Chemical Reactivity in Confined Systems

Chemical Reactivity in Confined Systems

Theory, Modelling and Applications

Edited by

Pratim Kumar Chattaraj
Department of Chemistry
Indian Institute of Technology Kharagpur
Kharagpur
India

Debdutta Chakraborty
Department of Chemistry
Katholieke Universiteit
Leuven
Belgium

WILEY

This edition first published 2021

Registered Offices
John Wiley & Sons, Inc., 111 River Street, Hoboken, NJ 07030, USA
John Wiley & Sons Ltd, The Atrium, Southern Gate, Chichester, West Sussex, PO19 8SQ, UK

Editorial Office
The Atrium, Southern Gate, Chichester, West Sussex, PO19 8SQ, UK

For details of our global editorial offices, customer services, and more information about Wiley products visit us at www.wiley.com.

Wiley also publishes its books in a variety of electronic formats and by print-on-demand. Some content that appears in standard print versions of this book may not be available in other formats.

For general information on our other products and services or for technical support, please contact our Customer Care Department within the United States at (800) 762-2974, outside the United States at (317) 572-3993 or fax (317) 572-4002.
Wiley also publishes its books in a variety of electronic formats. Some content that appears in print may not be available in electronic formats. For more information about Wiley products, visit our web site at ww.wiley.com.

Library of Congress Cataloging-in-Publication Data

Names: Chattaraj, Pratim Kumar, editor. | Chakraborty, Debdutta, editor.
Title: Chemical reactivity in confined systems : theory, modelling and applications / edited by Pratim Kumar Chattaraj, Debdutta Chakraborty.
Description: Hoboken, NJ : Wiley, 2021. | Includes bibliographical references and index.
Identifiers: LCCN 2021025002 (print) | LCCN 2021025003 (ebook) | ISBN 9781119684022 (cloth) | ISBN 9781119683384 (adobe pdf) | ISBN 9781119683230 (epub)
Subjects: LCSH: Reactivity (Chemistry)
Classification: LCC QD505.5 .C485 2021 (print) | LCC QD505.5 (ebook) | DDC 541/.39 – dc23
LC record available at https://lccn.loc.gov/2021025002
LC ebook record available at https://lccn.loc.gov/2021025003

Cover Design: Wiley
Cover Image: © Debdutta Chakraborty

Set in 9.5/12.5pt STIXTwoText by Straive, Chennai, India

10 9 8 7 6 5 4 3 2 1

Contents

Preface *xv*
List of Contributors *xvii*

1 Effect of Confinement on the Translation-Rotation Motion of Molecules: The Inelastic Neutron Scattering Selection Rule *1*
L. William Poirier
1.1 Introduction *1*
1.2 Diatomics in C_{60}: Entanglement, TR Coupling, Symmetry, Basis Representation, and Energy Level Structure *4*
1.2.1 Entanglement Induced Selection Rules *4*
1.2.2 $H@C_{60}$ *5*
1.2.3 $H_2@C_{60}$ *7*
1.2.3.1 Symmetry *7*
1.2.3.2 Spherical Basis and Eigenstates *7*
1.2.3.3 Energy Level Ordering with Respect to λ *8*
1.2.4 $HX@C_{60}$ *10*
1.3 INS Selection Rule for Spherical (K_h) Symmetry *11*
1.3.1 Inelastic Neutron Scattering *11*
1.3.2 Group Theory Derivation of the INS Selection Rule *12*
1.3.2.1 Group-Theory-Based INS Selection Rule for Cylindrical ($C_{\infty v}$) Environments *12*
1.3.2.2 Group-Theory-Based INS Selection Rule for Spherical (K_h) Environments *12*
1.3.3 Specific Systems, Quantum Numbers, and Basis Sets *13*
1.3.3.1 H@sphere *14*
1.3.3.2 H_2@sphere *14*
1.3.3.3 HX@sphere *15*
1.3.4 Beyond Diatomic Molecules *15*
1.3.4.1 H_2O@sphere *15*
1.3.4.2 CH_4@sphere *17*
1.3.4.3 Any Guest Molecule in any Spherical (K_h) Environment *18*
1.4 INS Selection Rules for Non-Spherical Structures *18*
1.5 Summary and Conclusions *20*
Acknowledgments *22*
References *22*

2 **Pressure-Induced Phase Transitions** *25*
Wojciech Grochala
2.1 Pressure, A Property of All Flavours, and Its Importance for the Universe and Life *25*
2.2 Pressure: Isotropic and Anisotropic, Positive and Negative *26*
2.3 Changes of the State of Matter *27*
2.4 Compression of Solids *30*
2.4.1 Isotropic or Anisotropic Compressibility *30*
2.4.2 Negative Linear Compressibility *30*
2.4.3 Negative Area Compressibility *31*
2.4.4 Anomalous Compressibility Changes at High Pressure *31*
2.5 Structural Solid-Solid Transitions *32*
2.5.1 Structural Phase Transitions Accompanied by Volume Collapse *32*
2.5.2 Effects of Volume Collapse on Free Energy *33*
2.5.3 Structure-Influencing Factors at Compression *34*
2.5.4 Changes in the Nature of Chemical Bonding upon Compression and upon Phase Transitions *35*
2.6 Selected Classes of Magnetic and Electronic Transitions *36*
2.6.1 High Spin–Low Spin Transitions *36*
2.6.2 Electronic Com- vs Disproportionation *37*
2.6.3 Metal-to-Metal Charge Transfer *37*
2.6.4 Neutral-to-Ionic Transitions *37*
2.6.5 Metallization of Insulators (and Resisting It) *38*
2.6.6 Turning Metals into Insulators *39*
2.6.7 Superconductivity of Elements and Compounds *39*
2.6.8 Topological Phase Transitions *41*
2.7 Modelling and Predicting HP Phase Transitions *41*
Acknowledgements *42*
References *42*

3 **Conceptual DFT and Confinement** *49*
Paul Geerlings, David J. Tozer, and Frank De Proft
3.1 Introduction and Reading Guide *49*
3.2 Conceptual DFT *50*
3.3 Confinement and Conceptual DFT *52*
3.3.1 Atoms: Global Descriptors *52*
3.3.2 Molecules: Global and Local Descriptors *56*
3.3.2.1 Electron Affinities *57*
3.3.2.2 Hardness and Electronic Fukui Function *59*
3.3.3 Inclusion of Pressure in the $E = E[N,v]$ Functional *63*
3.4 Conclusions *65*
Acknowledgements *65*
References *66*

4 Electronic Structure of Systems Confined by Several Spatial Restrictions *69*
Juan-José García-Miranda, Jorge Garza, Ilich A. Ibarra, Ana Martínez, Michael-Adán Martínez-Sánchez, Marcos Rivera-Almazo, and Rubicelia Vargas
4.1 Introduction *69*
4.2 Confinement Imposed by Impenetrable Walls *69*
4.3 Confinement Imposed by Soft Walls *72*
4.4 Beyond Confinement Models *74*
4.5 Conclusions *77*
References *77*

5 Unveiling the Mysterious Mechanisms of Chemical Reactions *81*
Soledad Gutiérrez-Oliva, Silvia Díaz, and Alejandro Toro-Labbé
5.1 Introduction *81*
5.1.1 Context *81*
5.1.2 Ideas and Methods *82*
5.1.3 Application *82*
5.2 Energy and Reaction Force *83*
5.2.1 The Reaction Force Analysis (RFA) *83*
5.2.2 RFA-Based Energy Decomposition *84*
5.2.3 Marcus Potential Energy Function *85*
5.2.4 Marcus RFA *86*
5.3 Electronic Activity Along a Reaction Coordinate *87*
5.3.1 Chemical Potential, Hardness, and Electrophilicity Index *87*
5.3.2 The Reaction Electronic Flux (REF) *88*
5.3.2.1 Physical Decomposition of REF *88*
5.3.2.2 Chemical Decomposition of REF *89*
5.4 An Application: the Formation of Aminoacetonitrile *90*
5.4.1 Energetic Analysis *91*
5.4.2 Reaction Mechanisms *91*
5.5 Conclusions *94*
Acknowledgments *95*
References *95*

6 A Perspective on the So-Called Dual Descriptor *99*
F. Guégan, L. Merzoud, H. Chermette, and C. Morell
6.1 Introduction: Conceptual DFT *99*
6.2 The Dual Descriptor: Fundamental Aspects *99*
6.2.1 Initial Formulation *99*
6.2.2 Alternative Formulations *100*
6.2.2.1 Derivative Formulations *100*
6.2.2.2 Link with Frontier Molecular Orbital Theory *101*
6.2.2.3 State-Specific Development *101*
6.2.2.4 MO Degeneracy *102*

6.2.2.5 Quasi Degeneracy *102*
6.2.2.6 Spin Polarization *103*
6.2.2.7 Grand Canonical Ensemble Derivation *105*
6.2.3 Real-Space Partitioning *105*
6.2.4 Dual Descriptor and Chemical Principles *106*
6.2.4.1 Principle of Maximum Hardness *106*
6.2.4.2 Local Hardness Descriptors *106*
6.2.4.3 Local Electrophilicity and Nucleophilicity *106*
6.2.4.4 Local Chemical Potential and Excited States Reactivity *107*
6.3 Illustrations *108*
6.3.1 Woodward Hoffmann Rules in Diels-Alder Reactions *108*
6.3.2 Perturbational MO Theory and Dual Descriptor *109*
6.3.3 Markovnikov Rule *109*
6.4 Conclusions *110*
References *111*

7 Molecular Electrostatic Potentials: Significance and Applications *113*
Peter Politzer and Jane S. Murray
7.1 A Quick Review of Some Classical Physics *113*
7.2 Molecular Electrostatic Potentials *113*
7.3 The Electronic Density and the Electrostatic Potential *114*
7.4 Characterization of Molecular Electrostatic Potentials *115*
7.5 Molecular Reactivity *116*
7.6 Some Applications of Electrostatic Potentials to Molecular Reactivity *118*
7.6.1 σ-Hole and π-Hole Interactions *118*
7.6.2 Hydrogen Bonding: A σ-Hole Interaction *119*
7.6.3 Interaction Energies *120*
7.6.4 Close Contacts and Interaction Sites *121*
7.6.5 Biological Recognition Interactions *124*
7.6.6 Statistical Properties of Molecular Surface Electrostatic Potentials *125*
7.7 Electrostatic Potentials at Nuclei *126*
7.8 Discussion and Summary *127*
References *127*

8 Chemical Reactivity Within the Spin-Polarized Framework of Density Functional Theory *135*
E. Chamorro and P. Pérez
8.1 Introduction *135*
8.2 The Spin-Polarized Density Functional Theory as a Suitable Framework to Describe Both Charge and Spin Transfer Processes *137*
8.3 Practical Applications of SP-DFT Indicators *141*
8.4 Concluding Remarks and Perspectives *145*
Acknowledgements *147*
References *147*

9 Chemical Binding and Reactivity Parameters: A Unified Coarse Grained Density Functional View *167*
Swapan K. Ghosh
9.1 Introduction *167*
9.2 Theory *169*
9.2.1 Concept of Electronegativity, Chemical Hardness, and Chemical Binding *169*
9.2.1.1 Electronegativity and Hardness *169*
9.2.1.2 Interatomic Charge Transfer in Molecular Systems *169*
9.2.1.3 Concept of Chemical Potential and Hardness for the Bond Region *170*
9.2.1.4 Spin-Polarized Generalization of Chemical Potential and Hardness *171*
9.2.1.5 Charge Equilibriation Methods: Split Charge Models and Models with Correct Dissociation Limits *172*
9.2.1.6 Density Functional Perturbation Approach: A Coarse Graining Procedure *173*
9.2.1.7 Atomic Charge Dipole Model for Interatomic Perturbation and Response Properties *174*
9.2.1.8 Force Field Generation in Molecular Dynamics Simulation *174*
9.3 Perspective on Model Building for Chemical Binding and Reactivity *175*
9.4 Concluding Remarks *175*
Acknowledgements *175*
References *175*

10 Softness Kernel and Nonlinear Electronic Responses *179*
Patrick Senet
10.1 Introduction *179*
10.2 Linear and Nonlinear Electronic Responses *181*
10.2.1 Linear Response Theory *181*
10.2.1.1 Ground-State *181*
10.2.1.2 Linear Responses [1] *181*
10.2.2 Nonlinear Responses and the Softness Kernel *182*
10.2.3 Eigenmodes of Reactivity *184*
10.3 One-Dimensional Confined Quantum Gas: Analytical Results from a Model Functional *185*
10.4 Conclusion *188*
References *188*

11 Conceptual Density Functional Theory in the Grand Canonical Ensemble *191*
José L. Gázquez, Marco Franco-Pérez, Paul W. Ayers, and Alberto Vela
11.1 Introduction *191*
11.2 Fundamental Equations for Chemical Reactivity *192*
11.3 Temperature-Dependent Response Functions *195*
11.4 Local Counterpart of a Global Descriptor and Non-Local Counterpart of a Local Descriptor *200*
11.5 Concluding Remarks *203*

Acknowledgements 204
References 204

12 **Effect of Confinement on the Optical Response Properties of Molecules** 213
Wojciech Bartkowiak, Marta Chołujv, and Justyna Kozłowska
12.1 Introduction 213
12.2 Electronic Contributions to Longitudinal Electric-Dipole Properties of Atomic and Molecular Systems Embedded in Harmonic Oscillator Potential 215
12.3 Vibrational Contributions to Longitudinal Electric-Dipole Properties of Spatially Confined Molecular Systems 218
12.4 Two-Photon Absorption in Spatial Confinement 219
12.5 Conclusions 220
References 221

13 **A Density Functional Theory Study of Confined Noble Gas Dimers in Fullerene Molecules** 225
Dongbo Zhao, Meng Li, Xin He, Bin Wang, Chunying Rong, Pratim K. Chattaraj, and Shubin Liu
13.1 Introduction 225
13.2 Computational Details 226
13.3 Results and Discussion 227
13.3.1 Changes in Structure 227
13.3.2 Changes in Interaction Energy 227
13.3.3 Changes in Bonding Energy 228
13.3.4 Changes in Energy Components 228
13.3.5 Changes in Noncovalent Interactions 229
13.3.6 Changes in Information-Theoretic Quantities 231
13.3.7 Changes in Spectroscopy 232
13.3.8 Changes in Reactivity 233
13.4 Conclusions 236
Acknowledgments 236
References 236

14 **Confinement Induced Chemical Bonding: Case of Noble Gases** 239
Sudip Pan, Gabriel Merino, and Lili Zhao
14.1 Introduction 239
14.2 Computational Details and Theoretical Background 241
14.3 The Bonding in He@$C_{10}H_{16}$: A Debate 243
14.4 Confinement Inducing Chemical Bond Between Two Ngs 244
14.5 XNgY Insertion Molecule: Confinement in One Direction 251
14.6 Conclusions 254
Acknowledgements 255
References 255

15 **Effect of Both Structural and Electronic Confinements on Interaction, Chemical Reactivity and Properties** 263
Mahesh Kumar Ravva, Ravinder Pawar, Shyam Vinod Kumar Panneer, Venkata Surya Kumar Choutipalli, and Venkatesan Subramanian
15.1 Introduction 263

15.2 Geometrical Changes in Small Molecules Under Spherical and Cylindrical Confinement *264*
15.3 Hydrogen Bonding Interaction of Small Molecules in the Spherical and Cylindrical Confinement *265*
15.4 Spherical and Cylindrical Confinement and Chemical Reactivity *267*
15.5 Concluding Remarks *268*
References *270*

16 Effect of Confinement on Gas Storage Potential and Catalytic Activity *273*
Debdutta Chakraborty, Sukanta Mondal, Ranjita Das, and Pratim Kumar Chattaraj
16.1 Introduction *273*
16.2 Endohedral Gas Adsorption Inside Clathrate Hydrates *274*
16.3 Hydrogen Hydrates *276*
16.4 Methane Hydrates *278*
16.5 Noble Gas Hydrates *279*
16.6 Confinement Induced Catalysis of Some Chemical Reactions *280*
16.7 Outlook *285*
Acknowledgements *285*
References *286*

17 Engineering the Confined Space of MOFs for Heterogeneous Catalysis of Organic Transformations *293*
Tapan K. Pal, Dinesh De, and Parimal K. Bharadwaj
17.1 Introduction *293*
17.2 Catalysis at the Open Metal Sites *293*
17.2.1 MOFs Endowed with Open Metal Site(s) *294*
17.2.2 Removal of Volatile Molecules From Metal Nodes to Perform Catalysis *297*
17.2.3 Catalysis at the Metal node Post Transmetalation *299*
17.3 Functionalization in the MOF to Furnish Catalytic Site *301*
17.3.1 Attaching the Catalytically Active Moieties to the Metal Nodes (SBU) *301*
17.3.2 Preconceived Catalytic Site into the Linker *301*
17.3.3 Post Synthetic Modification of the Linker *304*
17.3.4 MOFs with Linkers Having Coordinated Metal Ions (Metalloligands) *306*
17.4 MOFs as Bifunctional Catalyst *310*
17.5 Impregnation/Encapsulation of Nanoparticles in the MOF Cavity for Catalysis *317*
17.6 Engineering Homochiral MOFs for Enantioselective Catalysis *320*
17.7 Conclusion *325*
Acknowledgements *326*
References *326*

18 Controlling Excited State Chemistry of Organic Molecules in Water Through Incarceration *335*
V. Ramamurthy
18.1 Introduction *335*
18.2 Complexation Properties of OA *337*
18.3 Properties of OA capsule *339*
18.4 Dynamics of Encapsulated Guests *340*
18.5 Dynamics of Host-Guest Complex *346*
18.6 Room Temperature Phosphorescence of Encapsulated Organic Molecules *353*

18.7 Consequence of Confinement on the Photophysics of Anthracene *356*
18.8 Selective Photo-Oxidation of Cycloalkenes *358*
18.9 Remote Activation of Encapsulated Guests: Electron Transfer Across an Organic Wall *360*
18.10 Summary *362*
Acknowledgements *363*
References *363*

19 Effect of Confinement on the Physicochemical Properties of Chromophoric Dyes/Drugs with Cucurbit[*n*]uril: Prospective Applications *371*
J. Mohanty, N. Barooah, and A. C. Bhasikuttan
19.1 Introduction *371*
19.1.1 Confinement of Dyes/Drugs in Macrocyclic Hosts *372*
19.1.1.1 Cyclodextrins *372*
19.1.1.2 Calixarenes *373*
19.1.1.3 Cucurbiturils *373*
19.2 Confinement in Cucurbituril Hosts: Effects on the Physicochemical Properties of Guest Molecules – Advantages for Various Technological Applications *374*
19.2.1 Enhanced Photostability and Solubility of Rhodamine Dyes *375*
19.2.1.1 Water-Based Dye Laser *376*
19.2.2 Enhanced Luminescence and Photostability of $CH_3NH_3PbBr_3$ Perovskite Nanoparticles *377*
19.2.3 Enhanced Antibacterial Activity and Extended Shelf-Life of Fluoroquinolone Drugs with Cucurbit[7]uril *377*
19.2.4 Effect of Confinement on the Prototropic Equilibrium *379*
19.2.4.1 Salt-Induced pK_a Tuning and Guest Relocation *379*
19.2.5 Confinement in Cucurbit[7]uril-Mediated BSA: Stimuli-Responsive Uptake and Release of Doxorubicin *380*
19.2.6 Effect of Confinement on the Fluorescence Behavior of Chromophoric Dyes with Cucurbiturils *380*
19.2.6.1 Fluorescence Behavior of Chromophoric Dyes with Cucurbit[7]uril *381*
19.2.6.2 Fluorescence Behavior of Chromophoric Dyes with Cucurbit[8]uril *383*
19.2.7 Effect of Confinement on the Catalytic Performance within Cucurbiturils *386*
19.3 Conclusion *388*
Acknowledgement *389*
References *389*

20 Box-Shaped Hosts: Evaluation of the Interaction Nature and Host Characteristics of ExBox Derivatives in Host-Guest Complexes from Computational Methods *395*
Giovanni F. Caramori and Alvaro Muñoz-Castro
20.1 Introduction *395*
20.2 Noncovalent Interactions Through Energy Decomposition Analysis *396*
20.3 Ex^0Box^{4+} ($CBPQT^{4+}$) *398*
20.4 $ExBox^{4+}$ and Ex^2Box^{4+} *400*

20.5 Larger Boxes *406*
20.6 NMR Features *408*
20.7 All Carbon Counterpart *409*
20.8 Conclusions *409*
Acknowledgments *410*
References *411*

Index *417*

Preface

Confined systems often exhibit unusual behavior regarding their structure, stability, reactivity, bonding, interactions and dynamics. Quantization is a direct consequence of confinement. Confinement modifies the electronic energy levels, orbitals, electronic shell filling, etc., of a system thereby affecting its reactivity as well as various response properties as compared to the corresponding unconfined system. Confinement may enforce two rare gas atoms to form a partly covalent bond. Gas storage is facilitated through confinement and unprecedented optoelectronic properties are observed in certain cases. Some slow reactions get highly accelerated in an appropriate confined environment.

Analyzing the reactivity of atoms and molecules, present in a confined environment, by utilizing various theoretical and computational methods can unravel numerous new paradigms *vis-à-vis* physicochemical properties of the systems under consideration. Therefore, confined quantum systems have been extensively analyzed from both epistemological and applied points of view. The crucial point while analyzing quantum confined systems is to be able to construct an accurate theoretical model that takes into account changes in the electronic wave function due to the effect of confinement. To this end, model theoretical calculations could be conceived by suitable choice of the boundary condition. Host–guest complexes provide an ideal 'real' platform where the changes of reactivity and response properties of various systems could be understood. To this end, density functional theory (DFT) based calculation provides a cost-effective and reasonably accurate method.

On the other hand, experimental studies have helped to shed light on many fascinating aspects of confinement. The utility of confined systems could be gauged in various disciplines such as catalysis, gas storage, designing superior optoelectronic and magnetic materials, etc.

In this book, several experts, who have made seminal contributions in understanding the behavior of confined systems, have written authoritative accounts on state-of the-art research topics encompassing both theoretical as well as experimental endeavors. Hopefully, this book will be beneficial for graduate students in chemistry, materials science and physics in understanding the recent developments in this field.

Pratim Kumar Chattaraj
Debdutta Chakraborty

List of Contributors

Paul W. Ayers
McMaster University
Canada

Nilotpal Barooah
Bhabha Atomic Research Centre
Mumbai
India

Wojciech Bartkowiak
Wroclaw University of Science and Technology
Poland

Parimal Kanti Bharadwaj
Indian Institute of Technology Kanpur
India

A. C. Bhasikuttan
Bhabha Atomic Research Centre
Mumbai
India

Giovanni F. Caramori
Federal University of Santa Catarina – UFSC
Brazil

Eduardo Chamorro
Universidad Andrés Bello
Chile

Debdutta Chakraborty
Katholieke Universiteit
Leuven
Belgium

Pratim Kumar Chattaraj
Indian Institute of Technology Kharagpur
Kharagpur
India

Henry Chermette
Université de Lyon 1
France

Marta Chołuj
Wroclaw University of Science and Technology
Poland

Silvia Díaz
Pontificia Universidad católica de Chile
Chile

Frank De Proft
Durham University
Durham
UK

Marco Franco-Pérez
Universidad Nacional Autónoma de México
Mexico

Jorge Garza Olguín
UAM-Iztapalapa
Mexico

José Luis Gázquez
Universidad Autónoma
Metropolitana-Iztapalapa
Mexico

Paul Geerlings
Vrije Universiteit Brussel
Belgium

Frédéric Guégan
Université de Poitiers
France

Soledad Gutiérrez-Oliva
Pontificia Universidad Católica de Chile
Chile

Swapan Kumar Ghosh
University of Mumbai
India

Wojciech Grochala
University of Warsaw
Poland

Justyna Kozłowska
Wroclaw University of Science and Technology
Poland

Shubin Liu
University of North Carolina
USA

Gabriel Merino
Cinvestav
Mexico

Lynda Merzoud
Université de Lyon 1
France

Jyotirmayee Mohanty
Bhabha Atomic Research Centre
Mumbai
India

Christophe Morell
Université de Lyon 1
France

Alvaro Muñoz Castro
Universidad Autónoma de Chile
Chile

Sudip Pan
Philipps-Universität Marburg
Germany

Patricia Perez
Universidad Andrés Bello
Chile

L. William Poirier
Texas Tech University
USA

Peter Politzer
University of New Orleans
USA

Vaidhyanathan Ramamurthy
University of Miami
USA

Chunying Rong
Hunan Normal University
China

Patrick Senet
Université de Bourgogne Franche-Comté
(UBFC)
France

Venkatesan Subramanian
CSIR-Central Leather Research Institute
India

Alejandro Toro-Labbe
Pontificia Universidad Catolica de Chile
Chile

David J. Tozer
Durham University
Durham
UK

Rubicelia Vargas
Universidad Autónoma Metropolitana
Iztapalapa

Alberto Vela
Cinvestav
Mexico

Dongbo Zhao
Yunnan University
China

Lili Zhao
Nanjing Tech University
China

1

Effect of Confinement on the Translation-Rotation Motion of Molecules: The Inelastic Neutron Scattering Selection Rule

*L. William Poirier**

Texas Tech University, USA

1.1 Introduction

One of the core theoretical ideas used to understand the dynamics of free molecules is the simplifying notion that the overall (i.e., center-of-mass) translational motion can be cleanly separated from internal vibrations and rotations. Indeed, this separation is so universally applied – and the translational motion so easily dealt with – that it can be easy to "forget" that the latter even exists! On the other hand, if you place that same molecule in a nanoconfined environment, the situation can be vastly different. First and foremost, the continuum of states that characterizes translational motion for free molecules necessarily becomes *quantized* in the confined context. The quantum translational states for the trapped "guest" molecule can be "particle-in-a-box like" or more complicated, depending on the nature of the external field provided by the cage structure. Of course, larger cages give rise to smaller translational level spacings – which are in any event generally smaller than the rotational level spacings (and *much* smaller than the vibrational level spacings). If the confinement is very severe, however, then the translational and rotational level spacings can become comparable to each other – and even strongly coupled.

This is the situation for small molecules trapped inside C_{60} fullerene cages – e.g., $H_2@C_{60}$ [1–16], $HD@C_{60}$ [4, 7, 12, 15, 17, 18], $HF@C_{60}$ [14–16, 19], $H_2O@C_{60}$ [14–16, 20, 21], and $CH_4@C_{60}$ [15, 22], all of which will be considered in this chapter. Based on the physical size of the fullerene (diameter ≈ 7 Å), one might well imagine that a number of guest molecules could be crammed into a single C_{60} cage. In reality, the guest molecules are trapped via long-range van der Waals interactions that prevent them from getting closer than a few Å from the cage wall. Consequently, the *effective* cage size is much smaller – on the order of the size of the guest molecule itself. This implies that: (a) only one guest molecule can fit inside a single C_{60} cage; (b) the corresponding translational level spacing is comparable to the rotational level spacing. It is hardly surprising, then, to also find that translation and rotation are indeed strongly coupled in these systems. This complicates life from a theoretical/computational standpoint, for which one must adopt an exact, coupled quantum dynamical treatment, encompassing all relevant translation-rotation (TR) guest molecule degrees of freedom [3–5, 15, 16, 19, 23–25]. That said, the combined TR states that result are not necessarily entirely devoid of structure either – it is just not that of the standard form that one expects in terms of TR separability.

*Email: bill.poirier@ttu.edu

Chemical Reactivity in Confined Systems: Theory, Modelling and Applications,
First Edition. Edited by Pratim Kumar Chattaraj and Debdutta Chakraborty.

In addition to providing a fundamentally different spectroscopic picture, the strong confinement-induced TR coupling also gives rise to a remarkable physical effect that was once thought to be impossible – *selection rules* for inelastic neutron scattering (INS) [26–28]. INS is an experimental technique in which a beam of neutrons is scattered through nuclear force interactions with the nuclei of the target sample. It is an extremely useful tool for probing nanoconfined hydrogen, in large part because the H atom nucleus provides the *largest* neutron scattering cross section across the entire periodic table. Even that of the "second place" contender – i.e., deuterium – is more than an order of magnitude smaller. INS offers other advantages as well, such as the ability to examine transitions between individual quantum states – including ortho-para nuclear spin transitions in H_2, which would be forbidden, e.g., in an electric dipole or even Raman far-infrared (IR) spectrum. However, whereas such optically forbidden spectroscopic transitions certainly are allowed in INS, this is no guarantee that *all other* transitions are also necessarily allowed.

That INS spectroscopy does indeed have forbidden transitions, was first proposed and experimentally verified for $H_2@C_{60}$, in a stunning set of papers by Bačić, Horsewill, and coworkers [9, 10, 29, 30]. In this earlier work, in addition to computing the TR quantum eigenstates themselves (energy levels and wavefunctions) [3–5], an explicit numerical simulation of the experimental INS spectrum was also performed [9, 12, 29, 30]. These represent heroic calculations, applied to each TR transition individually, which also take various experimental circumstances into proper account. The end result is a reasonably accurate prediction of both INS transition energies and intensities (both "stick spectra" and more experimentally-relevant convolved spectra can be obtained). In this manner, it was discovered that some transition intensities for $H_2@C_{60}$ are vastly smaller than others – by four or more orders of magnitude. This provided excellent numerical evidence for a selection rule – which, indeed, was subsequently confirmed through actual INS experiments [7, 8, 10].

As impressive and unexpected as this discovery proved to be, one of the drawbacks of the above approach is that one must infer a general pattern for the selection rule, from amongst a necessarily limited set of specific transitions. Indeed, this led Bačić and coworkers to initially assume a selection rule for $H_2@C_{60}$ of the following form:

> **restricted INS selection rule for $H_2@C_{60}$ (from p-H_2 ground state)**: Transitions are forbidden to all states for which $(j + l - \lambda) = 1$ [note: all terms will be explained].

In fact, the correct rule for transitions starting from the $H_2@C_{60}$ ground state is more general:

> **correct INS selection rule for $H_2@C_{60}$ (from p-H_2 ground state)**: Transitions are forbidden to all states for which $(j + l - \lambda) =$ odd.

To the author's knowledge, the latter form above was first proposed by the author himself, at a scientific meeting in May, 2015. Since confirmed assignments for the experimental $H_2@C_{60}$ data did not exist beyond $(j + l - \lambda) = 2$, the theoretical simulations were not performed beyond this point either, and so the available information at the time was consistent with *either* of the two rules above. This state of affairs motivated the present author to develop a *group theoretical* derivation of the general INS selection rule [11]. In parallel with this effort, Bačić and coworkers extended the analytical part of their calculations of the transition integrals [12], to encompass all possible initial and final states. Both approaches then gave rise to the following,

most general INS selection rule for H_2@C_{60}: Transitions are forbidden between states for which $(j + l - \lambda)$ changes from even to odd (or vice-versa), and at least one $\lambda = 0$.

As will be described in this chapter, the group theory approach provides physical understanding, and also has the great advantage that it leads to the correct and completely general INS selection rule for the appropriate symmetry group, "all at once." However, this approach has some limitations and issues, as well, which will also be addressed. To begin with, it does not provide any intensity information for allowed transitions; for this, it is necessary to calculate transition integrals explicitly, as per Bačić et. al. Secondly there is the question of choosing the most correct symmetry group to work with, in terms of experimental relevance. This is a particularly important question for INS spectroscopy, for which there is a preferred direction or orientation – i.e., that of the *momentum transfer vector*, $\vec{k}$, of the scattered neutron beam. Thirdly, there are some group theoretical complications that arise, due to the fact that the INS interaction operator itself is incommensurate with most of the standard molecular point groups – a key difference, e.g., from optical spectroscopy. Finally, all true group-theory-based selection rules are expressed in terms of the *irreducible representations* (irreps) of the appropriate symmetry group. While this form of a selection rule is truly universal, for specific systems it is still necessary to assign irrep labels to individual quantum states and/or basis functions – thereby making an association with the pertinent *quantum numbers* for those systems. On balance, it is clear that *both* the group theory and explicit integral approaches are necessary for interpreting and predicting INS experiments, as they provide complementary understanding.

Since 2015, Bačić, Horsewill, Felker, and others have continued to explore the INS selection rule – together with other TR effects of nanoconfinement – through a fruitful synergy that has emerged between theory and experiment. These highly interesting TR effects have now been observed and/or predicted across a range of nanoconfined systems. Regarding the INS selection rule itself, the irrep-based version of [11] is in principle valid for any molecule in any spherical environment (section 1.3.2). In terms of specific basis sets and quantum states, Bačić and coworkers have generalized the rule for any *diatomic* molecule in a spherical environment, using their explicit integral approach. They have also applied this methodology to both HD@C_{60} [4, 7, 12, 15, 17, 18] and HF@C_{60} [14–16, 19], and discovered forbidden INS transitions in both of these nanoconfined systems. Experiments for H_2O@C_{60} [12, 20] suggest a selection rule for this system as well. All of these recent findings indicate that forbidden INS transitions may be much more prevalent than originally realized, with future studies potentially addressing larger cage systems such as H_2@C_{70} [5], $(H_2)_2$@C_{70} [31], etc. Other refinements include the development of improved potential energy surfaces (PESs) [32], and/or the explicit incorporation of vibrational effects.

Another important recent direction has been the consideration of the *larger environment* surrounding a given C_{60} cage, and the role of *intercage interactions* [14, 16, 19, 33, 34]. For example, Felker and Bačić have examined electric-dipole coupling between the two guest H_2O molecules in $(H_2O@C_{60})_2$ [33], whereas Roy and coworkers have studied the dipole interaction in $(HF@C_{60})_n$ when the n cages form a "peapod" arrangement [34]. One fascinating and related finding pertains to the small ($\sim$1 cm^{-1}) but puzzling level-splitting observed experimentally in the $j = 1$ states of H_2@C_{60}, HF@C_{60}, and H_2O@C_{60} [6, 13–16, 19, 20]. These splittings break the expected spherical or icosahedral level pattern, implying a reduction in symmetry caused by the larger environment. Whereas various explanations have been profferred (including the aforementioned guest–guest dipole interaction), the precise cause was recently definitively established [14–16] as being due to neighboring C_{60} cages in the solid – arranged in an alternating "P" orientation – which reduce the molecular point group to S_6.

From a group theory perspective, one of the most intriguing of the above recent results are the forbidden INS transitions that have been predicted for both HD@C_{60} [12, 15, 18] and HF@C_{60} [19]. Recall that neutrons only interact effectively with H atom nuclei – of which these two systems each contain only one. Yet, from group theory arguments [35, 36], it has been established that *there are no* single-atom wavefunctions that belong to the irreps necessary to effect a forbidden INS transition. At least two particles are necessary to realize the requisite irreps – in much the same way that in diatomic molecules, at least two electrons are required to realize an electronic state with Σ^- character [37, 38]. This state of affairs underscores the need for further theoretical development in a reconciling vein – thereby, in part, motivating the present and future effort. The resolution of the dilemma will be presented in section 1.2.1, but can be summarized in a single word: *entanglement*.

Evidently, recent developments in the study of quantum confinement effects in the TR dynamics of small molecules offer us plenty of material to ponder. In any event, the field is clearly proceeding at a very rapid pace, beyond what can be reviewed in detail in this single chapter. Accordingly, we choose to focus here primarily on the INS selection rule as it applies to diatomic molecules confined to spherical (or sphere-like) environments. As per the preceding paragraph, an important aim is to reconcile the group theory and explicit integral approaches, in a manner that will stimulate mutual developments going forward, and otherwise streamline future progress. Among other aspects, this effort necessarily requires that an explicit association be made between irrep labels and quantum states/basis functions/quantum numbers, for specific molecular systems. Finally, we address prospects for moving *beyond* all previous INS selection rule applications to date – by considering guest molecules that are larger than diatomic, and/or host cages other than spherical. Indeed, *we also derive here a quantum-number-based INS selection rule for* H_2O@C_{60}, and use it to provide unambiguous transition assignments for the experimental INS spectrum [12, 20] (section 1.3.4.1).

1.2 Diatomics in C_{60}: Entanglement, TR Coupling, Symmetry, Basis Representation, and Energy Level Structure

1.2.1 Entanglement Induced Selection Rules

In the usual group theory procedure, selection rules are determined by sandwiching the interaction operator between bra and ket system states, and decomposing the resultant triple-direct-product of irreps into its own irreducible form. (This procedure is worked out in detail for the case of INS spectroscopy in section 1.3.1.) Quantum mechanics, however, requires careful consideration of what, exactly, is meant by "the system." In particular, the approach above is only valid if the system is not appreciably entangled with its surroundings. Put another way, it presumes a separable product form for the "total system" wavefunction:

$$\Psi_{\text{total}} = \psi \times \psi_{\text{surroundings}}, \tag{1.1}$$

where ψ is the wavefunction of the system itself. Therefore, when defining which part of our experimental sample to regard as "the system," this should be taken as the smallest piece that includes all of the constituent particles of interest, *and also* leads to an unentangled state of the Eq. (1.1) form.

For the present purpose, we are considering a diatomic guest molecule trapped within a C_{60} cage. The strongest INS signals are observed when at least one of the two atoms is hydrogen. Let us therefore consider a diatomic of the "HX" variety, where X is any atom *other* than hydrogen. In the

HX@C_{60} system, one may presume that there is little entanglement between the host cage and the guest diatomic. Nevertheless, the H atom is *strongly* entangled with X; HX is a molecule, after all. Consequently, the smallest unentangled unit that can provide an INS signal is the HX molecule, which must be taken as "the system" for group theory purposes. We must work with quantum wavefunctions of the combined HX system – not of H itself – even though the nucleus X is mostly "dark," insofar as the neutron beam is concerned. This situation is reminiscent of the manner in which the existence of a black hole may be inferred from the observed behavior of its partner star, even though the black hole itself cannot be directly seen.

Including both the "bright" and entangled "dark" particles in the definition of "the system" can give rise to different selection rules than if just the bright particles alone were used. This rather intriguing situation gives rise to an effect that might be called *entanglement induced selection rules* (EISR). In this work, we demonstrate that the forbidden INS transitions predicted for HD@C_{60} and HF@C_{60} are indeed due solely to entanglement with the dark nucleus, X=D or F – and therefore disappear entirely for the system H@C_{60}. In principle, the EISR idea should not be restricted to INS spectroscopy, but could also be applied, e.g., in optical spectroscopy. So why is EISR evidently not discussed in the latter context? In fact, it turns out that the EISR effect sometimes *does* manifest in optical spectroscopy, but when it does so, it is known by other, specific names in specific instances – e.g., the "Herzberg-Teller Effect" [37]. Interestingly in the Herzberg-Teller context, EISR serves to *remove* an electric-dipole forbidden transition, whereas for HX@C_{60} INS spectroscopy, EISR does the opposite! More generally, this intriguing new phenomenon – or properly, new lens through which to interpret phenomena – promises to provide interesting insight and connections across a very broad range of applications. EISR therefore demands further study, and will serve as the focus of future investigations.

1.2.2 H@C_{60}

For H@C_{60}, the sole "bright" nucleus (in the INS sense) is that of the guest H atom, which will not be appreciably entangled with the C_{60} host. Accordingly, as per section 1.1, the relevant quantum system is just the H atom itself. The corresponding quantum Hamiltonian, $\hat{H}$, is constructed by fixing the C_{60} cage, and allowing the position of the H atom, $\vec{r} = (x, y, z)$, to vary within the cage. The explicit dimensionality of the quantum dynamical problem is thus three (3D), corresponding to pure translational motion only.

The symmetry group is defined as the set of symmetry operations that leave $\hat{H}$ invariant [37]. Note that the PES contribution to $\hat{H}$ (i.e. $\hat{V}$), describing the external field that is felt by the H atom, arises mainly from pair-wise van der Waals interactions with individual C atoms of the host cage, C_{60}. Consequently, keeping $\vec{r}$ fixed, but transforming the C_{60} under any of its own point group operations, clearly leaves the H@C_{60} $\hat{H}$ invariant. Conversely, keeping the C_{60} fixed but applying its point group operations to $\vec{r}$ also leaves $\hat{H}$ invariant. There are no other operations applied to $\vec{r}$ that also do so; consequently, the H@C_{60} symmetry group must be no larger than the C_{60} point group, I_h. It might in principle be a subgroup of I_h, if multiple I_h operations applied to $\vec{r}$ were to result in the same transformed point, $\vec{r}_t$. Whereas this is true for certain $\vec{r}$ points with special symmetric orientations, in the general case, each I_h operation yields a distinct $\vec{r}_t$ point – resulting in 120 such points in all. The H@C_{60} symmetry group is therefore I_h.

Although I_h is formally the correct symmetry group for H@C_{60}, I_h has the greatest symmetry of all finite point groups, and is otherwise a large subgroup of the *spherical point group*, K_h [35–37, 39]. More specifically, the structure of the C_{60} cage is itself nearly spherical; in solid angle space, one is never very far away from a C atom. Moreover, the van der Waals nature of the external field ensures that $\hat{V}$ is even *more* nearly perfectly spherical than the fullerene itself. This implies that the true energy eigenstate wavefunctions are very nearly spherically symmetric, and therefore well-described by a spherical basis of the form

$$\psi_{n,\lambda,m_\lambda}(r,\theta,\phi) = R_{n\lambda}(r)Y_{\lambda,m_\lambda}(\theta,\phi), \tag{1.2}$$

where $\vec{r} = (r, \theta, \phi)$ is now expressed in spherical coordinates. This leads to a natural state labeling scheme in terms of the quantum numbers (n, λ, m_λ), which can be directly related to irreps of the K_h point group [35, 37] (Table 1.1).

Note that the spherical eigenstates have a $(2\lambda + 1)$-fold degeneracy, whereas no irrep of I_h has a greater than five-fold degeneracy. Thus, to the extent that the true H@C_{60} eigenstates deviate from spherical symmetry, the energy level structure should exhibit a slight degeneracy-lifting, as will be discussed explicitly for H_2@C_{60} in section 1.2.3. We can use group theory – specifically, the correlation table between K_h and I_h, presented in Table 1.1 – to tell us precisely how this decomposition occurs, in terms of I_h irreps. Note that, whereas in general, the K_h irreps are labeled by both λ and (g/u) inversion parity, for the $Y_{\lambda,m_\lambda}(\theta,\phi)$ spherical harmonic basis functions, only *half* of these irreps are actually realized, because (g/u) parity is determined by the even/oddness of λ [35, 36]. This fact will become quite important in our discussion of INS selection rules (section 1.3).

Note also that the specific radial function in Eq. (1.2) is immaterial, with respect to the symmetry character of the corresponding eigenstate; in principle, $R_{n\lambda}(r)$ can take on any form. In all earlier studies of the H_2@C_{60} INS transition intensities, an explicit, isotropic harmonic oscillator basis was presumed [3–5, 9, 12, 14–16]. Such a choice might be expected to be appropriate for the very tight confinement in C_{60} – and indeed, turns out to be reasonable for all diatomic systems considered thus far, except for HF@C_{60}. For larger sphere-like cavities, however, the greatest density will lie

Table 1.1 Correlation table, indicating the multipole decomposition of K_h irreps, with respect to the $C_{\infty v}$ and I_h subgroups. Irreps that are *not* realized by the single-particle spherical harmonic basis functions, $Y_{\lambda,m_\lambda}(\theta,\phi)$, are listed in bold; note the alternating pattern with increasing λ.

		Correlates to:			Correlates to:	
λ	Irrep K_h	$C_{\infty v}$	I_h	Irrep K_h	$C_{\infty v}$	I_h
0	S_g	Σ^+	A_g	$\mathbf{S_u}$	Σ^-	A_u
1	$\mathbf{P_g}$	$\Sigma^- \oplus \Pi$	T_{1g}	P_u	$\Sigma^+ \oplus \Pi$	T_{1u}
2	D_g	$\Sigma^+ \oplus \Pi \oplus \Delta$	H_g	$\mathbf{D_u}$	$\Sigma^- \oplus \Pi \oplus \Delta$	H_u
3	$\mathbf{F_g}$	$\Sigma^- \oplus \Pi \oplus \Delta \oplus \Phi$	$T_{2g} \oplus G_g$	F_u	$\Sigma^+ \oplus \Pi \oplus \Delta \oplus \Phi$	$T_{2u} \oplus G_u$
4	G_g	⋮	$G_g \oplus H_g$	$\mathbf{G_u}$	⋮	$G_u \oplus H_u$
5	$\mathbf{H_g}$	⋮	$T_{1g} \oplus T_{2g} \oplus H_g$	H_u	⋮	$T_{1u} \oplus T_{2u} \oplus H_u$
6	I_g	⋮	$A_g \oplus T_{1g} \oplus G_g \oplus H_g$	$\mathbf{I_u}$	⋮	$A_u \oplus T_{1u} \oplus G_u \oplus H_u$
7	$\mathbf{J_g}$	⋮	$T_{1g} \oplus T_{2g} \oplus G_g \oplus H_g$	J_u	⋮	$T_{1u} \oplus T_{2u} \oplus G_u \oplus H_u$
8	K_g	⋮	$T_{2g} \oplus G_g \oplus 2H_g$	$\mathbf{K_u}$	⋮	$T_{2u} \oplus G_u \oplus 2H_u$
9	$\mathbf{L_g}$	⋮	$T_{1g} \oplus T_{2g} \oplus 2G_g \oplus H_g$	L_u	⋮	$T_{1u} \oplus T_{2u} \oplus 2G_u \oplus H_u$
10	M_g	⋮	$A_g \oplus T_{1g} \oplus T_{2g} \oplus G_g \oplus 2H_g$	$\mathbf{M_u}$	⋮	$A_u \oplus T_{1u} \oplus T_{2u} \oplus G_u \oplus 2H_u$
11	$\mathbf{N_g}$	⋮	$2T_{1g} \oplus T_{2g} \oplus G_g \oplus 2H_g$	N_u	⋮	$2T_{1u} \oplus T_{2u} \oplus G_u \oplus 2H_u$
12	O_g	⋮	$A_g \oplus T_{1g} \oplus T_{2g} \oplus 2G_g \oplus 2H_g$	$\mathbf{O_u}$	⋮	$A_u \oplus T_{1u} \oplus T_{2u} \oplus 2G_u \oplus 2H_u$
13	$\mathbf{P_g}$	⋮	$T_{1g} \oplus 2T_{2g} \oplus 2G_g \oplus 2H_g$	P_u	⋮	$T_{1u} \oplus 2T_{2u} \oplus 2G_u \oplus 2H_u$
14	Q_g	⋮	$T_{1g} \oplus T_{2g} \oplus 2G_g \oplus 3H_g$	$\mathbf{Q_u}$	⋮	$T_{1u} \oplus T_{2u} \oplus 2G_u \oplus 3H_u$
15	$\mathbf{R_g}$	⋮	$A_g \oplus 2T_{1g} \oplus 2T_{2g} \oplus 2G_g \oplus 2H_g$	R_u	⋮	$A_u \oplus 2T_{1u} \oplus 2T_{2u} \oplus 2G_u \oplus 2H_u$
⋮	⋮	⋮	⋮	⋮	⋮	⋮

not in the center of the cavity at $r = 0$, but at some nonzero radial value. For such systems, the harmonic basis will no longer be appropriate. On the other hand, neither is a harmonic basis necessary, in order for the INS selection rule to remain valid. This is an important group-theory-based generalization that goes beyond the early explicit-integral results, although the latter have since been generalized as well – e.g., for the HF@C_{60} system [19] (section 1.3.3.3).

1.2.3 H_2@C_{60}

1.2.3.1 Symmetry

For H_2@C_{60}, the two H atoms will interact strongly with an incident neutron beam. These atoms are also strongly entangled to each other – but again, are not expected to be appreciably entangled with the C_{60} host. The relevant quantum system is thus the H_2 guest molecule itself. The corresponding $\hat{H}$ is constructed by fixing the C_{60} cage, and allowing the positions of the two H atoms, $\vec{r}_1 = (x_1, y_1, z_1)$ and $\vec{r}_2 = (x_2, y_2, z_2)$, to vary. The explicit dimensionality of the quantum dynamical problem is thus six (6D), corresponding to three overall translational coordinates, plus two rotational coordinates, plus a single vibrational coordinate (whose inclusion would form the "TRV $\hat{H}$").

In practice, the H_2 vibrational motion is often frozen, so that $r = |\vec{r}_1 - \vec{r}_2| = \text{const}$ – thereby reducing the dimensionality of the resultant "TR $\hat{H}$" to five (5D) [3–5, 9]. The rationale for this simplification is that the fundamental vibrational frequency of (gas phase) H_2 is 4142 cm^{-1} – which is orders of magnitude larger than the typical nanoconfined TR level spacing, and the energy range of the INS experiments. Nevertheless, even the ground vibrational state of H_2@C_{60} is not very localized in r – which could lead to small corrections in the low-lying TR levels, thus providing some motivation for performing TRV calculations in full 6D. Here, however, we focus only on 5D TR calculations.

Following analogous arguments as in section 1.2.2, we find that the TR (or even TRV) $\hat{H}$ is again invariant with respect to all I_h symmetry operations, applied to both H atoms simultaneously. In addition, exchange of the two H atom positions – i.e., $\vec{r}_1 \leftrightarrow \vec{r}_2$ – also leaves $\hat{H}$ invariant. Since this (12) exchange permutation operation commutes with all of the I_h point operations, the H_2@C_{60} symmetry group is thus

$$I_h^{(12)} = I_h \otimes S_2 = I_h \otimes \{E, (12)\}. \tag{1.3}$$

If Γ is an I_h irrep, we denote the corresponding permutation symmetric/antisymmetric irrep of the H_2@C_{60} symmetry group as $\Gamma^{(s/a)}$.

Also as per section 1.2.2, the low-lying 5D TR eigenstates of H_2@C_{60} have been found to be nearly spherically symmetric [9] – leading again to a natural angular-momentum-based labeling scheme. Consider the usual center-of-mass coordinate transformation,

$$\vec{R} = (\vec{r}_1 + \vec{r}_2)/2 \quad ; \quad \vec{r} = (\vec{r}_1 - \vec{r}_2). \tag{1.4}$$

Working in spherical coordinates $\vec{R} = (R, \Theta, \Phi)$ and $\vec{r} = (r, \theta, \phi)$, and recalling that $r = \text{const}$, a natural uncoupled basis is thus

$$\mathbf{R}_{nl}(R) Y_{lm_l}(\Theta, \Phi) Y_{jm_j}(\theta, \phi). \tag{1.5}$$

1.2.3.2 Spherical Basis and Eigenstates

By combining the two angular momenta – i.e., the "orbital" $\vec{l}$ associated with translation, and the "rotational" $\vec{j}$ – one obtains a TR coupled basis, $|nj\lambda\, l\, m_\lambda\rangle = |nl\rangle |j\lambda\, l\, m_\lambda\rangle$, which can be given explicitly as follows:

$$\langle R, \Theta, \Phi, \theta, \phi | nj\lambda\, l\, m_\lambda \rangle = \mathbf{R}_{nl}(R) \sum_{m_l, m_j} C^{\lambda m_\lambda}_{l m_l j m_j} Y_{l m_l}(\Theta, \Phi) Y_{j m_j}(\theta, \phi). \tag{1.6}$$

In Eq. (1.6), the $C^{\lambda m_\lambda}_{l m_l j m_j}$ are the Clebsch-Gordan coefficients [40–42]. As in the case of $H@C_{60}$, the actual 5D TR eigenstate wavefunctions of $H_2@C_{60}$ closely resemble the spherical basis functions themselves [i.e., Eq. (1.2) for $H@C_{60}$, or Eq. (1.6) for $H_2@C_{60}$]. As a caveat, we again note that *any* $\mathbf{R}_{nl}(R)$ functions might be used to form a basis, or might emerge as the (nearly) spherical eigenstate solutions. As discussed in section 1.2.2, the choice of $\mathbf{R}_{nl}(R)$ has no impact on symmetry or INS selection rules – although to date, an isotropic harmonic oscillator form has always been used. Finally, taking nuclear spins into account, we note that exchange antisymmetry implies even/odd j values that correspond to (s/a) permutation symmetry – i.e., to para-(p-H_2) and ortho-(o-H_2), respectively.

In Table 1.2, we present the $H_2@C_{60}$ eigenstate energy levels, as calculated numerically in [9], and used as a basis for assessing experimental INS transitions [10]. Various labels are used, including the coupled spherical basis set quantum numbers, (n, j, λ, l), together with the corresponding K_h and (through Table 1.1) I_h irrep labels. Note that unlike the K_h labels, the I_h labels *exactly* match the observed level degeneracies. In cases where multiple I_h irreps correlate to a single K_h irrep (i.e., for $\lambda \geq 3$), the corresponding level splitting provides a quantitative measure of that eigenstate's deviation from true spherical symmetry – i.e., of the effect of icosahedral "corrugation." These cases are indicated in Table 1.2 using italicised pairs. From the table, these splittings are indeed seen to be very small indeed – just a few tenths or hundredths of a cm^{-1}, on an energy scale ranging over several hundred cm^{-1}. Not only are the true $H_2@C_{60}$ eigenstates very close to spherical as predicted – implying that λ and m_λ are excellent quantum numbers – but we also find that n, j, and l are good quantum numbers, as well.

The level structure as presented in Table 1.2 has much to tell us about the underlying TR quantum dynamics. A very detailed account of the physical significance – including, even, a discussion of the precise Hamiltonian source of the tiniest I_h level splittings discussed above [13, 15] – may be found in the aforementioned papers by Bačić, Horsewill, and coworkers. In broad terms, we see that the rotational level spacings are on the order of 100 cm^{-1}, whereas the translational level spacings are both smaller (in l) and larger (in n). Finally, for a given (n, j, l), we see that the λ level spacings are only around 10 cm^{-1} or less, with the levels for the *extremal* λ values [or the even $(j + l - \lambda)$ values] lying *highest* in energy.

1.2.3.3 Energy Level Ordering with Respect to λ

The last trend mentioned above is a bit puzzling, as it implies a *non*-monotonic energy level ordering with respect to λ. The situation bears further scrutiny, also, because it turns out to be related to the INS selection rule. One explanation can be provided through an analysis of Eq. (1.6). Let us redefine just the angular part of this basis as follows:

$$\mathbf{Y}_{j\lambda\, l\, m_\lambda}(\theta_1, \phi_1, \theta_2, \phi_2) = \sum_{m_l, m_j} C^{\lambda m_\lambda}_{l m_l j m_j} Y_{l m_l}(\theta_1, \phi_1) Y_{j m_j}(\theta_2, \phi_2). \tag{1.7}$$

Exchanging coordinates, $1 \leftrightarrow 2$, and also indices, $j \leftrightarrow l$,

$$\mathbf{Y}_{l\lambda\, j\, m_\lambda}(\theta_2, \phi_2, \theta_1, \phi_1) = \sum_{m_j, m_l} C^{\lambda m_\lambda}_{j m_j l m_l} Y_{j m_j}(\theta_2, \phi_2) Y_{l m_l}(\theta_1, \phi_1). \tag{1.8}$$

We next use the identity, $C^{\lambda m_\lambda}_{j m_j l m_l} = (-1)^{j+l-\lambda} C^{\lambda m_\lambda}_{l m_l j m_j}$, easily obtained from the well-known formula for exchanging two columns of the Wigner 3-j symbol [40–42]. Substituting into Eq. (1.8) above, and also flipping the order of the spherical harmonic factors, yields

$$\mathbf{Y}_{l\lambda\, j\, m_\lambda}(\theta_2, \phi_2, \theta_1, \phi_1) = (-1)^{j+l-\lambda} \mathbf{Y}_{j\lambda\, l\, m_\lambda}(\theta_1, \phi_1, \theta_2, \phi_2). \tag{1.9}$$

Table 1.2 Energy levels E and degeneracies g for $H_2@C_{60}$, from [9]. Energies are in cm^{-1}, relative to the p-H_2 ground state. Columns III and IV list (n, j, λ, l) and corresponding K_h irrep labels; Column V lists $I_h^{(12)}$ irrep labels. Italics indicate K_h irreps that correlate to multiple I_h levels. Column VI is the spectroscopic parity, $\sigma = (-1)^{j+l+\lambda} = (-1)^{j+l-\lambda}$, whereas Column VII is $(j + l - \lambda)$ itself. INS transitions from (0, 0, 0, 0) to the states listed in bold are forbidden, according to the restricted INS selection rule of [9] and section 1.1.

E	g	(n, j, λ, l)	K_h	$I_h^{(12)}$	σ	$(j + l - \lambda)$
0.00	1	(0, 0, 0, 0)	S_g	$A_g^{(s)}$	+1	0
116.53	3	(0, 1, 1, 0)	P_u	$T_{1u}^{(a)}$	+1	0
179.39	3	(1, 0, 1, 1)	P_u	$T_{1u}^{(s)}$	+1	0
290.63	**3**	**(1, 1, 1, 1)**	$\mathbf{P_g}$	$\mathbf{T_{1g}^{(a)}}$	**−1**	**1**
297.02	5	(1, 1, 2, 1)	D_g	$H_g^{(a)}$	+1	0
307.14	1	(1, 1, 0, 1)	S_g	$A_g^{(a)}$	+1	2
347.80	5	(0, 2, 2, 0)	D_g	$H_g^{(s)}$	+1	0
374.95	5	(2, 0, 2, 2)	D_g	$H_g^{(s)}$	+1	0
400.93	1	(2, 0, 0, 0)	S_g	$A_g^{(s)}$	+1	0
483.20	**5**	**(2, 1, 2, 2)**	$\mathbf{D_u}$	$\mathbf{H_u^{(a)}}$	**−1**	**1**
492.65	*4*	*(2,1,3,2)*	F_u	$G_u^{(a)}$	+1	0
493.16	*3*	*(2,1,3,2)*	F_u	$T_{2u}^{(a)}$	+1	0
494.66	3	(2, 1, 1, 2)	P_u	$T_{1u}^{(a)}$	+1	2
522.48	3	(2, 1, 1, 0)	P_u	$T_{1u}^{(a)}$	+1	0
522.56	**5**	**(1, 2, 2, 1)**	$\mathbf{D_u}$	$\mathbf{H_u^{(s)}}$	**−1**	**1**
528.49	*4*	*(1,2,3,1)*	F_u	$G_u^{(s)}$	+1	0
528.55	*3*	*(1,2,3,1)*	F_u	$T_{2u}^{(s)}$	+1	0
533.45	3	(1, 2, 1, 1)	P_u	$T_{1u}^{(s)}$	+1	2
583.78	*4*	*(3,0,3,3)*	F_u	$G_u^{(s)}$	+1	0
584.65	*3*	*(3,0,3,3)*	F_u	$T_{2u}^{(s)}$	+1	0
625.42	3	(3, 0, 1, 1)	P_u	$T_{1u}^{(s)}$	+1	0

The left and right sides of Eq. (1.9) involve an exchange of both indices and coordinates. We now consider the special case where the coordinates are *equal*, i.e. $(\theta_1, \phi_1) = (\theta_2, \phi_2) = (\theta, \phi)$. From Eq. (1.7), we see that

$$\mathbf{Y}_{j\lambda\, l\, m_\lambda}(\theta, \phi, \theta, \phi) = \mathbf{Y}_{l\lambda\, j\, m_\lambda}(\theta, \phi, \theta, \phi). \tag{1.10}$$

Substitution into Eq. (1.9) then yields

$$\mathbf{Y}_{j\lambda\, l\, m_\lambda}(\theta, \phi, \theta, \phi) = (-1)^{j+l-\lambda}\mathbf{Y}_{j\lambda\, l\, m_\lambda}(\theta, \phi, \theta, \phi). \tag{1.11}$$

Note that the two functions on each side of Eq. (1.11) are *identical*, implying that the function itself must be *zero* when $(j + l - \lambda)$ is odd. Consequently, the eigenstate wavefunction *vanishes* when $\vec{R}$ and $\vec{r}$ point in the same direction [43].

Now, the geometries that correspond to collinear $\vec{R}$ and $\vec{r}$ vectors are those for which the H_2 molecule points radially outward from the center of the C_{60} cage. For fixed $R = |\vec{R}|$ and $r = |\vec{r}|$, these are the geometries for which one H atom gets closest to the C_{60} wall – and which therefore have the highest energy. Since they also correspond to even values of $(j+l-\lambda)$, it follows (in a kind of "Hund's Rule" sense) that these states should lie higher in energy than the odd-$(j+l-\lambda)$ states. The mathematical derivation provided above is typical of – and generally much simpler than – the sorts of spherical basis manipulations that are applied in the explicit integral context, for which lengthy appendices are often required. In contrast, a *much* simpler explanation for the λ trend – even than that provided above – can be obtained from a simple symmetry argument (section 1.3.3.2).

1.2.4 HX@C_{60}

Finally, we consider the HX@C_{60} case, which is perhaps the most interesting from the INS selection rule perspective. As discussed in section 1.2.4, due to entanglement, we must treat the entire HX molecule as the relevant quantum system – even though the X nucleus has essentially zero interaction with the incident neutron beam. Even so, its presence serves to induce new selection rules, not observed for H@C_{60}, through EISR.

Roughly speaking, we can follow the same procedure as in section 1.2.3. The biggest difference is that we no longer have (12) exchange permutation symmetry, as a consequence of which the HX@C_{60} symmetry group is just I_h. There is also an important difference in the center-of-mass coordinate transformation:

$$\vec{R} = (m_1\vec{r}_1 + m_2\vec{r}_2)/(m_1 + m_2) \qquad ; \qquad \vec{r} = (\vec{r}_1 - \vec{r}_2). \tag{1.12}$$

Working in spherical coordinates as before, we can generate a suitable uncoupled basis of the same form as Eq. (1.5), and a coupled basis of the form of Eq. (1.6).

Here, however, we encounter a big difference with H_2@C_{60}, which is that the HX@C_{60} eigenstates are *not* generally well described by Eq. (1.6) [4, 12, 15, 16, 19]. From a group theory perspective, whereas the near perfect spherical symmetry of $\hat{H}$ ensures that λ and m_λ will be good quantum numbers, *there are no such assurances with regard to n, j and l.* Indeed, through explicit calculation, Bačić and coworkers first demonstrated that *j* is *not* a good quantum number for HD@C_{60} [4, 12, 18], – and very recently, that both *l and j* are not good quantum numbers for HF@C_{60} [19]. For HF@C_{60}, the isotropic harmonic oscillator *n* was observed not to be good either, although this is no guarantee that some other choice might not work better.

Despite the fact that *j* and *l* are not good quantum numbers, Bačić and coworkers have shown that the "most general INS selection rule" from section 1.1 is still valid, even in this more general context. This approach works, because in the Eq. (1.6) basis set expansions of the true eigenstates, one finds that only even or odd $(j+l)$ values are included – at least for HD@C_{60} and HF@C_{60}. But what about more general HX@C_{60} systems? What guarantee is there that the property holds for *all* such systems?

This example underscores the advantages of a group theory approach – and more properly, of working with the *symmetry characters* or irreps of quantum states, rather than with a particular choice of basis set or quantum numbers. The former are *always* reliable, and can be universally applied; the latter, as we have seen above, are not. In this case, it turns out that the general procedure proposed above *is*, in fact always valid, because the even/odd restriction of $(j+l)$ is always enforced by the *inversion parity* of the corresponding irreps. Further discussion will be deferred to section 1.3.

1.3 INS Selection Rule for Spherical (K_h) Symmetry

1.3.1 Inelastic Neutron Scattering

As discussed in section 1.1, INS (and also quasielastic neutron scattering) has proven to be an extremely powerful and precise tool for analyzing nanoconfined hydrogen [7, 8, 10, 12, 24, 26, 28, 44, 45]. In INS experiments the "bright" hydrogen atom nuclei interact strongly with the incident monochromatic neutron beam, resulting in inelastic scattering of the latter. A comparison of incident and scattered beam energies then provides transition energies between the initial and final quantum states of the target.

An important feature of the INS interaction operator $\hat{M}$ is that it has a *preferred spatial direction*, defined by the "neutron momentum transfer vector," $\vec{k} = (\vec{k}_{\text{scattered}} - \vec{k}_{\text{incident}})$. However, for target systems with (nearly) perfect spherical symmetry, which serve as the main focus of this chapter, all directions are (essentially) equivalent. The orientation of the beam is thus largely immaterial in practice, although we shall revisit this issue in section 1.4.

The INS interaction itself involves the nuclear force, and operates on a scale of 10^{-14}–10^{-15} m. On the other hand, the wavelength of the neutron beam is on the atomic or molecular scale – i.e., typically around 10^{-10} m. To compute scattering cross sections, it is therefore appropriate to use S wave scattering and the first Born approximation, in terms of which the INS interaction operator takes the form [9, 26, 28–30]

$$\hat{M} = \sum_n \hat{b}_n \hat{I}_n = \sum_n \hat{b}_n \exp\left(i\vec{k} \cdot \vec{r}_n\right). \tag{1.13}$$

In Eq. (1.13) above, the summation runs only over the *bright* nuclei (i.e., H atoms) in the quantum system.

The $\hat{b}_n$ operators in Eq. (1.13) are "scattering length operators." They consist of both spin and scalar contributions, corresponding respectively to coherent and incoherent INS scattering cross sections. In practical terms, insofar as determining the forbidden (i.e. zero-intensity) transitions is concerned, it suffices to ignore the nuclear spin states altogether and concentrate solely on the *spatial* wavefunctions of the quantum system – as are presumed throughout this work. A transition from initial state $|\Psi_i\rangle$ to final state $\langle\Psi_f|$ is then forbidden if and only if the transition integral with the spatial part of the INS interaction operator, $\hat{I}_n$, is zero for all n:

$$I_n^{f\leftarrow i} = \langle\Psi_f| \exp\left(i\vec{k} \cdot \vec{r}_n\right) |\Psi_i\rangle = 0 \qquad \text{for all } n \text{ implies forbidden } f \leftarrow i \text{ transition} \tag{1.14}$$

For H_2 and HX, the individual $\hat{I}_n$ interactions can be rewritten in terms of the center-of-mass coordinate vectors $\vec{R}$ and $\vec{r}$, using Eqs (1.4) and (1.12):

$$\hat{I}_n = \exp\left(i\vec{k} \cdot \vec{R}\right) \exp[i(-1)^{n-1}\, \vec{k} \cdot \vec{r}/2] \qquad \text{for } H_2 \text{ guest} \tag{1.15}$$

$$\hat{I}_1 = \hat{I} = \exp\left(i\vec{k} \cdot \vec{R}\right) \exp\left(i\alpha\vec{k} \cdot \vec{r}\right) \qquad \text{for HX guest} \tag{1.16}$$

In Eq. (1.16) above, $\alpha = m_1/(m_1 + m_2)$, where $m_1 = m_H$ and $m_2 = m_X$. The differences between Eqs (1.15) and (1.16), and between $\hat{I}_n$ and $\hat{I}_1$ are minor; in truth, they all behave identically, insofar as the INS selection rule of Eq. (1.14) is concerned. It therefore suffices to consider just $\hat{I}_1 = \hat{I}$ with $\alpha = 1/2$, without loss of generality.

1.3.2 Group Theory Derivation of the INS Selection Rule

1.3.2.1 Group-Theory-Based INS Selection Rule for Cylindrical ($C_{\infty v}$) Environments

Standard group theory dictates [37] that the vanishing of the INS transition matrix element of Eq. (1.14) can be determined from the triple-direct-product representation formed from the irreps of $\langle\Psi_f|$, $\hat{I}$, and $|\Psi_i\rangle$, in the symmetry group for the quantum system. The product representation is generally not irreducible, but can be reduced to a direct sum of an integer number of copies, a_j, of every group irrep, Γ^j:

$$\Gamma = \Gamma(\Psi_f) \otimes \Gamma(\hat{I}) \otimes \Gamma(\Psi_i) = a_0\Gamma^0 \oplus a_1\Gamma^1 \oplus \cdots \tag{1.17}$$

The transition intensity is then identically zero provided that $a_0 = 0$ – where by convention, Γ^0 labels the totally symmetric irrep.

At this point we encounter an immediate technical difficulty for spherical systems [11], which is that *the INS interaction $\hat{I}$ does not belong to any K_h irrep!* This stems from the aforementioned fact that $\hat{I}$ has a preferred direction – i.e., that of the vector $\vec{k}$. (In contrast, the familiar electric dipole operator from optical spectroscopy has no preferred direction, and can therefore be represented in any molecular point group.) To rectify the situation, one strategy is to replace K_h with the largest subgroup in which $\hat{I}$ can be represented. Taking $\vec{k}$ to define the positive z axis, the largest point group with a preferred direction is clearly the cylindrical group, $C_{\infty v}$. In this group, all of the various $\hat{I}_n$ forms of section 1.3.1 belong to the totally symmetric irrep, Σ^+.

The fact that $\hat{I}$ belongs to Γ^0 is a "lucky break," in that we need only consider the simpler direct product representation, $\Gamma(\Psi_f) \otimes \Gamma(\Psi_i)$ in Eq. (1.17). The right-hand side of Eq. (1.17) is thus provided by the $C_{\infty v}$ direct product table, Table 1.3. *Only* products that contain at least one copy of Σ^+ correspond to *allowed* transitions. From the table, these are found to correspond to the diagonal entries. This leads to the following, cylindrical version of the INS selection rule:

> **most general INS selection rule for any guest molecule in any cylindrical ($C_{\infty v}$) environment**: Transitions for which $\Gamma(\Psi_f) = \Gamma(\Psi_i)$ are allowed; all others are forbidden.

From a group theory standpoint, the key symmetry operation associated with the above INS selection rule is the *vertical reflection* operation. The character of a given irrep or quantum state under vertical reflection is called the *(e/f)* or *spectroscopic parity*, σ [37, 38]. The singly degenerate irreps $\Sigma^\pm$ have definite $\sigma = \pm 1$ character under vertical reflection. The remaining irreps, which are all doubly-degenerate, have zero σ character – meaning that one +1 state and one −1 state can always be constructed through suitable linear combinations of any given degenerate pair.

1.3.2.2 Group-Theory-Based INS Selection Rule for Spherical (K_h) Environments

The above is all fine and good, for a cylindrical system that is perfectly aligned with $\vec{k}$; however, we must still deal with the fact that in reality, our Ψ_i and Ψ_f states belong to the K_h group, not to $C_{\infty v}$. This situation is addressed using the correlation table between K_h and $C_{\infty v}$ – which is also presented in Table 1.1. From the table, we see that every degenerate ($\lambda > 0$) irrep of K_h contains one copy of the $C_{\infty v}$ irrep, Π. Thus, the direct product of *any* two (i.e. not necessarily the same) degenerate K_h irreps will include at least one copy of Σ^+, through $\Pi \otimes \Pi = \Sigma^+ \oplus \Sigma^- \oplus \Delta$ (where the latter decomposition comes from Table 1.3). Therefore, all transitions between degenerate K_h irreps are allowed.

For the case where at least one of the $\Gamma(\Psi_i)$ and $\Gamma(\Psi_f)$ K_h irreps is singly degenerate (i.e., $\lambda = 0$), we first note that every K_h irrep decomposition in Table 1.1 contains exactly one copy of either Σ^+

Table 1.3 Direct product table for the $C_{\infty v}$ point group. The $\Gamma(\Psi_f)$ irreps are listed in the first column; the $\Gamma(\Psi_i)$ irreps are listed in the first row. The totally symmetric irrep, $\Gamma^0 = \Sigma^+$, is realized in the $\Gamma(\Psi_f) \otimes \Gamma(\Psi_i)$ direct product only when $\Gamma(\Psi_f) = \Gamma(\Psi_i)$.

	$\mathbf{\Sigma^+}$	$\mathbf{\Sigma^-}$	$\mathbf{\Pi}$	$\mathbf{\Delta}$	$\cdots$
Σ^+	Σ^+	Σ^-	Π	Δ	$\cdots$
Σ^-	Σ^-	Σ^+	Π	Δ	$\cdots$
Π	Π	Π	$\Sigma^+ \oplus \Sigma^- \oplus \Delta$	$\Pi \oplus \Phi$	$\cdots$
Δ	Δ	Δ	$\Pi \oplus \Phi$	$\Sigma^+ \oplus \Sigma^- \oplus \Gamma$	$\cdots$
$\vdots$	$\vdots$	$\vdots$	$\vdots$	$\vdots$	$\ddots$

or Σ^-. According to Table 1.3, forbidden transitions will therefore be those for which one of the two K_h irreps contains Σ^+, and the other contains Σ^-. Note that vertical reflection is again the key; this symmetry operation belongs to the K_h group as much as to the $C_{\infty v}$ group. Furthermore, the character of a given K_h irrep under vertical reflection is either ± 1, so that we can continue to take this as our definition of σ, the spectroscopic parity. In Table 1.1, the K_h irreps with $\sigma = -1$ (i.e., those that contain Σ^-) are presented in bold, whereas those with $\sigma = +1$ are not in bold.

Note the alternating pattern, which results in the following useful relation between σ, and the K_h irrep labels, λ and p:

$$\sigma = p(-1)^{\lambda} = p(-1)^{-\lambda} \tag{1.18}$$

In Eq. (1.18) above, λ is the "total angular momentum," and $p = \pm 1$ is the total or (g/u) inversion parity. It must be stressed that Eq. (1.18) is a *universal relation*, since all of the quantities are group-theory based. Note that both forms given in Eq. (1.18) are equivalent, since λ is an integer. However, both forms are useful; the second connects with the earlier selection rules as presented in section 1.1, whereas the first generalizes in the case where there are half-integer angular momenta quantities due to spin.

Putting together the various pieces above, in terms of spectroscopic parity, we obtain the following completely general spherical INS selection rule:

> **most general INS selection rule for any guest molecule in any spherical (K_h) environment**: Transitions for which $\Gamma(\Psi_f)$ and/or $\Gamma(\Psi_i)$ are singly degenerate, and for which $\sigma_f \sigma_i = -1$, are forbidden; all others are allowed.

From Table 1.1, a version in terms of explicit K_h irreps can also be provided:

> **most general INS selection rule for any guest molecule in any spherical (K_h) environment**: Transitions between S_g and $\{S_u, P_g, D_u, F_g, G_u, \ldots\}$ are forbidden; transitions between S_u and $\{S_g, P_u, D_g, F_u, G_g, \ldots\}$ are forbidden; all other transitions are allowed.

1.3.3 Specific Systems, Quantum Numbers, and Basis Sets

The INS selection rules given at the end of section 1.3.2.2 are *always correct* for spherical environments, regardless of the specific details of the host cage or guest molecule. Nevertheless, it can also be useful to have specific forms that are tailored to the quantum numbers and basis sets used in specific applications. Note that p and λ are *always* well-defined quantum numbers for any K_h system; moreover, these quantities are usually straightforward to obtain for a given explicit eigenstate or basis representation. Using Eq. (1.18), it is then straightforward to obtain a system- and basis-set-specific form of σ. The individual cases are presented below, and in section 1.3.4.

1.3.3.1 H@sphere

For H@sphere systems, all eigenstate wavefunctions are necessarily of the form of Eq. (1.2), for which λ is specified. Furthermore, $p = (-1)^\lambda$ [35, 36], as can be shown by transforming the solid angle coordinates and spherical harmonics under the inversion symmetry operation:

$$\theta \to (\pi - \theta) \quad ; \quad \phi \to (\pi + \phi) \quad ; \quad Y_{\lambda,m_\lambda}(\theta, \phi) \to (-1)^\lambda \, Y_{\lambda,m_\lambda}(\theta, \phi) \tag{1.19}$$

In any event, from Eq. (1.18), we find that *all* H@sphere states have $\sigma = +1$. Thus, only the K_h irreps that are *not* in bold in Table 1.1 are physically realized. More importantly, there are no transitions that can change the sign of σ; consequently, *all transitions are allowed.*

1.3.3.2 H_2@sphere

For $H_2@C_{60}$ (and likely other sphere-like cavities), the true eigenstates are well-approximated by a coupled basis of the form of Eq. (1.6) – for which λ again directly appears as a quantum number. As for p, let us first consider the uncoupled basis of Eq. (1.5). Here, the total inversion symmetry operation applies separately to both (θ, ϕ) and (Θ, Φ). Accordingly, the total inversion parity is just the product of the parities for each individual spherical harmonic factor – i.e., $p = (-1)^j(-1)^l = (-1)^{j+l}$. Importantly, Eq. (1.6) does not combine uncoupled basis functions of different j or l values. Thus, j and l are good quantum numbers for the coupled basis (and for the eigenstates themselves) and so the above expression for p is still valid.

From Eq. (1.18), this leads at once to

$$\sigma = (-1)^{j+l+\lambda} = (-1)^{j+l-\lambda}. \tag{1.20}$$

The sign of σ is thus determined by the even/oddness of $(j + l - \lambda)$. From the end of section 1.3.2.2, the forbidden transitions are thus those that change this character, and for which at least one of the two states is singly degenerate. This gives rise to the *most general INS selection rule for* $H_2@C_{60}$ as given in section 1.1.

As special cases, we can consider forbidden transitions starting from the p-H_2 and o-H_2 ground states. The p-H_2 ground state has S_g character; thus, transitions to all states with $\sigma = -1$ are forbidden. These are the states that correspond to the first list in the second INS selection rule in section 1.3.2.2 – or alternatively, the K_h irreps listed in bold in Table 1.1. In Table 1.2, the $\sigma = -1$ energy levels of $H_2@C_{60}$ are also listed in bold. INS transitions from the $\sigma = +1$ p-H_2 ground state $(0, 0, 0, 0)$, to the three bold states indicated in the table, are thus *forbidden*.

The o-H_2 case is more interesting. Here, the ground state has P_u character, so that we obtain:

INS selection rule for H_2 in any spherical (K_h) environment (from o-H_2 ground state): Transitions are forbidden to all S_u states; all other transitions are allowed.

Note that for H_2, *there are no* S_u states, and therefore no forbidden transitions from the o-H_2 ground state. Nevertheless, the above form may be useful in more general situations.

Finally, we revisit the trend discussed at the end of section 1.2.3, whereby the even $(j + l - \lambda)$ states lie higher in energy than the odd $(j + l - \lambda)$ states. A simple symmetry-based explanation is presented as follows. Consider those geometries for which $\vec{r}$, $\vec{R}$, and $\vec{k}$ are all coplanar (note the generalization beyond the *collinear* geometries of section 1.2.3). For these geometries, both of the H atoms also lie within the common plane, which is vertical. These are also the geometries with the highest energies. Quantum states with $\sigma = -1$ change sign under reflection through this vertical plane, and must therefore have zero amplitude at all such coplanar geometries. Consequently, their energies lie lower than the $\sigma = +1$ states.

1.3.3.3 HX@sphere

For HX@sphere systems, the uncoupled and coupled basis sets used to represent the TR $\hat{H}$ are identical to those of H_2, as provided in Eqs (1.5) and (1.6). Moreover, p and γ are still guaranteed by group theory to be good quantum numbers. Generally, speaking, however, j and l need not be good quantum numbers – or alternatively, the true eigenstates do not need to adhere to the form of Eq. (1.6).

In such cases, the true eigenstates can be expressed as expansions over the coupled basis functions – which, by spherical symmetry, must not combine different values of λ and p. Since $p = (-1)^{j+l}$, it is only the even or oddness of $(j + l)$ that matters, not the value itself. Thus, a single expansion taken over different j and l values is fine, provided that $(j + l)$ is always even or always odd – exactly as discussed in section 1.2.4.

Note that similar comments apply also to the spectroscopic parity, $\sigma = (-1)^{j+l-\lambda}$, which is also well defined over the eigenstate expansions described above, even if j and/or l are not. This means that $(j + l - \lambda)$ can continue to be used in the formulation of the *corresponding INS selection rule for HX@sphere systems, which is exactly as given in sections* 1.1 *and* 1.3.3.2. In particular, HX@sphere systems are also predicted to have forbidden INS transitions, although these have yet to be fully confirmed experimentally for HD@C_{60} and HF@C_{60} [12, 18, 19].

1.3.4 Beyond Diatomic Molecules

1.3.4.1 H_2O@sphere

For bent triatomic molecules in sphere-like cages – e.g. H_2O@C_{60} – a suitable TR $\hat{H}$ can still be defined by freezing the triatomic structure. A similar procedure to that used for H_2@C_{60} can be applied here as well, with the primary complication being that asymmetric rotor basis functions, $\langle \phi, \theta, \chi | j_{k_a k_c} m_j \rangle$, must replace the rotational spherical harmonics $Y_{jm_j}(\theta, \phi)$ in Eq. (1.5). As per the discussion in section 1.3.3.2, the total inversion parity is $p = p_j(-1)^l$, where p_j is the inversion parity of the asymmetric rotor state, $|j_{k_a k_c} m_j\rangle$. There are actually several different possible relations for p_j, depending on the type of molecule and the choice of body-fixed frame. For H_2O, it can be shown [46] that the correct relation is $p_j = (-1)^{k_c}$. From the relation for p above, together with Eq. (1.18), we are thus immediately led to the following

> **most general INS selection rule for** H_2O@C_{60}: Transitions are forbidden between states for which $(k_c + l + \lambda)$ changes from even to odd (or vice-versa), and at least one $\lambda = 0$.

It should be stressed that despite the simplicity of the above derivation, the conclusion is completely rigorous, and not merely an "educated guess." We do assume, however, that the true eigenstates are well described by the coupled form of Eq. (1.6) as modified for asymmetric rotor functions – i.e., by the basis functions, $|n j_{k_a k_c} \lambda\, l\, m_\lambda\rangle$. That this is the case for H_2O@C_{60} has been very well established [15, 16, 21].

In comparison with the corresponding $H_2@C_{60}$ INS selection rule of section 1.1, one of the most intriguing new features for $H_2O@C_{60}$ is the appearance of *forbidden transitions for pure rotational transitions*. Thus, TR coupling is not strictly necessary. For rotational transitions from the the p-H_2O ground state, we have

INS selection rule for pure rotational transitions in $H_2O@C_{60}$ (from p-H_2O ground state): Pure rotational transitions are forbidden to all states for which $(j + k_c) = \text{odd}$.

Although no previous quantum-number-based INS selection rule for $H_2O@C_{60}$ has ever been devised to our knowledge – neither through group theory nor the explicit integral approach – some important prior progress in this direction has nevertheless been made. In 2014, Horsewill and coworkers obtained an experimental INS spectrum, and were able to assign the pure rotational transitions from the p-$H_2O@C_{60}$ ground state [20], based on gas-phase rotational spectra. It thus became clear that certain states were missing, and that some sort of INS selection rule might be at play [12, 20]. A bit later, Felker and Bačić made further progress by computing the energy levels, quantum numbers, and K_h and I_h irreps for the $H_2O@C_{60}$ TR eigenstates [21]. Based on these calculations, they were able to make tentative assignments for several experimental transitions to translationally excited states – although no explicit INS transition integrals *per se* were computed. The need to make further progress was acknowledged.

We can now report that all *pure rotational transitions* that have been previously unambiguously assigned to the experimental INS spectrum of $H_2O@C_{60}$ are indeed *allowed*, according to the INS selection rule above. Moreover, several low-lying rotational transitions that are predicted to be *forbidden* (e.g., to 1_{10}) are conspicuous by their absence from the experimental spectrum. Similar comments apply to the translationally excited assignments of [21]; however, whereas the method of that study could not generate λ labels, we can now assign λ values using the new INS selection rule. In addition, there are two previous assignments that were ambiguous, in the sense that two distinct possibilities were provided [21]. In each case, only one of the two possibilities is allowed by the INS selection rule – again, allowing us to make definitive assignments for the first time. In Table 1.4 below, we present new assignments for just those transitions where the previous assignments were ambiguous and/or incomplete. Further details will be provided in an upcoming publication.

Table 1.4 Selected experimental transitions from the ground state of p-$H_2O@C_{60}$ [20], together with final state labels determined using the new INS selection rule. Only those transitions for which the previous assignments [21] were ambiguous and/or incomplete are included. Experimental transition energies (in meV) are provided in Column I; new assignment labels, in the form $(n, j_{k_a k_c}, l)$, $\lambda = \ldots$, are provided in Column II.

experimental transition energy	final state assignment
16.6	$(0,3_{03},0)$, $\lambda = 3$
22.8	$(1,1_{01},1)$, $\lambda = \{0, 2\}$
30.8	$(1,2_{12},1)$, $\lambda = \{1, 3\}$
39.4	$(1,2_{21},1)$, $\lambda = 2$
45.1	$(2,1_{10},2)$, $\lambda = 2$

1.3.4.2 CH_4@sphere

The $CH_4@C_{60}$ complex has not yet been experimentally realized, except in open fullerene form [15, 22]. However, such structures may well be synthesized in the future – if not for C_{60}, then perhaps for larger spherical cages. From a group theory standpoint, the procedure used for treating the CH_4@sphere TR eigenstates is very similar to that for the H_2O@sphere system. In particular, since the vibrational motion is frozen, the coordinates describing TR motion are exactly the same, despite the added number of atoms. Indeed, CH_4 is in some respects *simpler* than H_2O, because the rotor is *spherical*, and hence well-described by the symmetric rotor or Wigner rotation basis functions, $D^{j*}_{m_j k_j}(\phi, \theta, \chi)$. Of course, one complicating factor is the additional H atom permutation symmetry. In practice, this will restrict the set of K_h irreps that can be physically realized by the spatial wavefunctions, although this restriction should have no direct impact on the INS selection rule itself. Accordingly, permutation symmetry is ignored in the present analysis. The effect of the inversion symmetry operation acting on individual Wigner rotation states can be determined from a consideration of the manner in which the Euler angle rotational coordinates themselves are transformed:

$$\phi \to (\pi + \phi) \quad ; \quad \theta \to (\pi - \theta) \quad ; \quad \chi \to (\pi - \chi) \quad ;$$
$$D^{j*}_{m_j k_j}(\phi, \theta, \chi) \to (-1)^{j+k_j} \, D^{j*}_{m_j -k_j}(\phi, \theta, \chi) \tag{1.21}$$

Based on Eq. (1.21), a *parity-adapted* Wigner basis, $\overline{D}^{jp_j}_{m_j \overline{k}_j}(\phi, \theta, \chi)$, can be easily constructed from suitable linear combinations of the $\pm k_j \neq 0$ states. Here, $\overline{k}_j = |k_j|$, and $p_j = \pm 1$ is the requisite inversion parity. For a given j, each $\overline{k}_j > 0$ gives rise to one copy each of both the $(\lambda = j, p_j = +1)$ and $(\lambda = j, p_j = -1)$ K_h irreps. Thus, both g and u inversion parities are realized for *any* $j > 0$, unlike for spherical harmonics. For $k_j = 0$, the $D^{j*}_{m_j 0}$ behave exactly like the spherical harmonics themselves – i.e., $p_j = (-1)^j$. Thus, the $(2j+1)^2$ Wigner functions associated with a given value of j decompose into j copies of the $\lambda = j$ irreps of both parities, as per Table 1.5. Note that all K_h irreps are realized except S_u.

The irreps in each decomposition in Table 1.5 can be labelled by j, $\overline{k}_j$, and p_j. These labels are thus good quantum numbers for the coupled basis as well as the uncoupled basis. Conversely, the index m_j still labels individual degenerate states within a given irrep, and so gets summed over when forming the coupled basis. Each coupled basis function is thus labelled $|nj\overline{k}_j p_j \lambda \, l \, m_\lambda\rangle$; such states are anticipated to provide a good description of the true TR eigenstates of $CH_4@C_{60}$.

Table 1.5 Irrep decomposition of the $D^{j*}_{m_j k_j}(\phi, \theta, \chi)$ Wigner rotation function representation of the K_h symmetry group. For all $j > 0$, both g and u inversion parities are realized.

j	K_h irrep decomposition
0	S_g
1	$P_g \oplus 2P_u$
2	$3D_g \oplus 2D_u$
3	$3F_g \oplus 4F_u$
4	$5G_g \oplus 4G_u$
5	$5H_g \oplus 6H_u$
⋮	⋮

From Table 1.5, it is convenient to introduce another parity quantity, $\epsilon = p_j(-1)^j$, which is essentially the spectroscopic parity of the *rotor state alone* (note that $\epsilon = 1$ for all $H_2@C_{60}$ TR states). When $\epsilon = +1$, the corresponding k_j values range from 0 to j; otherwise they range from 1 to j. In terms of ϵ, the total spectroscopic parity becomes $\sigma = \epsilon(-1)^{j+l-\lambda}$. This leads to the following

> **most general INS selection rule for $CH_4@C_{60}$:** Transitions are forbidden between states for which at least one $\lambda = 0$ and *either* (a) $(j + l - \lambda)$ changes even/oddness and ϵ stays the same, *or* (b) the $(j + l - \lambda)$ even/oddness stays the same and ϵ changes sign.

Again, further details will be provided in an upcoming publication.

1.3.4.3 Any Guest Molecule in any Spherical (K_h) Environment

Because it is entirely group theory based, the completely general INS selection rules provided at the end of section 1.3.2.2 are applicable to *any* guest molecule in *any* spherical environment. Moreover, insofar as the TR $\hat{H}$ eigenstates are concerned, the basis sets as described in Secs. 1.3.4.1 and 1.3.4.2 can be universally applied. More generally, however, one may wish to include the vibrational motions explicitly – thus working with the TRV $\hat{H}$ and eigenstates. If the total number of atoms in the guest molecule is less than four, then the vibrational states are completely unaffected by inversion parity, and will have no effect on the INS selection rule. However, for four or more atoms, the inversion symmetry operation affects vibrational as well as rotational states. Since the INS selection rule depends on spectroscopic parity, which in turn depends on inversion parity, it will become necessary to assess the impact of inversion – first on the vibrational coordinates themselves, and then on the corresponding basis functions.

1.4 INS Selection Rules for Non-Spherical Structures

The analysis provided in section 1.3 – particularly in section 1.3.2.2 – is completely general, except in one important respect: it presumes a (nearly) spherical cage environment. For C_{60} complexes such as $H_2@C_{60}$, very substantial evidence has been amassed that the true quantum eigenstates are indeed very nearly spherical. Some of this evidence is presented in sections. 1.1 and 1.2; more is presented in earlier work [3, 9, 12, 14–16]; still more has been gathered independently by the author, through group-theory-based numerical investigations of the I_h decompositions of the low-lying coupled basis functions of Eq. (1.6), as well as the $\hat{I}_n$ INS interaction operator itself. Of course, the fact that all C_{60} INS experiments conducted to date (i.e. for $H_2@C_{60}$, $HD@C_{60}$, $HF@C_{60}$, and $H_2O@C_{60}$) appear to be consistent with the spherical INS selection rule (even though complete validation has thus far only been achieved for $H_2@C_{60}$) lends further credence to the use of the K_h symmetry group in this context.

Yet, we know that in reality, the *true* symmetry group for these C_{60} systems (ignoring nuclear permutation symmetry) is I_h, not K_h. Thus, the results obtained using K_h are only true in an *approximate sense* – i.e., the "forbidden" transitions are actually found to have very tiny intensities, in direct proportion to the extent to which the true system deviates from perfect sphericality. All of this begs the rather obvious question: why not simply work with the I_h group directly? How do we know that the observed experimental results are not *also* consistent with an I_h-based explanation? These are important questions, that also bring up the prospect of moving *beyond* spherical symmetry – so as to be able to derive new INS selection rules, not only for I_h, but also more general cage structures, in principle belonging to any other point group. All of these points will be addressed in this section.

We begin by demonstrating that the K_h INS selection rules that have been actually observed experimentally in $H_2@C_{60}$ are *inconsistent* with an I_h-based explanation. In Table 1.2, the $H_2@C_{60}$ levels listed in bold – i.e., T_{1g}, and H_u (in the I_h point group) – are experimentally inaccessible from the p-H_2 ground state, which has (I_h) A_g character (note that (12) labels are dropped here, because: (a) they do not apply to $HX@C_{60}$ systems; (b) they have no bearing on the INS selection rule). A consistent I_h INS selection rule will thus require that *all* transitions from A_g to T_{1g} or A_g to H_u be forbidden. Conversely, all transitions to I_h irreps that are *not* in bold in Table 1.2 – i.e., A_g, T_{1u}, T_{2u}, H_g, and G_u – must be allowed. However, this contradicts the K_h prediction from Table 1.1, which states that all transitions to (K_h) G_u states are forbidden – even though (K_h) G_u contains one copy of (I_h) G_u.

To understand why the spherical selection rule is the most correct one for interpreting C_{60} experiments – even though the symmetry group is not strictly K_h but I_h – we have to consider further the experimental details. To begin with, the target samples used are powders, whose grains are themselves polycrystalline. Given that neutron beam sizes are typically on the order of a few mm^2, this means that the orientations of the host C_{60} cages must generally be regarded as *random*, relative to the incident neutron beam [7–10, 27]. This state of affairs might not be especially significant for say, optical spectroscopy, but it has important ramifications for INS spectroscopy because of the preferred spatial direction.

For systems with (nearly) perfect spherical symmetry, the orientation of the cage structure relative to the neutron beam momentum transfer vector $\vec{k}$ does not (much) matter; for a sphere, all directions are equivalent, and so the system can accommodate any such $\vec{k}$. Specifically, in group theory terms, this means *restricting* the set of all symmetry operations from the original group, down to just those operations that leave $\vec{k}$ invariant. In the case of K_h, once this restriction is applied, the symmetry operations that remain are found to comprise the *cylindrical subgroup*, $C_{\infty v}$ – with the vertical or z axis naturally aligned with the vector $\vec{k}$, even though this direction effectively varies randomly from one cage to the next. For any point group other than K_h, however, a randomly-oriented $\vec{k}$ will generally *not* be aligned with the z axis – nor with any other symmetry axis. The subsequent restriction of the group's symmetry operations will then leave *nothing at all* remaining, apart from the identity operation. As a consequence, there will be no forbidden INS transitions.

The above argument serves to explain why *there are no INS selection rules for non-sphere-like systems* – at least not for powder samples. Historically speaking, this may, in turn, help explain why so few forbidden INS transitions have been observed in the past, and why their existence was not even *suspected* until relatively recently (although another possible explanation is that the level structure is often too fine to completely resolve individual transitions [47]). On the other hand, there is evidently some preliminary indication that $H_2@C_{70}$ may admit INS selection rules [48]. More specifically, when simulating the INS spectrum of $H_2@C_{70}$ using explicit calculations of INS transition integrals, an *orientational averaging* was applied, in order to mimic the experimental conditions (as was also done in all previous calculations with C_{60} cages [9, 10, 12, 13, 15, 29]). Since intensities are nonnegative, the orientational average can be (close to) zero if and only if the intensity is zero for every possible $\vec{k}$ direction. If this were true for a non-sphere-like system such as $H_2@C_{70}$, it would imply a bona-fide *non-spherical* INS selection rule that holds for all $\vec{k}$ orientations – for which group theory, evidently, provides no explanation.

Conversely, we can "turn the problem around," by using group theory analysis to actually *suggest new experiments* for which non-spherical INS selection rules *are* predicted. Evidently, what is required is a means of consistently orienting $\vec{k}$ across every cage structure throughout the sample. This could be difficult to set up in practice, as it obviously entails the use of large crystal samples instead of powders, although it might be possible, according to experimentalists [47]. As an added technical challenge, even crystalline forms will have more than one standard orientation for the cage structures. For example, recent work suggests that for crystalline C_{60}, the S_6 P form is prevalent [14–16], for which there are two distinct and alternating cage orientations. On the other hand, even pure crystalline C_{60} need not in principle exhibit perfect orientational ordering [49].

As a final technical challenge, we point out that the direction of $\vec{k}$ is *not* actually that of the incident neutron beam itself, although they are obviously related. Furthermore, whereas in principle the entire momentum transfer vector $\vec{k}$ can be measured experimentally, in practice, many newer neutron facilities focus solely on the transfer *energy*, $(\hbar^2/2m)\vec{k}\cdot\vec{k}$ – sacrificing directional information for improved signal [47, 50]. Evidently, there are still some active facilities where $\vec{k}$ can be measured, and it would be quite an achievement to see such "direction-specific INS selection rules" actually realized in the laboratory.

In principle, forbidden INS transitions of the direction-specific variety could emerge for almost *any* symmetric environment – e.g., not just I_h structures such as C_{60}, but also D_{5h} structures such as C_{70}. A full description of the group theoretical analysis for each case is beyond the purview of this chapter, but will form the focus of a future paper. Here, we simply sketch the general procedure to be followed. For the most part, this is analogous to that described above and in section 1.3.2. The first and most important step is to restrict the original group symmetry operations to just those that leave $\vec{k}$ invariant. This defines the appropriate *subgroup* to work with – i.e., the analog of $C_{\infty v}$, in the case of the K_h group. Since the INS interaction operator belongs to the totally symmetric irrep, Γ^0, of the resultant restricted subgroup, the forbidden transitions are those for which $a_0 = 0$ in the Eq. (1.17) expansion of $\Gamma(\Psi_f) \otimes \Gamma(\Psi_i)$.

Of course, this analysis pertains to the *subgroup* irreps; to connect with the original group irreps, we must use the appropriate correlation table, as in Table 1.1. As a final wrinkle, let us not forget that the resultant INS selection rules – and even the restricted subgroup itself – depend not only on the point group of the cage structure, but also on the direction of $\vec{k}$. Indeed, a "random" $\vec{k}$, not chosen along any symmetry axis, results in the trivial subgroup and no forbidden INS transitions, as already discussed. If $\vec{k}$ is chosen to lie along a symmetry axis other than the principal (z) axis, then the restricted group will necessarily be a proper subgroup – although there can be a labeling confusion, associated with the fact that the subgroup labels correspond to the direction of $\vec{k}$ rather than z. In any event, in this manner, the appropriate group-theory-based direction-specific INS selection rules may be obtained.

1.5 Summary and Conclusions

When small molecules are confined inside nanoscale cages, many interesting quantum dynamical effects emerge that are not seen in the gas phase. In particular, translational states become quantized and strongly coupled to the rotational states. In spherical or sphere-like caging environments, the situation is well-described using spherical harmonic basis functions for the translational "orbital angular momentum" states, and the appropriate rigid rotor states for the "rotational angular momentum." The TR coupling then emerges from the familiar angular momentum addition machinery. The latter step is not just a reasonable guess; it is, in fact, *required* by K_h group theory, as it generates the three rigorously good quantum numbers, λ, m_λ, and p. In any event, the final result is a coupled basis expansion of the form of Eq. (1.6), described (at a minimum) by the quantum numbers $(n, j, \lambda, l, m_\lambda)$ (with p determined by the other quantum numbers).

Ideally, the TR quantum eigenstates of the trapped guest molecule will conform to the coupled basis above, in which case, all of the above quantum numbers are good. However, group theory provides no assurances for n, j, or l, which may or may not be good, depending on the particular system. In practice, it appears that these are generally good numbers for systems with H atom exchange symmetry – e.g., $H_2@C_{60}$ and $H_2O@C_{60}$. On the other hand, strong asymmetry – e.g., for $HD@C_{60}$, and especially $HF@C_{60}$ – leads to more complicated eigenstates for which a superposition over multiple coupled basis functions is required. Here too, however, group theory provides us with a simplification, based on inversion parity. Specifically, $p = p_j(-1)^l$ is *always* true, and can be used to restrict the l and p_j values that appear in any given expansion (e.g., $(j + l)$ = even or odd, for the $HX@C_{60}$ systems).

The above overview represents the kind of synergy that can emerge from what we believe to be a winning general strategy: combining the group theory and explicit basis set approaches. The two approaches are inherently complementary – very much representing "top-down" and "bottom-up" philosophies, respectively. As discussed, the group theory approach is universally applicable and always rigorous, but is couched in an abstract language that does not always relate directly to the problem at hand. The explicit basis set approach, on the other hand, is built up from the pieces of the actual system. Those pieces (and their combinations) may vary substantially from system to system, and do not always provide good overall quantum numbers. However, they do offer specific, concrete information about the particular system. We argue not only that both viewpoints are essential in and of themselves, but that each is also necessary to advance the other. Without a consideration of the specific system and experimental details, it might not be clear which group one should even be working with. Conversely, without guidance from group theory, explicit basis sets for complex systems can quickly lead to a morass of possibilities from which there may be no clear path forward. A combined strategy is best, but it requires that the results from each approach be "translated" into the language of the other – as we have sought to achieve throughout this chapter.

If the combined strategy is beneficial when dealing with basis sets and eigenstates, it is considerably *more* useful in the context of spectroscopic selection rules – where one has not only both bra and ket states to contend with, but also an interaction operator that may (as in the case of INS) come with its own peculiar quirks. In the explicit basis set approach to computing explicit transition intensities, it becomes necessary to perform complex integrations that are *especially* tedious, given that each coupled basis function must be expanded into uncoupled form in order to reduce the dimensionality to a manageable level. Insofar as determining selection rules is concerned, the general group theory decomposition procedure of Eq. (1.17) is certainly far simpler – and has the great advantage that it need only be performed "once for all time," for a given point group. Moreover, the results are very straightforward to apply to specific systems and basis representations, once proper irrep labels have been assigned to individual quantum states.

For INS spectroscopy, for example, we find that all that matters for the selection rule is the *spectroscopic parity* σ, given by Eq. (1.18) for all sphere-like systems. *Every quantum-number-based selection rule presented in this chapter was derived simply by determining* σ *and* p *as a function of the relevant quantum numbers for particular applications*, which in general is quite straightforward. The system-specific INS selection rules that result from the above simple procedure can serve as a *very* powerful tool for analyzing experimental INS spectra for nanoconfined molecules – providing information that is complementary to that obtained from gas-phase spectra, and TR quantum eigenstate calculations. A great example here is the experimental INS spectrum for $H_2O@C_{60}$ (section 1.3.4.1), for which the latter two tools were not, even in conjunction, able to provide completely unambiguous transition assignments. That our newly derived INS selection rule *does* make this possible (Table 1.4) represents a triumph of the method of course, but is also something of a "happy accident." In truth, the complete characterization of any given INS spectrum *will* require that explicit transition integrals be performed; what is actually observed in the lab, after all, are the *allowed* transitions, for which group theory provides no information about intensities.

We conclude with a list of valuable lessons learned, in no particular order.

- *Forbidden INS transitions* not only exist, they appear to be *much more common* than previously realized. They do not even seem to always require TR coupling, so that for some systems *even pure rotational transitions are forbidden* (section 1.3.4.1).
- In both INS and optical spectroscopy, *entanglement induces selection rules* (EISR). Entanglement can either *create or destroy forbidden transitions*, depending on circumstances. This effect should be explored more thoroughly in future.
- (Corollary to above): In INS spectroscopy, *dark particles matter*. A system with a single H atom has no forbidden INS transitions – *unless* that H atom is entangled with a dark particle.
- If the experimental samples are randomly oriented powders, *forbidden INS transitions can be observed only for sphere-like systems*. Conversely, for non-spherical systems, a whole slew of *direction-specific INS selection rules* might emerge – if the relative beam orientation can be controlled throughout the sample.
- For sphere-like cages such as C_{60}, *INS selection rules exist for larger-than-diatomic guest molecules*. For all such systems, an explicit basis or quantum-number-based INS selection rule is *straightforward to obtain using group theory*.
- *Top down* (group theory) and *bottom up* (explicit basis) approaches *must inform each other*, if progress is to be made in interpreting experimental INS spectra.

Acknowledgments

The work by the author discussed in this chapter was conducted with research grant support from both The Robert A. Welch Foundation (D-1523) and the National Science Foundation (CHE-1665370). Z. Bačić, C. A. Chatzidimitriou-Dreismann, J. Eckert, P. Felker, A. J. Horsewill and P.-N. Roy are also acknowledged, for many stimulating discussions.

References

1 K. Komatsu, M. Murata, and Y. Murata, Science **307**, 238 (2005).
2 M. Murata, Y. Murata, and K. Komatsu, J. Am. Chem. Soc. **128**, 8024 (2006).
3 M. Xu, F. Sebastianelli, Z. Bačić, R. Lawler, and N. J. Turro, J. Chem. Phys. **128**, 011101 (2008).
4 M. Xu, F. Sebastianelli, Z. Bačić, R. Lawler, and N. J. Turro, J. Chem. Phys. **129**, 064313 (2008).
5 M. Xu, F. Sebastianelli, B. R. Gibbons, Z. Bačić, R. Lawler, and N. J. Turro, J. Chem. Phys. **130**, 224306 (2009).
6 S. Mamone, M. Ge, D. Hüvonen, U. Nagel, A. Danquigny, F. Cuda, M. C. Grossel, Y. Murata, K. Komatsu, M. H. Levitt, T. Rõõm, and M. Carravetta, J. Chem. Phys. **130**, 081103 (2009).
7 A. J. Horsewill, S. Rols, M. R. Johnson, Y. Murata, M. Murata, K. Komatsu, M. Carravetta, S. Mamone, M. H. Levitt, J. Y.-C. Chen, J. A. Johnson, X. Lei, and N. J. Turro, Phys. Rev. B **82**, 081410(R) (2010).
8 A. J. Horsewill, K. S. Panesar, S. Rols, J. Ollivier, M. R. Johnson, M. Carravetta, S. Mamone, M. H. Levitt, Y. Murata, K. Komatsu, J. Y.-C. Chen, J. A. Johnson, X. Lei, and N. J. Turro, Phys. Rev. B **85**, 205440(R) (2012).

9 M. Xu, S. Ye, A. Powers, R. Lawler, N. J. Turro, and Z. Bačić, J. Chem. Phys. **139**, 064309 (2013).

10 M. Xu, M. Jiménez-Ruiz, M. R. Johnson, S. Rols, S. Ye, M. Carravetta, M. S. Denning, X. Lei, Z. Bačić, and A. J. Horsewill, Phys. Rev. Lett. **113**, 123001 (2014).

11 B. Poirier, J. Chem. Phys. **143**, 101104 (2015).

12 M. Xu, S. Ye, and Z. Bačić, J. Phys. Chem. Lett. **6**, 3721 (2015).

13 S. Mamone, M. Jiménez-Ruiz, M. R. Johnson, S. Rols, and A. J. Horsewill, Phys. Chem. Chem. Phys. **18**, 29369 (2016).

14 P. M. Felker, V. Vlček, I. Hietanen, S. Fitzgerald, D. Neuhauser, and Z. Bačić, Phys. Chem. Chem. Phys. **19**, 31274 (2017).

15 Z. Bačić, J. Chem. Phys. **149**, 100901 (2018).

16 Z. Bačić, V. Vlček, D. Neuhauser, and P. M. Felker, Faraday Discuss. Chem. Soc. **212**, 547 (2018).

17 M. Ge, U. Nagel, D. Hüvonen, T. Rõõm, S. Mamone, M. H. Levitt, M. Carravetta, Y. Murata, K. Komatsu, X. Lei, and N. J. Turro, J. Chem. Phys. **135**, 114511 (2011).

18 M. Xu, S. Ye, R. Lawler, N. J. Turro, and Z. Bačić, Phil. Trans. R Soc. A **371**, 20110630 (2013).

19 M. Xu, P. M. Felker, S. Mamone, A. J. Horsewill, S. Rols, R. J. Whitby, and Z. Bačić, J. Phys. Chem. Lett. **10**, 5365 (2019).

20 K. S. K. Goh, M. Jiménez-Ruiz, M. R. Johnson, S. Rols, J. Ollivier, M. S. Denning, S. Mamone, M. H. Levitt, X. Lei, Y. Li, N. J. Turro, Y. Murata, and A. J. Horsewill, Phys. Chem. Chem. Phys. **16**, 21330 (2014).

21 P. M. Felker and Z. Bačić, J. Chem. Phys. **144**, 201101 (2016).

22 K. E. Whitener, Jr., R. J. Cross, M. Saunders, S. Iwamatsu, M. Murata, M. Mizorogi, and S. Nagase, J. Am. Chem. Soc. **131**,6338 (2009).

23 M. Xu, F. Sebastianelli, and Z. Bačić, J. Chem. Phys. **128**, 244715 (2008).

24 L. Ulivi, M. Celli, A. Giannasi, A. J. Ramirez-Cuesta, D. J. Bull, and M. Zoppi, Phys. Rev. B **76**, 161401(R) (2007).

25 J. L. McAfee and B. Poirier, J. Chem. Phys. **134**, 074308 (2011).

26 G. L. Squires, Introduction to the Theory of Thermal Neutron Scattering (Dover Publications, New York, 1996).

27 G. J. Kearley and M. R. Johnson, Vib. Spect. **53**, 54 (2010).

28 P. C. H. Mitchell, S. F. Parker, A. J. Ramirez-Cuesta, and J. Tomkinson, Vibrational Spectroscopy with Neutrons (World Scientific, Singapore, 2005).

29 M. Xu, L. Ulivi, M. Celli, D. Colognesi, and Z. Bačić, Phys. Rev. B **83**, 241403(R) (2011).

30 M. Xu and Z. Bačić, Phys. Rev. B **84**, 195445 (2011).

31 M. Murata, S. Maeda, Y. Morinaka, Y. Murata, and K. Komatsu, J. Am. Chem. Soc. **130**, 15800 (2008).

32 Y. N. Kalugina and P.-N. Roy, J. Chem. Phys. **147**, 244303 (2017).

33 P. M. Felker and Z. Bačić, J. Chem. Phys. **146**, 084303 (2017).

34 T. Halverson, D. Iouchtchenko, and P.-N. Roy, J. Chem. Phys. **148**, 074112 (2018).

35 B. E. Douglas and C. A. Hollingsworth, Symmetry in Bonding and Spectra: An Introduction (Academic Press, Orlando, FL, 1985).

36 J. M. Hollas, High Resolution Spectroscopy (Elsevier, New York, 1982).

37 P. F. Bernath, Spectra of Atoms and Molecules (Oxford University Press, Oxford, 1995).

38 *Advances in Quantum Chemistry*, edited by J. Sabin and M. Zerner (Academic Press, San Diego, 1997), Vol. 28.

39 E. B. Wilson et al,Molecular Vibrations (Dover Publications, New York, NY, 1980).

40 M. E. Rose, Elementary Theory of Angular Momentum (John Wiley & Sons, Inc., New York, 1957).

41 R. N. Zare, Angular Momentum (John Wiley & Sons, Inc., New York, 1988).

42 W. J. Thompson, Angular Momentum (Wiley-VCH, Weinheim, 1994).

43 In [40], it is shown (p. 61 Eq. 4.32) that Eq. (1.10) is proportional to $C^{\lambda 0}_{j0l0} Y_{\lambda m_\lambda}(\theta, \phi)$. The Clebsch-Gordan coefficient $C^{\lambda 0}_{j0l0}$, which vanishes when $(j + l - \lambda) = 0$ (p. 42 Eq. 3.22), is known as the "parity C coefficient" or "parity-conservation coupling coefficient" ([42], p. 274).

44 J. M. Nicol, J. Eckert, and J. Howard, J. Phys. Chem. **92**, 7117 (1988).

45 P. A. Georgiev, A. Albinati, B. L. Mojet, J. Ollivier, and J. Eckert, J. Am. Chem. Soc. **129**, 8086 (2007).

46 P. R. Bunker and P. Jensen, Molecular Symmetry and Spectroscopy (National Research Council of Canada, Ottawa, 1998).

47 C. A. Chatzidimitriou-Dreismann, J. Eckert, and A. J. Horsewill, (private communications).

48 Z. Bačić, (private communication).

49 W. I. F. David, R. M. Ibberson, J. C. Matthewman, K. Prassides, T. J. S. Dennis, J. P. Hare, H. W. Kroto, R. Taylor, and D. R. M. Walton, Nature **353**, 147 (1991).

50 Indeed, one such facility is the IN1 Lagrange Spectrometer at the Institut Laue-Langevin in Grenoble, where the original $H_2@C_{60}$ experiments were performed [7, 8, 10].

2

Pressure-Induced Phase Transitions

Wojciech Grochala

Center of New Technologies, University of Warsaw, Żwirki i Wigury 93 02089, Warsaw, Poland

2.1 Pressure, A Property of All Flavours, and Its Importance for the Universe and Life

Temperature and pressure are the two key parameters which influence the fate of a chemical system of a given constant composition. Importantly, also living organisms are greatly influenced by even subtle changes in both temperature and pressure. Take the entire stratum of our gaseous atmosphere, which exerts a pressure on living things at Earth's surface of about 1 atm. Even small variations of this value, of the order of 1% of what we call "atmospheric pressure", lead to undesired meteopathic reactions in many humans. Or take the oceanic depths such as Mariana Trench, with the pressure at its bottom exceeding 1000 atm (or 0.1 GPa); benthonic organisms living in such abyss differ greatly from their oceanic shelf counterparts, and they reached the resistance to high pressure via evolution. Indeed, it is currently believed that hydrothermal vents are one possible location linked to the origin of organic life on our planet [1, 2]. Extremophiles living in total darkness, such as high-temperature and high-pressure resistant bacteria and archaea, which utilize chemosynthesis rather than photosynthesis, are believed to be the very first one-cell organisms developed in the process of evolution. Many others are easily killed by elevated pressure, and this is used nowadays more and more often for sterilization of food.

Applying (increasing) pressure in nearly all cases leads to compression of matter, i.e. to reduction of its volume. On the other hand, increasing temperature usually has an inverse effect, i.e. of thermal expansion. Although pressure is usually treated as totally independent from temperature in terms of effects it may exert, yet in experimental reality the two may influence (or compensate) each other. This is nicely exemplified by ideal gas law, which states, in its equation of state (EOS) form, that:

$$V_m = R\,T\,p^{-1} \tag{2.1}$$

where V_m is a molar volume of gas, R is the gas constant, T is temperature, and p is pressure. Clearly, an external observer may not notice any change of the volume of gas contained inside a balloon if temperature and pressure change simultaneously by the same factor. The isochore of an ideal gas is just a perfect hyperbola. For that reason, effects of pressure may be considered to some extent *inverse to the effects of temperature*. Increase of T tends to thermally expand and melt a

*Corresponding Author: W.GROCHALA@CENT.UW.EDU.PL

Chemical Reactivity in Confined Systems: Theory, Modelling and Applications,
First Edition. Edited by Pratim Kumar Chattaraj and Debdutta Chakraborty.

solid, thermally expand and vaporize a liquid, thermally expand a gas and turn it into a plasma, and eventually break even all nuclei into an assembly of elementary particles. The effects of pressure are just reciprocal; increase of p tends to bump elementary particles and small nuclei into each other, thus producing larger nuclei in nuclear fusion process, condense plasma back to the atomic or molecular (gas) form, compress gas and liquify it, compress liquid and solidify it, and ultimately compress solid and induce phase transitions in it. The entire evolution of our Universe from the Big Bang and via diverse stadia:

- *mathematical* (fluctuation and/or inhomogeneity leading to the Bang at 13.6 billions years ago),
- *physical* (since the Bang, via the Planck Era with the resulting Planck pressure of 4.6×10^{108} atm, to the formation of the first atoms at 380,000 years after the Bang),
- *chemical* (since the Era of Recombination i.e. 380,000 years after the Bang),
- *biological* (since ca. 3.6 billions years ago), and so on,

testifies to the importance of these two key parameters: temperature and pressure. Whenever one factor yields, the other predominates and influences the fate of physical, chemical, and biological systems.

As follows from Eq. (2.1), at a constant temperature, volume of a gas is inversely proportional to pressure. Similar reciprocity (albeit with more complex power dependence) is also seen in all empirical equations of state of a solid matter, and known as "*inverse relationship rule*" [3]. Obviously, volume of a solid changes less dramatically than that of a gas, if the same pressure is applied to both. Notably, if only the largest-power terms are kept in the Birch–Murnagham EOS [4, 5], one obtains:

$$V \sim p^{-1/3} \quad (2.2)$$

Since a solid is usually much less compressible than a gas, a much larger pressure must be applied to a solid to achieve similar volume reduction as for a gas. Nevertheless, the least densely packed atomic or molecular solids (He or H_2 and the like), with their components interacting only by weak van der Waals interactions, may be pushed to quite substantial volume reduction at 100 GPa pressure range. For example, H_2 compressed dynamically at ca. 250 GPa (phase IV) exhibits volume of the unit cell of ca. 6.4 $Å^3$ at room temperature [6]. This implies the reduction by the factor of 11 as compared to that for solid hydrogen at 1 atm and 4.2 K (70.7 $Å^3$) [7]. The said volume reduction (or density increase) translates to the "average H…H distance" [8] reduction by the cubic root of that factor, i.e. 2.2-fold. This is an immense number, since huge changes to electronic features of materials occur even at 10% compression of chemical bonds (ca. 1.1-fold compression).

2.2 Pressure: Isotropic and Anisotropic, Positive and Negative

So far, we have silently assumed in the discussion that the pressure was isotropic and thus describable by a single scalar parameter. In experimental reality this assumption is, however, crude, in most cases. Leaving away the technicalities of "isotropic pressure medium" (and how much isotropic it may really be in the conditions of uniaxial stress of the diamond anvil cell, for example) one immediately realizes that non-isotropic compression is present anyway in the majority of chemical systems studied. This is because a *non-cubic* macroscopic crystal subjected to ideally isotropic pressure transmits *non-isotropic* pressure anyway so that stress is anisotropic in its bulk. This is easily described by a stress tensor.

External pressure, be it iso- or anisotropic, has by definition a positive value; null pressure corresponds to vacuum conditions but negative pressure is unphysical. Nevertheless, diverse tricks may be played in order to mimic (and also simulate in theory) the said "negative pressure" regime:

(i) When a non-isotropic external pressure is applied deliberately in a smart way to a non-isotropic system, some components of the stress tensor may become negative. Imagine a crystal built of molecular wires packed in a parallel fashion with respect to each other. When a uniaxial stress is applied along the direction of propagation of these wires, a crystal will tend to expand in the other two dimensions, i.e. the corresponding elements of the stress tensor will correspond formally to negative pressure components. Similar arguments apply to two-dimensional (layered) systems, etc.

(ii) A somewhat similar situation takes place when a thin layer of a compound is grown in vacuum condition from the gas phase (e.g. via chemical vapor deposition or laser beam deposition) on a different substrate. The epitaxial growth on a substrate which is isostructural with the deposited crystal but has a slightly smaller lattice constant, corresponds to positive stress in two directions. On the other hand, growth on a substrate with a slightly larger lattice constant that of the deposited crystal mimics a negative pressure (in two dimensions), with the other dimension accommodating accordingly.

(iii) Needless to say, a negative pressure may be applied to any solid using theoretical modelling. While for small negative pressure very interesting crystal structures are obtained [9], the fate of any chemical system at very large negative pressure seems to be obvious: it should undergo transitions with progressively decreasing coordination number. For example, ZnO crystal which crystallizes in NaCl polytype (CN=6) would supposedly undergo transition to CN=4 coordinated polymorphs (wurtzite, sphalerite) [10], and then via CN=3 reach ultimately the gaseous state with isolated ZnO molecules (CN=1). In some systems, this will take place via cluster stage [9]. Even as "simple" systems as alkali metal halides, oxides, or nitrides may show a diversity of low-density forms [11, 12], and some were already prepared [13].

(iv) Consequently, when a chemist performs reactions on a molecular system dissolved in a solvent, or in a gas phase, rather than in the condensed one, he or she formally uses conditions similar to those corresponding to full dissociation of a crystal to individual molecules. However, even at high vacuum conditions, most crystals will not decompose to molecules, since they are bound by at least the van der Waals forces. Thus, a solution chemistry (reactive molecules in isolation from each other) is essentially a "negative pressure regime". Most organic chemistry performed is *de facto* "negative pressure" one. In inorganic chemistry, careful deintercalation of formerly intercalated systems may preserve the existing heavy element framework and thus lead to new low-density polymorphic forms [14]. Smart chemistry may break down the macroscopic alkali metal halide to nanocrystalline derivatives [15] and potentially also clusters [16]. All these correspond to "negative pressure regime" as well.

Following this brief introduction of the notion of pressure, let us now look at how it may affect the states of matter, structure, as well as magnetic and electronic transitions.

2.3 Changes of the State of Matter

Since the phase diagram is in its most common meaning a graph in which a property or a state of matter is shown in the function of pressure and temperature (sometimes chemical composition is added, as well), and since pressure acts, as described above, seemingly as an inverse of temperature, then it comes as no surprise that a phase diagram at a given chemical composition most often contains boundaries which, upon proper scaling, resemble straight lines (or nearly so), separating the three common states of matter pairwise (Figure 2.1). In other words, if the system is in equilibrium

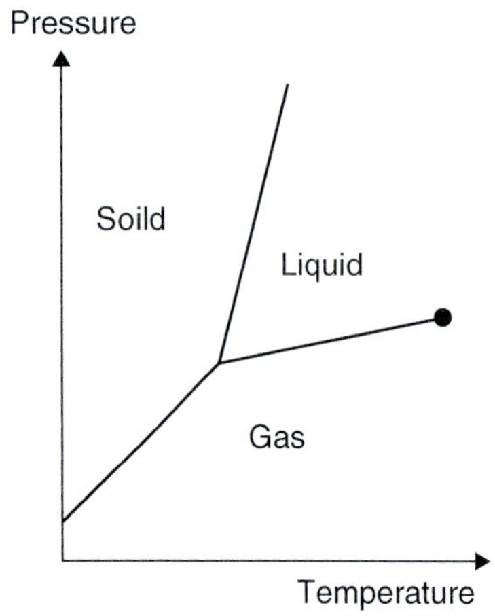

Figure 2.1 A typical phase diagram at a constant chemical composition. Dot marks critical point. Three lines meet at a triple point.

(let us say, water at the melting point of ice at ambient pressure conditions) and both temperature and pressure will be increased by certain related increments (as follows from Eq. 1.1, Eq. 1.2, and the like), one may still stay at the melting curve and preserve the equilibrium situation.

The most common behavior is that a state of matter at both low temperature and high pressure corresponds to a solid state, but when either temperature is raised and/or pressure decreased, transitions to liquid and gaseous state may occur. This is a most typical behavior. For example, krypton is a prototypical unreactive gas at ambient (p,T) conditions but at pressure of 60 GPa it is a solid, and it takes nearly 3000 K to melt it down [17]. On the other hand, helium does not solidify if pressure is not applied; several atmospheres are needed so that a perennial liquid becomes a solid at sufficiently low temperature. Interestingly, hydrogen was suggested to become a quantum liquid at sufficiently high pressure [18].

As a rule of thumb, pressure will have an immense impact on melting point if the molar volumes of a solid and a liquid will differ a lot. This is often connected with the change of the coordination number (just like during some solid-solid phase transitions). An especially dramatic melting is seen for aluminum chloride, which is a densely packed ionic solid with the CN of 6 (and density of 2.48 g cm^{-3}), but quasi-molecular Al_2Cl_6 units with CN of 4 in a liquid phase (density of 1.78 g cm^{-3}); there is a huge concomitant volume increase at melting of close to 40%. Moreover, the melting point is anomalously low in the series of Group 13 chlorides. It is naturally expected that pressure would have immense impact on the melting curve of $AlCl_3$. Indeed, the experimental data at high pressure up to 6 GPa (Figure 2.2) indicate that the melting temperature changes enormously as the pressure is applied up to 4 GPa, with the nearly constant increment exceeding 250 K / 1 GPa [19].

One of the most interesting phenomena related to phase transitions at elevated pressure is that of anomalous melting, i.e. when either a melting curve shows a sudden change upon a structural phase transition (the higher-pressure form melting at lower temperatures than a low-pressure form) and/or that of the presence of the maximum at the melting curve. Exceeding that maximum one finds a state where a continuous decrease of the melting point is seen when pressure is systematically increased. This is exemplified, e.g. by graphite (Figure 2.3), which at 4200 K and 1 GPa is a liquid. A compression of that liquid to ca. 1.5 GPa results in solidification, but an even further compression up to over 10 GPa results in re-melting [20]. The presence of such re-entrant phase transition in the phase diagram testifies to interesting changes in molar volume (and entropy) of a liquid versus a solid phase, which result from compression. Ice VII has been predicted to show such re-entrant behavior, as well.

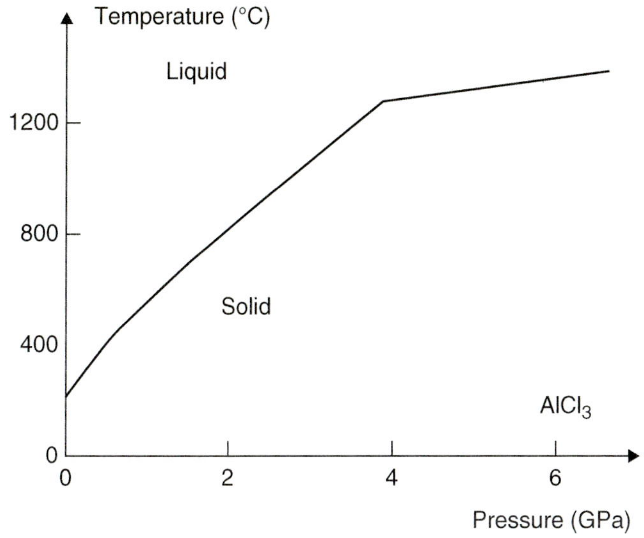

Figure 2.2 Phase diagram of $AlCl_3$ up to 6 GPa and 1300 K. Redrawn after [19] with changes.

Figure 2.3 Semi-logarithmic phase diagram of carbon with a re-entrant transition marked by arrow. Redrawn after [20] with changes. The point marks the case when volume of liquid equals volume of solid graphite.

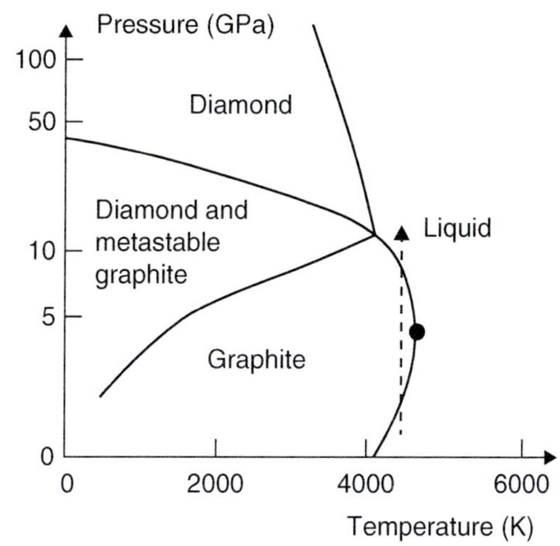

Another interesting feature is shown by diamond (Figure 2.3), which in a temperature range of ca. 3000–4000 K and at pressures from ca.10 GPa to at least 150 GPa shows anomalous melting. Here, again, the liquid is denser than the solid when they coexist. The same trend was observed or suggested for hydrogen [18, 21], sodium [22], lithium [23], and more. Plain hexagonal ice shows the same property, and the formation of a thin layer of water when ice is pressurized by a skate is the reason why skating is facile. Not only is ice less dense than liquid water at 1 atm (recall the Titanic case…), but liquid water shows another strange property: it is the densest at 4 °C. Had it not been the case, lakes would freeze completely during winter, as ice would be falling to their bottom. Instead, bottom parts of lakes host relatively warm water which permits many organisms to survive and stay in lethargic state over winter. The physical peculiarities of water related to molecular volume and pressure effects simply sustain organic life!

Having discussed impact of pressure on changes of physical state of matter, let us turn to effects of compression on solids.

2.4 Compression of Solids

2.4.1 Isotropic or Anisotropic Compressibility

Solids undergoing compression will react by reducing of their volume, just like gas and liquid does. And for most the volume reduction is achieved by simultaneous reduction of all lattice constants. It may (for isotropic crystals) be identical in all three dimensions, but as a rule compressibility in each dimension may be quite different. This most common phenomenon will be exemplified here by graphite (Figure 2.4) [24].

Inspection of Figure 2.4 immediately shows that the hexagonal parameter of graphite (**a**), is rather weakly compressible, and compressibility is nearly linear in the pressure region of up to 25 GPa. The total reduction of the hexagonal lattice constant **a** is ca. 1.5% at this pressure. On the other hand, the lattice constant **c** is quite compressible in the low-pressure region (up to 10 GPa), but progressively stiffens at higher pressure (10–25 GPa). It is reduced by as much as ca. 16 % at 25 GPa as compared to the 1 atm value. Although the qualitative behavior of both lattice parameters is identical (they decrease with pressure increase), yet quantitatively they differ by one order of magnitude. This is not surprising, of course, as lattice constant **a** reflects the stiff chemical bonding (with bond order of ca. 1.5) between carbon atoms, while lattice constant **c** is related to weak van der Waals interactions between graphitic sheets.

2.4.2 Negative Linear Compressibility

While graphite exemplifies the most typical behavior of a solids under compression, there are interesting qualitative exceptions from the scenario where all three lattice parameters decrease when pressure is applied. In some cases, while two lattice parameters decrease, one grows; this is called negative linear compressibility, NLC [25]. Recently, a seemingly simple three-dimensional metal–organic framework material, [Ag(ethylenediamine)]NO_3, has been shown to exhibit giant

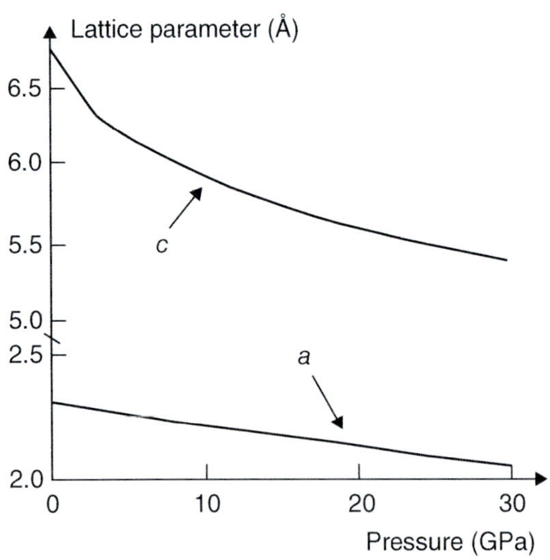

Figure 2.4 Pressure-induced reduction of lattice constants of graphite. Redrawn after [24] with changes.

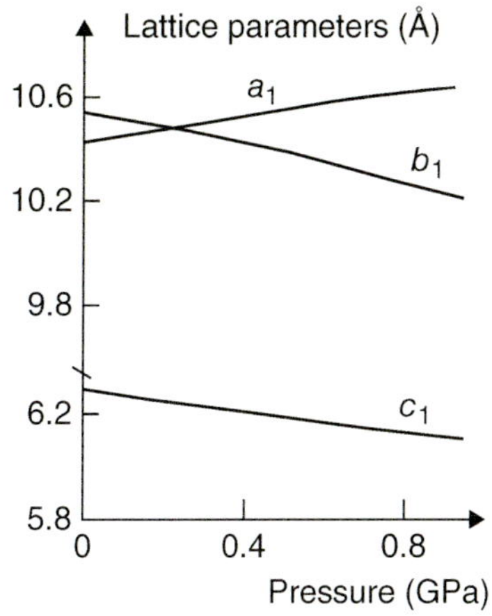

Figure 2.5 Evolution of lattice constants of [Ag(ethylenediamine)] NO_3 (phase I) with pressure, up to the first structural transition. Redrawn after [26] with changes.

NLC [26]. When squeezed up to ca. 1 GPa, lattice constant **a** of this material exhibits an increase of over 2% (Figure 2.5). Simultaneously, two other lattice parameters show a comparable decrease, hence volume effectively decreases with pressure. Interestingly, the third-order Birch–Murnaghan equation of state gives the pressure-derivative of the bulk modulus B′ of −1.6(31) at zero pressure which indicates that the material becomes *increasingly softer under pressure*. This is quite an anomalous behavior.

2.4.3 Negative Area Compressibility

A phenomenon analogous to NLC, where, however, only one lattice parameter decreases with pressure, but the remaining two grow, is called negative area compressibility, NAC. Recently a dramatic manifestation of NAC was observed for a crystal of $[Zn(L)_2(OH)_2]_n$·Guest (where L is 4-(1H-naphtho[2,3-d]imidazol-1-yl)benzoate; Guest is water or methanol) [27]. The first high-pressure phase of this material (where Guest is water) in the pressure range of ca. 1–2.6 GPa exhibits growth of two lattice constants (one by ca. 1% and another by ca. 5%) while the third constant decreases by as much as 17% (Figure 2.6). Overall this leads to a systematic pressure-induced volume reduction, albeit the immense mechanical anisotropy of this system is fascinating. It is worthwhile to study the behavior of this system under anisotropic stress as well. It also remains to be seen how the NLC and NAC may become useful in diverse piezo devices and mechanical energy transformers.

2.4.4 Anomalous Compressibility Changes at High Pressure

As a rule of a thumb, compressibility of a solid will decrease when pressure is increased. That is, compressibility will be larger and bulk modulus will be smaller at a low pressure but compressibility will decrease and bulk modulus will increase at higher pressure. While this is certainly the most common scenario (cf. Figure 2.4), remarkable exceptions have been noticed from this rule. For example, $Zn(CN)_2$ compressed to 0.5 GPa has been shown to exhibit the reverse behavior [28]. This also leads to pressure enhancement of the negative thermal expansion in this compound. ScF_3 is yet another peculiar system of the same sort [29]; other exotic compounds showing similar behavior are now extensively sought for.

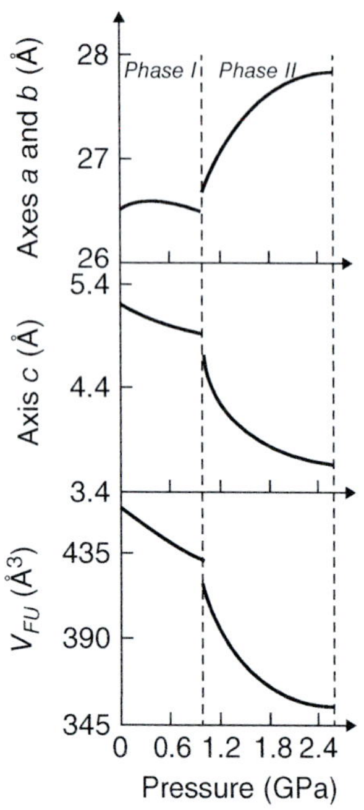

Figure 2.6 Evolution of lattice constants of $[Zn(L)_2(OH)_2]_n$·Guest (where L is 4-(1H-naphtho[2,3-d]imidazol-1-yl)benzoate; Guest is water). Redrawn after [27] with changes.

2.5 Structural Solid-Solid Transitions

2.5.1 Structural Phase Transitions Accompanied by Volume Collapse

A compression of a solid may lead not only to smooth changes of its lattice constants, but also to a more dramatic event: a structural phase transition. The transition most often is accompanied by a substantial volume drop, more or less sudden in terms of a pressure range needed for accomplishing of the transition. Quite often the severe reconstruction of the crystalline lattice is seen, either of the light element sublattice, or both the light and heavy element ones.

Just as a solidification of a liquid (section 3 above) results in a better packing, the solid-solid phase transitions lead to even more densely packed polymorphic forms. This is exemplified by the classical NaCl- to CsCl-type transition (Figure 2.7), here shown for the parent rock salt [30].

A discontinuity of volume is seen at ca. 20 GPa, amounting to about 5% in terms of initial volume at 1 atm conditions. Size and compressibility of an anion is usually larger than that of a cation, hence packing of anions often determines the crystal structure. Hence, the ratio of ionic radii of anion and cation drops upon squeezing so that more anions may be more effectively packed around a cation. Simultaneously, two spheres of different size may pack much better than just one type of a sphere. Therefore, the observed transition is that from ccp (fcc) to Po (bcc) type rather than the reverse one.

The case of diamond and graphite discussed above (Figure 2.3) indeed exemplifies the allotropy of chemical elements with *the largest* known difference of densities of the two forms. The experimental atomic volumes are as follows: 5.671 Å^3 for diamond (at 10 K) [31] and 8.730 Å^3 for graphite (at 4.2 K) [32]. This implies a drop of volume upon phase transition of about 35%! In other words, diamond is 1.54-fold better packed than graphite! This is immense factor, larger than the ratio of 1.42 between the packing fraction of the hexagonal close packing (0.74) and that for beta-polonium (simple cubic) structure (0.52).

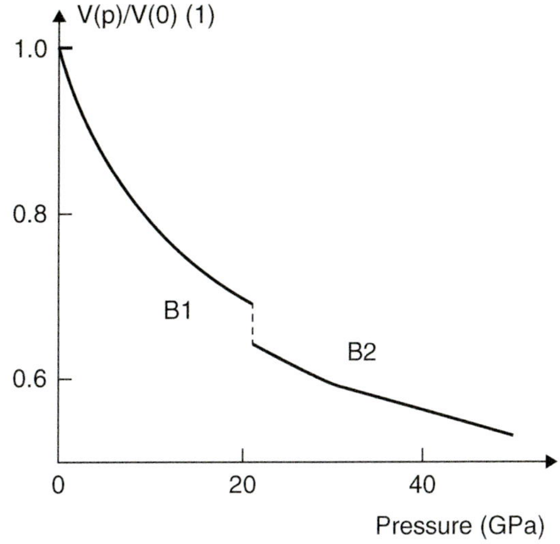

Figure 2.7 Equation of state and the signature of the structural phase transition for NaCl. Volume has been normalized to that at ambient pressure. Redrawn after [30] with changes.

2.5.2 Effects of Volume Collapse on Free Energy

While diamond and graphite have nearly identical free energy at p↔ 0 GPa, and T ↔ 0 K [33, 34], yet this situation changes dramatically at elevated pressure. For example, the difference of atomic volumes between diamond and graphite of ca. 3.06 $Å^3$, translates at a mere 1 GPa to the differential pV factor of ca. 1.84 kJ/mol or 19.1 meV per atom (in favor of diamond). These values are actually comparable to a typical free energy difference between diverse polymorphic forms of elements and compounds; hence, even a small pressure may impact relative pressure of diverse polymorphs. Consequently, electronically soft systems, which are built of large compressible atoms, may exhibit many consecutive phase transitions even at a rather narrow pressure range of several GPa; this is exemplified by cesium, which shows four different allotropic forms at pressures not exceeding 5 GPa [35].

Pressure has an impact on yet another important factor, i.e. absolute entropy of solids, and this also affects the free energy at finite temperatures. The same example of diamond and graphite will be used to illustrate this. According to simplified yet powerful considerations [36] there exists a direct proportionality between the formula unit volume, V_{FU}, and absolute entropy at 298 K:

$$S^{298}\ K(J\ mol^{-1}\ K^{-1}) = 1.757 \times V_{FU}\ (Å^3) \tag{2.3}$$

In the said case of diamond and graphite, the difference of their atomic volumes of ca. 3.06 $Å^3$, translates at 298 K to the differential ST factor of ca. 1.60 kJ/mol or 16.6 meV per atom (in favor of graphite). The actual experimental estimates are smaller, but of the same order, i.e. 1.00 kJ/mol, or 10.4 meV [33]. Again, these values are of the order of typical differences in free energy of polymorphs at ambient (p,T) conditions, and this shows that any change of formula unit volume (e.g. pressure-driven) will impact the relative stability of polymorphs via not only the pV but also ST factor. A subtle balance between these two factors, as well as internal energies (including zero-point vibrational energies), will determine the fate of the chemical system at elevated pressure [33]. Because of this subtle balance, even an extremely *exotic* behavior may be theorized, that of *increase of the unit cell volume upon application of small external pressure*, mostly due to prevalent entropy factor.

2.5.3 Structure-Influencing Factors at Compression

Having discussed the basics of the most common solid-solid structural phase transitions, let us turn (to the factors which determine crystal structures at elevated pressure. From what was said before, packing is definitely the key factor. However, at a relatively small pressures it may not be decisive, especially that reduction of formula unit volume works against stability at finite temperatures (in terms of entropy). The fact that different polymorphic forms may exhibit different compressibility, adds to the complexity of the pressure-induced phenomena; for example, graphite is much more compressible than diamond, hence the huge difference of atomic volumes at 1 atm becomes much smaller at elevated pressure, with consequences for differential PV and ST factors.

From our own experience [37, 38] it follows that minimization of bonding pair and/or lone pair repulsions is a very important structure-influencing factor. In subsequent theoretical studies of silane, SiH_4, and related germane, GeH_4, we have frequently seen that any modification of the crystal structures which decreased repulsion of hydride terminals, has led to improved enthalpy of the system. This general feature has later been seen to apply to other systems; for example, AgF_2 is known to exhibit two consecutive pressure-induced phase transitions [39]. The one at ca. 8.5 GPa leading to the HP-I structure is characterized by the disappearance of local inversion center at Ag(II) site. This may be thought of as Nature's way to decrease the lone pair repulsion Ag(dz^2) ... F^- where fluorine atoms come from adjacent layers, by substantially distorting the secondary interactions (Figure 2.8). In addition, this helps to increase the coordination number of Ag(II) from 6 to 7, and thus to pack better. The second phase transition at ca. 14 GPa leading to HP-II structure leads to a counter-intuitive dramatic reduction of dimensionality [39, 40]. From a quasi-2D layered coordination polymer, AgF_2 turns into a unique quasi-molecular nanowire. While at the first sight this might be interpreted as leading to worse packing, actually the reverse is true. The HP-II form benefits from even larger reduction of the Ag(dz^2) ... F^- repulsion where fluorine atoms come from adjacent nanowires. Here, no apical Ag(II)...F contact is no longer present, and the lone pair Ag(dz^2) points comfortably to the void between two secondary F atoms (Figure 2.8).

Although molecular crystal of XeF_2 is so different from polymeric AgF_2, the fate of secondary bonding in XeF_2 upon compression is very similar. Theoretical calculations [41] meant initially to interpret experimental data [42], point out at entirely different interpretation of experimental results than the one presented in [42]. Theory suggests that the linear dumbbell F-Xe-F unit subjected to 105 GPa should bend the F-Xe-F angle, and then at 200 GPa transform to essentially ionic

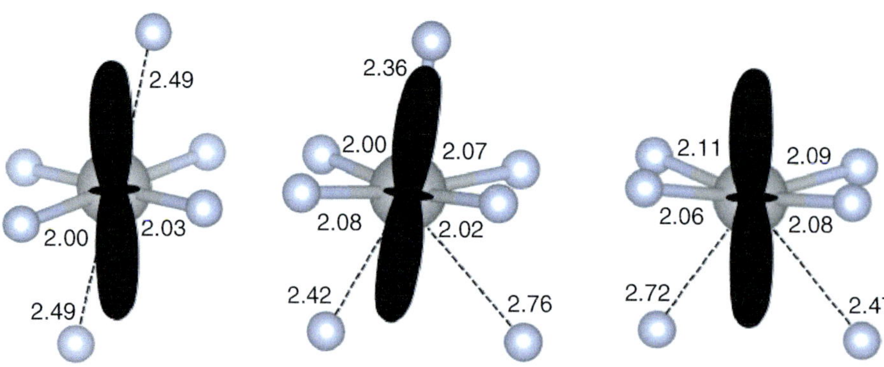

Figure 2.8 Local coordination of Ag(II) site in: layered AgF_2 at 8.5 GPa (left), HP-I structure at 13.5 GPa (middle), and HP-II structure at 16.9 GPa (right). Lone pair Ag(dz^2) has been symbolically drawn to emphasize the Ag(dz^2) ... F^- repulsion. Structure drawings based on cif data from [39].

$(FXe^+)(F^-)$ system. Upon autodissociation one rather covalent Xe-F bond is substituted by two more ionic contacts. This helps to decrease the repulsion between the lone pairs at Xe(II) center and those at F^- site. More recent calculations confirm this scenario [43].

Even the formation of "exotic" electrides at high pressure [44] may be seen as a manifestation of the rule of minimizing the inter-electron repulsions. Thus, it seems that, just like at ambient pressure, the valence shell electron pair repulsion (VSEPR) [45] influences the behaviour of chemical systems at elevated pressure. In other words, for fermionic systems, this is the Pauli exclusion principle as well as purely Coulombic forces, which are a powerful driving force behind structural transitions.

2.5.4 Changes in the Nature of Chemical Bonding upon Compression and upon Phase Transitions

A chemical bond is a compromise between different types of attractions and repulsion, with the electron pair(s) (for even-electron systems) playing the key role in minimizing the repulsion between nuclei (or atomic cores in general). Therefore, when the chemical bond is elongated, some share of electron density sitting between nuclei is transferred back to its parent atomic cores. When the bond is heterolytic, this leads to increase of bond iconicity as measured by a local dipole moment. On the other hand, when the chemical bond is compressed, there is usually a pressing need for more electron density in the internuclear region, and therefore bond covalence increases. This is the most typical scenario at high pressure as exemplified e.g. by quite robust N_2 molecule [46]. This situation, typical of an isolated molecule, is different for a solid at the stage, when N_2 molecules come in closer contact; now, intermolecular internuclear repulsions must also be softened by transferring electron density between molecule; this ultimately results in building new chemical bonds at the expense of the old ones, and N_2 polymerizes; the polymer exhibits three single bonds around each N atom.

When a bulk heteronuclear system, such as the said NaCl (Figure 2.6) is compressed, covalence of shrinking Na...Cl bonding systematically increases. Then, upon the structural phase transition, six shorter bonds are substituted by eight longer ones. Hence, iconicity of each of the eight bonds must now be larger than that of each of the former six bonds (on the per bond basis). Further compression of the CsCl polytype results again in shortening of each of the eight bonds, and therefore in increasing covalence. NaCl exemplifies the most typical behavior of inorganic solids, with systematic increase of bonding covalence upon compression, and then sudden increase of iconicity at phase transitions which increase the coordination number.

One spectacular example of increase of bond iconicity is provided by O-H...O hydrogen bond system in water. Common hexagonal ambient-pressure molecular ice shows an alternation of short covalent O-H bonds and longer ionic H...O contacts [47]; this is feature typical of weak or moderately strong asymmetric hydrogen bonding. On the other hand, ice X forming at pressures above 62 GPa features a fully symmetric O...H...O bonding arrangement [48]; this corresponds to so-called strong hydrogen bonding but also implies a substantially increased bond iconicity in this non-molecular form as compared to common more covalent short O-H terminal contact.

Volume reduction at high pressure may not just change the existing chemical bonds, but affect significantly the stoichiometry, and therefore the bonding pattern. Formation of Fe_3S_2 from FeS and Fe at elevated pressure conditions [49] is one good example. Before trying to understand the bonding and electronic structure of Fe_3S_2, first attempt to assign the formal oxidation states in this compound!

2.6 Selected Classes of Magnetic and Electronic Transitions

2.6.1 High Spin–Low Spin Transitions

The spin crossover phenomenon has now been known for nearly a century [50]. The most common type of phase transition is from the LS to HS upon increase of temperature, or the reverse transition upon application of external pressure [51]. These transitions may easily be understood either ligand field theory; here, temperature acts to increase unit cell volume, hence increase the metal–ligand distance; pressure does just the opposite. In consequence, ligand field strength decreases in the first case but increases in the latter. This leads to instability for systems at the verge of the transition. Since temperature and pressure act in opposite directions, increase of temperature may be compensated by the increase of pressure, as exemplified by $Fe(phenantroline)_2(NCS)_2$ complex (Figure 2.9).

While the spin crossover phenomenon is most frequently seen for the first raw TM systems, yet those bearing selected elements from the second raw are supposed to also exhibit this behavior. For example, although HS state of Pd(II) in PdF_2 seems to be quite robust at 5 GPa [52] yet it is naturally expected to turn to LS one upon further compression.

The pressure-induced SC is not limited to TM compounds but also exhibited by elemental TMs. Although magnetic moment in elemental iron is rather small compared to the large moments seen in iron(II) or iron(III) compounds, yet the itinerant magnetism of iron is quite persistent. It turns out that it takes pressure of 10 GPa to turn iron into a nonmagnetic form [53].

Obviously, it is much more difficult to kill paramagnetism in inner-transition metals (lanthanides) and their compounds than in outer-transition metals (i.e. d-block elements). This is because unpaired electrons in the former sit in quasi-core orbitals which are poorly accessible to chemistry and weakly influenced by external conditions. However, altering magnetism of lanthanides is not impossible, as exemplified by the case of elemental europium metal. At normal (p,T) conditions metal cores are found in the most typical divalent state with a half-filled f shell (J=7/2), resulting in an antiferromagnetic state. However, when subjected to pressure of 80 GPa, europium core dissociates one more electron to the electron plasma and changes its valence state to Eu(III) (J=0). The non-magnetic europium metal results [54]. Recent structural measurements shed more light on the behavior of Eu and they suggest that some divalent character is preserved even at 92 GPa [55].

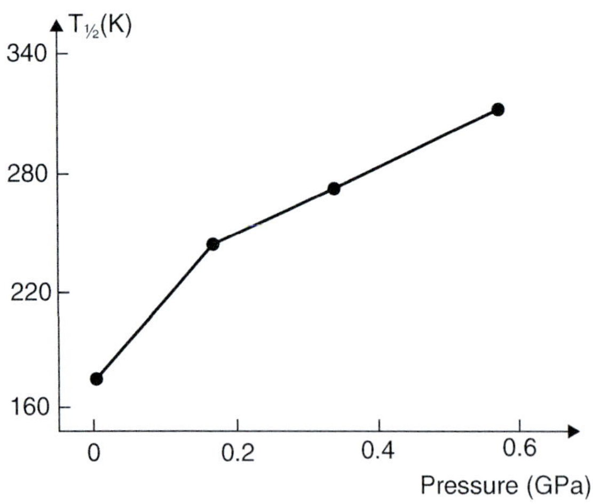

Figure 2.9 The temperature $T_{1/2}$ at which 50% of the complex is in the LS state, versus the external pressure, for $Fe(phenantroline)_2(NCS)_2$; redrawn from [51] with changes.

Similar behavior is expected for ytterbium systems. For example, it turns out that Yb metal exhibits some mixed-valence character of its atomic cores under pressure up to 179 GPa [56]. Recent calculations suggest fluctuation valence scenario [57]. Manipulations of valence state of Yb in its alloys are rather straightforward [58] but pressure effects have been scarcely studied.

2.6.2 Electronic Com- vs Disproportionation

The mixed valence compounds are of great interest to chemists and physicists alike. One key question is whether, and at what external conditions, the system will preserve its mixed valence (i.e. charge density wave), or rather would become comproportionated. Examples may be shown of CDW to single valence as well as reverse transitions, tuned by external pressure. For example, $BiNiO_3$ containing Ni(II) is disproportionated and at ambient pressure two distinct bismuth sites are seem, corresponding to Bi(III) and Bi(V). However, at pressure increase to ca. 4 GPa, this compound shows comproportionation to single valence Bi(IV), with substantial mixing between Bi(6s) and O(2p) states [59]. Quite similar behavior has been seen for $SrCu(II)_3Fe_4O_{12}$. At ambient (p,T) conditions this compound is disproportionated, with distinct Fe(III) and Fe(V) sites; however, pressure increase melts down the differences between iron sites and at 28 GPa charge distribution is uniform [60]. $CsAuCl_3$ is yet another system which comproportionates at elevated pressure; here, Au(I) and Au(III) sites exchange electron and both become Au(II) [61]. Analogous transitions were seen for bromide and iodide derivatives. In all cases described above, comproportionation is associated by transformation from insulating or semiconducting state to a metal.

Behaviour of $BaBiO_3$ and AgO remains in striking contrast to that of their $BiNiO_3$ and $CsAuCl_3$ analogues, respectively. Calculations suggest that both compounds would preserve mixed valence character up to at least 100 GPa [62, 63]. Even more complex pressure-induced behavior has been predicted for AuO [64]. This compound is expected to be disproportionated at ambient (p,T) conditions, but turns into a comproportionated $AuSO_4$-related form at 80 GPa, and then becomes disproportionated again at p>105 GPa, and finally it metallizes at 329 GPa in structure bearing only one type of gold site.

The complex case of AuO definitely merits an experimental study; theoretical results show that behavior at high pressure is influenced by many factors pulling the system back and forth to and from disproportionation; while packing of different size spheres may be beneficial at elevated pressure, for some systems it comes with a too large energy penalty. On the other hand, some systems show robust diamagnetic configurations (such as Ag(I) and low spin Ag(III) in AgO) and they remain disproportionated up to very high pressures.

2.6.3 Metal-to-Metal Charge Transfer

In some systems, pressure-induced electron transfer may occur not between two cations of the same element but of two different elements. This is exemplified by $EuMnO_3$, which was predicted to transform from the Eu(III)Mn(III) formulation to the Eu(II)Mn(IV) one, with obvious consequences for electronic and magnetic properties [65]. Many more such examples may be shown; in principle, systems which contain two elements belonging to redox pairs with similar redox potentials will exhibit such behavior more often.

2.6.4 Neutral-to-Ionic Transitions

Electrons may also be transferred between two different neutral organic molecules leading to ionic formulation, resembling that of inorganic ionic salt. Some of these systems exhibiting alternating

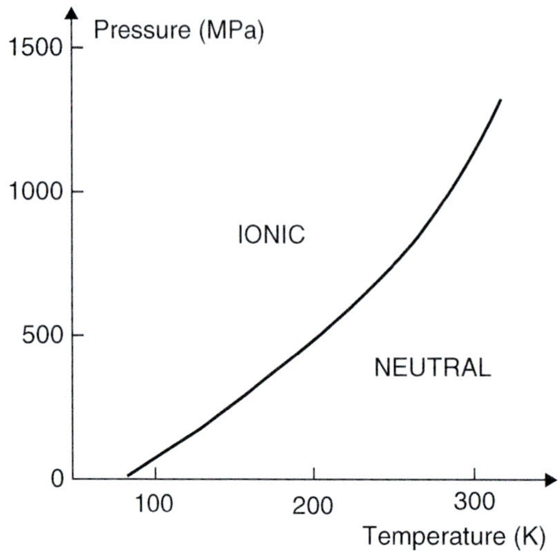

Figure 2.10 Phase boundary between neutral and ionic forms of TTF/CA system. Redrawn in a simplified form after [67].

stack (donor-acceptor-donor-acceptor etc.) structure are on the verge of the transition from neutral to ionic state, and pressure or temperature may greatly affect the state of the system. The first system of this kind based on tetrathiafulvalene (TTF) and chloranile (CA) was found by Torrance et al. [66]. The phase boundary between neutral and ionic (ferroelectric) form in a (p,t) diagram exhibits a familiar quasi-linear dependence, thus again pointing out the reverse effects of pressure and temperature, which may compensate for each other (Figure 2.10).

2.6.5 Metallization of Insulators (and Resisting It)

Just as progressive forcing one H atom into other in the H_2 molecule leads to an increasing splitting between the bonding and antibonding combination of atomic orbitals, so forcing of atoms into one another in a solid state leads to electronic bands broadening. Ultimately, this may result in metallization of a non-metal via the band overlap [68]. Numerous examples are known of this important phenomenon, and some may be predicted with surprisingly good accuracy using a simplistic model of metallization by Herzfeld and Goldhammer [69, 70]; the model uses only the electronic dipole polarizability and atomic volume as parameters. As a rule of thumb, compression leads to volume decrease and thus metallization [71]. Examples of pressure-induced metallization are given in reviews devoted to high pressure physics and chemistry [46], with metallization pressures (for static and not dynamic compression experiments) ranging from a few GPa range (for I_2, Si etc.) via hundreds of GPa predicted theoretically for metallization of Kr [72], or Ar [73] and up to tens of TPa needed to metallize helium according to theoretical calculations [74]. Indeed, since kinetic energy of electrons scales as $V^{-2/3}$ while Coulombic interactions as $V^{-1/3}$, it may be expected that at sufficiently high pressure (possibly with some support from increase of temperature) any chemical matter would become a free-electron metal.

One particular type of inorganic compounds is worth mentioning here due to their particular resistance to metallization; these are systems exhibiting Jahn-Teller effects at the metal site. Many such systems have been studied in the past in the function of external pressure. While generally they show substantial plasticity of the coordination sphere of a metal where short/strong and longer/weak bonds are mutually interrelated [75] yet pressure naturally tends to make all bonds similar to each other; in other words, it may suppress the Jahn-Teller effect [76]. Importantly, if the Jahn-Teller was entirely eliminated, orbital degeneracy would appear thus likely corresponding to a metallic system. However, experimental studies have showed that many Jahn-Teller active systems are remarkably resistant towards metallization; this is the case of Cu(II) chloride derivatives [76], CuF_2 [77] or AgF_2 discussed above [39, 40]. As already mentioned the theoretical prediction places the metallization threshold for hypothetical AuO at nearly 330 GPa [64], which is an immense pressure, indeed.

2.6.6 Turning Metals into Insulators

Turning a metal into an insulator seems to be just the reverse of the previously described metallization. However, there is rather little symmetry between the two. The fascinating sudden decrease of optical reflectivity in metallic sodium at 118 GPa has been observed in 2009 [78, 79], and band gap opening at 178 GPa has been observed in the coming years [80]. Simultaneously, similar behavior was seen for lithium at around 80 GPa [81]. Interestingly, drop of electric conductivity for Ca in fcc structure at ca. 14 GPa was found as early as in 1963 [82]. This behavior thought to be anomalous, has been rationalized by observing that electron localization in interatomic sites takes place in these materials [83, 84], thus providing the conceptual link [44] to the long-known electrides [85]. The Pauli exclusion principle is a nice tool to explain the appearance of electride phases in "simple" metals. An alternative explanation is based on orthogonality of atomic orbitals. In H atom, all eigenfunctions must be orthogonal to each other; this is impossible to achieve if 2s orbital would occupy similar space as the 1s one. While the orthogonality criterion is no longer strict for non-H-atom-like species, yet it approximately holds. The application of external pressure on, say, lithium, results in compression of predominantly the valence 2s shell, which becomes more similar to the 1s one and the non-orthogonality might appear. One good way to avoid this is to place valence electrons somewhere else, i.e. in structure interstices, far away from atomic cores.

2.6.7 Superconductivity of Elements and Compounds

Transition to the superconducting state is yet another important type of phase transition which is markedly affected by pressure. Take elemental lithium; it superconducts at ambient pressure below the critical temperature, T_C, of 0.4 mK [86] but after a series of phase transitions at elevated pressure the T_C value rises to 20 K at 48 GPa [87, 88]. This is a 50,000-fold increase. Calcium, which does not superconduct at ambient pressure at any measurable low temperature, has been turned into a superconductor by use of high pressure [89]. Schilling periodically updates information about the state-of-the-art in superconductivity studies at elevated pressure [90]; the current record of T_C amongst confined elements is 29 K for calcium at 216 GPa [91].

Pressure markedly affects T_C also for high-T_C oxocuprate superconductors [92]. As early as in 1987, application of external pressure has helped to raise T_C value for $(La,Ba)_2CuO_{4-x}$ superconductor first up to 40 K (Figure 2.11) [93] and then up to 52.5 K, the record value at that time [94]. In 1994 the new record was set for $HgBa_2Ca_2Cu_3O_{8+delta}$ of 164 K at 31 GPa [95]. To date this remains the highest T_C value for oxocuprate material.

Following Ashcroft's early suggestion that superconductivity might be observed in compressed hydrogen [96] (later extended to hydrogen-dominant metallic alloys [97]), a series of theoretical and experimental studies were carried out. These were initiated by the theoretical study for silane [37] and an experimental one for the same system [98]. The topic is too broad to be extensively presented in this short account, so only the most striking recent results will be discussed.

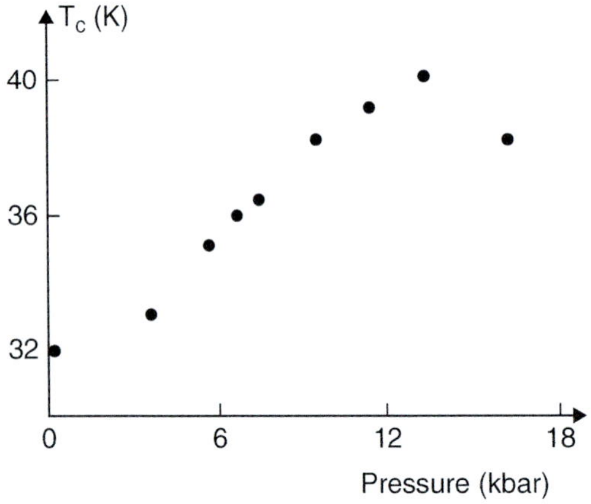

Figure 2.11 Pressure dependence T_C of the HTS superconductor $(La,Ba)_2CuO_{4-x}$. Redrawn after [93] with changes.

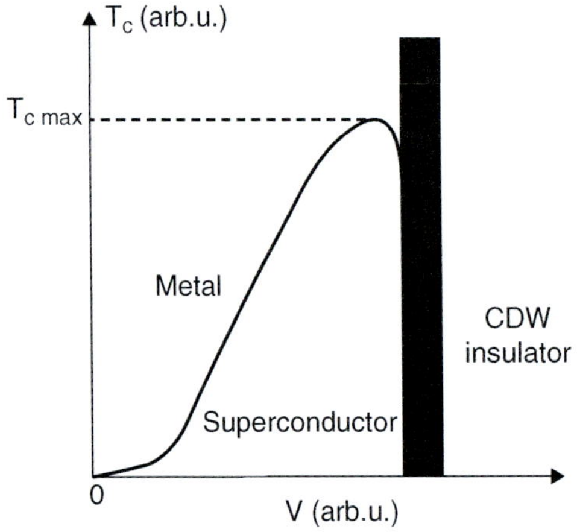

Figure 2.12 Phase diagram of the electron-phonon coupling constant, V, vs superconducting critical temperature, T_C. Redrawn after [105] with changes.

In 2015 superconductivity at 203 K has been discovered for H_xS samples, where x is close to 3, at pressures approaching 150 GPa [99]. More recently, samples of LaH_x, where x is close to 10–11, have been shown to exhibit T_C of 250–260 K at pressures close to 200 GPa [100, 101]. The samples turn out to be multi-phase with some components exhibiting T_C value in excess of 273 K i.e. 0 °C [102]. Thus, the ambient temperature superconductivity is now a fact. And even higher T_C values up to 326 K have been predicted for compressed related yttrium hydride system [103], with even higher values predicted for ternary hydride [104]. However, despite all beauty and thrill of these discoveries it must be noticed that hydrogen rich materials exhibit high T_C superconductivity at unpractically high pressures, and quenching to ambient pressures with preservation of superconductivity is simply impossible.

The fact that ultra-high pressure is needed to prevent lattice distortions which open the fundamental gap, can be easily rationalized considering traditional arguments stemming from BCS theory of superconductivity. The electron-phonon coupling constant, which is a measure of the strength of the coupling which generates superconductivity, cannot be raised infinitely, as it will ultimately result in an insulating state (Figure 2.12). In consequence, high-T_C superconductivity at temperatures exceeding boiling point of N_2 but without the need for external pressure, is immensely rare, i.e. limited to oxocuprates.

Maximum Hardness Principle [106] may be used to rephrase the explanation of the imminent rarity of high-T_C families. Nature avoids metals with large density of states at the Fermi level i.e. with large carrier density available for pairing up; Nature's way is to remove that density of states by distortions which minimize electronic softness. It turns out that doped oxocuprate superconductors are unstable towards phase separation to undoped and fully doped phases [107]; it is only the way of their preparation which prevents phase separation from taking place. Synthesis of metastable systems seems to be the key strategy to bypass the Maximum Hardness Principle and achieve high-T_C superconductivity, just like in the case of doped oxocuprates. Epitaxial growth of thin layers of superconductors on proper substrates, which result in 2D strain, is yet another option.

2.6.8 Topological Phase Transitions

The complexity of pressure-induced phase transitions is additionally enriched by electronic topological (Lifshitz) transitions, with marked qualitative changes of the Fermi surface. For example, elemental tellurium compressed at 2 GPa shows the topological transition from a semiconductor to a Weyl semimetal [108]. MgB_2 shows a phonon-coupled electronic topological phase transition at ca. 20 GPa and this affects the T_C dependence on pressure [109]. Last but not the least, topological transitions are also seen for oxocuprate superconductors leading to quite unusual pressure dependence of T_C (Figure 2.13) [110]. More such interesting examples are known.

2.7 Modelling and Predicting HP Phase Transitions

Together with the systematic increase of the available computer power, theoretical modelling has become an important tool to explain experimental data as well as to predict new systems and phenomena. This is an immense field *per se* which has been described in numerous reviews (e.g. [111–114] to give just a few). More recently, predictive theoretical quest for high-T_C superconductivity in diverse hydride materials has been very intense [37, 103, 104, 115–117]. Some other directions pursued during the last two decades encompass for example quantum liquids [18], melting behaviour [118], new carbon allotropes [119], superhard materials [120], unprecedented compounds of noble gases [121] and new polymorphs of their known compounds [41], electride phases of alkali metals [83, 84], novel stoichiometries [122], sometimes with peculiar oxidation states [123], processes of auto-dissociation [124] and more.

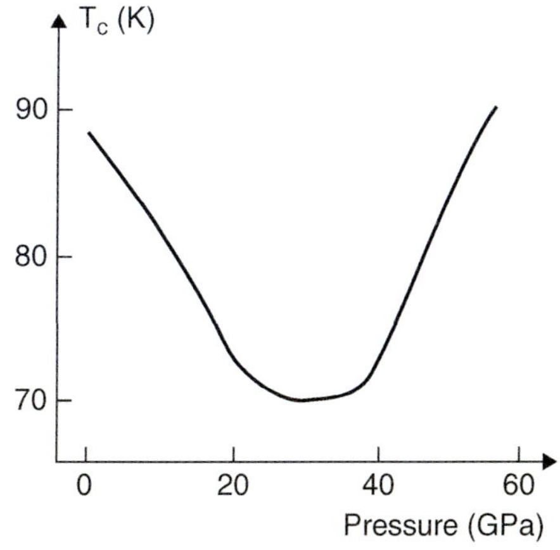

Figure 2.13 Pressure dependence of T_c in slightly overdoped Bi2212. Redrawn after [110] with changes.

Acknowledgements

Generous financial support over the years from the Polish National Science center is acknowledged, notably for the project Harmonia "HP" 2012/06/M/ST5/ 00344.

References

1 Wachtershauser, G. (1990). Evolution of the first metabolic cycles. Proc. Nat. Acad. Sci. USA 87: 200–204.
2 Colín-García, M. et al. (2016). Hydrothermal vents and prebiotic chemistry: a review. Bol. Soc. Geol. Mexic. 68: 599–620.
3 Hazen, R.M. and Finger, L.W. (1982). Comparative Crystal Chemistry: Temperature. Composition and the Variation of Crystal Structure. John Wiley & Sons, New York: Pressure.
4 Birch, F. (1947). Finite elastic strain of cubic crystals. Phys. Rev. 71: 809–824.
5 Murnaghan, F.D. (1944). The compressibility of media under extreme pressures. Proc. Nat. Acad. Sci. USA 30: 244–247.
6 Ji, C. et al. (2019). Ultrahigh-pressure isostructural electronic transitions in hydrogen. Nature 573: 558–562.
7 Barrett, C.S. et al. (1966). Crystal structure of solid hydrogen and deuterium, and of neon-hydrogen and neon-deuterium mixtures. J. Chem. Phys. 45: 834–837.
8 Obviously, both the inter- and intra-molecular H…H distances undergo compression albeit they are affected to a different degree.
9 Wilson, M. and McMillan, P.F. (2003). Crystal-liquid phase relations in silicon at negative pressure. Phys. Rev. Lett. 90: 135703.
10 Sangthong, W. et al. (2010). Predicting transition pressures for obtaining nanoporous semiconductor polymorphs: oxides and chalcogenides of Zn, Cd and Mg. Phys. Chem. Chem. Phys. 12: 8513–8520.
11 Sangthong, W. et al. (2008). Stable nanoporous alkali halide polymorphs: a first principles bottom-up study. J. Mater. Chem. 18: 5871–5879.
12 Zwijnenburg, M.A. et al. (2010). Apparent scarcity of low-density polymorphs of inorganic solids. Phys. Rev. Lett. 175503: 104.
13 Liebold-Ribeiro, Y. et al. (2008). Experimental substantiation of the "energy landscape concept" for solids: synthesis of a new modification of LiBr. Angew. Chem. Int. Ed. 47: 4428–4431.
14 Kim, D.Y. et al. (2015). Synthesis of an open-framework allotrope of silicon. Nat. Mater. 14: 169–173.
15 Guanyun, Q.S. et al. (2016). Alkali halide cubic cluster anions ($[Cs_8X_{27}]^{19-}$, X = Cl, Br) isolated from water. Inorg. Chem. 55: 11125–11130.
16 Bickelhaupt, F.M. et al. (2007). Table salt and other alkali metal chloride oligomers: structure, stability, and bonding. Inorg. Chem. 46: 5411–5418.
17 Boehler, R. et al. (2001). High-pressure melting curves of argon, krypton, and xenon: deviation from corresponding states theory. Phys. Rev. Lett. 86: 5731–5734.
18 Bonev, S.A. et al. (2004). A quantum fluid of metallic hydrogen suggested by first-principles calculations. Nature 431: 669–672.
19 Brazhkin, V.V. et al. (2005). Structural studies of phase transitions in crystalline and liquid halides ($ZnCl_2$, $AlCl_3$) under pressure. J. Exp. Theor. Phys. Lett. 82: 713–718.

20 dos Santos, J.L. et al. (2017). A new dynamic powder consolidation technique using shock waves. Mat. Res. 20: 260–264.

21 Liu, H. et al. (2013). Anomalous melting behavior of solid hydrogen at high pressures. J. Phys. Chem. C 117: 11873–11877.

22 Gregoryanz, E. et al. (2005). Melting of dense sodium. Phys. Rev. Lett. 185502: 94.

23 Cf. S. F. Elatresh, et al., Role of quantum ion dynamics in the melting of lithium. Phys. Rev. B 94: 104107 (2016), and references therein.

24 Wang, Y. et al. (2012). Crystal structure of graphite under room-temperature compression and decompression. Sci. Rep. 2: 520.

25 Baughman, R.H. et al. (1998). Materials with negative compressibilities in one or more dimensions. Science 279: 1522–1524.

26 Weizhao, C. and Katrusiak, A. (2014). Giant negative linear compression positively coupled to massive thermal expansion in a metal–organic framework. Nat. Commun. 5: 4337.

27 Weizhao, C. et al. (2015). Giant negative area compressibility tunable in a soft porous framework material. J. Am. Chem. Soc. 137: 9296–9301.

28 Chapman, K.W. and Chupas, P.J. (2007). Pressure enhancement of negative thermal expansion behavior and induced framework softening in zinc cyanide. J. Am. Chem. Soc. 129: 10090–10091.

29 Greve, B.K. et al. (2010). Pronounced negative thermal expansion from a simple structure: cubic ScF_3. J. Am. Chem. Soc. 132: 15496–15498.

30 Jain, A. and Dixit, R.C. (2018). Pressure induced structural phase transition from NaCl-type (B1) to CsCl-type (B2) structure in sodium chloride. AIP Conf. Proc. 1953: 040023.

31 Haruna, K. et al. (1992). Thermal expansion coefficient of synthetic diamond single crystal at low temperatures. Jpn. J. Appl. Phys. 31: 2527–2529.

32 Baskin, Y. and Meyer, L. (1955). Lattice constants of graphite at low temperatures. Phys. Rev. 100: 544–544.

33 Grochala, W. (2014). Diamond: the electronic ground state of carbon at temperatures approaching 0 K. Angew. Chem. Int. Ed. Engl. 53: 3680–3683.

34 Lechner, C. et al. (2016). First-principles study of the structural, electronic, dynamic, and mechanical properties of HOPG and diamond: influence of exchange–correlation functionals and dispersion interactions. J. Phys. Chem. C 120: 5083–5100.

35 Young, D.A. (1975). Phase Diagrams of the Elements. UCRL-51902 prepared for the U.S. Energy Research and Development Administration under contract No. In: W-7405-Eng-48.

36 Jenkins, H.D.B. and Glasser, L. (2003). Inorg. Chem. 42: 8702–8708.

37 Feng, J. et al. (2006). Prediction of superconductivity in SiH_4 at high pressure – en route to "metallic hydrogen". Phys. Rev. Lett. 017006: 96.

38 Martinez-Canales, M. et al. (2006). Pressure induced metallization of germane. J. Phys. Chem. Sol. 67: 2095–2099.

39 Grzelak, A. et al. (2017). High pressure behavior of silver fluorides up to 40 GPa. Inorg. Chem. 56 (23): 14651–14661.

40 Grzelak, A. et al. (2017). Metal fluoride nanowires featuring square-planar building blocks in a high-pressure polymorph of AgF_2. Dalton Trans. 46: 14742–14745.

41 Kurzydłowski, D. et al. (2011). Freezing in resonance structures for better packing: XeF_2 becomes $(XeF^+)(F^-)$ at large compression, Inorg. Chem. 50: 3832–3840.

42 Kim, M. et al. (2010). Two- and three-dimensional extended solids and metallization of compressed XeF_2. Nat. Chem. 2: 784–788.

43 Wu, G. et al. (2017). Confirmation of the structural phase transitions in XeF_2 under high pressure. J. Phys. Chem. C 121: 6264–6271.

44 Miao, M.-S. and Hoffmann, R. (2014). High pressure electrides: a predictive chemical and physical theory. Acc. Chem. Res. 47: 1311–1317.

45 Sidgwick, N.V. and Powell, H.M. (1940). Stereochemical types and valency groups (Bakerian Lecture). Proc. Roy. Soc. A 176: 153–180.

46 W. Grochala W, et al., The chemical imagination at work in very tight places. Angew. Chem. Int. Ed. Engl. 46: 3620–3642 (2007).

47 Bernal, J.D. and Fowler, R.H. (1933). A theory of water and ionic solution, with particular reference to hydrogen and hydroxyl Ions. J. Chem. Phys. 1: 515–548.

48 Hirsch, K.R. and Holzapfel, W.B. (1984). Symmetric hydrogen bonds in ice X. Phys. Lett. A 101: 142–144.

49 Fei, Y. et al. (1997). High-pressure iron-sulfur compound, Fe_3S_2, and melting relations in the Fe-FeS system. Science 275: 1621–1623.

50 Cambi, L. and Szegö, L. (1931). Über die magnetische Susceptibilität der komplexen Verbindungen. Chem. Ber. Dtsch. Ges. 64: 2591–2598.

51 Wentzcovitch, R.M. et al. (2009). Anomalous compressibility of ferropericlase throughout the iron spin cross-over. Proc. Natl. Acad. Sci. USA 106: 8447–8452.

52 Demazeau, G. et al. (1980). Physical and chemical properties of some palladium fluorides under high pressures. In: Proc. 7th Int. AIRAPT Conf. High Press. Sci. Techn. 1: 579–582.

53 Cort, G. et al. (1982). Search for magnetism in hcp ε-Fe. J. Appl. Phys. 53: 2064–2065.

54 Debessai, M. et al. (2009). Pressure-induced superconducting state of europium metal at low temperatures. Phys. Rev. Lett. 197002: 102.

55 Bi, W. et al. (2011). Pressure-induced structural transitions in europium to 92 GPa. Phys. Rev. B 104106: 83.

56 Song, J. et al. (2018). Pressure-induced superconductivity in elemental ytterbium metal. Phys. Rev. Lett. 037004: 121.

57 Ylvisaker, E.R. et al. (2009). Charge fluctuations and the valence transition in Yb under pressure. Phys. Rev. Lett. 246401: 102.

58 Svane, A. et al. (2000). Ab initio theory of valency in ytterbium compounds. Phys. Rev. B 62: 13394.

59 See: I. Leonov, et al., Unusual Mott transition associated with charge-order melting in $BiNiO_3$ under pressure. Phys. Rev. B 100: 161112(R) (2019), and references therein.

60 Kawakami, T. et al. (2019). Pressure-induced intersite charge transfer in $SrCu_3Fe_4O_{12}$. J. Phys. Soc. Jap. 88: 064704.

61 Matsushita, N. et al. (2007). Single crystal X-ray diffraction study of a mixed-valence gold compound, $Cs_2Au^IAu^{III}Cl_6$ under high pressures up to 18 GPa: Pressure-induced phase transition coupled with gold valence transition. J. Solid State Chem. 180: 1353–1364.

62 Smolyanyuk, A. et al. (2017). Ab initio prediction of the high-pressure phase diagram of BaBiO3. Phys. Rev. B 035103: 96.

63 Grzelak, A. et al. (2017). Persistence of mixed valence in the high-pressure structure of silver (I,III) oxide AgO: a combined Raman, XRD and DFT study. Inorg. Chem. 56: 5804–5812.

64 Hermann, A. et al. (2016). AuO: evolving from dis- to comproportionation, and back again. Inorg. Chem. 55: 1278–1286.

65 Aschauer, U. et al. (2019). Strain-induced heteronuclear charge disproportionation in $EuMnO_3$. Phys. Rev. Mater. 013601: 3.

66 Torrance, J.B. et al. (1981). Discovery of a neutral-to-ionic phase transition in organic materials. Phys. Rev. Lett. 46: 253–257.
67 Cailleau, H. et al. (1997). Physics of neutral-to-ionic phase transition in organic charge transfer semiconducting compounds. Acta Phys Polon A 92: 597–608.
68 Imada, M. et al. (1998). Metal-insulator transitions. Rev. Mod. Phys. 70: 1039–1263.
69 D. A. Goldhammer, Dispersion und Absorption des Lichtes in ruhenden isotropen Körpern. Ed. E. Jahnke, vol. 16, Teubner, Leipzig-Berlin (1913).
70 Herzfeld, K.F. (1927). On atomic properties which make an element a metal. Phys. Rev. 29: 701–705.
71 Hensel, F. et al. (2015). On the occurrence of metallic character in the periodic table of the chemical elements. Phil. Trans. Roy. Soc. A 20140477: 373.
72 Hama, J. and Suito, K. (1989). Equation of state and metallization in compressed solid krypton. Phys. Lett. A 140: 117–121.
73 Ishikawa, T. et al. (2015). Superconductivity of compressed solid argon from first principles. Phys. Rev. B 064512: 91.
74 Khairallah, S.A. and Militzer, B. (2008). First-principles studies of the metallization and the equation of state of solid helium. Phys. Rev. Lett. 101: 106407.
75 Gažo, J. et al. (1976). Plasticity of the coordination sphere of copper(II) complexes, its manifestation and causes. Coord. Chem. Rev. 19: 253–297.
76 Aguado, F. et al. (2012). Pressure effects on Jahn-Teller distortion in perovskites: The roles of local and bulk compressibilities. Phys. Rev. B 100101 (R): 85.
77 Kurzydłowski, D. (2018). The Jahn-Teller distortion at high pressure: the case of copper difluoride. Crystals 8: 140.
78 Lazicki, A. et al. (2009). Anomalous optical and electronic properties of dense sodium. Proc. Nat. Acad. Sci. USA 106: 6525–6528.
79 Ma, Y. et al. (2009). Transparent dense sodium. Nature 458: 182–185.
80 Marques, M. (2011). Optical and electronic properties of dense sodium. Phys. Rev. B 83: 184106.
81 Matsuoka, T. and Shimizu, K. (2009). Direct observation of a pressure-induced metal-to-semiconductor transition in lithium. Nature 458: 186–189.
82 Stager, R.A. and Drickamer, H.G. (1963). Effect of temperature and pressure on the electrical resistance of four alkaline earth metals. Phys. Rev. 131: 2524–2527.
83 Gatti, M. et al. (2010). Sodium: a charge-transfer insulator at high pressures. Phys. Rev. Lett. 216404: 104.
84 Rousseau, B. et al. (2011). Exotic high pressure behavior of light alkali metals, lithium and sodium. Eur. Phys. J. B 81: 1–14.
85 Dye, J.L. (1990). Electrides: Ionic salts with electrons as the anions. Science 247: 663–668.
86 Tuoriniemi, J. et al. (2007). Superconductivity in lithium below 0.4 millikelvin at ambient pressure. Nature 447: 187–189.
87 Shimizu, K. et al. (2002). Superconductivity in compressed lithium at 20 K. Nature 419: 597–599.
88 Struzhkin, V.V. et al. (2002). Superconductivity in dense lithium. Science 298: 1213–1215.
89 Yabuuchi, T. et al. (2006). Superconductivity of Ca exceeding 25 K at megabar pressures. J. Phys. Soc. Jap. 083703: 75.
90 https://web.physics.wustl.edu/jss/NewPeriodicTable.pdf
91 Sakata, M. et al. (2011). Superconducting state of Ca-VII below a critical temperature of 29 K at a pressure of 216 GPa. Phys. Rev. B 220512 (R): 83.

92 B. Lorenz, C. W. Chu, High pressure effects on superconductivity. In: Frontiers in Superconducting Materials, ed. A. V. Narlikar, Springer, Berlin pp. 459–497 (2005).

93 Chu, C.W. et al. (1987). Evidence for superconductivity above 40 K in the La-Ba-Cu-O compound system. Phys. Rev. Lett. 58: 405–407.

94 Chu, C.W. et al. (1987). Superconductivity at 52.5 K in the lanthanum-barium-copper-oxide system. Science 235: 567–569.

95 L. Gao, et al., Superconductivity up to 164 K in $HgBa_2Ca_{m-1}Cu_mO_{2m+2+delta}$ (m=1, 2, and 3) under quasihydrostatic pressures. Phys. Rev. B, 50: 4260(R)–4263(R) (1994).

96 Ashcroft, N.W. (1968). Metallic hydrogen: a high-temperature superconductor? Phys. Rev. Lett. 21: 1748–1749.

97 Ashcroft, N.W. (2004). Hydrogen dominant metallic alloys: high temperature superconductors? Phys. Rev. Lett. 92: 187002.

98 Sun, L. (2006). High pressure studies on silane to 210 GPa at 300 K: optical evidence of an insulator–semiconductor transition. J. Phys.: Condens. Matter 18: 8573–8580.

99 Drozdov, A.P. et al. (2015). Conventional superconductivity at 203 kelvin at high pressures in the sulfur hydride system. Nature 525: 73–76.

100 Drozdov, A.P. et al. (2019). Superconductivity at 250 K in lanthanum hydride under high pressures. Nature 569: 528–531.

101 Somayazulu, M. et al. (2019). Evidence for superconductivity above 260 K in lanthanum superhydride at megabar pressures. Phys. Rev. Lett. 027001: 122.

102 V. V (Dec 2019). Struzhkin. In: private communication to the author.

103 Liu, H. et al. (2017). Potential high-TC superconducting lanthanum and yttrium hydrides at high pressure. Proc. Nat. Acad. Sci. USA 114: 6990–6995.

104 Sun, Y. et al. (2019). Route to a superconducting phase above room temperature in electron-doped hydride compounds under high pressure. Phys. Rev. Lett. 097001: 123.

105 Akimitsu, J. (2019). Towards higher-TC superconductors. Proc. Jap. Acad. Series B 95: 321–342.

106 Grochala, W. (2017). The Maximum Hardness Principle revisited and applied to solids (Part 2). Phys. Chem. Chem. Phys. 19: 30984–31006.

107 Grochala, W. and Derzsi, M. (2018). High-temperature superconductivity as viewed from the Maximum Hardness Principle. J. Mol. Model. 24: 233.

108 Ideue, T. et al. (2019). Pressure-induced topological phase transition in noncentrosymmetric elemental tellurium. Proc. Nat. Acad. Sci. USA 116: 25530–25534.

109 Struzhkin, V.V. et al. (2001). Phonon-assisted electronic topological transition in MgB_2 under pressure. In: arXiv preprint cond-mat/0106576.

110 Deng, L. et al. (2019). Higher superconducting transition temperature by breaking the universal pressure relation. Proc. Nat. Acad. Sci. USA 116: 2004–2008.

111 Zurek, E. and Grochala, W. (2015). Predicting crystal structures and properties of matter at extreme conditions via quantum mechanics: the pressure is on. Phys. Chem. Chem. Phys. 17: 2917–2934.

112 Wang, H. et al. (2018). Hydrogen-rich superconductors at high pressures. WIREs Comput. Molec. Sci. e1330: 8.

113 Needs, R.J. and Pickard, C.J. (2016). Perspective: Role of structure prediction in materials discovery and design. APL Mater. 4: 053210.

114 Chakraborty, D. and Chattaraj, P.K. (2019). Bonding, reactivity, and dynamics in confined systems. J. Phys. Chem. A 123: 4513–4531.

115 Zurek, E. et al. (2009). A little bit of lithium does a lot for hydrogen. Proc. Nat. Acad. Sci. USA 106: 17640–17643.

116 Li, Y. et al. (2014). The metallization and superconductivity of dense hydrogen sulfide. J. Chem. Phys. 174712: 140.

117 Peng, F. et al. (2017). Hydrogen clathrate structures in rare earth hydrides at high pressures: possible route to room-temperature superconductivity. Phys. Rev. Lett. 107001: 119.

118 O. R. Smits, et al., A hundred-year-old experiment re-evaluated: accurate ab initio Monte Carlo simulations of the melting of radon. Angew. Chem. Int. Ed. Engl., 57: 9961–9964 (2018).

119 Zhu, Q. et al. (2012). Systematic search for low-enthalpy sp^3 carbon allotropes using evolutionary metadynamics. Phys. Rev. B 201407: 85.

120 Kvashnin, A.G. et al. (2019). Computational discovery of hard and superhard materials. J. Appl. Phys. 040901: 126.

121 Grochala, W. (2007). Atypical compounds of gases, which have been called 'noble'. Chem. Soc. Rev. 36: 1632–1655.

122 Liu, Z. et al. (2018). Reactivity of He with ionic compounds under high pressure. Nat. Commun. 9: 951.

123 J. Botana, et al., Mercury under pressure acts as a transition metal: calculated from first principles. Angew. Chem. Int. Ed. Engl., 54: 9280–9283 (2015).

3

Conceptual DFT and Confinement

Paul Geerlings[1,], David J. Tozer[2], and Frank De Proft[1]*

[1] *General Chemistry (ALGC), Vrije Universiteit Brussel, Brussels, Belgium*
[2] *Department of Chemistry, Durham University, Durham, UK*

3.1 Introduction and Reading Guide

The study of chemical reactivity is at the heart of chemistry: can we foresee that molecule A will react with molecule B, and if so, which position will be preferred? Modern quantum-chemical methods allow us to calculate reaction paths for small and even medium-sized molecules, revealing thermodynamic and kinetic aspects of a reaction (a glimpse at the abstract books of recent WATOC conferences is convincing). However, in order to avoid such a – still computing-expensive – behavior and moreover to gain more *insight* into its result, approaches are still developed and used which, on the basis of a number of relatively simple concepts and principles, explain and interpret experimental and theoretical data on reactivity and in a later stage turn to a predictive mode.

Of the multitude of approaches that were presented in the literature hitherto, the wave function approach, with its central icon, the orbital concept, stands out: Hückel 's π- electron theory (for an authoritative account, see [1]), Fukui's frontier MO theory [2], and the celebrated Woodward-Hoffmann rules for pericyclic reactions [3] have been at the forefront for decades. In the 1980s an interesting alternative was offered by Parr and co-workers, in the context of a subbranch of Kohn's Density Functional Theory [4], which at that time started to pervade chemistry and which considers the electron density and no longer the wavefunction as the basic carrier of information. Conceptual DFT (CDFT) [5–8] was born in which the E=E[N, v] functional stands central. Here E is the energy of the system, N its number of electrons and v the external potential, i.e. the potential, due to the nuclei, felt by the electrons. Upon a chemical reaction N and/or v are changing and inspection of the response of the energy of the atom, molecule, surface … under consideration of these perturbations describes the evolution of the system at the onset of the reaction. Adopting Klopman's rule [9], which holds when the reaction profiles of similar reactions do not cross between the reactant and the transition state, information then becomes available on the kinetic, possibly thermodynamic, preference of one reaction to another one by considering the reaction at the onset. This onset should therefore be characterized by a minimal and well-chosen response functions telling us how the system "answers" to the aforementioned

* Corresponding Author: pgeerlin@vub.be

Chemical Reactivity in Confined Systems: Theory, Modelling and Applications,
First Edition. Edited by Pratim Kumar Chattaraj and Debdutta Chakraborty.

perturbations in N and/or v. The study, including their numerical evaluation, of these response functions has therefore been crucial in the years behind. In section 3.2 a very short overview of this endeavor is given, which can also be of help when reading other chapters in this book.

Referring, however, to the title of this book, "Chemical Reactivity in Confined Space", the question arises if, between other issues/approaches described in this book, CDFT can/has been refurbished to describe reactivity in confined space [10, 11]. The answer is "yes", although the literature is not that extensive yet and a lot of work still has to be done. In section 3.3 we will give an overview and critical analysis of this literature. As we will see, the larger part of these studies has been devoted to atoms, undoubtedly because for symmetry reasons confinement is easier to tackle mathematically in these systems of spherical symmetry than for a molecule with strongly reduced symmetry or even, in many cases, deprived of any symmetry. Nevertheless, avenues were set out, but in section 3.3 we start with atoms anyway. Apart from their energy, polarizability is one of the atomic properties which has stood at the forefront since the early days of confinement, going back to the work by Michels, De Boer, and Bijl in 1937 [12]. Ionization energies and, to a lesser extent, electron affinities were also studied [13]. In section 3.3.1, devoted to atoms, it will be seen that polarizability, certainly for atoms, bears a proportionality to the well-known CDFT descriptor softness [14]. So the work on polarizability of atoms in confinement can be considered as a forerunner of later work on global CDFT descriptors on atoms in confinement. A similar issue appears for the Ionization energy and the Electron affinity which, in the finite difference approach to the electronic chemical potential (or electronegativity) [15] and the hardness [14], can be considered as the building blocks of the latter two quantities. So, when considering CDFT descriptors such as softness, hardness, and electronegativity, it will be with a hindsight to (mostly earlier) results on ionization energies and electron-affinities. In section 3.3.2 where molecules are at stake we start with a special section, a little bit aside from the main line of the chapter, but absolutely relevant in the context of this book, on electron affinities for which confinement is *used* as an effective way out to calculate accurate electron affinities for metastable ions [16]. The central part of that paragraph is devoted to a possible solution to discuss a general confinement procedure applicable to any kind of molecules with an application for both a global (hardness) and – for the first time – a local CDFT descriptor, the electronic Fukui function [17] and its companion the Molecular Electrostatic Potential [18, 19]. Finally the link between confinement and pressure is exploited to propose, in line with recent evolutions in CDFT, an extension of the E=E[N,v] functional with a pressure variable to cope with recent evolutions in high-pressure chemistry [20].

3.2 Conceptual DFT

As stated in the section 3.1, the E =E [N,v] functional and its various partial, functional, and mixed derivatives are at the heart of Conceptual DFT. In Scheme 1 the definition of the derivatives the derivatives $\partial^n E/\partial^m N\, \delta v(\mathbf{r}_1) \ldots \delta v(\mathbf{r}_{m'})$ $(n = m + m')$ is given for $n \leq 2$, as well as their connections. One recognizes that the first-order derivative $(\partial E/\partial N)_v$ is, in analogy with macroscopic thermodynamics, termed the electronic chemical potential μ, identified by Parr and coworkers [15] as minus the Mulliken electronegativity χ_M, given by I +A with an arbitrary factor ½ in front [21]. The other first derivative, the functional derivative of E with respect to v, $(\delta E/\delta v(\mathbf{r}))_N$, is easily seen to be equal to the electron density $\rho(\mathbf{r})$ (see e.g. [7]). In this way two fundamental quantities of the variational equation of time independent DFT [5]

$$\mu + \delta F_{HK}/\delta\rho(\mathbf{r}) = v(\mathbf{r}) \tag{3.1}$$

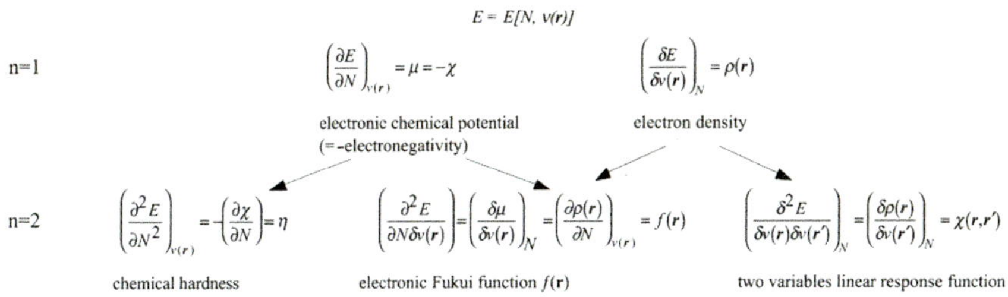

Scheme 3.1 The response function tree of Conceptual Density Functional Theory up to n=2 (see text).

are retrieved in CDFT. Here F_{HK} is the Hohenberg-Kohn functional and μ is the Lagrangian multiplier introduced in the variational procedure to ensure that ρ (**r**) is normalized to N at all times. *As such CDFT is part of DFT*.

In the third line in Scheme 3.1 one recognizes the second derivative of E with respect to N, identified by Parr and Pearson [14] as Pearson's chemical hardness [22], its inverse being baptized as the chemical softness S. In a finite difference approximation, its value can, just as is the case for the electronic chemical potential, be easily extracted from the knowledge of the ionization energy and the electron affinity as I -A. with sometimes a factor ½ put in front for convenience. The mixed derivative $\partial^2 E/\partial N\, \delta v(\mathbf{r})$ is termed the Fukui function f (**r**); it is a generalization of Fukui's frontier MO (density) concept: in a finite difference/frozen orbital approximation it indeed reduces to the density of the HOMO or LUMO for ΔN = −1 or +1 respectively. Finally the second functional derivative $(\delta^2 E/\delta v(\mathbf{r})\, \delta v(\mathbf{r}'))_N$, called the linear response function $\chi(\mathbf{r},\mathbf{r}')$, was recently shown to bear important chemical information retrieving inductive and mesomeric effects, and aromaticity. This might be inferred from the fact that it can be written as $(\delta\, \rho(\mathbf{r})/\delta\, v(\mathbf{r}'))_N$, expressing the response of the density ρ at a given point **r** to a perturbation δv in the external potential at a different point **r′** ubiquitous in any chemical reaction [23]. Obviously the "tree" in Figure 3.1 can and has been extended for higher n. For the

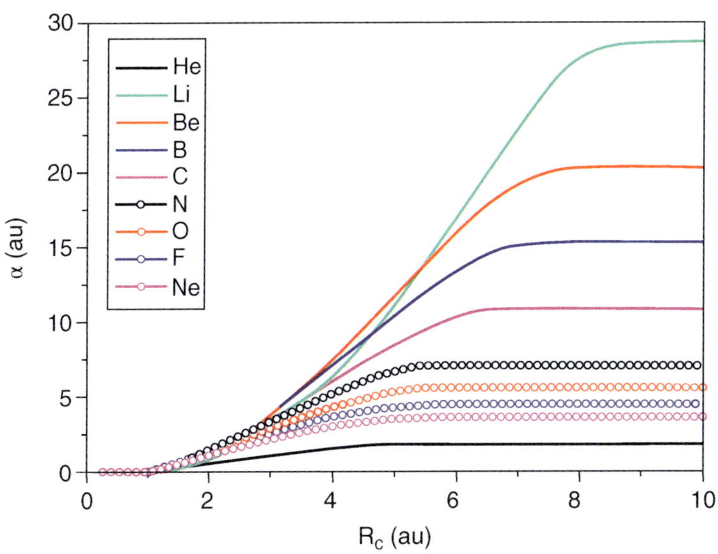

Figure 3.1 Plot of the polarizability α (a.u.) versus the cut-off radius (R_C,au) for atoms confined in a spherical box (reprinted with permission from J. Phys. Chem. A, 107, 4877–4882, Copyright (2003) American Chemical Society).

sake of brevity we mention that, until now, the n=3 and certainly n=4 cases received relatively minor attention (they also become more awkward to calculate and to interpret), except for one case: the N derivative of the Fukui function, called the dual descriptor $f^{(2)}(\mathbf{r}) = (\partial^2 \rho(\mathbf{r})/\partial N^2)_v = (\partial f(\mathbf{r})/\partial N)_v$, which has proven to be extremely useful for giving a one-shot representation of nucleophilic and electrophilic regions in a molecule [24, 25]. This "regional" aspect brings us to the well-known classification of the response functions into global, local and non- local descriptors depending on whether the quantity considered is **r**-independent (global), **r**-dependent (local) or is a function of two or more position vectors ($\mathbf{r}$, $\mathbf{r}'$..) (non-local). In the present work, also as a consequence of the existing literature, we will mainly focus on global descriptors (μ, η, S) and their "building blocks" I and A, or their relative, α, be it that in the section on molecules some results on the Fukui function and its relative, the Molecular Electrostatic Potential, will be discussed as examples of local descriptors. We again stress that particular attention will be devoted to the electron affinity because in a slightly different context confinement offers a way out to obtain high-quality values for negative electron affinities at stake when studying metastable anions.

Besides the abovementioned global descriptors we will also report on (confinement effects on) the electrophilicity ω [26]. Recently a status report on CDFT was published insisting that CDFT descriptors should be based on three fundamental precepts: observability, universality, and mathematical rigor [8]. In this context it was recognized that response functions, as energy derivatives, perfectly meet these requirements. Another class of descriptors, that are derived from the E=E[N,v] functional and exploit its characteristics are also perfectly acceptable. The most notorious member of this class is the electrophilicity ω introduced by Parr, von Szentpaly, and Liu, which Chattaraj and coworkers have shown to be exceptionally useful across a broad range of applications both in its global and local forms [27]. In its global form, at stake in this contribution, the electrophilicity is defined as the energy associated to a maximal flow of electrons transferred from a sea of free electrons at 0K and zero chemical potential to an electrophilic system. It corresponds to the minimum of the E = E (N) curve at constant v. It can be proven that the expression for ω boils down to a simple combination of two of the afore mentioned global descriptors

$$\omega = \mu^2/\eta \tag{3.2}$$

A final, almost philosophical, comment at the end of this short introduction to CDFT. In the past CDFT has been used primarily in the interpretation of both experimental and theoretical results. It has been extensively commented on in the recent status report [8] that, when following the guidelines for physically acceptable CDFT descriptors (e.g. the response functions), when respecting and adopting the correct mathematical framework (e.g. when facing for example of the E = E(N) discontinuity problem)) and when thoughtfully using the descriptors upon applying certain principles such as Pearson's HSAB principle [22], CDFT can and should turn into a powerful predictive modus. In this context introducing/combining confinement into CDFT may offer interesting perspectives for studying atomic and molecular behavior and reactivity at high pressure. In the next paragraph an overview will be given of what the combination of CDFT and confinement has led to until now.

3.3 Confinement and Conceptual DFT

3.3.1 Atoms: Global Descriptors

As stated above, polarizability was one of the first properties addressed in the context of confinement, and concentrated, for obvious reasons of computational simplicity, on atoms. Already in

1937 Michels et al. [12] presented numerical results on the decrease of (Kirkwood's [28] approximate expression for the) polarizability of the H- atom under pressure realized by confining the atom in a sphere of a given, finite radius, the confinement radius, often denoted as R_C. In the following decades many more refined studies on H and also He appeared (for comprehensive accounts see [10] and [11]) unambiguously showing a monotonous decrease of the polarizability with decreasing confinement radius R_c, approaching zero for very small R_c [29–31]). Extension to the case of the hyperpolarizability γ by Waugh in 2010 [31] revealed that for He γ also undergoes a monotonous decreasing trend. However, the decrease in γ is much more rapid as compared to the polarizability and with increasing compression the hyperpolarizability even undergoes sign inversion as already noticed by Banerjee for hydrogen [29]), a finding of high relevance in non-linear optics in which third harmonic generation and degenerate four wave mixing involve γ. In a reactivity context, trends along the periodic table are of primary interest, and sporadically a result for a many-electron atom was reported as by Von Faassen for Ne [30] using the cut-off function defined by Ludena: $f_k = (1-r/R)^k$ for $r \leq R$ and 0 for $r>R$) [13, 32]. A major breakthrough was then realized by Chattaraj and coworkers in 2003 [33–35], who performed numerical Hartree-Fock calculations on the He-Ne series (note that Ludena already in 1978 presented Hartree Fock studies on compressed H, He, Li, Be, B, C, Ne focusing, however, only on total and orbital energies) [13]). Chattaraj performed numerical Hartree Fock calculations with Dirichlet boundary conditions by multiplying the SCF wavefunction by a step function of the type $\Theta = \exp((-r/R_c)^{\lambda})$ where R_c is the cut-off radius of the spherical box on whose surface the wavefunction vanishes and λ is a parameter usually taken to be equal to 20 as suggested by Boeyens in 1994 [36].

Plotting α vs the cut-off radius for these atoms shows a decrease as a function of the confinement radius. The polarizability decreases monotonously and approaches zero for very small radii (Figure 3.1).

Note that except for special cases the curves do not cross so that the sequence in α values at zero pressure remains the same under pressure. The plot of the softness in Figure 3.2 is not unexpected in

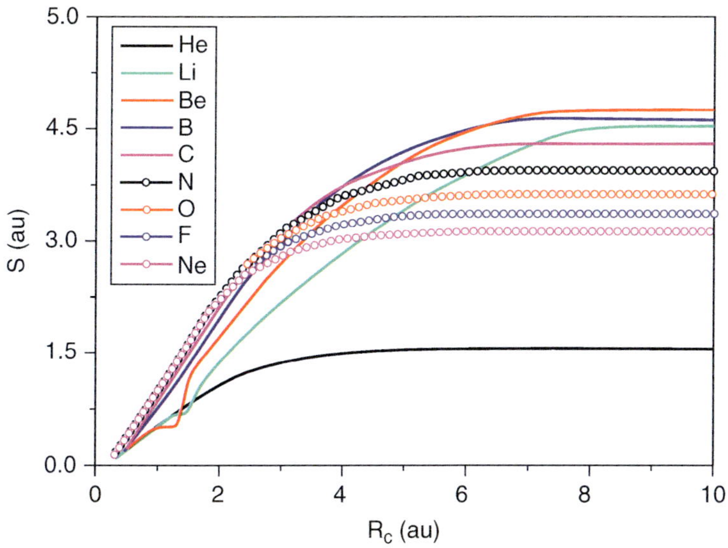

Figure 3.2 Plot of softness S (a.u.) versus cut-off radius R_C (a.u.) for atoms confined in a spherical box (reprinted with permission from J. Phys. Chem. A, 107, 4877–4882, Copyright (2003) American Chemical Society).

view of the well-established proportionality between α and S^3 for atoms [37]: upon decreasing cutoff radius the softness decreases continuously with hardly any crossings observed between curves for different atoms.

In the same vein Garza, Vargas, Aquino, and Sen [38], using the Dirichlet approach as proposed by Garza and Vela in 1998 [39] for confining, investigated the influence of confinement on the softness of the families IA, IIA, VA, and VIIIA of the periodic table. Their conclusions are in line with the previous works: 1° quasi- monotonic decrease of S upon increasing r_c and 2° crossings occur but not systematically ; remarkably, for some atoms, when extremely confined, an increment in S is observed due to changes in the electronic configuration of the atoms. The hardness shows, as expected, the inverse behavior and increases with pressure be it not infinitely due to the aforementioned changes in electronic configuration.

Later on Chattaraj et al. compared the Boeyens and Dirichlet boundary conditions [40]. In the first method the step function approach described above is used. In the second method the confinement is incorporated by using appropriate Dirichlet conditions so that the wave function goes to zero at the cut-off radius R_c. As the boundary condition is part of the variational optimization the latter method should lead to superior results than the former. It turned out that for *qualitative* trends differences in formulas used for the calculation of reactivity indices are much more influential than the difference between the use of the Dirichlet conditions and the cut-off method (see also next paragraph).

Concerning the electronic chemical potential or the electronegativity less studies are available. In 2003 Chattaraj and Sarkar [34] observed that χ is not very sensitive to confinement except for very small R_c, at which it shoots up to a high value (Figure 3.3).

Crossings between atoms are rarely observed. These results are, however, strongly different from those of Garza, Vargas, Aquino, and Sen [38] who observed later (in 2005) a decrease in electronegativity upon increasing confinement, attaining even negative values downwards from a certain radius, attributing this behavior to the negative I and A values below that radius, as mentioned

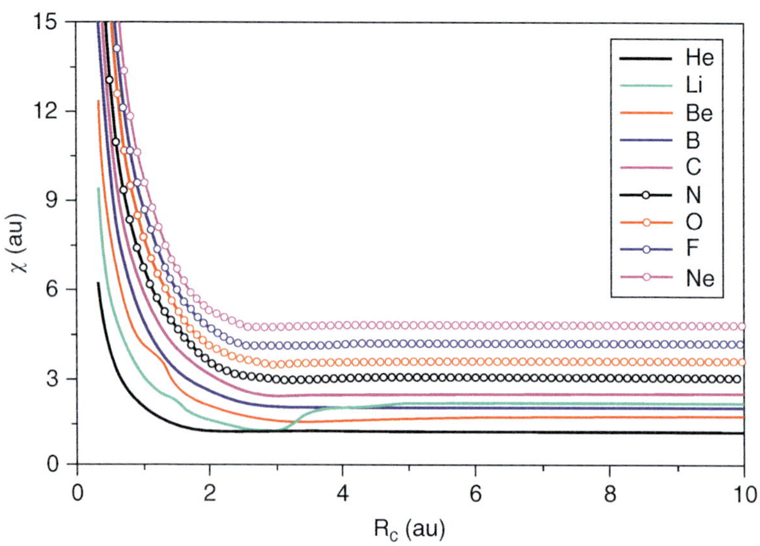

Figure 3.3 Plot of the electronegativity χ (in a.u.) versus cut-off radius R_C (a.u.) for atoms confined in a spherical box (reprinted with permission from J. Phys. Chem. A, 107, 4877–4882, Copyright (2003) American Chemical Society).

in section 3.2 the building blocks of μ or χ. They scrutinized the fundamental difference of their results with those in Figure 3.3 and attributed it to the way of imposing the confinement and the way of evaluating N derivatives of the energy in view of the derivative discontinuity at integer N.

At this moment it is advisable to reflect, with these results at hand, on the behavior of the two components of χ (and also of η), I and A.

The behavior of the ionization energy as a function of confinement was already addressed by Ludena in 1978 [13]. A typical trend for all atoms was revealed where this quantity diminishes upon confinement and becomes negative when lowering R_c to a critical value R_I, meaning that the confined atom prefers the ionized state in these circumstances. Expressed in terms of kinetic energy, the collapse of the electronic structure under pressure will take place when the outermost electron has enough kinetic energy to leave the atom. One can therefore define an atomic critical radius or ionization radius R_I, defined as the R value for which I is zero. Ludena et al. discussed this issue in their first series of SCF Hartree-Fock studies on H, He, Li, Be, B, C, and Ne and revealed some of the most salient features of the effect of pressure on the electronic structure of atoms. They tabulate the pressure at which the ionization of these different atom occurs, which is of the order of 10^6 atm or larger in order to produce a collapse of the electronic structure of atoms. High pressures are clearly necessary, which can be found in the interior of stellar bodies. Note that this behavior was discussed by Boeyens in 1994 [36] for the complete periodic table (using a step function confinement approach): a periodic behavior of the Ionization Radius R_I for the atoms He-Lr was retrieved, which, according to the author, could serve in quantum-chemical studies of reactivity. Note, however, that the critical radii values in the works by Garza [38], Ludena [13], and also Sen [41] are closer to each other as compared to those published by Boeyens (Table 3.1), probably due to the non-use of the Dirichlet boundary conditions. Sen et al., moreover, depicted a clear periodic behavior for the atoms He-Ca, with outspoken minima at the noble gases (He, Ne, Ar) and maxima for the alkali atoms (Li, Na, K) [41]. Anyway, an overall typical behavior of main group elements is that R_I increases when going down in the periodic table and decreases along a period, be it with a discontinuity when a subshell is filled.

Returning now to electronegativity and concentrating on the study by Garza et al., they depicted the behavior of both the ionization energy and the electron affinity of Kr (note that to the best of our knowledge this is the most detailed study of a multi-electron atom electron affinity under confinement). In Figure 3.4a it is seen that I becomes negative for a R_C value of about 3.3 a.u. (cf. Table 3.1) and then continuously drops to very low negative values for decreasing confinement radius. When comparing this curve with the corresponding curve for the electron affinity it is seen that A drops faster to highly negative values for small R_C than I. This results in an overall decrease of χ (obtained as the average of I and A), with decreasing confinement radius and a negative χ value below a R value of the order of 5 a.u. The increasing hardness upon confinement can then easily be traced back to the more negative A values as compared to I when both are inserted in the I-A expression.

The behavior is nicely formulated by Garza et al. stating that "when an atom is confined it starts to lose its capacity to attract electrons and, after that, the confined atom ejects an electron since the ionization radius is less than that R_c where χ starts to attain negative values."

Note that very recently Hoffmann et al. [42] obtained comparable results for the electronegativity. They explored atoms under isotropic compression in a non- reactive neon-like environment. Though using a different definition of electronegativity, χ^* (the average of the orbital energies of the occupied orbitals [43]) implying that electronegative atoms take more negative (or less positive) values of χ^*, they invariably found in their extensive study comprising the complete Table a decrease

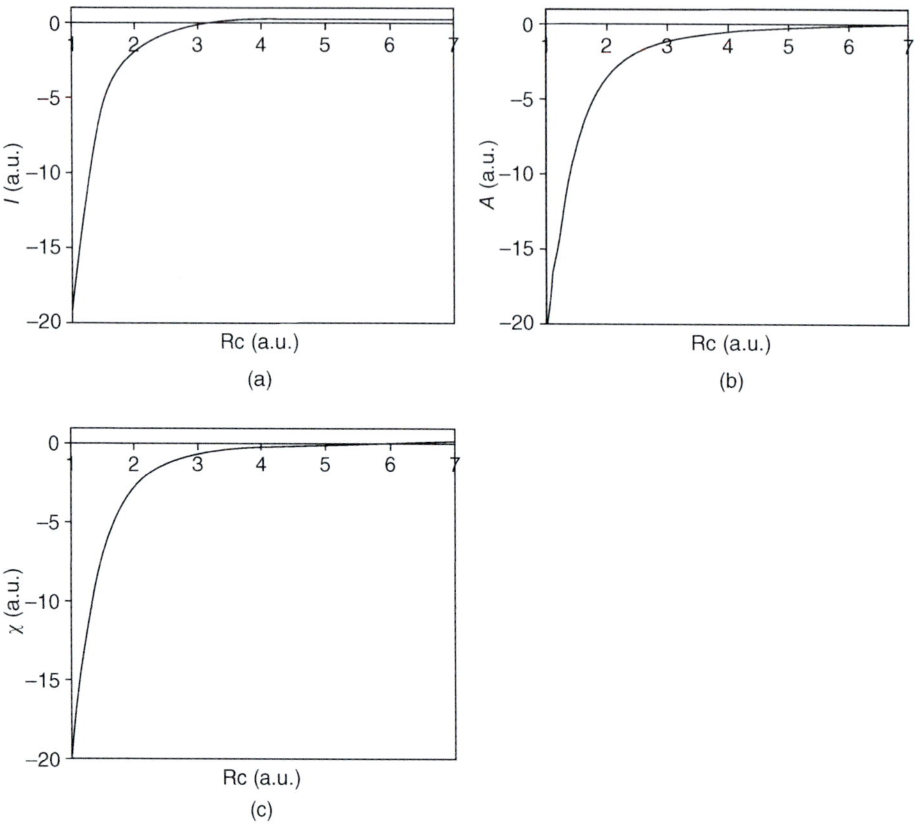

Figure 3.4 Plot, for Kr, of (a) the ionization energy I, (b) the electron affinity A, and (c) the electronegativity χ as a function of the confinement radius R_C (reprinted with permission from J. Chem. Sci. 117, 379–386 Copyright (2005) Springer Nature).

in electronegativity under pressure. Indeed, the increase in orbital energies upon increasing pressure, leads, via the aforementioned definition of χ^* to a decrease in its value. This result parallels the early findings by Ludena in 1978 [13], who in the first SCF calculation on first row atoms under compression always observed an increase in orbital energies under compression.

Anyway, in view of the fundamentally different results, this issue deserves further attention and a systematic study with different methods both for confinement on a series of preferably main group atoms (in order not to complicate things too much with too many spin issues) should be undertaken to provide a final answer.

We finally devote some place to electrophilicity. To the best of our knowledge its behavior under confinement has only be scrutinized by Chattaraj and co-workers [34]. A relative insensitivity at larger R_c value was observed followed by a decrease around 3 au and then a strong increase for R_c approaching 0. *In globo* ω is not very sensitive to confinement except for small R_c (Figure 3.5).

3.3.2 Molecules: Global and Local Descriptors

The work on the behavior of CDFT descriptors of molecules is rather limited. Concerning global reactivity descriptors, the main interest was given to chemical hardness. As expected, in view of their broad use in the literature, the first treatise on local descriptors in the molecular field pertain

Table 3.1 Atomic Ionization Radii: comparison for group IA, IIA, VA, and VIIIA atoms between different sources. All distances in A (taken from Garza et al) [38].

Atom	Garza et al. [38]	Previous works
Li	2·17	2·26 [41] Sen et al.
		1·25 [36] Boeyens
		2.21 [13] Ludena
Na	2·26	2·39 [41]
		2·73 [36]
K	2·71	2·88 [41]
		3·74 [36]
Be	1·70	1·70 [41]
		1·09 [36]
Mg	1·98	2·02 [41]
		2·36 [36]
Ca	2·54	2·52 [41]
		3·26 [36]
N	1·21	1·29 [41]
		1·56 [36]
P	1·69	1·79 [41]
		2·20 [36]
As	1·77	
Ne	0·97	0·98 [41]
		1·20 [36]
		0·97 [13]
Ar	1·36	1·38 [41]
		1·8 [36]
Kr	1·68	

to the electronic Fukui function and its companion, the Molecular Electrostatic Potential. However, as stated in the Introduction, confinement has been investigated in some detail when looking for an efficient computational strategy for the obtention of the (negative) electron affinities of metastable anions. We therefore start this paragraph with reporting results on electron affinity, as stated above, one of the building blocks of the electronic chemical potential and the chemical hardness.

3.3.2.1 Electron Affinities

It is well known that many neutral molecules do have a negative first vertical electron affinity Electron Transmission Spectroscopy [16] offers a way to measure these values, related to unstable, temporary anions, with an energy above that of the neutral molecule and unstable with respect to electron loss. Aiming at calculating these quantities hurts a lot of obstacles and so the evaluation of μ and η in a finite difference approach. When extended diffuse basis sets are used for an optimal description of the outer regions of the molecule the electron shows tendency to move to the outermost diffuse orbitals and to leave the molecule. The use of a compact basis set is a way out

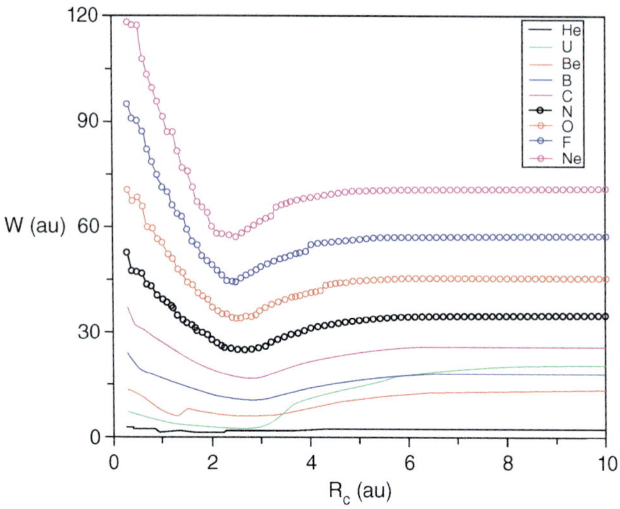

Figure 3.5 Plot of the electrophilicity index ω (a.u.) versus cut-off radius R_C (a.u.) for atoms confined in a spherical bow (reprinted with permission from J. Phys. Chem. A, 107, 4877–4882, Copyright (2003) American Chemical Society).

(for an early discussion see [44]), but its selection is a delicate task and its compactness evidently is harmful for an overall adequate description of the electronic structure of the molecule.

A solution was put forward by Puiatti et al. [45] by stabilizing the unstable anion in solvents with high dielectric constant ε and extrapolating a series of results with different ε to the ε=1 case This approach was followed in a study on a series of organic molecules, all of them featuring a negative electron affinity, by calculating their electron affinities in a set of 15 solvents of decreasing polarity and then plotting A vs 1/ε to yield a trustworthy value for A as

$$A = \lim(\varepsilon \to 1)\, \Delta E(1/\varepsilon) \tag{3.3}$$

where ΔE is the energy difference anion—neutral for a given ε value.

The same ansatz was more recently followed by one of the present authors in an endeavor to obtain an internally consistent scale of noble gas electronegativities and hardnesses, evaluated in the finite different approach from ionization energies and electron affinities [46]. All A and I values were evaluated in the same set of solvents as Puiatti, with ε varying from 78.39 (water) to 1.43 (argon). In this way both an electronegativity and a hardness scale could be constructed for the He-Rn noble gases, indicating a high electronegativity, second to the halogens and slightly above the chalcogens, and extremely high hardness values, way above the halogens and the chalcogens. These results were presented as a key to understand the chemistry of the noble gases. Indeed, a CDFT picture emerges on the "why" of the chemical inertness of the noble gases (their extremely high hardness) and the "when" about the situations in which it can be overruled (lower hardness for high Z values) and the polarity of the resulting Ng-halogen or Ng-oxygen bonds.

Two of the present authors presented some years ago [47] a way to artificially bind the extra electron in metastable anions in a different way closer to the confinement approach as sketched so far in this chapter. Their aim was to obtain an electron affinity value, starting from the conventional expression

$$A = E_N - E_N + 1 \tag{3.4}$$

as close as possible to the value determined via an alternative way they presented before as

$$A = -(\varepsilon_{HOMO} + \varepsilon_{LUMO}) - I \tag{3.5}$$

where ε_{HOMO} and ε_{LUMO} are the KS orbital energies from a local exchange-correlation functional, to tackle the problem of negative electron affinities by considering the integer discontinuity [48]. The excess electron is bound by the inclusion of a potential wall in the Kohn-Sham orbital equation in which context the calculations are done. Specifically a sphere of radius R_A is placed around each constituent atom A in the anion with value given by

$$R_A = \lambda\ B_A \tag{3.6}$$

where λ is a positive dimensionless parameter and B_A is the Bragg Slater radius [49] of atom A Within any sphere the exchange correlation potential v_{xc} in the anion unrestricted KS orbital equation is defined to be the conventional potential. Outside the sphere, however, the potential is set equal to a spin dependent constant

$$\mu^{\alpha} = \varepsilon^{\alpha}_{HOMO,N+1} - \varepsilon_{LUMO,N} \tag{3.7}$$

$$\mu^{\beta} = \varepsilon^{\beta}_{HOMO,N+1} - \varepsilon_{HOMO,N} \tag{3.8}$$

where μ^{α} and μ^{β} define the height of the potential wall and also control the decay of the α and β anion orbitals. In expressions (3.7) and (3.8) $\varepsilon^{\alpha}{}_{HOMO,N+1}$ and $\varepsilon^{\beta}{}_{HOMO,N+1}$ are the α and β HOMO energies of the anion in the SCF procedure and $\varepsilon_{LUMO,N}$ and $\varepsilon_{HOMO,N}$ are the LUMO and HOMO energies respectively of the neutral system, obtained from a separate conventional DFT calculation. This choice ensures that in the limit of a complete basis-set, the α and β HOMOs of the anion decay as the LUMO and HOMO of the neutral respectively [47]. On the basis of a training set an optimal value for λ was determined to be 3.4. With this value a correlation of 97% was obtained between experimental and theoretical electron affinity values for a series of 34 organic molecules, all of them displaying a negative experimental A value [50].

$$A_{\lambda} = 1.23\ A_{exp} - 0.22\ (R^2 = 0.97) \tag{3.9}$$

It also indicates that all electronic properties, as CDFT descriptors, for a temporary anion can be determined from a single DFT calculation on that anion using a large diffuse basis-set as (vide supra) required for these systems. With the $\lambda = 3.4$ value excellent results were obtained in a comparative study on the negative electron affinities of a series of fluoro-ethenes and on the role of d-orbitals on the bonding characteristics in the heavy analogues of pyridine and furane radical anions [51, 52].

3.3.2.2 Hardness and Electronic Fukui Function

The spherical confinement ansatz was applied to evaluate CDFT descriptors at the molecular level, either using the approach summarized in Eqs (3.7) and (3.8) or by taking the potential outside the spheres equal to a very large, constant, value to ensure the appropriate asymptotic density decay achieving confinement. Primary attention has been given to the global hardness and to the electronic Fukui function, the latter being the first local CDFT descriptor discussed in the context of confinement.

For ethylene the change in the HOMO-LUMO gap, used as an approximation for the hardness, and evaluated with the large potential outside the spheres, as a function of the confinement parameter λ is depicted in Figure 3.6 [53].

Upon reduction of λ from $\lambda = 10$ to $\lambda = 3$ a gradual increase in the hardness of 0.03 au is observed. The results are confirmed by a planar confinement model which is interesting in

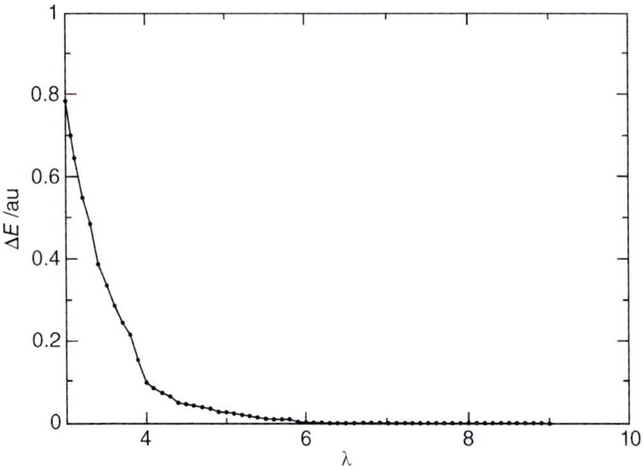

Figure 3.6 Change in the HOMO-MUMO gap as a function of λ for spherically confined ethylene (reproduced from [53], Copyright (2008) Royal Society of Chemistry).

comparative studies on planar hydrocarbons. In that case the potential in the Kohn-Sham equations is unchanged within a distance λB_C above and below the molecular plane (where B_C is the Bragg Slater radius of the Carbon atom) but set equal to a large constant at larger distances from the molecular plane. The results of increasing hardness with increasing confinement are in line with the studies on atoms discussed in section 3.3.1 and were also confirmed in the present approach by some preliminary atomic calculations on He, Ne, and Mg. Similar results were obtained for three more hydrocarbons: benzene, toluene, and naphthalene all showing a steady increase in hardness with decreasing λ. Note that the overall results were in contradiction with earlier work by Marquez et al. [54] on the influence of confinement on photophysical properties of naphthalene in a zeolitic environment. A bathochromic shift of the 0-0 transition was observed, concomitant with a decrease of the naphthalene band-gap or hardness, as was also found in periodic HF and KS calculations. However, these studies explicitly included the interaction between the molecule and the confining chemical medium whereas our studies only focus on the pure effect of confinement. The present studies eliminate specific interactions between the molecule and the confining medium, thereby exclusively focusing on the effect of confinement.

Turning to the case of the Fukui function, in analogy with the equation for the electron affinity

$$A_\lambda = E_N - E^\lambda_{N+1} \tag{3.10}$$

where λ is the confinement parameter, the working equation can be written in a finite difference approximation as [47]

$$f^+_\lambda(\mathbf{r}) = \rho^\lambda_{N+1}(\mathbf{r}) - \rho_N(\mathbf{r}) \tag{3.11}$$

where $\rho^\lambda{}_{N+1}$ $(\mathbf{r})$ is the electron density of the anion determined from orbitals evaluated in the presence of the potential wall and ρ_N $(\mathbf{r})$ is the conventional density of the neutral system. The evaluation of $f^-{}_\lambda$ $(\mathbf{r})$ is completely analogous. This technique was first presented by two of the present authors in 2007, using the ansatz based on Eqs (3.7) and (3.8), with λ selected on the basis of the best overall performance in the calculation of the (negative) electron affinities for H_2CO, C_2H_4, NH_3, and H_2O, the value being of the order of 3.5. The aim of that study was mainly to test if the use of a

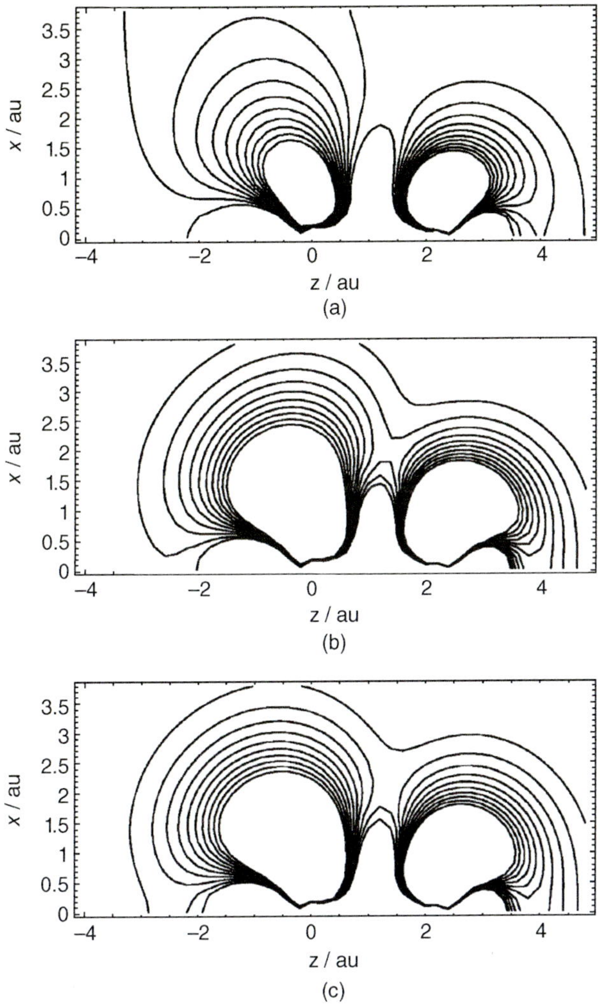

Figure 3.7 H_2CO Fukui function f^+ in a plane perpendicular to the molecular plane: influence of basis-set and potential wall (see text) (reprinted with permission from J. Chem. Phys. 127, 034108, Copyright (2007) AIP Publishing).

potential wall could afford a way out to cope with the insufficient description of temporary anions with extended basis sets, rather than investigate the influence on confinement on reactivity. With that value the H_2CO Fukui function f^+ for a nucleophilic attack was calculated in the plane perpendicular to the molecular plane. Figure 3.7a represents a conventional DFT calculation with no potential wall, but with a diffuse, extended basis aug-cc-pVTZ, Figure 3.7b again without a potential wall but with a compact basis 6-311G* and finally Figure 3.7c combines the potential wall with the extended basis.

It is clear that when using large, diffuse bases without applying a potential wall, the excess electron which is tending to leave the molecule occupies an orbital with a significant contribution from diffuse functions centered on the Carbon atom. In Figure 3.6b the tendency is removed by using a compact basis at the price of a less adequate description of the electronic structure of the anion. Finally in Figure 3.7c the introduction of a wall permits the use of an extended, diffuse basis

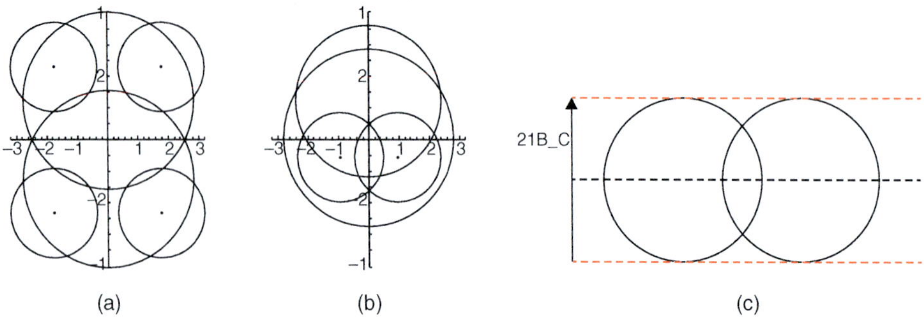

Figure 3.8 Spherical (a)(b) and planar confinement (c) of ethylene and formaldehyde (reproduced from [56] Copyright (2009) Royal Society of Chemistry).

set leading to an almost identical plot as in Figure 3.6b. Consistent with experimental data and many theoretical studies among others earlier Fukui function calculations [5, 55] the C atom is identified as the preferential site for nucleophilic attack as evidenced by the more extended contours around it.

In a follow-up paper [56] the confinement technique with very large constant potential outside the spheres was applied to ethylene and again to formaldehyde but with the hindsight of investigating the relation confinement-reactivity.

By choosing λ =4 (a little bit higher than the aforementioned 3.5) it is seen in Figure 3.8 that both molecules are effectively placed in a cavity which is similar to an ellipsoid with dimensions that correspond to (some cases) in real zeolites. In Figure 3.9 one notices that f^+, the Fukui function for nuclear attack, involving the calculation of the metastable ethylene-anion, is drastically affected by confinement, whereas f^- involving the ethylene cation is hardly affected, a tendency in line with what was explained before on the artificial binding of the excess electron in metastable anions.

These results are a very clear indication that *reactivity* can be strongly influenced, and, as a consequence, also be tuned, by spatial restrictions, and that the chemistry of anions (e.g. when encapsulated in a zeolite framework) is expected to be highly sensitive to confinement.

In the case of formaldehyde (Figure 3.10) the same striking differences between the influence of confinement on f^+ and f^- are observed. Clearly more detailed and quantitative studies should be undertaken to quantify these effects, to further investigate changes in overall reactivity and in regioselectivity and to put them in a perspective with "real" confinement situation.

A step in this direction has been taken by investigating a companion descriptor of the Fukui function, the Molecular Electrostatic Potential (cf. Chapter 7), which is preferred to be used instead of the Fukui function in hard-hard interactions (for detailed discussions see for example [57–59]). Comparison of the two MEP plots for H_2CO on a density iso-surface showcases (Figure 3.11) the influence of confinement on its values and so on the reactivity, in this case of the oxygen towards a hard electrophile (in principle a proton).

The color code clearly shows that the oxygen region becomes more negative upon confinement i.e. more prone to interact with a hard electrophile, with an appreciable shift of the MEP minimum in the oxygen region from −0.056 to −0.081 a.u. A more detailed inspection of the evolution of the density under confinement shows that confinement pushes the oxygen lone pair closer to the nucleus with concomitant deepening of the minimum of the MEP. Also in this context more comparative qualitative and quantitative studies should be undertaken.

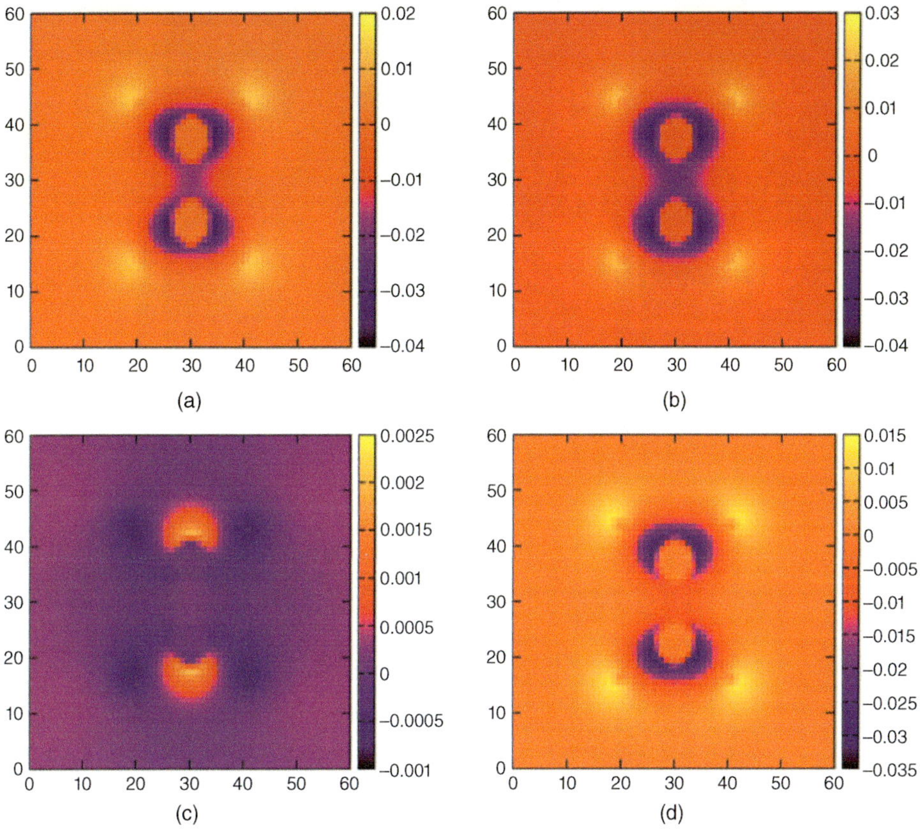

Figure 3.9 The Fukui function plotted in the plane of ethylene (reproduced from [56], Copyright (2009) Royal Society of Chemistry).

3.3.3 Inclusion of Pressure in the E = E [N,v] Functional

As stated in section 3.2, the fundamental quantity in CDFT is the E[N,v] functional. In the development of CDFT avenues were presented to include, besides N and v, other variables in this functional such as spin, electric and magnetic fields, and temperature to cope with a larger variety of reaction conditions. For a recent account on this issue we refer to [8] and for an up to date vision on the subject which in recent years was by far most frequently addressed, namely the inclusion of temperature, see [60]. Very recently the introduction of mechanical forces was described by two of the present authors in order to bridge CDFT and mechanochemistry [61, 62]. The confinement idea and the techniques described in this contribution might be a way out to extend the series of variables with pressure. Crucial is here the establishment of clear cut relations between (the degree of) confinement and the pressure, as has already been successfully done in early work on atoms in spherical confinement by Michels [12] (see also Key-Loo and Rubinstein [63], Ludena [32], Suryanarayana and Weil [64], Aquino et al. [65, 66]). Based on these relations working equations can be formulated to compute and interpret response functions involving pressure. In this way CDFT may enter the exciting field of (very) high pressure chemistry of widespread interest nowadays, both experimentally and theoretically (for the former aspect see for example a Chem. Soc. Rev. Special Issue [67], for the latter aspect for example the recent work by Cammi and Hoffmann [20, 68, 69]) a domain where simulation is a particular asset in view of the extreme experimental conditions.

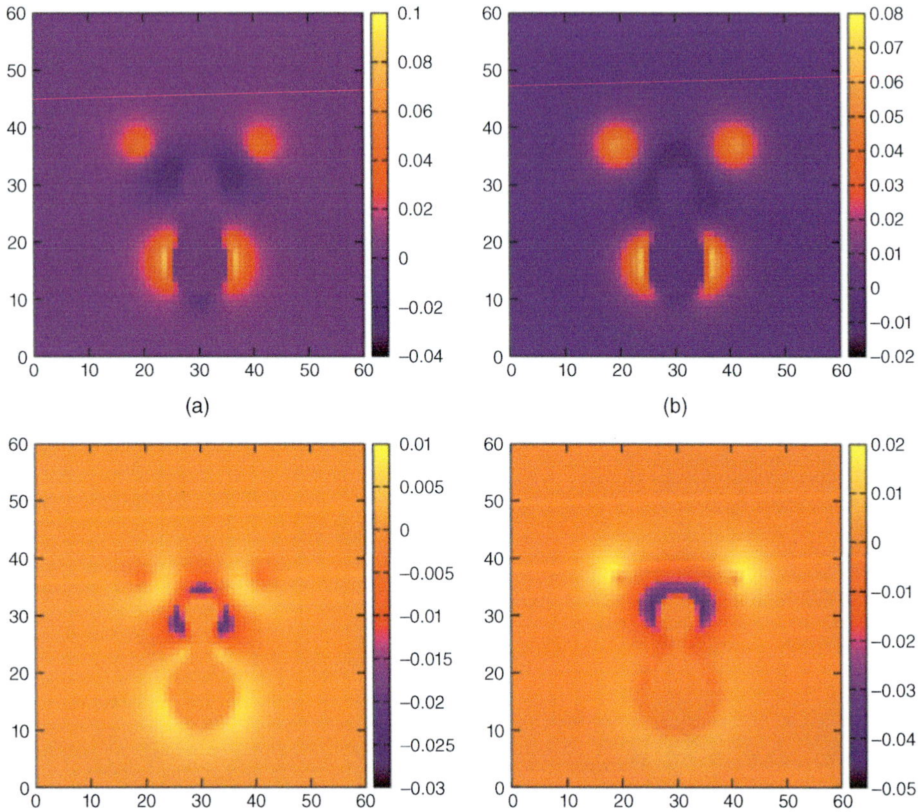

Figure 3.10 The Fukui function plotted in the plane of formaldehyde (reproduced from [56], Copyright (2009) Royal Society of Chemistry).

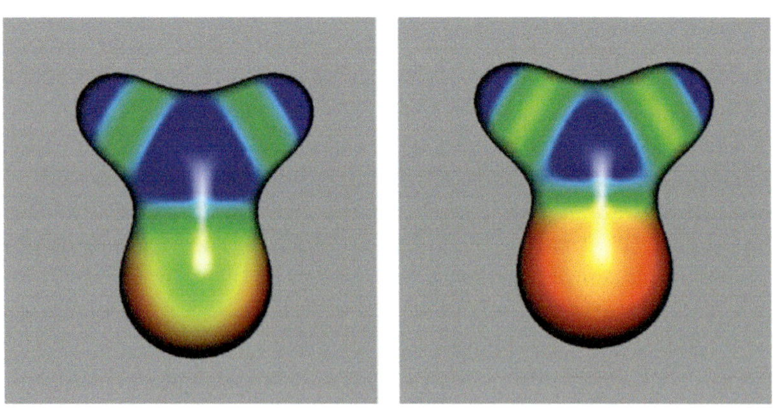

Figure 3.11 The MEP of (a) unconfined and (b) confined formaldehyde on an iso-surface of the density (reproduced from [56], Copyright (2009) Royal Society of Chemistry).

3.4 Conclusions

Conceptual Density Functional Theory has been offering an elegant way to interpret chemical reactivity in the past decades. If based on an adequate choice of descriptors, e.g. response functions, and within the correct mathematical framework, the thoughtful use of this ansatz may further serve as an interpretational tool but should be turned, in the very near future, in a predictive mode. The choice of descriptors, relatively small in number (electronic chemical potential, hardness, softness, Fukui function, linear response function, dual descriptor, and electrophilicity), made in the CDFT section deliberately fits the requirements formulated above. Their role in studying atomic or molecular properties under confinement as a proxy for chemistry high pressure is scrutinized on the basis of the existing literature, far more extensive for atoms, due to symmetry reasons, than for molecules. In line with early work on polarizability, softness is seen to be invariably decreasing upon increasing confinement, the opposite behavior, obviously, being noticed for the hardness. The situation for the electronic chemical potential, or the electronegativity, is less clear-cut, but the majority of works strongly indicate a decrease of electronegativity, even turning to negative values, upon increasing confinement. These results could be rationalized via the building blocks of these descriptors in the finite difference approximation, the ionization energy and the electron affinity. Whereas the latter has been studied extensively in the literature, introducing the concept of ionization radius, numerical data on the latter property are scarce but indicate that this quantity should decrease faster to deep negative values than the ionization energy upon decreasing confinement radius. For molecules two research lines can be distinguished: (1) the successful use of confinement to generate accurate negative electron affinities using extended bases, thereby avoiding the use of small, inaccurate basis-sets to prevent the escape of the outermost electron, and (2) the use of confinement to study now not only global but also local CDFT quantities. Using a similar technique as in (1), namely a spherical confinement for each atom in the molecule incorporated in DFT Kohn-Sham approach, the hardness (and by extension the softness) for simple test molecules as unsaturated hydrocarbons and formaldehyde was shown to behave analogously to the atomic case. The Fukui function shows marked differences in its sensitivity for electrophilic and nucleophilic attacks respectively. The results for the nucleophilic attack on H_2CO are in line with the literature for unconfined systems. Its companion for "hard" reagents (e.g. a proton), the Molecular Electrostatic Potential, shows a clear and substantial evolution towards higher reactivity of the oxygen region upon confinement.

Further research is at stake to set things straight in the case of atomic electronegativity, and to investigate a larger variety of molecules and a larger variety of global and local descriptors, to judge overall and local reactivity, including regioselectivity, under confinement as a proxy for reactions under high pressure.

Acknowledgements

The authors warmly thank the editors for their kind invitation. FDP and PG thank the VUB for continuous support and in particular for offering to the ALGC-group a second five-year period of a Strategic Research Program.

References

1 Streitwieser, A. (1961). Molecular Orbital Theory for Organic Chemists. New York: Wiley.
2 Fukui, K. (1975). Theory of Orientation and Stereoselection. Berlin: Springer Verlag.
3 Woodward, R.B. and Hoffmann, R.W. (1971). The Conservation of Orbital Symmetry. Verlag Chemie: Weinheim.
4 Hohenberg, P. and Kohn, W. (1964). Phys. Rev. B136: 864–871.
5 Parr, R.G. and Yang, W. (1989). Density Functional Theory of Atoms and Molecules. Oxford: Oxford University Press.
6 Parr, R.G. and Yang, W. (1995). Annu. Rev. Phys. Chem. 46: 701–728.
7 Geerlings, P., De Proft, F., and Langenaeker, W. (2003). Chem. Rev. 103: 1793–1873.
8 Geerlings, P., Chamorro, E., Chattaraj, P.K., De Proft, F., Gazquez, J.L., Liu, S., Morell, C., Toro-Labbé, A., Vela, A., Ayers, P. (2020) Theor. Chem. Acc., 139:36.
9 Klopman, G. in (1974) Chemical Reactivity and Reaction Paths. Klopman, G. Editor, John Wiley. New York Chapter 1.
10 Sen, K.D. (2014). Electronic Structure of Quantum Confined Atoms and Molecules. Springer.
11 Theory of Confined Systems – Part One (2009). Advances in Quantum Chemistry, 57. Sabin, J.R. and Brandas: E.J. Editors.
12 Michels, A., Deboer, J., Bijl, A. (1937) Physica IV, 981–994.
13 Ludena, E.V. (1978). J. Chem. Phys. 69: 1770–1775.
14 Parr, R.G. and Pearson, R.G. (1983). J. Am. Chem. Soc. 105: 7512–7516.
15 Parr, R.G., Donnelly, R.A., Levy, M., and Palke, W.E. (1978). J. Chem. Phys. 68: 3801–3807.
16 Jordan, K.D. and Burrow, P.D. (1987). Chem. Rev. 87: 557–588.
17 Parr, R.G. and Yang, W. (1984). J. Am. Chem. Soc. 106: 4049–4050.
18 Bonaccorsi, R., Scrocco, E., and Tomasi, J. (1970). J. Chem. Phys. 52: 5270–5284.
19 Molecular Electrostatic Potentials: Concepts and Appplication, Theoretical and Computational Chemistry, (1996) Volume 3, Murray, J.S. and Sen, K.D. Editors, Elsevier, Amsterdam.
20 Grochala, W., Hoffmann, R., Feng, W., and Ashcroft, N.W. (2007). Angew. Chem. Int. Ed. 46: 3620–3642.
21 Mulliken, R.S. (1934). J. Chem. Phys 2: 782–793.
22 Pearson, R.G. (1997). Chemical Hardness: Applications from Molecules to the Solid State. Weiheim, Germany: Wiley-VCH Verlag.
23 Geerlings, P., Fias, S., Boisdenhien, Z., De Proft, F. (2014) Chem. Soc. Rev., 43, 4989–5008.
24 Morell, C., Grand, A., and Toro-Labbé, A. (2005). J. Phys. Chem. A 109: 205–212.
25 Morell, C., Grand, A., and Toro-Labbé, A. (2006). Chem. Phys. Lett. 425: 342–346.
26 Parr, R.G., Von Szentpaly, L., and Liu, S. (1999). J. Am. Chem. Soc. 121: 1922–1924.
27 Chattaraj, P.K., Sarkar, U., and Roy, D.R. (2006). Chem. Rev. 106: 2065–2091.
28 Kirkwood, J.G. (1932). Phys. Z. 33: 57.
29 Banerjee, A., Sen, K.D., Garza, J., and Vargas, R. (2002). J. Chem. Phys. 116: 4054–4057.
30 Van Faassen, M. (2009). J. Chem. Phys. 131 (104108).
31 Waugh, S., Chowdury, A., and Banerjee, A. (2010). J. Phys. B 43: 225002.
32 Ludena, E.V. (1977). J. Chem. Phys. 66: 468–470.
33 Chattaraj, P.K. and Sarkar, U. (2003). Chem. Phys. Lett. 372: 805–809.
34 Chattaraj, P.K. and Sarkar, U. (2003). J. Phys. Chem. A 107: 4877–4882.
35 Chattaraj, P.K., Maiti, B., Sarkar, U. (2003) Proc. Indian Acad. Sci. (Chem. Sci.), 115, 195–218.
36 Boeyens, J.C.A. (1994). J. Chem. Soc. Faraday Trans. 90: 3377–3381.
37 Simon-Manso, Y. and Fuentealba, P. (1998). J. Phys. Chem. A 102: 2029–2032.

38 Garza, J., Vargas, R., Aquino, N., and Sen, K.D. (2005). J. Chem. Sci. 117: 379–381.
39 Garza, J., Vargas, R., and Vela, A. (1998). Phys. Rev. E 58: 3949–3954.
40 Sarkar, U., Giri, S., and Chattaraj, P.K. (2009). J. Phys. Chem. A 113: 10759–10766.
41 Sen, K.D., Garza, J., Vargas, R., and Vela, A. (2000). Chem. Phys. Lett. 325: 29–32.
42 Rahm, M., Cammi, R., Ashcroft, N.W., and Hoffmann, R. (2019). J. Am. Chem. Soc. 141: 1025–1027.
43 Rahm, M., Zeng, T., and Hoffmann, R. (2019). J. Am. Chem. Soc. 141: 342–351.
44 Guerra, M. (1990). Chem. Phys. Lett. 167: 315–319.
45 Puiatti, D., Mariano, D., Vera, A., and Pirini, A.B. (2009). PCCP 11: 9013–9024.
46 Furtado, J., De Proft, F., and Geerlings, P. (2015). J. Phys. Chem. A 119: 1339–1346.
47 Tozer, D.J., De Proft, F. (2007) J. Chem. Phys., 127, 034108.
48 Tozer, D.J. and De Proft, F. (2005). J. Phys. Chem. A. 109: 8923–8929.
49 Slater, J.C. (1964). J. Chem. Phys. 41: 3199.
50 Sablon, N., De Proft, F., Geerlings, P., and Tozer, D.J. (2007). PCCP 9: 5880–5884.
51 Peach, M.J., De Proft, F., and Tozer, D.J. (2010). J. Phys. Chem. Lett. 1: 2826–2831.
52 Hajgato, B., De Proft, F., Szieberth, D. et al. (2011). PCCP 13: 1663–1668.
53 Borgoo, A., Tozer, D.J., Geerlings, P., and De Proft, F. (2008). PCCP 10: 1406–1410.
54 Marquez, F., Zicovich-Wilson, C.M., Corma, A. et al. (2001). J. Phys. Chem. B 105: 9973–9997.
55 Langenaeker, W., De Decker, M., Geerlings, P., and Raeymaekers, P. (1990). J. Mol. Struct. (Theochem) 207: 115–130.
56 Borgoo, A., Tozer, D.J., Geerlings, P., and De Proft, F. (2009). PCCP 11: 2862–2868.
57 Langenaeker, W., De Proft, F., and Geerlings, P. (1995). J. Phys. Chem. 99: 6424–6431.
58 Geerlings, P., Langenaeker, W., De Proft, F., and Baeten, A. (1996). in Molecular Electrostatic Potentials: Concepts and Applications, Theoretical and Computational Chemistry, Vol, 3. Amsterdam, The Netherlands: Murray, J.S. and Sen, K.D. Editors, Elsevier Science.
59 Anderson, J.S.M., Melin, J., and Ayers, P.W. (2007). J. Chem. Theor. Comp. 3: 375–389.
60 Gazquez, J.L., Franco-Perez, M., Ayers, P.W., and Vela, A. (2019). Int. J. Quant. Chem. 119: e25797.
61 Bettens, T., Alonso, M., Geerlings, P., and De Proft, F. (2019). PCCP 21: 7378–7388.
62 Bettens, T., Alonso, M., Geerlings, P., and De Proft, F. (2020). Chem. Sci. 11: 1431–1439.
63 LeyKoo, E. and Rubinstein, S. (1979). J. Chem. Phys. 71: 351–357.
64 Suryanarayana, D. and Weil, J.A. (1976). J. Chem. Phys. 64: 510–513.
65 Aquino, N. (1994). Int. J. Quant. Chem. 54: 107–115.
66 Aquino, N., Campoy, G., and Montgomery, H.E. (2007). Int. J. Quant. Chem. 197: 1548–1155.
67 McMillan, P.F. (2006). Chem. Soc. Rev. 35: 855–857.
68 Chen, B., Hoffmann, R., and Cammi, R. (2017). Angew. Chem. 56: 11126–11142(2017).
69 Cammi, R., Chen., B., Rahm, M., (2018) J. Comp. Chem., 39, 2243–2250.

4

Electronic Structure of Systems Confined by Several Spatial Restrictions

Juan-José García-Miranda[1], Jorge Garza[1], Ilich A. Ibarra[2], Ana Martínez[2], Michael-Adán Martínez-Sánchez[1], Marcos Rivera-Almazo[1], and Rubicelia Vargas[1*]*

[1] *Departamento de Química, División de Ciencias Básicas e Ingeniería, Universidad Autónoma Metropolitana-Iztapalapa, 09340, Iztapalapa, Ciudad de México, San Rafael Atlixco 186, Col. Vicentina, México*
[2] *Instituto de Investigaciones en Materiales, Universidad Nacional Autónoma de México, 04510, Ciudad Universitaria, Ciudad de México, Circuito Exterior s/n, México*

4.1 Introduction

The electronic structure of atoms and molecules confined by spatial restrictions exhibits some characteristics drastically different to those observed for atoms and molecules without spatial restrictions. This conclusion had been verified experimentally and theoretically: in the 90s people found, experimentally, that some alkali metals present electron configurations similar to those of transition metals when these alkali metals were exposed to high pressure. Thus, extreme pressure induced electronic transitions in some metals. Another example where extreme conditions have a crucial impact on the electronic structure of atoms is the plasma. For these situations, some confinement models have been generated to analyse the effects of extreme conditions on the electronic structure of atoms and molecules [1]. Moreover, fullerenes have been used to confine atoms and molecules where the guest system is not submitted to extreme conditions [2]. Recently, metal-organic frameworks (MOFs) appeared in scene as new materials with potential technology applications [3, 4]. Similar to fullerenes, the confinement imposed on molecules embedded in MOFs cannot be associated to extreme conditions and consequently simple potential models are not easy to construct. Evidently, the importance to study confinements effects is crucial in the design of MOFs as trapper agents. In this chapter we do discuss different confinements models and their pertinence on some systems. For this discussion, Hartree-Fock (HF) [5] and Kohn-Sham (KS) [6] methods are central to obtain the corresponding electronic structure.

4.2 Confinement Imposed by Impenetrable Walls

The confinement imposed by impenetrable walls was proposed by Michels et. al. [7]. In this model, one atom is centered in a sphere with a surface of infinite potential. With this restriction the wave-function or the electron density must satisfy the condition

$$\psi(\mathbf{r}) = 0 \text{ for } r \geq r_c, \tag{4.1}$$

*Corresponding Author: R. Vargas; ruvf@xanum.uam.mx. J. Garza; jgo@xanum.uam.mx

Chemical Reactivity in Confined Systems: Theory, Modelling and Applications, First Edition. Edited by Pratim Kumar Chattaraj and Debdutta Chakraborty.

or

$$\rho(\mathbf{r}) = 0 \text{ for } r \geq r_c, \tag{4.2}$$

where r_c represents the radius of the sphere that imposes the confinement. In this way, Michels et. al. tried to simulate high pressure over the hydrogen atom. It is worth noting that the hydrogen atom represents a workhorse for confinement models, since many characteristics observed for this system are present in many-electron atoms [8–10]. The treatment of this system is straighfortward. First, the wave-function is written as $\psi(\mathbf{r}) = R(r)Y(\theta, \phi)$ and the radial contribution of the Schrödinger equation must satisfy the equation

$$\left[-\frac{1}{2}\frac{\partial^2}{\partial r^2} - \frac{1}{r}\frac{\partial}{\partial r} + \frac{\ell(\ell+1)}{2r^2} - \frac{1}{r}\right] R(r) = \epsilon R(r). \tag{4.3}$$

In this chapter we do use atomic units (au) such that the charge and the mass of the electron are equal to 1, and the Hartree is the unit for the energy. The solution to this equation is

$$R(r) = Ar^{\ell} \sum_{i=0}^{\infty} c_i r^i, \tag{4.4}$$

where c_is have the recurrence relation

$$c_j = -\frac{2(c_{j-1} + \epsilon c_{j-2})}{j(j + 2\ell + 1)}, \tag{4.5}$$

with $c_1 = -\frac{1}{\ell+1}c_0$ and $c_0 = 1$. To solve this equation, we fix r_c, l, and the number of terms in the sum (we can not use an infinite number of terms in a computer). From here we search the zero values of $R(r_c)$ since they represent the eigenvalues ϵ. By using 30 terms in the sum of Eq. (4.4) we found the graphs presented in Figure 4.1 for several values of ℓ and r_c. From this figure the hydrogen atom confined by impenetrable walls exhibits an important result, which is exhibited by many-electron atoms confined by this kind of spatial restriction: there are crossings between

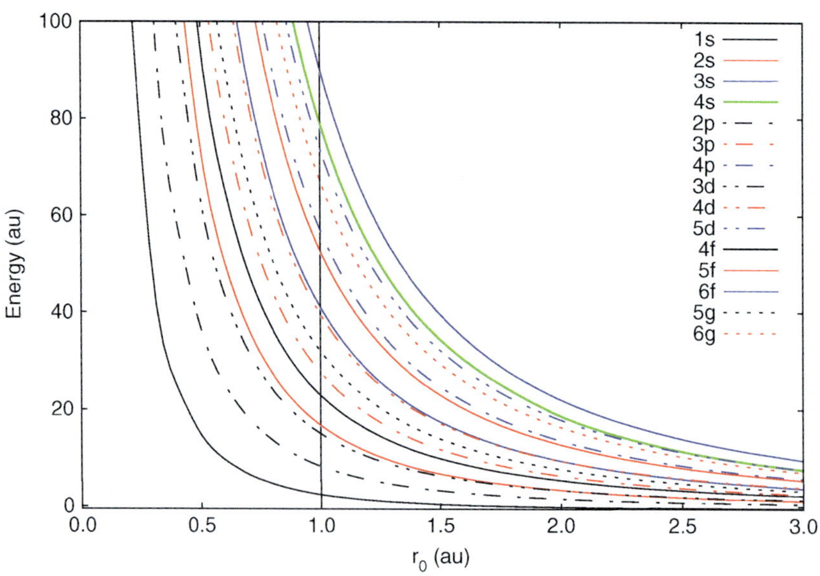

Figure 4.1 Orbital energies for the hydrogen atom confined by impenetrable walls as function of the confinement radius r_c.

orbitals with different ℓ. For example, for $r_c = 1.0$ au (vertical line in Figure 4.1) the orbital energy ordering is $1s2p3d2s4f3p5g$.., which is quite different to the ordering showed by the free hydrogen atom where the energy for orbitals with the same principal number is the same for different values of ℓ. Thus, the energy degeneration presented in the hydrogen atom is broken when this system is confined by spatial restrictions.

For many-electron atoms, the solution of HF or KS equations requires new software, since widely used programs for atoms, molecules and solids are unable to take into account the spatial restrictions imposed by equations (4.1) and (4.2). The MEXICA-C code [11, 12] solves HF and KS equations for atoms enclosed by hard walls using a basis set ad-hoc to satisfy the Dirichlet boundary conditions; there are alternative efforts to solve HF and KS equations for atoms confined by impenetrable walls [13–16].

The orbital energy crossing is one effect of hard walls over the electronic structure of many-electron atoms. In particular, such an effect have been observed for alkali and alkaline earth atoms where we have found the $s - d$ transition observed experimentally. Under the KS approach, some alkali and alkaline earth atoms have been confined by hard walls, the corresponding results predict crossing among orbital energies of different principal quantum number and consequently electronic transitions are observed. Some of these results are in agreement with experimental information [17, 18].

For a confinement imposed by rigid walls we expect localization of the electron density when the confinement radius is reduced. This effect has been explored by the Shannon (S) entropy defined in configuration space [19]. For atoms confined by impenetrable walls, the changes of the Shannon entropy, defined as $\Delta S = S_{conf} - S_{free}$, are always negative, indicating that the Shannon entropy is reduced when an atom is under a confinement imposed by rigid walls [20, 21]. It is worth noting that all conclusions delivered by the HF method applied over many-electron atoms confined by hard walls are reached also by the KS method. In fact, some exchange functionals give almost the same description as the HF method [22].

The confinement by impenetrable walls has been also applied over some molecules. However, the solution of HF or KS equations for these systems requires of new codes, as we have mentioned above. Fortunately, the H_2^+ molecule admits the exact solution of the Schrödinger equation [23], and with the exact solution we can build the electron density [24]. From this property, the Shannon entropy in configuration space delivers the same behavior exhibited by atoms confined by hard walls; ΔS is negative when this molecule is confined. Thus, confinements by rigid walls always localize the electron density. However, this molecule has the electron density localized in different regions, which depend on the confinement stages [24]. This property is presented in Figure 4.2 for different confinements, where the bold line corresponds to the free molecule and the dashed line is associated to extreme confinement.

The plot exhibited in Figure 4.2 is not expected since in previous publications (textbooks and papers) the electron density decays exponentially. In this figure we are using a logarithm scale and for this reason the electron density exhibits a linear behavior when this quantity goes to zero and we have a free molecule. The electron density of the free molecule is localized close to the nuclei. When the system is confined, the electron density augmented with maxima values localized at the nuclei. However, there is a confinement where the electron density is constant around the nuclei, and after this point presents a maximum at the middle of the nuclei. One of the consequences of this behavior of the electron density is that the Quantum Theory of Atoms in Molecules (QTAIM) [25, 26] fails, since under extreme conditions the QTAIM predicts an attractor in the middle of the nuclei, which apparently is not correct. The high values of kinetic energy are responsible of this effect since under extreme confinements the electron density avoids the surface

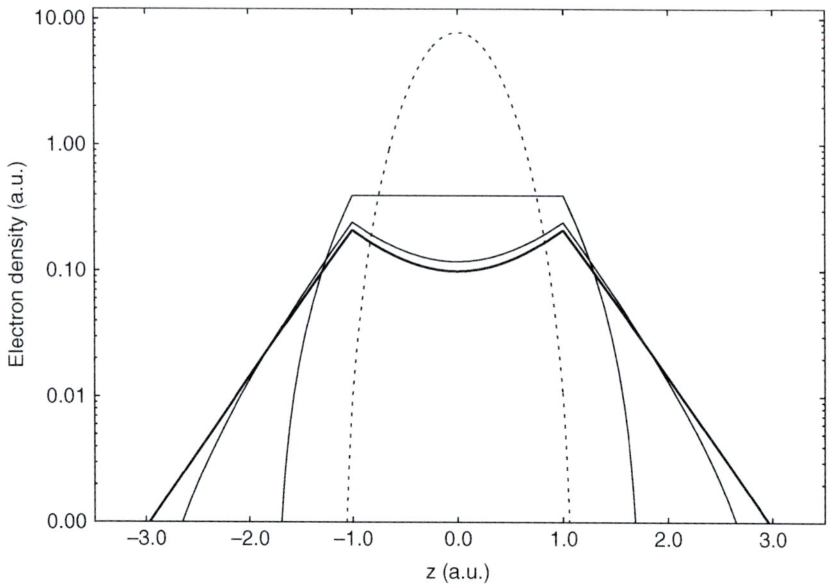

Figure 4.2 Electron density, in logarithm scale, along the internuclear axis (z) of the confined H_2^+ molecule. The bold line corresponds to the free molecule.

of the container and the middle of the molecule is the furthest region of the surface [24]. Although the H_2^+ is a small molecule we think that the main features of the chemical bond under extreme pressure will be preserved for many-electron molecules. For these molecules with many electrons we expect changes on the electron configuration for confinement imposed by impenetrable walls. Currently, there are a few reports of small molecules confined by spatial restrictions. However, the QTAIM has not been applied to investigate such systems. Thus, there are many opportunities to discover new features of the matter under extreme pressure by using a confinement imposed by rigid walls.

4.3 Confinement Imposed by Soft Walls

The confinement imposed by rigid walls gives many interesting features of the electronic structure of atoms and molecules when they are submitted under spatial restrictions. However, this confinement does not allow penetration of the electron density to regions where the confinement is imposed. Let us think of an atom or molecule trapped within a crystal structure. In this case, the electron density of the guest will penetrate to the crystal or will be mixed with the electron density of the crystal. Thus, impenetrable walls are not adequate to simulate this system. For this reason, there are models to take into account soft confinements. In general terms, a confinement potential over an atom can be expressed as

$$v(r) = \begin{cases} -\dfrac{Z}{r} & 0 \leq r < r_c \\ V_c(r) & r_c \leq r < \infty \end{cases}. \tag{4.6}$$

Ley-Koo and Rubinstein proposed [27]

$$V_c(r) = U_0, \tag{4.7}$$

where U_0 represents a constant potential. To solve the Schrödinger equation for the hydrogen atom under this confinement, the radial equation is divided into two regions: I) Inside the cavity, $r < r_c$. II) Outside the cavity, $r > r_c$. For region I, the solution is obtained in the same way to that used for hard walls through Eq. (4.3). For the region II, the radial equation has the form

$$\left[-\frac{1}{2}\frac{\partial^2}{\partial r^2} - \frac{1}{r}\frac{\partial}{\partial r} + \frac{\ell(\ell+1)}{2r^2} + V_c(r)\right] R_{\text{ext}}(r) = \xi R_{\text{ext}}(r). \tag{4.8}$$

Martínez-Sánchez et al. found that for different forms of V_c this equation can be transformed to the Kummer's equation [28]

$$\left[x\frac{\partial^2}{\partial x^2} + (b - x)\frac{\partial}{\partial x} - a\right] f(x) = 0, \tag{4.9}$$

where a, b and x depend on the specific form of V_c, and $f(x)$ is represented by a confluent hypergeometric function of the first or second kind. The technique used in this approach has been applied on the hydrogen atom confined by several soft potentials, which is not limited for a constant potential [28].

The hydrogen atom confined by a constant potential shows results different to those found when impenetrable walls are used as spatial restrictions. A possible ionization for small confinement radii is the most important result observed for this spatial restriction [27]. Another example of this behavior is shown in Figure 4.3. The shell structure obtained from the radial distribution function (RDF) corresponding to the free and confined lithium atom is depicted in Figure 4.3. From here we observe that there is a critical confinement radius ($r_c = 2.31$ au) where the external shell disappears. Precisely for this confinement, the energy of the highest occupied atomic orbital is zero. This result has important consequences. For example, for some confinement radii the electron density is delocalized and the Shannon entropy presents important increments, contrary to that observed for confinements imposed by hard walls [29]. Thus, confinements imposed by penetrable walls induce interesting responses on the electronic structure of the hydrogen atom.

Gorecki and Byers-Brown showed that the helium atom confined by a penetrable potential reproduces experimental values of this atom under high pressures. Thus, a penetrable potential is a good representation of some physical situations [30]. In fact, Duarte et al. found that the helium atom in

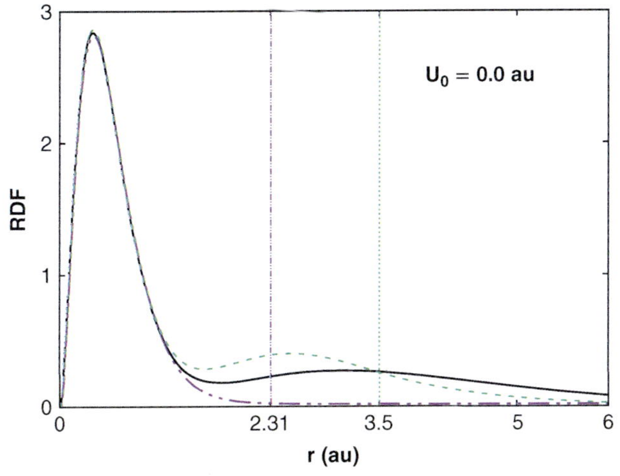

Figure 4.3 Radial distribution function (RDF) of the free lithium atom (solid line) and confined by penetrable walls ($U_0 = 0.0$ au) with $r_c = 3.5$ au (dashed line) and $r_c = 2.31$ au (dot-dot-dashed line).

a crystal structure can be simulated reasonably well by using a finite potential [22]. We must insist that with the potential represented in Eq. (4.7) there are difficulties to solve HF or KS equations and new codes are necessary to study many-electron systems. Rodriguez-Bautista et al. reported one numerical approach to solve the HF equations for atoms confined by a constant potential. In this work, a new basis set was proposed to solve the HF equations [12]. Such a basis set was inspired in the solution obtained by Ley-Koo and Rubinstein for the hydrogen atom. With this new basis set and the MEXICA-C code several many-electron atoms were studied under a confinement imposed by penetrable walls [11]. Several results for impenetrable walls are not found for soft confinements, in agreement with results obtained for the hydrogen atom. For example, for small confinement radii there are critical values where an atom ejects an electron by the action of the confinement. This result has important physical implications since right before the critical radii the electron density is spread over a large region, in particular, over classical forbidden regions. Connected with this result, the Shannon entropy shows a contrary behavior to that observed for impenetrable walls for small confinement radii $\Delta S > 0$. This result indicates that when the electron density is non-localized over particular regions then the corresponding Shannon entropy will be bigger than that observed for the free system [20].

Martínez-Sánchez et al. implemented the new basis set in the MEXICA-C code to solve KS equations [31]. Thus, this computational program has the possibility to solve HF and KS equations for atoms confined by impenetrable or penetrable walls. The study of the KS approach on confined atoms has given insight about the performance of exchange-correlation functionals to describe correctly atoms under these circumstances. For example, the delocalization of the electron density observed for atoms enclosed by soft walls cannot be predicted well by exchange functionals based on the local density approximation (Dirac functional) or on the generalized gradient approximation (Becke88 and PBE). Some problems observed for LDA and GGA are partially solved when a fraction of the exact exchange is included in a exchange functional [22]. Thus, hybrid exchange functionals are necessary to describe reasonably well systems confined by soft walls. This conclusion was obtained by using a set of many-electron atoms, however, this is connected with reports for molecules submitted to high pressures within crystal structures. Therefore the confined atom model can be used as a computational tool to obtain the performance of new exchange functionals.

The next step of this topic is the implementation of the new basis set on correlated methods like coupled cluster, multiconfiguration or configuration interaction to obtain details of the correlation energy in regions where an atom is almost ionized.

4.4 Beyond Confinement Models

So far we have discussed potentials to simulate spatial restrictions. However, in a real situation atoms or molecules are not encaged by perfect spheres of ellipsoids. Thus, another way to take into account the confinement is by inserting a molecular system within a crystal or inside in a large molecule. In this section we report two examples: a molecule inside the fullerene and a molecule within a MOF.

Let us start with the CH_4 molecule embedded in the C_{60}. In this case we want to observe changes on the CH_4 due to the confinement imposed by the fullerene C_{60}. In Figure 4.4 two possible interactions involved in the $CH_4@C_{60}$ system are presented. The bond paths obtained through the QTAIM are shown in magenta on the left-hand side. These bond paths indicate that the hydrogen atoms in the CH_4 interact with carbon atoms of the C_{60}, two of these contacts present an electron density at

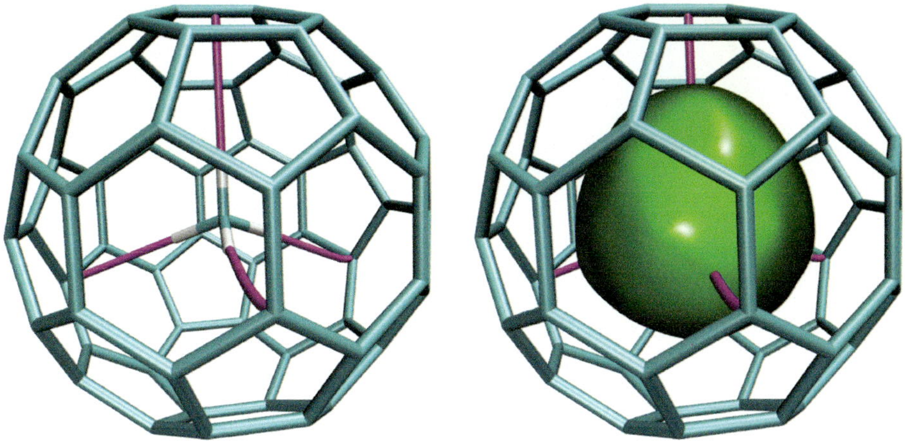

Figure 4.4 Bond paths, in magenta color, between C_{60} and CH_4 (left), and non-covalent interactions index in a green surface (right). Full optimization and QTAIM analysis done by using the PBE0-D3/6-311++G** method. Source: Based on E.R. Johnson, S. Keinan, P. Mori-Sánchez, J. Contreras-García, A.J. Cohen, and W. Yang. Revealing non-covalent interactions. J. Am. Chem. Soc., 132:6498–6506, 2010.

the bond critical point (ρ_{BCP}) of 0.1256 au, and the remained two contacts have $\rho_{BCP} = 0.1216$ au. If we compare the C-H bond length in the CH_4 between isolated and within the fullerene, we find a reduction in this property since it goes from 1.09 Å to 1.08 Å. This small decrement in the C-H bond length is reflected in ρ_{BCP} since it goes from 0.272 to 0.277 au. These results indicate that the CH_4 is compressed by the fullerene and the possible contacts with the CH_4 do not stabilize the full system. The right-hand side of Figure 4.4 shows the non-covalent interactions index (NCI) [32]. The color code used in this chapter represents strong attractive interactions in blue color, non-attractive interactions in red and weak attractive interactions in green color. From this figure it is clear that in the $CH_4@C_{60}$ system, weak interactions like van der Waals interactions form the attractive contacts between both fragments. To obtain this figure we subtracted the NCI corresponding to the free C_{60}, for that reason we do not observe isosurfaces in red color.

There are reports where the potential generated by the fullerene is modeled by a shell with small negative values [33–35]. From the discussion presented for the $CH_4@C_{60}$ system, we conclude that such models are well justified since carbon atoms in the fullerene keep their positions with regard to the free system [2]. This is not true when a guest molecule interacts in an important way with the atoms involved in the cavity.

To analyze the second example, a molecule within a MOF, we present in Figure 4.5 the Molecular Electrostatic Potential (MEP) mapped over the electron density of the MOF entitled Mg-CUK-1 [36]. This MOF has been experimentally evaluated for the capture of CO_2 and H_2S molecules. However, it is not clear how these guest molecules interact with the confinement cavity.

From this figure it is evident that this MOF exhibits large cavities with positive and negative regions of the MEP. Thus, we observe several regions where, for example, one H_2S molecule could interact with the cavity. Naturally, this target needs a deeper analysis and that is not discussed in this chapter. Instead for this purpose, we allocate in the center of a cavity three H_2S molecules to explore the effect of the cavity on this trimer. As starting point we centered three molecules with the same structure optimized in the gas phase, forming a triangle, and we relaxed these molecules and the hydrogen atoms were assigned by single crystal X-ray diffraction to the MOF. This procedure was done by using the CRYSTAL v14 code with the B3LYP-D3/POB-TZVP method.

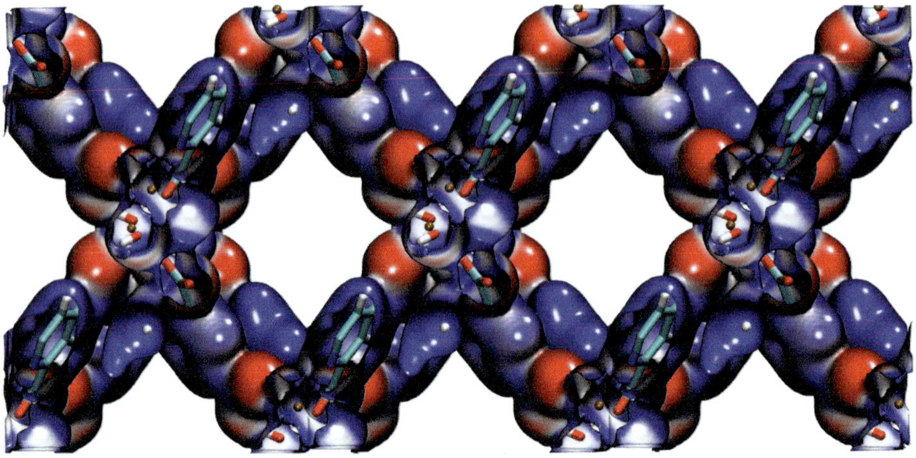

Figure 4.5 Molecular electrostatic potential of the Mg-CUK-1 mapped over the electron density. For this figure the electron density has an isosurface of 0.01 atomic units. Negative values in red and positive in blue. Source: Based on E. Sánchez-González, Paulo G. M. Mileo, M. Sagastuy-Breña, J.-Raziel Alvarez, J.-E. Reynolds, A. Villarreal, A. Gutiérrez-Alejandre, J. Ramírez, J. Balmaseda, E. González-Zamora, Guillaume Maurin, S.-M. Humphrey, and I.-A. Ibarra. Highly reversible sorption of H_2S and CO_2 by an environmentally friendly Mg-based MOF. J. Mater. Chem. A, 6:16900–16909, 2018.

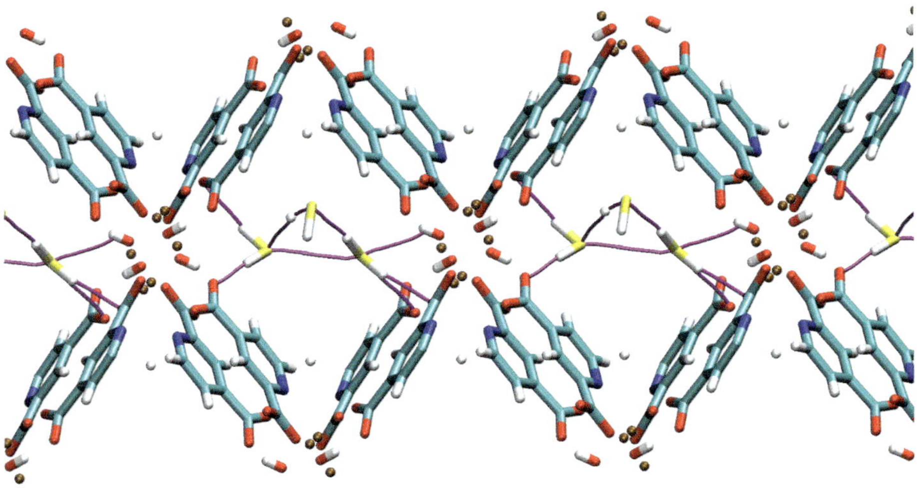

Figure 4.6 Bond paths in magenta involved between the H_2S trimer and the Mg-CUK-1 MOF. Magnesium, oxygen, carbon, nitrogen and hydrogen atoms in brown, red, cyan, blue and white colors, respectively.

After the optimization process, possible interactions between the H_2S trimer and Mg-CUK-1 found by the QTAIM are presented in Figure 4.6.

In this figure we clearly observe the S-H…O, S…S, and S…O interactions. Such interactions are responsible of distortions on the triangle defined by the H_2S trimer, with regard to the free trimer where there are only S…H intermolecular contacts. However, within Mg-CUK-1 the contacts between the H_2S molecules present two S…H contacts and one S…S contact. The S-H…O hydrogen bond distance is changed from 2.784 Å (in the free system) to 2.612 Å and 2.437 Å. It means that the MOF shrinks two hydrogen bonds, and the third one disappears. Obviously, there are consequences over the ρ_{BCP} corresponding to the hydrogen bonds involved with the H_2S trimer. This quantity goes from 0.0103 au (in the free system) to 0.0141 au and 0.0205 au, respectively. In this example, the confinement imposed by Mg-CUK-1 induces guest-cavity interactions in such a way that one S…H contact disappears. The triangle within this MOF is more compact than the one obtained without confinement. It seems then, that the H_2S cluster keeps its shape and the MOF is capable of compressing it.

4.5 Conclusions

In this chapter we have presented two different confinements: model potentials and explicit cavities. For the first case, we have discussed confinement potentials with hard and soft walls. For hard walls, the electron density of atoms or molecules has no possibilities to penetrate the potential barrier. The soft walls offer the possibility that the distribution of charge explores classically forbidden regions, which is appropriate to simulate several physical situations. These potentials exhibit well defined shapes, which is not necessarily true for real systems. For this reason, sometimes it is necessary to consider explicitly the atoms that conform the cavity. This strategy is useful to study cavity-guest systems but it is computationally expensive and the computational resources or software are not always available to perform this task. When it is possible to investigate these systems, the expected results are related to the nature of the cavity and the guest. The two systems presented in this chapter have non-covalent interactions between the cavity and the guest and we found that the cavities act as a confinement agent that compresses the guest.

References

1 E. Ley-Koo. Recent progress in confined atoms and molecules: Superintegrability and symmetry breakings. Rev. Mex. Fis., 64:326–363, 2018.

2 S. Bloodworth, G. Sitinova, S. Alom, S. Vidal, G.-R. Bacanu, S.-J. Elliot, M.-E. Light, J.-M. Herniman, G.-J. Langley, M.-H. Levitt, and R.-J. Whitby. First synthesis and characterization of CH_4@C_{60}. Angew. Chem. Int. Ed., 58:5038–5043, 2019.

3 Q. Wang and D. Astruc. State of the art and prospects in metal-organic framework (mof)-based and mof-derived nanocatalysis. Chem. Rev., 120:1438–1511, 2020.

4 M. Sánchez-Serratos, J. R. Álvarez, E. González-Zamora, and I. A. Ibarra. Porous coordination polymers (PCPs): New platforms for gas storage. J. Mex. Chem. Soc., 60:43–57, 2016.

5 A. Szabo and N. S. Ostlund. Modern Quantum Chemistry: Introduction to Advanced Electronic Structure Theory. Dover, New York, 1996.

6 R. G. Parr and W. Yang. Density-Functional Theory of Atoms and Molecules. Oxford University Press, Oxford, 1994.

7 A. Michels, J. De Boer, and A. Bijl. Remarks concerning molecular interaction and their influence on the polarisability. Physica, 4:981–994, 1937.

8 A. Sommerfeld and H. Welker. Artificial limiting conditions in the kepler problem. Ann. Phys., 32:56–65, 1938.

9 N. Aquino. Accurate energy eigenvalues for enclosed hydrogen atom within spherical impenetrable boxes. Int. J. Quantum Chem., 54:107–115, 1995.

10 E. V. Ludeña. SCF calculations for hydrogen in a spherical box. J. Chem. Phys., 66:468–470, 1977.

11 J. Garza, J.-M. Hernández-Pérez, J.-Z. Ramírez, and R. Vargas. Basis set effects on the Hartree-Fock description of confined many-electron atoms. J. Phys. B: At. Mol. Opt. Phys., 45:015002, 2012.

12 M. Rodriguez-Bautista, C. Díaz-García, A. M. Navarrete-López, R. Vargas, and J. Garza. Roothaan's approach to solve the Hartree-Fock equations for atoms confined by soft walls: Basis set with correct asymptotic behavior. J. Chem. Phys., 143:34103, 2015.

13 P. K. Chattaraj and U. Sarkar. Effect of spherical confinement on chemical reactivity. J. Phys. Chem. A, 107:4877–4882, 2003.

14 S. A. Cruz. Thomas-Fermi-Dirac-Weizsäcker density functional formalism applied to the study of many-electron atom confinement by open and closed boundaries. Adv. Quantum Chem., 57:255–283, 2009.

15 J. Garza, R. Vargas, and A. Vela. Numerical self-consistent-field method to solve the Kohn-Sham equations in confined many-electron atoms. Phys. Rev. E, 58:3949–3954, 1998.

16 M. van Faassen. Atoms in boxes: From confined atoms to electron-atom scattering. J. Chem. Phys., 131:104108, 2009.

17 D. Guerra, R. Vargas, P. Fuentealba, and J. Garza. Modeling pressure effects on the electronic properties of Ca, Sr, and Ba by the confined atoms model. Adv. Quantum Chem., 58:1, 2009.

18 K. D. Sen, J. Garza, R. Vargas, and A. Vela. Effective pressure induced electronic transition in spherically confined alkali metal atoms. Proc. Indian Nat. Sci. Acad., 70A:675–681, 2004.

19 C. E. Shannon. A mathematical theory of communication. Bell Syst. Tech., 27:379–423, 1948.

20 M. Rodriguez-Bautista, R. Vargas, N. Aquino, and J. Garza. Electron-density delocalization in many-electron atoms confined by penetrable walls: A Hartree-Fock study. Int. J. Quantum Chem., 118: e25571, 2018.

21 K. D. Sen. Characteristic features of Shannon information entropy of confined atoms. J. Chem. Phys., 123:074110, 2005.

22 F.-A. Duarte-Alcaráz, M.-A. Martínez-Sánchez, M. Rivera-Almazo, R. Vargas, R. A. Rosas-Burgos, and J. Garza. Testing one-parameter hybrid exchange functionals in confined atomic systems. J. Phys. B: At., Mol. Opt. Phys., 52:135002, 2019.

23 E. Ley-Koo and S. A. Cruz. The hydrogen-atom and the H_2^+ and HeH^{++} molecular ions inside prolate spheroidal boxes. J. Chem. Phys., 74: 4603–4610, 1981.

24 R. Hernández-Esparza, B. Landeros-Rivera, R. Vargas, and J. Garza. Electron density analysis for the H_2^+ system confined by hard walls: The chemical bond under extreme conditions. Ann. Phys. (Berl.), 531:1800476, 2019.

25 R. F. W. Bader. Atoms in molecules: A Quantum Theory. Oxford University Press, New York, 1990.

26 R. F. W. Bader. Principle of stationary action and the definition of a proper open system. Phys. Rev. B, 49:13348–13356, 1994.

27 E. Ley-Koo and S. Rubinstein. The hydrogen atom within spherical boxes with penetrable walls. J. Chem. Phys., 71:351–357, 1979.

28 M.-A. Martínez-Sánchez, R. Vargas, and J. Garza. Asymptotic Behavior for the Hydrogen Atom Confined by Different Potentials, Chapter 3, pages 101–132. Mathematics Research Developments. Nova Science Pub Inc, New York, 2020.

29 M.-A. Martínez-Sánchez, R. Vargas, and J. Garza. Shannon entropy for the hydrogen atom confined by four different potentials. Quantum Rep., 1:208–218, 2019.

30 J. Gorecki and W. Byers-Brown. Padded-box model for the effect of pressure on helium. J. Phys. B: At. Mol. Opt. Phys., 21:403–410, 1988.

31 M.-A. Martínez-Sánchez, M. Rodriguez-Bautista, R. Vargas, and J. Garza. Solution of the Kohn-Sham equations for many-electron atoms confined by penetrable walls. Theor. Chem. Acc., 135:207, 2016.

32 E.R. Johnson, S. Keinan, P. Mori-Sánchez, J. Contreras-García, A. J. Cohen, and W. Yang. Revealing non-covalent interactions. J. Am. Chem. Soc., 132:6498–6506, 2010.

33 J. P. Connerade, V. K. Dolmatov, P. A. Lakshmi, and S. T. Manson. Electron structure of endohedrally confined atoms: atomic hydrogen in an attractive shell. J. Phys. B: At. Mol. Opt. Phys., 32:L239–L245, 1999.

34 A. Cortés-Santiago, R. Vargas, and J. Garza. Noble gases encaged by the C-60 increase their chemical reactivity. J. Mex. Chem. Soc., 56:270–274, 2012.

35 A. S. Baltenkov, S. T. Manson, and A. Z. Msezane. Jellium model potentials for the C60 molecule and the photoionization of endohedral atoms, A@C_{60}. J. Phys. B: At. Mol. Opt. Phys., 48:185103, 2015.

36 E. Sánchez-González, Paulo G. M. Mileo, M. Sagastuy-Breña, J.-Raziel Álvarez, J.-E. Reynolds, A. Villarreal, A. Gutiérrez-Alejandre, J. Ramírez, J. Balmaseda, E. González-Zamora, Guillaume Maurin, S.-M. Humphrey, and I.-A. Ibarra. Highly reversible sorption of H_2S and CO_2 by an environmentally friendly mg-based mof. J. Mater. Chem. A, 6: 16900–16909, 2018.

5

Unveiling the Mysterious Mechanisms of Chemical Reactions

Soledad Gutiérrez-Oliva, Silvia Díaz, and Alejandro Toro-Labbé

Laboratorio de Química Teórica Computacional (QTC), Facultad de Química y de Farmacia. Pontificia Universidad Católica de Chile., Santiago, Chile

5.1 Introduction

5.1.1 Context

It can be hypothesized that chemical reactions take place through a sequence of events involving structural and electronic rearrangements; these events take the reactants into the products, thus achieving the chemical transformation [1]. The ensemble of these events defines the mechanism of a chemical reaction and its knowledge is very important since it may lead to control complex processes such as catalytic reactions which are a fundamental component of modern chemistry [2–4]. For example, conversion of raw materials into higher-value products materials through new designed and efficient chemical processes may produce large social benefits and discloses the fundamental link between basic chemistry and technology.

Selectivity, energetic saving, and reduction of environmental impact associated with chemical processes have become main subjects of research in chemistry, the design of new routes and methods aimed at the chemical transformation of unwanted products is a very important issue that is increasingly addressed by researchers [5]. Although generally the mechanisms of chemical reactions are very complicated and the reactivity of chemical species is very difficult to foresee, theoretical and computational chemistry provide reliable tools, reactivity and selectivity descriptors, to predict chemical behavior, pathways, and rates of chemical reactions [6–8]. Complex reactions most frequently present a large number of reactive degrees of freedom that may lead to a huge number of possible reaction paths that have to be considered to obtain a complete description of the dynamic of a chemical reaction. Such a complete description involves the mapping of multidimensional potential energy surfaces, making this approach quite expensive computationally and difficult to handle. Instead, a simplified method computing the minimum energy path relying reactants and products is frequently used [9–11]. Based on this approach and having reaction paths at hand, in the last years we have developed concepts and computational technologies aimed at characterizing the mechanisms of chemical reactions and the properties of the chemical species that participate in those reactions [12–14]. Modern mathematical methods and computational techniques, such as the intrinsic reaction coordinate method (IRC) [9], opened the possibility of computing reliable

Chemical Reactivity in Confined Systems: Theory, Modelling and Applications,
First Edition. Edited by Pratim Kumar Chattaraj and Debdutta Chakraborty.

reaction paths connecting reactants, transition states, and products for quite complex chemical reactions. The IRC method thus provides, at high level of quantum chemical calculations, a direct way to determine the profiles along the reaction coordinate of energy and other chemical properties that, together, give insights not only on the kinetic and thermodynamic outcomes, but also on the mechanism of a chemical reaction.

Selectivity is one of the most important factors to be controlled in chemistry; among the several ways to obtain high selectivity for a desired product are induced changes in the molecular and electronic structures of the reacting species, changes that can be computer-designed to help guide the synthesis of appropriate chemicals for specific reactions. Energetic saving and reduction of the environmental impact will result from the optimization of processes and the discovery of new, more efficient methods of synthesis. In both issues, the knowledge of structure-activity relations of the reacting species and the mechanisms of the reactions is necessary.

5.1.2 Ideas and Methods

In section 5.2 of this chapter, we are going to review and discuss few concepts and methodologies that have been developed to get insights into reaction mechanisms. Special attention will be paid to the energies involved and to the electronic activity that is ultimately responsible for driving chemical reactions. The reaction force analysis (RFA) [2, 12, 13, 15] will be exposed as a quite useful tool to describe chemical events taking place within different regions along the reaction coordinate. These chemical events have been characterized as being of structural and/or electronic nature and they drive a chemical reaction in a cooperative way. Within this framework, structural and energetic characterization of transition states is a crucial issue to formulate reliable reaction mechanisms. We will take advantage of the natural decomposition of the activation and reaction energies provided by the RFA putting forward the so-called reaction works, which are basically associated with electronic and structural effects. Activation and reaction energies are then written in terms of reaction works and the physical nature of these energies will be characterized through the relative weight of the structural and electronic effects. Moreover, we will use the RFA energy decomposition in the framework of the Marcus potential energy, an extension of the well-known Marcus equation, to produce analytic forms for reaction works [16].

On the other hand, the reaction electronic flux (REF) [1, 14, 17] has proven to be a quite good and reliable descriptor of the electronic activity taking place during a chemical reaction. This activity is rationalized in terms of bond strengthening/formation and/or bond weakening/breaking processes and has been linked to the change of specific local electronic populations. In section 5.3 of this chapter we will expose two complementary ways to characterize and rationalize the REF, a physical perspective in which the REF is expressed in terms of polarization and electron transfer phenomena and a chemical approach in which the REF is expressed in terms of the chemical reactivity descriptors, molecular hardness, and electrophilicity index.

5.1.3 Application

To illustrate these ideas, in section 5.4 we will use the combination of RFA and REF to analyze the two-steps mechanism of the formation of aminoacetonitrile (NH_2CH_2CN) from methanimine ($NHCH_2$) and hydrogen isocyanide (CNH). The interest of this reaction comes from the fact that aminoacetonitrile is a direct precursor of glycine and the reaction may take place in the interstellar medium, where both reactants ($NHCH_2$ and CNH) have been observed [18, 19]. This might be an interesting input for intriguing questions about the origin and evolution of life on prebiotic earth.

5.2 Energy and Reaction Force

5.2.1 The Reaction Force Analysis (RFA)

It can be noticed in Figure 5.1, left panel, that the energy profile E(ξ), representing the minimum energy path connecting the transition state (**TS**) with the reactants (**R**) and the products (**P**) of a generic elementary step along a reaction coordinate ξ = IRC displays three key points needed to define the activation and reaction energies. These are the positions of **R**, **TS**, and **P**, ξ_R, ξ_{TS}, and ξ_P, respectively. It is important to mention that in real-life computational studies E(ξ) will be a numerical function obtained through the IRC procedure. Unfortunately E(ξ) does not produce direct evidence on the way the reaction takes place; however, key information on the reaction mechanism emerges from its derivative, the reaction force [12, 13, 20]:

$$F(\xi) = -\left(\frac{dE}{d\xi}\right) \tag{5.1}$$

It can be seen in the reaction force profile depicted in the right panel of Figure 5.1 that the activation process is associated with a retarding reaction force (F(ξ) < 0) that prepares the reactants to continue the reaction through structural changes, mostly taking place within the reactant region, and electronic reordering, mainly at the transition state region. Analytic observation of many different types of reactions has led to the conclusion that at the reaction force minimum (ξ_1) the activated reactant has absorbed more than 50% of the activation energy and, most likely, will reach the transition state and continue toward the product of the reaction. On the other hand, the relaxation process is carried out with positive reaction force that mostly involves electronic reordering in the TS region and structural relaxation at the product region. Most of the energy delivered in the relaxation process takes place in the product region, thus being a structural energy. Besides helping characterize the nature of the energies involved at the different steps of the reaction and giving information on the mechanisms responsible for the chemical change, F(ξ) provides a general frame to analyze chemical reactions through a natural partition of the reaction coordinate into the so-called reaction regions in which different driving mechanisms might be operating. It has

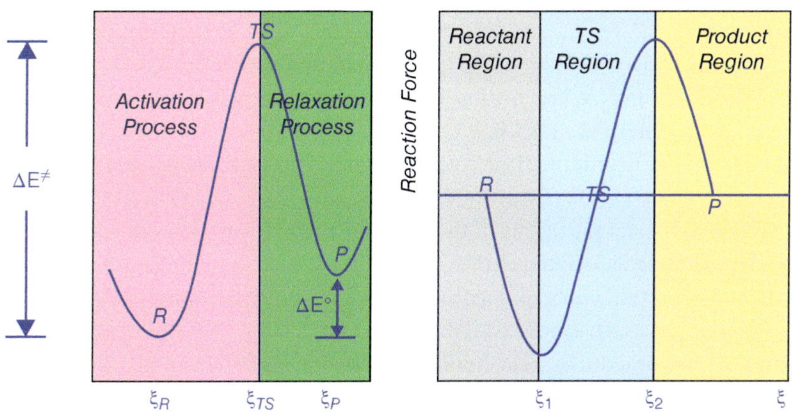

Figure 5.1 Generic profiles of energy and reaction force along the reaction coordinate ξ. On the energy profile the three key points needed to determine activation and reaction energies, ξ_R, ξ_{TS}, and ξ_P, are indicated. On the reaction force profile, two extra points, ξ_1 and ξ_2, at the force minimum and maximum, become relevant to define the three reaction regions depicted. Figure taken from [16].

been shown that any elementary step reaction can be analyzed through the demarcation of three reaction regions: the reactants, transition state, and products regions that are defined using the critical points of the reaction force profile, as indicated in Figure 5.1. For reactions taking place through a sequence of elementary steps, more reaction regions will arise accordingly [2]. To quantify the energetic intake/release taking place at any stage of the process, reaction works (W_i) can be defined through integration of the reaction force:

$$W_i = -\int_{\xi_i}^{\xi_i+\delta\xi_i} F(\xi)d\xi. \tag{5.2}$$

In summary, reaction work is the energy the system absorbs or deliver at different stages of the reaction. In particular, specific reaction works can be defined within each reaction region to quantify structural and electronic energy involved in activation and relaxation processes:

$$W_1 = -\int_{\xi_R}^{\xi_1} F(\xi)d\xi > 0 \qquad W_2 = -\int_{\xi_1}^{\xi_{TS}} F(\xi)d\xi > 0$$
$$W_3 = -\int_{\xi_{TS}}^{\xi_2} F(\xi)d\xi < 0 \qquad W_4 = -\int_{\xi_2}^{\xi_P} F(\xi)d\xi < 0. \tag{5.3}$$

Since W_1 and W_4 are defined in the reactant and product regions, respectively, they are mostly accounting for structural rearrangements, whereas W_2 and W_3, defined within the **TS** region, are mostly due to electronic effects.

5.2.2 RFA-Based Energy Decomposition

In the context of the RFA, a rational phenomenological decomposition of the reaction and activation energies emerge [13, 16]:

$$\Delta E^\circ = \int_{\xi_R}^{\xi_P} F(\xi)d\xi = \left(W_1 + W_2 + W_3 + W_4\right), \tag{5.4}$$

$$\Delta E^{\neq} = \int_{\xi_R}^{\xi_{TS}} F(\xi)d\xi = \left(W_1 + W_2\right). \tag{5.5}$$

The RFA takes advantage of the above partition of the activation and reaction energies in terms of structural $\{W_1,W_4\}$ and electronic $\{W_2,W_3\}$ contributions to identify the physical nature, structural and/or electronic, of these energies. This is a crucial issue in modern chemical catalysis because the above equations may help guiding the choice of catalysts indicating the need for structural and/or electronic activity.

Equations (5.4) and (5.5) are rational partitions of the reaction and activation energies with the extra feature of the physical meaning associated to the prevalence of structural or electronic effects within the region in which the reaction works are defined, thus unveiling their physical nature. A partition of the reaction energy, Eq. (5.4), is now available through the RFA. The amount of energy absorbed during activation, or released during relaxation, can now be explored from the perspective of structural and electronic effects that involves both, equilibrium and non-equilibrium configurations. Although there are many interesting activation energy decompositions based on specific physical models [21, 22], most of them intend to explain its physical nature through highlighting specific interactions or contributions to the total energy. In contrast to this, the reaction force partition given in Eq. (5.5) provides a simple way to rationalize activation energies in terms of structural and electronic effects. Although RFA is based on the functional analysis of a numerical energy profile, previously obtained through the IRC procedure, it can be applied to get information from an actual analytic potential energy function.

5.2.3 Marcus Potential Energy Function

The Marcus potential energy function (MPE) formulated recently [16] is an extension of the Marcus formula for the activation energy of an elementary step reaction [23, 24]; it is a very simple two-parameters potential function that, besides complying with the Marcus equation, can be used to rationalize experimental and computational data. MPE is based on a two-state approximation illustrated in Figure 5.2, an analytic potential function can be obtained from the combination of local potentials associated with the reactants and products, it is given by:

$$E(\xi) = \left[\omega_R(\xi)E_R(\xi) + \omega_P(\xi)E_P(\xi)\right], \tag{5.6}$$

where $\{\omega_R(\xi), \omega_P(\xi)\}$ are the so-called conformational functions defining the statistical weight of the reactant and product reference potentials along the reaction coordinate, and they comply with $\left[\omega_R(\xi) + \omega_R(\xi)\right] = 1$ [25]. The functions $\{E_R(\xi), E_P(\xi)\}$ are harmonic potentials localized at ξ_R and ξ_P, as indicated in Figure 5.2. Among the possible mathematical models used to define the conformational and the local potentials functions, we use a cosine expansion of a torsional potential expressed as a limited Fourier series that complies with the Marcus equation for the activation energy [25]. In this context the reaction coordinate spans the interval {0,2} and the resulting analytic potential energy function is a two-parameters function that defines the variation of the energy along the reaction coordinate. We have named it the Marcus potential energy function, and it is given by [16]:

$$E^{(m)}(\xi) = -\frac{1}{4}\Delta E_0^{\neq}\xi^4 + \Delta E_0^{\neq}\xi^2 + \frac{1}{4}\Delta E^{\circ}\xi^2 \tag{5.7}$$

where $\Delta E_0^{\neq}$ is the so-called Marcus intrinsic activation energy and ΔE° is the reaction energy. These two energies are input parameters necessary to construct the potential function. It is important to note that these energetic parameters may come from experiment or computation. This opens a way to define experimental potential functions that might be useful to better rationalize experimental data and to explore non-equilibrium states using realistic potential functions within computer simulation schemes.

Within this representation of the potential function, the position of the transition state is given by:

$$\left(\frac{dE^{(m)}}{d\xi}\right)_{\xi_{TS}} = 0 \Rightarrow \xi_{TS}^2 = 2 + \frac{1}{2}\left(\frac{\Delta E^{\circ}}{\Delta E_0^{\neq}}\right). \tag{5.8}$$

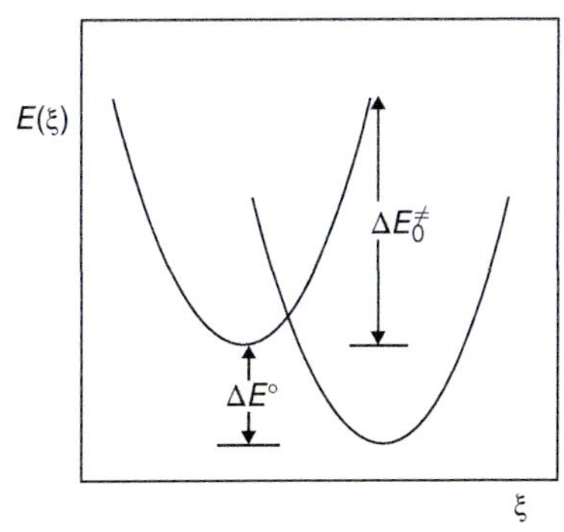

Figure 5.2 Schematic illustration of a two-state reaction represented by local harmonic potentials centered at the reactants and products. Figure taken from [16]. © 2018, Springer Nature.

Evaluating Eq. (5.7) at ξ_{TS} produces the well-known Marcus equation to estimate the energy barrier of an elementary chemical process:

$$\Delta E^{\neq} = \Delta E_0^{\neq} + \frac{1}{2}\Delta E^\circ + \frac{1}{16}\frac{(\Delta E^\circ)^2}{\Delta E_0^{\neq}} \tag{5.9}$$

Equation (5.9) is then a particular case of Eq. (5.7); consistency between MPE and the Marcus equation for the activation energy is then reached. As already mentioned, the MPE, Eq. (5.7), is a two-parameter function; it depends on the Marcus intrinsic activation energy and the reaction energy, $\{\Delta E_0^{\neq}, \Delta E^\circ\}$. Note that provided the reaction and activation energies are available, $\Delta E_0^{\neq}$ can be determined numerically by solving a simple second-degree equation formulated from Eq. (5.9).

5.2.4 Marcus RFA

From Eq. (5.7) an analytic form of the corresponding reaction force, $F^{(m)}(\xi)$, emerges:

$$F^{(m)}(\xi) = -\left(\frac{dE^{(m)}}{d\xi}\right) = \Delta E_0^{\neq}\xi^3 - 2\Delta E_0^{\neq}\xi - \frac{1}{2}\Delta E^\circ \xi. \tag{5.10}$$

Analytic formulae for the reaction works are then obtained straightforwardly:

$$\left.\begin{aligned}
W_1^{(m)} &= \tfrac{5}{9}\Delta E_0^{\neq} + \tfrac{5}{18}\Delta E^\circ + \tfrac{5}{144}\tfrac{(\Delta E^\circ)^2}{\Delta E_0^{\neq}},\\
W_2^{(m)} &= \tfrac{4}{9}\Delta E_0^{\neq} + \tfrac{4}{18}\Delta E^\circ + \tfrac{4}{144}\tfrac{(\Delta E^\circ)^2}{\Delta E_0^{\neq}},\\
W_3^{(m)} &= \tfrac{5}{9}\Delta E_0^{\neq} + \tfrac{5}{18}\Delta E^\circ + \tfrac{10}{144}\tfrac{(\Delta E^\circ)^2}{\Delta E_0^{\neq}},\\
W_4^{(m)} &= \tfrac{4}{9}\Delta E_0^{\neq} - \tfrac{14}{18}\Delta E^\circ - \tfrac{1}{144}\tfrac{(\Delta E^\circ)^2}{\Delta E_0^{\neq}}.
\end{aligned}\right\} \tag{5.11}$$

Note in Eq. (5.11) that $W_1 > W_2$ and a closer analysis of the Marcus reaction works gives insights on the nature of $\Delta E^{\neq}$ since 56% of it corresponds to W_1 and 44% to W_2, indicating that the activation energy is mostly due to structural energy effects quantified by W_1.

It is important to stress the fact that Eq. (5.7) can be integrated over any interval on the reaction coordinate, allowing the energy involved in that specific portion of the reaction coordinate to be quantified. Of course, higher-order derivatives of Eq. (5.7) are straightforward; the second-order derivative leads to analytic reaction force constant that has been defined few years ago and is used to characterize synchronicity of the chemical events driving the reaction [26]:

$$\kappa^{(m)}(\xi) = -\left(\frac{dF^{(m)}}{d\xi}\right) = -3\Delta E_0^{\neq}\xi^2 + 2\Delta E_0^{\neq} + \frac{1}{2}\Delta E^\circ. \tag{5.12}$$

$E^{(m)}(\xi)$ describes correctly the activation process that takes the reactants to transition state; however, its qualitative behavior for the relaxation process lacks precision because of the absence of an inflection point after the transition state, and this is the reason why $F^{(m)}(\xi)$ and $\kappa^{(m)}(\xi)$ diverge after the TS is reached. In this context, the reaction works $W_3^{(m)}$ and $W_4^{(m)}$, given in Eq. (5.11), have been estimated assuming that the reverse activation energy behaves as it does the direct one. On the other hand, it should be emphasized that obtaining "experimental" potential and reaction force functions is now possible provided experimental values of activation and reaction energies. We believe this is a very important progress in the characterization of reaction mechanisms and in the generation of experimentally parameterized potential functions that would be used in computer simulations and spectroscopy characterization, to better rationalize experimental data and to get new insight on the properties of transient structures appearing during the advancement of a chemical reaction.

5.3 Electronic Activity Along a Reaction Coordinate

5.3.1 Chemical Potential, Hardness, and Electrophilicity Index

Once the energetics of a chemical process have been characterized, one can ask about how the process took place, and what the actual mechanism is that brings the reactants into the products. To answer these questions it is necessary to look at the electronic changes the chemical species suffer during the transformation. Chemical species can be characterized through their electronic structure that, in turn, define their reactivity properties and how they react in front of another specie [7, 8, 27, 28]. Reactivity properties are classified in terms of their nature global or local. Global properties of a molecule are those describing changes taking place on the whole system, whereas local properties describe local changes taking place in a given region of the molecular topology. In this chapter we are going to deal with reactivity descriptors that come out from density functional theory (DFT) in which the energy of a molecular system can be written as a function of the total number of electrons N and as a functional of the external potential $\upsilon(\vec{r})$, so that $E \equiv E[N, \upsilon(\vec{r})]$. Global properties are obtained by successive derivative of the energy with respect to N, chemical potential (μ) [29, 30] and hardness (η)[31] corresponds to the first and second derivative, respectively:

$$\mu = \left(\frac{dE}{dN}\right)_{\upsilon(\vec{r})} \approx -\frac{1}{2}(I + A) \approx -\frac{1}{2}\left(\varepsilon_h + \varepsilon_\ell\right); \tag{5.13}$$

$$\eta = \left(\frac{d^2E}{dN^2}\right)_{\upsilon(\vec{r})} \approx \frac{1}{2}(I - A) \approx -\frac{1}{2}\left(\varepsilon_h - \varepsilon_\ell\right). \tag{5.14}$$

It has been demonstrated that the chemical potential is the negative of the absolute electronegativity, so it is frequently used to characterize the primary interaction between two chemical species: electrons will flow from the system with higher chemical potential to the one with lower value of μ. Therefore, just by comparing the μ values of interacting molecules it is possible to identify the donor and acceptor species. On the other hand, by virtue of the maximum hardness principle, hardness can be seen as a resistance of the system to change its electronic structure, in short, η represents the resistance to charge transfer. The second and third terms of the right-hand side of Eqs. (5.13) and (5.14) are approximate formulae obtained using the finite difference approximation and the Koopmans' theorem, respectively [32]. These formulae express the chemical potential and hardness in terms of the ionization potential and electron affinity $\{I, A\}$ and the energy of the frontier molecular orbitals HOMO(Highest Occupied Molecular Orbital) and LUMO (Lowest Unoccupied Molecular Orbital), $\{\varepsilon_h, \varepsilon_\ell\}$. Local properties such as the electron density ($\rho(\vec{r})$) or the Fukui functions ($f(\vec{r})$) are obtained through derivation of the energy with respect to the external potential $\upsilon(\vec{r})$ [8, 27].

The capability of a molecule to accept electrons from a donor is given by the electrophilicity index which is a global property that can be obtained from the combination of μ and η [33]:

$$\omega = \left(\frac{\mu^2}{2\eta}\right) \tag{5.15}$$

and the maximum number of electrons the molecule may accept is given by:

$$\Delta N_{max} = -\left(\frac{\mu}{\eta}\right). \tag{5.16}$$

5.3.2 The Reaction Electronic Flux (REF)

The electronic activity taking place along a reaction coordinate during a chemical reaction can be quantified through the reaction electronic flux, which is defined as [1]:

$$J(\xi) = -\left(\frac{d\mu}{d\xi}\right). \tag{5.17}$$

The REF is interpreted in terms of the spontaneity of the electronic activity, when $J(\xi) > 0$ the electronic activity is spontaneous and it is driven by bond formation/strengthening processes; when $J(\xi) < 0$ then the electronic activity is non-spontaneous and it is driven by bond breaking/weakening processes. In this chapter we will explore two ways to characterize the REF, and both of them imply a partition of $J(\xi)$: (i) physical partition, in which the REF is expressed in terms of polarization and electron transfer phenomena; (ii) chemical partition, in which the REF is expressed in terms of reactivity descriptors, molecular hardness, and electrophilicity index.

5.3.2.1 Physical Decomposition of REF

The physical phenomena underlying any kind of electronic activity are polarization and transfer. To comply with this phenomenological view, the REF can be expressed as:

$$J(\xi) = \left[J_p(\xi) + J_t(\xi)\right] \tag{5.18}$$

where $J_p(\xi)$ and $J_t(\xi)$ are, respectively, polarization and transfer contributions. Evaluating these components will produce insights on the way the chemical transformation takes place. The polarization flux, $J_p(\xi)$, can be determined, making a fragmentation of the reactive complex such that the fragment polarization flux, $J_p^{(i)}(\xi)$, can be obtained using different methodologies available for evaluating the behavior of atoms or fragments in a molecular system. The counterpoise method [34, 35] is our preferred methodology for this task even though this procedure is computationally expensive because at each point along ξ counterpoise calculations must be performed to obtain the fragment's chemical potential $\mu_i(\xi)$; this leads to:

$$J_p^{(i)}(\xi) = -\left(\frac{N_i}{N}\right)\frac{d\mu_i}{d\xi}, \tag{5.19}$$

where N_i is the number of electrons of fragment i and N is the total number of electrons of the supermolecular system. In this way the contribution to $J_p(\xi)$ coming from each fragment remains proportional to its own number of electrons. Then, total polarization flux will be the sum of the fragments' fluxes:

$$J_p(\xi) = \sum_{i=1}^{n} J_p^{(i)}(\xi), \tag{5.20}$$

where n is the number of chemical fragments (atoms or groups of atoms) in which the reactive system has been partitioned. It is important to make clear that the assumption of separated chemical fragments underlies the calculation of $J_p(\xi)$ using Eq. (5.20).

Transfer reaction electronic flux can be obtained by making the difference between $J(\xi)$, calculated using the supermolecule approximation, and $J_p(\xi)$, determined from the isolated fragments through the use of the counterpoise method:

$$J_t(\xi) = [J(\xi) - J_p(\xi)] = -\left(\frac{d\mu^\circ}{d\xi}\right) + \sum_{i=1}^{n} \left(\frac{N_i}{N}\right) \frac{d\mu_i}{d\xi}, \tag{5.21}$$

where μ° corresponds to the chemical potential of the reactive supermolecular system. It is possible to obtain the fragment contribution to $J_t(\xi)$ and $J(\xi)$. To do so the principle of equalization of the chemical potential has to be invoked [36, 37]: since the chemical potential is a global property of the system, then its value is preserved all over it, then the chemical potential of the supermolecule (μ°) should be equal to the μ_i°, the chemical potentials associated to the different topological regions on the unfragmented supermolecule. Therefore we can write:

$$\mu^\circ(\xi) = \sum_{i=1}^{n} \left(\frac{N_i}{N}\right) \mu_i^\circ(\xi), \tag{5.22}$$

and Eq. (5.17) can be written as:

$$J(\xi) = -\left(\frac{d\mu^\circ}{d\xi}\right) = -\sum_{i=1}^{n} \left(\frac{N_i}{N}\right) \frac{d\mu_i^\circ}{d\xi} = -\sum_{i=1}^{n} \left(\frac{N_i}{N}\right) \frac{d\mu^\circ}{d\xi} = \sum_{i=1}^{n} J^{(i)}(\xi). \tag{5.23}$$

Finally, using Eqs (5.19), (5.21), and (5.23), the electron transfer flux is given by:

$$J_t(\xi) = \sum_{i=1}^{n} \left(\frac{N_i}{N}\right) \frac{d}{d\xi}(\mu_i - \mu_i^\circ) = \sum_{i=1}^{n} J_t^{(i)}(\xi). \tag{5.24}$$

Equation (5.24) features a difference of chemical potentials ($\mu_i - \mu_i^\circ$), which is a clear signature of electronic transfer. It is important to emphasize the fact that the $\mu_i(\xi)$ come from counterpoise calculations of the fragments using the geometry they would have within the supermolecule at every point ξ, whereas $\mu^\circ(\xi)$ results from the supermolecule calculation and $\mu_i^\circ(\xi)$ is a product of a dummy partitioning of $\mu^\circ(\xi)$. Therefore, the total REF can be expressed in terms of fragment's polarization and transfer, as:

$$J(\xi) = [J_p(\xi) + J_t(\xi)] = \sum_{i=1}^{n} \left[J_p^{(i)}(\xi) + J_t^{(i)}(\xi)\right] = \sum_{i=1}^{n} J^{(i)}(\xi). \tag{5.25}$$

By virtue of the Sanderson's equalization priciple [1, 36, 37] Eq. (5.25) only has a symbolic value since $J(\xi) = J^{(i)}(\xi)$ for all ξ. However, it puts forward the validity of the partition of Eq. (5.18) at a fragment level:

$$J^{(i)}(\xi) = \left[J_p^{(i)}(\xi) + J_t^{(i)}(\xi)\right], \tag{5.26}$$

in this way the reaction electronic flux admits the double and simultaneous scrutiny of the physical phenomena, polarization, and transfer, and the chemical partitioning of the supermolecule.

5.3.2.2 Chemical Decomposition of REF

Expressing $J(\xi)$ in terms of the changes of hardness and alectrophilicity gives new insights on this property and produces a chemical partition of it. Taking advantage of the definition of ω, the REF becomes:

$$J(\xi) = \frac{1}{2}\Delta N_{max}(\xi)\left(\frac{d\eta}{d\xi}\right) + \frac{1}{\Delta N_{max}(\xi)}\left(\frac{d\omega}{d\xi}\right) = [J_\eta(\xi) + J_\omega(\xi)] \tag{5.27}$$

it is interesting to observe from the above equation that electronic activity comes out from a balance between the change of the resistance to electronic transfer ($J_\eta(\xi)$) and the capability of the system to accept electrons($J_\omega(\xi)$). In this approach, the electronic activity results from a sequence of electronic donation and capture processes taking place along the reaction coordinate and within different regions of the molecular topology. It is worth mentioning that to identify specific regions on the molecule that are prone to change their local electronic structure, it is necessary to take a look at local properties such as atomic charges, bond orders, Fukui and electron localization functions, etc. [2, 3, 8]. Since the link between local and global properties is not a matter for this chapter, we refer the reader to references [2, 15, 20] for an overview of this issue. It is possible to qualitatively correlate the physical concepts involved in Eq. (5.18) and the chemical description of Eq. (5.27); the first term of the right-hand side of the latter equation contains the change of the resistance to charge transfer and so it must be related to $J_t(\xi)$ whereas the second term would be more related to $J_p(\xi)$.

5.4 An Application: the Formation of Aminoacetonitrile

In this section a complete characterization of the reaction mechanism of the formation of aminoacetonitrile is presented, in particular the isomerization HCN/CNH with methanimine acting as a catalyst. The interest of this reaction comes from the fact that methanimine and hydrogen cyanide, two well known interstellar molecules, have been postulated to react in the interstellar medium to form aminoacetonitrile, which is a direct precursor of glycine, recently detected in this medium [18, 19, 38].

Since the exact composition of the prebiotic Earth's atmosphere is still a matter of debate, the study of the existence of biomolecules in the interstellar medium (ISM), the vast space between stars, is crucial and the use of computational chemistry has been a great contribution to the characterization of molecular systems and reactions that occur in this medium, which is a huge reservoir of molecular material ranging from simple diatomic molecules to more complex systems. One of the motivating factors for the study of biological molecules found in interstellar space is the hypothesis that the basic constituents of amino acids may have come from space.

Following the prebiotic Strecker's synthesis [39], methanimine ($NHCH_2$) may have reacted in the ISM with CNH to form aminoacetonitrile (NH_2CH_2CN), a direct precursor of glycine, which has recently been detected in the ISM. The reaction scheme which is shown in Figure 5.3 consists of a series of chemical reactions which synthesizes an amino acid from an aldehyde in the presence of aqueous ammonia (NH_3) and hydrogen cyanide, generating aminoacetonitrile which is subsequently hydrolyzed to give the desired amino acid. The formation reaction of aminoacetonitrile from methanimine and hydrogen cyanide, framed in red in Figure 5.3, proceeds in two steps that are displayed in Figure 5.4. In this study we want to put forward the significance of the reaction

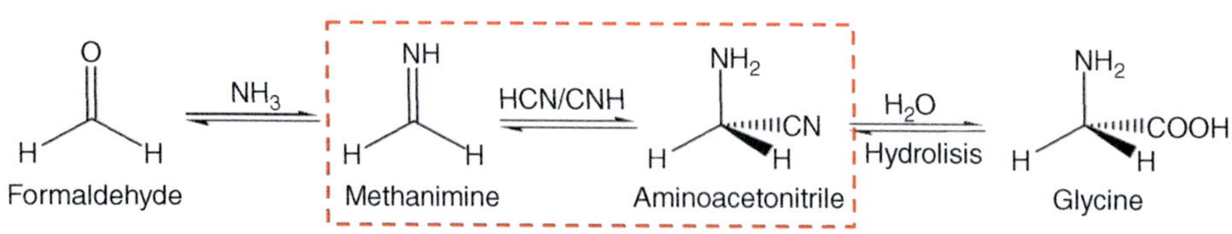

Figure 5.3 Schematic view of the main steps of the Strecker prebiotic synthesis for the formation of glicine. The characterization of the mechanism of the key step, framed in red, is presented in this chapter.

electronic flux and its representations in the characterization of a reaction mechanism. All calculations were performed at the MP2/6-31+g(d,p) level of theory using the Gaussian 09 package [40]. The fragmentation of the whole reactive system used to determine $J_p(\xi)$ in the succesive steps of the reactions was $NHCH_2$ + HCN in **R1** and $NHCH_2$ + HNC in **R2**.

5.4.1 Energetic Analysis

Table 5.1 displays the energetic data associated with the reactions under study. Both reactions present quite high energy barriers that are reasonably well described by the Marcus potential function. For the IRC calculation the nature of the energy barrier of **R1** is mainly structural ($W_1 > W_2$) whereas in **R2** it is mostly electronic ($W_1 < W_2$). On the other hand, the reaction works obtained using the Marcus analytic potential are reasonably close to the values obtained from the numerical IRC procedure. By construction of the analytic MPE function, $W_1^{(m)} > W_2^{(m)}$, it can be concluded that the model potential MPE describes reasonably well the energetic data of **R1**. In the case of **R2**, the MPE description is not as good as for **R1**, though it accounts for the main energetic trends of the reaction.

5.4.2 Reaction Mechanisms

Figure 5.4 displays the two-steps reaction to form acetonitrile. The chemical events taking place along the reaction **R1** are summarized by the cleavage of the C_7H_6 bond and the formation of N_2H_6 and N_8H_6, these events take place in a concerted stepwise fashion. We also note that the

Table 5.1 Energetic data for reactions **R1** and **R2**. All values are given in Kcal/mol

Parameter	R1	R2
ΔE°	13.51	−31.67
$\Delta E^{\neq}$	35.82	29.21
$\Delta E_0^{\neq}$	28.66	43.60
W_1	21.77 (61%)	12.59 (43%)
$W_1^{(m)}$	20.06 (56%)	16.36 (56%)
W_2	14.05 (39%)	16.62 (57%)
$W_2^{(m)}$	15.76 (44%)	12.85 (44%)

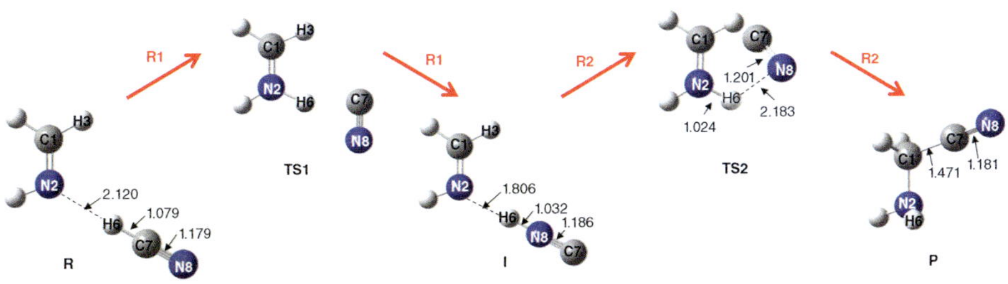

Figure 5.4 Structural data for the molecular systems involved in **R1** and **R2**.

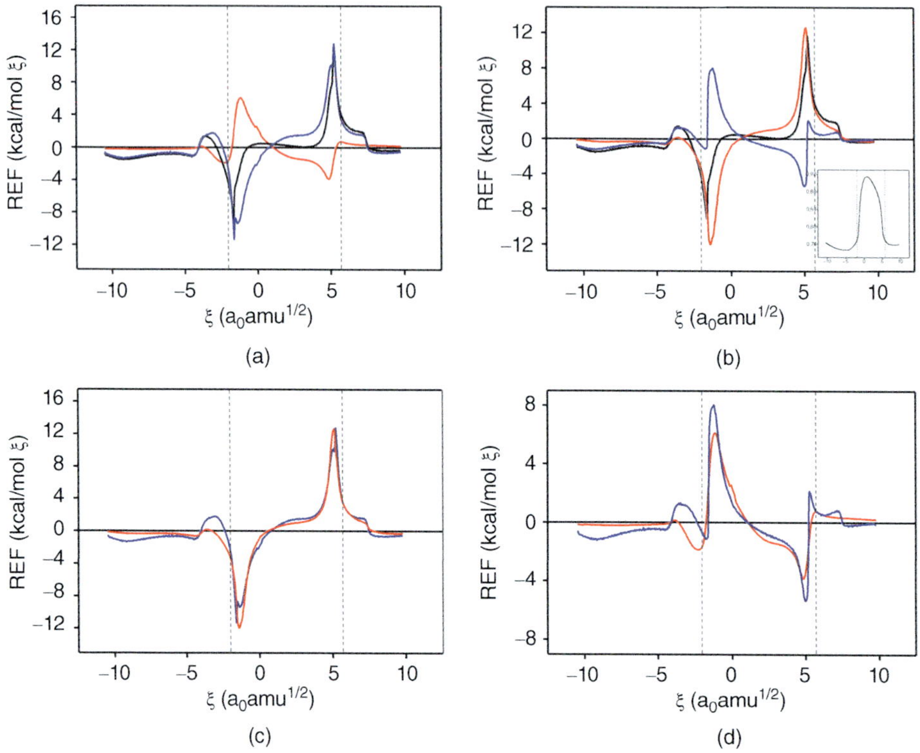

Figure 5.5 Physical and chemical representations of the REF for **R1**. (a) Red and blue curves correspond to $J_p(\xi)$ and $J_t(\xi)$, respectively; black curve is the total REF: $J(\xi) = J_p(\xi) + J_t(\xi)$. (b) Red and blue curves correspond to $J_\eta(\xi)$ and $J_\omega(\xi)$, respectively; black curve is the total REF: $J(\xi) = J_\eta(\xi) + J_\omega(\xi)$. The inset on the figure shows the variation of $\Delta N_{max}(\xi)$. (c) Comparison between $J_t(\xi)$ and $J_\eta(\xi)$. (d) Comparison between $J_p(\xi)$ and $J_\omega(\xi)$.

transition state is mainly characterized by the transfer of the H_6 proton from C_7 to N_2 and therefore the formation of the N_2H_6 bond and the cleavage of the C_7H_6 bond drive the reaction.

Figures 5.5 and 5.6 display the REF for reactions **R1** and **R2**, respectively, the figures contain the physical and chemical representations of the REF (panels (a) and (b)) and a comparison between the two representations (panels (c) and (d)). The main features of $J(\xi)$, the black curve in Figures 5.5(a) and 5.5(b) are the two peaks within the transition state region, indicating that most electronic activity takes place within this region. The negative peak indicates that non spontaneous bond breaking/weakening processes drive the reaction when entering the transition state region; then the positive peak evidence that spontaneous bond forming/strengthening processes takes over when leaving this region. It can be noticed in Figure 5.5(a) that $J(\xi)$ follows quite closely the trend exhibited by the transfer flux $J_t(\xi)$; polarization flux, $J_p(\xi)$, presents an opposite trend featuring low-intensity positive and negative peaks.

A closer view of the REF shows that electronic activity initiate with a tiny and broad positive peak which is mostly due to electronic transfer effects, although polarization effects are also present at that interval of the reaction coordinate. Then a large negative peak develops when entering the transition state region; again, transfer effects are responsible for this manifestation of electronic activity. The negative peak entering the transition state region in the REF profile (black curve) is attributed to the breaking of the C_7H_6 bond together with weakening of the CN triple bond, whereas

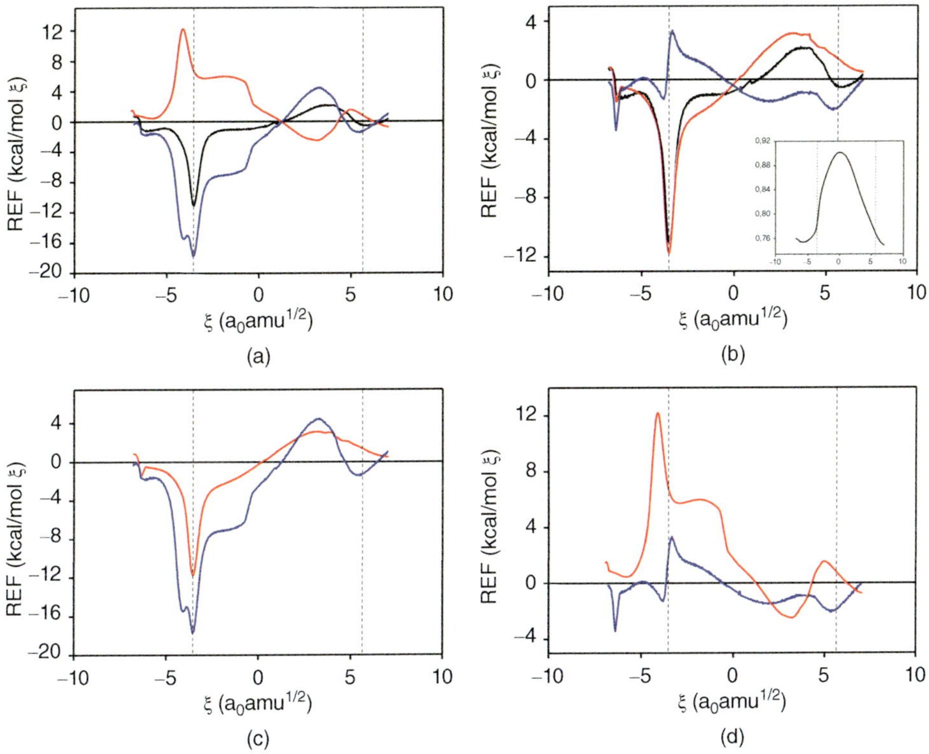

Figure 5.6 Physical and chemical representations of the REF for **R2**. (a) Red and blue curves correspond to $J_p(\xi)$ and $J_t(\xi)$, respectively; black curve is the total REF: $J(\xi) = J_p(\xi) + J_t(\xi)$. (b) Red and blue curves correspond to $J_\eta(\xi)$ and $J_\omega(\xi)$, respectively; black curve is the total REF: $J(\xi) = J_\eta(\xi) + J_\omega(\xi)$. The inset on the figure shows the variation of $\Delta N_{max}(\xi)$. (c) Comparison between $J_t(\xi)$ and $J_\eta(\xi)$. (d) Comparison between $J_p(\xi)$ and $J_\omega(\xi)$.

the positive peak developed when leaving this region should be assigned to the formation of the N_8H_6 bond. This assignment of the electronic activity to specific events taking place as the reaction proceed can be confirmed when studying local electronic properties.

The components of the chemical representation of the REF show a very interesting behavior (Figure 5.5(b)). The red curve associated with $J_\eta(\xi)$ initiates with a negative and intense peak indicating non-spontaneous electronic activity. This is consistent with the resistance to change the electronic structure represented by the hardness. The blue curve on Figure 5.5(b) represents $J_\omega(\xi)$, the term related to the change of electrophilicity. It exhibits a positive peak followed by a negative one, indicating a complementary opposite behavior to the hardness term (red curve), a change of the resistance of electron transfer comes together with a change of the electrophilicity index.

It is interesting to mention that at the minimum of the red negative peak more than 80% of the activation energy has been absorbed by the system and the reaction is already triggered. Simultaneously to the development of the red peak, just before reaching the transition state, $J_\omega(\xi)$ (blue curve) presents a positive peak indicating a large electronic reordering, an observation that is confirmed by the variation of $\Delta N_{max}(\xi)$ displayed in the inset of Figure 5.5(b). This quantity remains quite constant at the reactant and product regions, it is turned on at the transition state region, showing a positive variation of about 0.20 electrons which are transferred from one atom to another in the course of the bond breaking and formation processes that drive the reaction.

In the search for consistency between the two representations of the REF and its components, we have compared in Figures 5.5(c) and 5.5(d) the individual elements given in Eqs (5.18) and (5.27). In this way we expect to shed more light on the physical interpretation of the reaction electronic flux. We observe in these figures that $J_p(\xi)$ and $J_t(\xi)$, determined using Eqs (5.20) and (5.24), respectively, are amazingly close to $J_\omega(\xi)$ and $J_\eta(\xi)$ of Eq. (5.27). This seems to be a very important result because, although these quantities were determined using very different protocols, the pairs $\{J_p, J_\omega\}$ and $\{J_t, J_\eta\}$ appear to be well correlated, thus giving new conceptual insights on the physical meaning of the REF and its components. It appears that $J_\eta(\xi)$ contains a great deal of information on the electronic transfer phenomena while $J_\omega(\xi)$ explains most of polarization effects.

Figure 5.6 displays the REF and its components for reaction **R2**. The total REF profile (black curve in Figures 5.6(a) and 5.6(b)) shows a negative peak that starts to develop before entering the transition state region having a negative minimum at the location of the force minimum. This is evidence of the early break of the N_8H_6 bond. The broad low-intensity maximum of $J(\xi)$ observed just after the TS should be associated to the rotation of the C_7N_8- moiety that promotes the hydrogen transfer for the formation of the N_2H_6 bond. Qualitative trends of $J_p(\xi)$, $J_t(\xi)$ and $J(\xi)$ are similar to their counterparts in **R1**. However, the development of the first negative peak of $J(\xi)$, mainly explained by transfer effects, takes place before leaving the reactant region, where the energy absorbed by the system to produce the reaction is barely around 30% of the activation energy. This results from the early resistance to electronic activity opposed by the hardness term (red curve in Figure 5.6(b)), which is activated more rapidly in **R2** than in **R1**. As already mentioned, the REF at the second part of the reaction is characterized a broad low-intensity positive peak which is mainly due to the transfer/hardness terms, although polarization/electrophilicity are also present to a lesser extent. Consistently, in this case the variation of $\Delta N_{max}(\xi)$ is activated before entering the transition state region.

In contrast to the excellent results concerning consistency between the physical and chemical representations of the REF obtained for **R1**, results in **R2** are not as good. Comparison of the REF's components in **R2**, displayed in Figures 5.6(c) and 5.6(d), is fairly good for the pair $\{J_p(\xi), J_\eta(\xi)\}$ but it shows only qualitative agreement for the pair $\{J_p(\xi), J_\omega(\xi)\}$. The most plausible reason to explain the large deviation between $J_p(\xi)$ and $J_\omega(\xi)$ may be the choice of the fragments used to compute the polarization term of Eq. (5.24). The simultaneity of different kind of strong interactions involved in the different structures along the reaction coordinate (see Figure 5.4) suggests the use of various fragmentation models as the reaction advances.

5.5 Conclusions

We have presented a battery of conceptual and computational tools aimed at the characterization of reaction mechanisms and the energetic cost associated with local processes that drive chemical reactions as they advance. It has been shown that the reaction force analysis provides a natural framework to define reaction regions in which different mechanisms might be operating. These mechanisms are driven by specific local interactions and the energy involved as they progress is determined by reaction works defined in Eq. (5.2). Moreover, the RFA offers a natural decomposition of the activation and reaction energies in terms of structural and electronic effects, which allows their physical nature to be characterized. In this context, a two-parameters analytic potential function complying with the well-known Marcus equation has been formulated and reviewed. This leads to the possibility of defining analytic versions of the reaction works anywhere along the reaction coordinate.

On the one hand, two representations for the reaction electronic flux have been presented and analyzed. A physical representation involving polarization and electronic transfer effects allows a phenomenological rationalization of the electronic activity taking place during a chemical reaction. On the other hand, a chemical representation based upon well established reactivity descriptors hardness and electrophilicity index was drawn up, which gives insight into the chemical nature of the electronic activity that drives the reaction. Consistency between both approaches was reached for the case of the formation reaction of aminoacetonitrile. A very good correspondence between the pairs $\{J_p, J_\omega\}$ and $\{J_t, J_\eta\}$ was obtained. This gives new conceptual insights on the physical interpretation of the REF and its components. Our results indicate that $J_\eta(\xi)$ contains a great deal of information on the electronic transfer phenomena while $J_\omega(\xi)$ can be explained in terms of polarization effects.

Acknowledgments

The authors are indebted to Professor Pratim K. Chattaraj for his kind invitation to participate in this book. The authors wish to give thanks for financial support from FONDECYT (ANID-Chile) through Projects Nos 1181072 and 1201617.

References

1 María Luisa Cerón, Eleonora Echegaray, Soledad Gutiérrez-Oliva, Bárbara Herrera, and Alejandro Toro-Labbé. The reaction electronic flux in chemical reactions. Sci. China Chem., 54(12):1982–1988, 2011.

2 Daniela E. Ortega, Soledad Gutiérrez-Oliva, Dean J. Tantillo, and Alejandro Toro-Labbé. A detailed analysis of the mechanism of a carbocationic triple shift rearrangement. Phys. Chem. Chem. Phys., 17:9771–9779, 2015.

3 Daniela E. Ortega, Q.N.N. Nguyen, Dean J. Tantillo, and Alejandro Toro-Labbé. The catalytic effect of the NH_3 base on the chemical events in the caryolene-forming carbocation cascade. J. Comp. Chem., 37:1068–1081, 2016.

4 N. Villegas-Escobar, M.H. Larsen, S. Gutiérrez-Oliva, A.S.K. Hashmi, and A. Toro-Labbé. Double gold activation of 1-ethynyl-2-(phenylethynyl)benzene toward 5-exo-dig and 6-endo-dig cyclization reactions. Chem. Eur. J., 20(7):1901–1908, 2014.

5 A.S.K. Hashmi, I. Braun, P. Nosel, J. Schadlich, M. Wieteck, M. Rudolph, and F. Rominger. Simple gold-catalyzed synthesis of benzofulvenes-gem-diaurated species as instant dual-activation precatalysts. Angew. Chem. Int. Ed., 51(18):4456–4460, 2012.

6 Weitao Yang and Wilfried J. Mortier. The use of global and local molecular parameters for the analysis of the gas-phase basicity of amines. J. Am. Chem. Soc., 108(19):5708–5711, 1986.

7 Robert G. Parr and Weitao Yang. Density-Functional Theory of Atoms and Molecules. Oxford University Press, New York, 1989.

8 Paul Geerlings, Frank De Proft, and Wilfried Langenaeker. Conceptual density functional theory. Chem. Rev., 103(5):1793–1874, 2003.

9 Kenichi Fukui. The path of chemical reactions. The IRC approach. Acc. Chem. Res., 14(12):363–368, 1981.

10 H.P Hratchian and H.B. Schlegel. Accurate reaction paths using a Hessian based predictor-corrector integrator. J. Chem. Phys., 120:9918–9924, 2004.

11 H.P Hratchian and H.B. Schlegel. Using Hessian updating to increase the efficiency of a Hessian based predictor-corrector reaction path following method. J. Chem. Theory Comput., 1:61–69, 2005.

12 Alejandro Toro-Labbé. Characterization of chemical reactions from the profiles of energy, chemical potential, and hardness. J. Phys. Chem. A, 103(22):4398–4403, 1999.

13 Peter Politzer, Alejandro Toro-Labbé, Soledad Gutiérrez-Oliva, Bárbara Herrera, Pablo Jaque, Mónica C. Concha, and Jane S. Murray. The reaction force: Three key points along an intrinsic reaction coordinate. J. Chem. Sci., 117(5):467–472, 2005.

14 Eleonora Echegaray and Alejandro Toro-Labbé. Reaction electronic flux: A new concept to get insights into reaction mechanisms. study of model symmetric nucleophilic substitutions. J. Phys. Chem. A, 112(46):11801–11807, 2008.

15 Cristina Ortega-Moo, Rocio Durán,Bárbara Herrera, Soledad Gutiérrez-Oliva, Alejandro Toro-Labbé, and Rubicelia Vargas. Study of antiradical mechanisms with dihydroxybenzenes using reaction force and reaction electronic flux. Phys. Chem. Chem. Phys., 19:14512–14519, 2017.

16 S. Gutiérrez-Oliva, B. Herrera, and A. Toro-Labbé. An extension of the Marcus equation: the Marcus potential energy function. J. Mol. Model., 24:104, 2018.

17 Stefan Vogt-Geisse and Alejandro Toro-Labbé. The mechanism of the interstellar isomerization reaction $HOC^+ \rightarrow HCO^+$ catalyzed by H_2: New insights from the reaction electronic flux. J. Chem. Phys., 130(24):244308, 2009.

18 J.M. Hollis, J.A. Pedelty, D.A. Boboltz, S.Y. Liu, L.E. Snyder, P. Palmer, F.J. Lovas, and P.R. Jewell. Kinematics of the Sgr B2(N-LMH) molecular core. Astrophys. J. Lett., 596:L235, 2003.

19 C. De Duve. Origins life evolution. Biospheres, 33:559, 2003.

20 J.P. Hernández-Mancera, F. Núñez-Zarur, S. Gutiérrez-Oliva, A. Toro-Labbé, and R. Vivas-Reyes. Diels-Alder reaction mechanisms of substituted chiral anthracene: A theoretical study based on the reaction force and reaction electronic flux. J. Comp. Chem., 41(23):2022–2032, 2020.

21 Lili Zhao, Moritz von Hopffgarten, Diego M. Andrada, and Gernot Frenking. Energy decomposition analysis. WIREs Computational Molecular Sciences, 8(3):e1345, 2018.

22 F.M. Bickelhaupt and K.N. Houk. Analyzing reaction rates with the distortion/interaction-activation strain model. Angewandte Chemie, 56(34):10070–10086, 2017.

23 Rudolph A. Marcus. Theory, experiment, and reaction rates. a personal view. J. Phys. Chem., 90(16):3460–3465, 1986.

24 Rudolph A. Marcus. Electron transfer reactions in chemistry. theory and experiment. Rev. Mod. Phys., 65(3):599–610, 1993.

25 Gloria I. Cárdenas-Jirón, Alejandro Toro-Labbé, Charles W. Bock, and Jean Maruani. Characterization of rotational isomerization processes in monorotor molecules. In Y.G. Smeyers, editor, Structure and Dynamics of Non-Rigid Molecular Systems, pages 97–120. Kluwer Academic Publishers, 1995.

26 Diana Yepes, Julen Munarriz, Daniel l'Anson, Julia Contreras-Garcia, and Pablo Jaque. Real-space approach to the reaction force: Understanding the origin of synchronicity/nonsynchronicity in multibond chemical reactions. J. Phys. Chem. A, 124(10):1959–1972, 2020.

27 Henry Chermette. Chemical reactivity indexes in density functional theory. J. Comp. Chem., 20(1):129–154, 1999.

28 Paul Geerlings, Eduardo Chamorro, Pratim K. Chattaraj, Frank De Proft, José Luis Gázquez, Shubin Liu, Christophe Morell, Alejandro Toro-Labbé, Alberto Vela, and Paul W. Ayers. Conceptual density functional theory: Status, prospects, issues. Theor. Chem. Acc., 139:36, 2020.

29 Robert G. Parr, Robert A. Donnelly, Mel Levy, and William E. Palke. Electronegativity: The density functional viewpoint. J. Chem. Phys., 68(8):3801–3807, 1978.
30 Ralph G. Pearson. Absolute electronegativity and absolute hardness of Lewis acids and bases. J. Am. Chem. Soc., 107(24):6801–6806, 1985.
31 Ralph G. Pearson. Chemical hardness and density functional theory. J. Chem. Sci., 117(5):369–377, 2005.
32 Tjalling Koopmans. Uber die Zuordnung von Wellenfunktionen und Eigenwerten zu den einzelnen Elektronen eines Atoms. Physica, 1:104–113, 1934.
33 Robert G. Parr, Lázló v. Szentpánly, and Shubin Liu. Electrophilicity index. J. Am. Chem. Soc., 121:1922–1924, 1999.
34 S.F. Boys and F. Bernardi. The calculation of small molecular interactions by the differences of separate total energies. some procedures with reduced errors. Mol. Phys., 19:553–566, 1970.
35 B Brauer, M.K. Kesharwani, and J.M.L. Martin. Some observations on counterpoise corrections for explicitly correlated calculations on noncovalent interactions. J. Chem. Theory Comput., 10:3791–3799, 2014.
36 R.T. Sanderson. An interpretation of bond lengths and a classification of bonds. Science, 114:670–672, 1951.
37 R.T. Sanderson. Partial charges on atoms in organic compounds. Science, 121:207–208, 1955.
38 J.M. Hollis, L.E. Snyder, R.D. Suenram, and F.J. Lovas. A search for the lowest energy conformer of interestellar glycine. Astrophys J., 241:1001, 1980.
39 Jun Wang, Xiaohua Liu, and Xiaoming Feng. Asymmetric strecker reactions. Chem. Rev., 111(11):6947–6983, 2011.
40 M. J. Frisch, G. W. Trucks, H. B. Schlegel, G. E. Scuseria, M. A. Robb, J. R. Cheeseman, G. Scalmani, V. Barone, B. Mennucci, G. A. Petersson, H. Nakatsuji, M. Caricato, X. Li, H. P. Hratchian, A. F. Izmaylov, J. Bloino, G. Zheng, J. L. Sonnenberg, M. Hada, M. Ehara, K. Toyota, R. Fukuda, J. Hasegawa, M. Ishida, T. Nakajima, Y. Honda, O. Kitao, H. Nakai, T. Vreven, J. A. Montgomery, Jr., J. E. Peralta, F. Ogliaro, M. Bearpark, J. J. Heyd, E. Brothers, K. N. Kudin, V. N. Staroverov, R. Kobayashi, J. Normand, K. Raghavachari, A. Rendell, J. C. Burant, S. S. Iyengar, J. Tomasi, M. Cossi, N. Rega, J. M. Millam, M. Klene, J. E. Knox, J. B. Cross, V. Bakken, C. Adamo, J. Jaramillo, R. Gomperts, R. E. Stratmann, O. Yazyev, A. J. Austin, R. Cammi, C. Pomelli, J. W. Ochterski, R. L. Martin, K. Morokuma, V. G. Zakrzewski, G. A. Voth, P. Salvador, J. J. Dannenberg, S. Dapprich, A. D. Daniels, O. Farkas, J. B. Foresman, J. V. Ortiz, J. Cioslowski, and D. J. Fox. Gaussian 09 Revision B.01. Gaussian Inc. Wallingford CT 2009.

6

A Perspective on the So-Called Dual Descriptor

F. Guégan[1], L. Merzoud[2], H. Chermette[2], and C. Morell[2]

[1] *IC2MP UMR 7285, Université de Poitiers – CNRS, 4, rue Michel Brunet TSA 51106–86073 Cedex 9 Poitiers, France*
[2] *Université de Lyon, Institut des Sciences Analytiques, UMR 5280, CNRS, Université Lyon 1, ENS Lyon - 5, rue de la Doua, F-69100, Villeurbanne, France*

In this chapter, the so-called dual descriptor from conceptual DFT framework is reviewed. The power of the dual descriptor is highlighted by selected examples.

6.1 Introduction: Conceptual DFT

Chemistry can be defined as a science dedicated to the study of the transformations of matter. Its central concern is thus to understand and predict reactivity and selectivity during reactions [1, 2]. Theoretical chemistry offers in this respect a wonderful tool to achieve first-principle rationalizations, and thus it is not surprising that many different theoretical frameworks were proposed over time. Among these one may find Conceptual Density Functional Theory (C-DFT) [3–8]. The basic assumption in C-DFT is that the first Hohenberg-Kohn theorem [9] is satisfied, and thus that energy is a unique functional of the electron density. This implies that, if all accessible information on the system is contained within the wavefunction, it is also contained within the electron density. C-DFT then focuses on extracting this information, by the calculation of relevant quantities designed from first principles, often responses of the system to a (chemical) perturbation. The dual descriptor is one example of such descriptors, designed to provide insight into the chemical reactivity of a system.

In the following sections, we will present how it has initially been formulated. This first approach already revealed a strong connection between the dual descriptor and reactivity, which was further established by alternative derivations of this quantity, revealing deeper links with other reactivity descriptors and principles. Then, its efficiency in predicting reactivity and selectivity will be illustrated on simple yet insightful classical reactions in organic chemistry.

6.2 The Dual Descriptor: Fundamental Aspects

6.2.1 Initial Formulation

The dual descriptor was first derived in 2005, through a study of the energy variation two reagents undergo when put in interaction [10]. Decomposing this interaction into covalent, electrostatic,

Chemical Reactivity in Confined Systems: Theory, Modelling and Applications,
First Edition. Edited by Pratim Kumar Chattaraj and Debdutta Chakraborty.

and polarization contributions (or, in a HSAB framework, into soft-soft, hard-hard, and hard-soft contributions), it could be shown that maximal covalent interaction with a nucleophile (respectively an electrophile) is expected when the quantity

$$\Delta f(\mathbf{r}) = f^{+}(\mathbf{r}) - f^{-}(\mathbf{r}) \tag{6.1}$$

is maximal (respectively minimal). In this expression, f^+ and f^- are, respectively, the electrophilic and nucleophilic Fukui functions [11] (right and left derivatives of the electron density with respect to the number of electrons). Since each Fukui Function is normalized to one, the dual descriptor is also normalized:

$$\int_{\mathbb{R}^3} \Delta f(\mathbf{r}) d\mathbf{r} = 0 \tag{6.2}$$

Thus, localization of both the most electrophilic and nucleophilic sites of a given reagent are possible by a simple study of the extrema of a single quantity, Δf, hence the name "dual descriptor".

But beside this search for the most reactive regions, one may expect that Δf will allow to characterize all reactive regions in the system. Indeed, in any electrophilic area, one expects that $f^+(\mathbf{r}) > f^-(\mathbf{r})$, thus $\Delta f(\mathbf{r}) > 0$. Conversely, in nucleophilic regions $f^+(\mathbf{r}) < f^-(\mathbf{r})$ and $\Delta f(\mathbf{r}) < 0$. One may thus approach local philicities by the sign of the dual descriptor, and compare site reactivities by comparing their dual descriptor values. This is illustrated in section 6.3.

6.2.2 Alternative Formulations

6.2.2.1 Derivative Formulations

Before moving to the illustrations, further arguments can be provided to illustrate the strong connection between Δf and reactivity/selectivity. It can indeed be shown that Δf represents the local variation of hardness η due to a change in the external potential (approach of a reagent) $\delta v(\mathbf{r})$:[1]

$$\Delta f(\mathbf{r}) = \left(\frac{\delta \eta}{\delta v(\mathbf{r})}\right)_N . \tag{6.3}$$

From Maxwell identities, it can then be shown that

$$\Delta f(\mathbf{r}) = \left(\frac{\partial^2 \rho(\mathbf{r})}{\partial N^2}\right)_{v(\mathbf{r})} . \tag{6.4}$$

As one may note, Eq. 6.1 can be retrieved from Eq. 6.4 using finite difference approximation (unitary variation of N). This latter formulation furthermore stresses an important feature of the dual descriptor: formally, this quantity is a high-order response of the system (second-order response of the electron density and third-order response of the energy). Nevertheless, it presents a significant advantage over the first-order response of ρ: it is uniquely defined, contrarily to the Fukui functions (left and right derivatives being inequivalent). Thus one may catch both nucleophilicity and electrophilicity through a single object, which furthermore may offer deeper insight on local reactivity than the Fukui functions. There may indeed be regions of space where $f^+(\mathbf{r})$ is rather large and $f^-(\mathbf{r})$ too. If one restricts oneself to a study of either Fukui function, these regions could be deemed rather electrophilic or nucleophilic, while Δf suggests these regions are merely "electron interchanges". One may, however, note that this deeper insight is offered at a certain expense. Δf is indeed small either when both Fukui functions are large or small. These two situations are chemically inequivalent: in the first case, a marked polarisability or local softness is expected,

1 Note that, formally, identity between these two quantities is only met if temperature equals 0 K, which is an implicit assumption in most DFT calculations.

since electrons are easily mobilized, while in the second case a marked local hardness (or weak polarisability) is expected.

Overall, Δf appears to be a suitable descriptor for the characterization of local reactivity and selectivity, but not for local polarisability/softness. Actually, the definition of local softness and hardness were quite recently revisited (*vide infra*) and show, as expected from the previous discussion, an explicit dependence over both the dual descriptor and the Fukui functions (thus regaining information relative to their magnitude).

6.2.2.2 Link with Frontier Molecular Orbital Theory

The strong insight on reactivity Δf offers is further stressed out by a common working approximation. Using finite differentiation, Eq. 6.4 rewrites:

$$\Delta f(\mathbf{r}) \approx \rho_{N+1}(\mathbf{r}) + \rho_{N-1}(\mathbf{r}) - 2\rho_N(\mathbf{r}) \tag{6.5}$$

with ρ_j the electron density of the system with j electrons (at constant geometry). If orbitals are frozen, that is they remain unchanged by addition or subtraction of electrons, one obtains

$$\Delta f(\mathbf{r}) \approx \rho_{LU}(\mathbf{r}) - \rho_{HO}(\mathbf{r}) \tag{6.6}$$

with ρ_{HO} and ρ_{LU}, respectively, the highest occupied and lowest unoccupied Molecular Orbital densities. This result is in line with the Frontier Molecular Orbital approximation of Fukui [12, 13]. In this framework, electrophilic regions correspond to places where the LUMO develops, while nucleophilic areas are characterized by the HOMO. This is quite obviously retrieved here, since $\Delta f(\mathbf{r}) > 0$ where the LUMO develops and $\Delta f(\mathbf{r}) < 0$ where the HOMO develops.

6.2.2.3 State-Specific Development

So far, we did not discuss the "computability" of the dual descriptor, and only focused on formal expressions and approximations for this quantity. Yet there may be cases where our previous expressions cannot be computed, or at least cannot afford a satisfactory picture. For instance, when the FMO theory is known to fail (which is expected when orbital relaxation plays an important role), or when frontier orbitals are (quasi)-degenerate, Eq. 6.6 can hardly provide a clear and relevant insight on reactivity.

In such cases, one could still rely on Eq. 6.5, but issues can also arise. In the case of anionic systems, for instance, it is not always possible to attach an additional electron, and thus ρ_{N+1} cannot be computed. Furthermore, in the case of transition-metal containing systems, the proper spin-state of the $N + 1/N - 1$ species is often not trivially guessed.

From this, it appeared necessary to propose alternative ways to compute Δf, or generalization of this quantity. In 2013, [14] a formulation relying on excited states was proposed. One may note that Pearson already proposed in 1988 to analyse the ground state reactivity of molecules through a careful analysis of their first excitation [15]. The central idea was that one could express the ground state electron density of a chemical system undergoing a perturbation as an expansion on the set of ground and excited state electron densities,

$$\rho_{pert}(\mathbf{r}) = \sum_i \alpha_i \rho_i(\mathbf{r}), \ \sum_i \alpha_i = 1 \tag{6.7}$$

i indexing the electronic states (0 meaning ground state). From this, the ground state electron density reshuffling under perturbation can be expressed as

$$\Delta\rho(\mathbf{r}) = \rho_{pert}(\mathbf{r}) - \rho_0(\mathbf{r}) = \sum_{i>0} \alpha_i \left(\rho_i(\mathbf{r}) - \rho_0(\mathbf{r})\right) = \sum_{i>0} \alpha_i \Delta f_i(\mathbf{r}) \tag{6.8}$$

In the previous equation, Δf_i is called the ith state-specific dual descriptor. It may indeed be shown, under the frozen orbital hypothesis, that Δf_1 equals the FMO approximation of the dual descriptor. The sum over all excited states is then called an "extended dual descriptor".

Now a central question remains: how can one express the weighting parameters α_i? So far, no formula could be proposed, but in line with the sum-over-states formula one may encounter for instance in the expression of the molecular static polarisabilities, it may be surmised that α_i decays with excitation energies. Thus, it seems reasonable to restrict the calculation to the first few excited states, which are expected to convey most information on the ground state reactivity. This approach has been successfully applied to some organic reactions whose outcome cannot be predicted by Frontier Molecular Orbital Theory [14] and to the trans effect in octahedral transition metal complexes [16].

6.2.2.4 MO Degeneracy

The state-specific development offers a solution for the computation of the dual descriptor when classical approximations are impossible to use, especially when the addition of one electron is impossible (self-ionizing anion) or when a near-degeneracy is observed around the Fermi level. However, in this latter case computations are still rather cumbersome, since no algebraic formula can be provided for the weighting parameters in Eq. 6.8; at best, one may expect here to reach a qualitative level of description, but no quantification is at hand…

J. Martinez-Araya proposed in 2016 a way to address the problem of orbital degeneracy [17]. More specifically, he proposed a general operational formula for Δf in closed-shell molecules presenting q-fold degenerate HOMO and p-fold degenerate LUMO. This proposal relies on Eq. 6.1, in which f^+ and f^- are reformulated to take the orbital degeneracy into account. Let us take the example of f^+. Usually, addition of one electron in a chemical system results in addition of a supplementary molecular orbital in the Slater determinant describing its ground state wavefunction (addition of the newly occupied LUMO). However, if the LUMO is p-fold degenerate, p different determinants can be build, corresponding to addition of the electron into a given LUMO. All these determinants should offer an equally valid representation of the state of the anion. Thus one should consider a linear combination of all these determinants as ground-state wavefunction. Obviously this is a tedious task for large p values. J. Martinez-Araya thus proposed a simplification: instead of considering p configurations associated to a one-electron addition, one could consider a unique configuration associated to the addition of p electrons at once (one electron per LUMO).

Following the same line of argument in the case of the q-fold degenerate HOMO (one configuration associated to the loss of q electrons), he then proposed that:

$$\Delta f(\mathbf{r}) = \frac{q \cdot \rho_{N+p}^{p+1}(\mathbf{r}) - (p+q) \cdot \rho_N^1(\mathbf{r}) + p \cdot \rho_{N-q}^{q+1}(\mathbf{r})}{pq} \tag{6.9}$$

where the subscripts indicate the total number of electrons and the superscript the spin multiplicity to consider in the calculation (as suggested by Hund's rule). Quite obviously, if $q = p = 1$ one retrieves Eq. 6.5. This operational formula then proved very efficient in the rationalization of the chemical properties of molecular systems displaying a genuine MO degeneracy.

6.2.2.5 Quasi Degeneracy

However, it may be surmised that the same approach will provide a less accurate picture in the case of near-degeneracy. Let us consider here the case of a system displaying an occupied MO close in energy to the HOMO, but not exactly degenerate with it, and a vacant MO close to the LUMO but here also not degenerate with it. In such case, the FMO approximation is expected to fail, as both the frontier and "near frontier" levels should contribute to reactivity. Yet these contributions should be quantitatively different, and thus Eq. 6.9 (which implicitly assumes all contributions are equivalent) is not expected to perform well either.

Qualitatively, one may expect that the further away a MO is from the frontier levels, the less it will contribute to the ground state reactivity. Following this idea, R. Pino-Rios and co-workers recently proposed a MO-weighted dual-descriptor, which may be seen as an extension of the FMO approximation of Δf [18].

More specifically, they proposed to define orbital-weighted Fukui functions:

$$f_w^-(\mathbf{r}) = \sum_{i=1}^{HOMO} \omega_i^- \rho_i(\mathbf{r}),\ \omega_i^- = \frac{\exp\left[-\left(\frac{\mu-\epsilon_i}{\Delta}\right)^2\right]}{\sum_{i=1}^{HOMO} \exp\left[-\left(\frac{\mu-\epsilon_i}{\Delta}\right)^2\right]} \tag{6.10}$$

$$f_w^+(\mathbf{r}) = \sum_{i=LUMO}^{N'} \omega_i^+ \rho_i(\mathbf{r}),\ \omega_i^+ = \frac{\exp\left[-\left(\frac{\mu-\epsilon_i}{\Delta}\right)^2\right]}{\sum_{i=LUMO}^{N'} \exp\left[-\left(\frac{\mu-\epsilon_i}{\Delta}\right)^2\right]} \tag{6.11}$$

where i indexes a MO (assuming the system is represented by N' MO), μ is the chemical potential (average of the HOMO and LUMO energy in the FMO approximation), ϵ_i is the MO energy, ρ_i the associated electron density, and Δ a width parameter. From these weighted Fukui functions, they then proposed to design a weighted dual descriptor as

$$\Delta f_w(\mathbf{r}) = f_w^+(\mathbf{r}) - f_w^-(\mathbf{r}). \tag{6.12}$$

As one may note from f_w^- and f_w^+ definitions, contributions from all occupied and vacant MOs are taken into account. Nevertheless, the further away from the frontier levels, the lower the contribution, as a consequence of the Gaussian weighting. It may also be noted that perfect degeneracy can still be taken into account here, resulting as expected in similar weights for the degenerate levels.

Eq. 6.12 thus provides an interesting way to evaluate Δf in various situations, noticeably when degeneracy is present. Yet, it must be noted the Gaussian weighting scheme is an ansatz (since it does not stem from a first principle derivation). Thus, even though it presents desirable features, especially regarding normalization and convergence, it is not theoretically motivated. Additionally, the choice of the width parameter Δ is also not trivial from first principles, and thus calculations of Δf_w may be quite cumbersome in the general case.

6.2.2.6 Spin Polarization

Up to now, we have made only limited mention of the problem of the spin-state and spin polarization in the calculation of Δf, and we have especially implicitly assumed the system under study bears a closed shell. This is of course often the case for organic compounds, but not in coordination or organometallic chemistry.

It is an understatement to say that extension of reactivity models to open-shell systems is not trivial: new concepts should indeed be developed, to account for the peculiar properties of unpaired electrons.[2] In the frame of C-DFT, this can be achieved by a study of the responses of the system to

2 One cannot expect to describe a unique electron as being halfway between an electron pair and a vacancy, as is unfortunately implicitly assumed in some approaches, such as in the definition of the radical Fukui function $f^0 = \frac{1}{2}(f^+ + f^-)$.

a change in the spin number N_S, or equivalently to a change in the total number of α and β spins, and to variations an external magnetic field $\mathbf{B}(\mathbf{r})$.

Thus, several different descriptors can be derived from the dual descriptor in spin-polarized DFT [19]. Recalling that

$$\rho(\mathbf{r}) = \left(\frac{\delta E}{\delta v(\mathbf{r})}\right)_{\mathbf{B}(\mathbf{r}),N,N_S}, \quad \rho_S(\mathbf{r}) = \left(\frac{\delta E}{\delta \mathbf{B}(\mathbf{r})}\right)_{v(\mathbf{r}),N,N_S} \tag{6.13}$$

with ρ_S the spin-density, in analogy with Eq. 6.4 one may derive six different spin-polarized dual descriptors:

$$\Delta f_{N,NN}(\mathbf{r}) = \left(\frac{\partial^2 \rho(\mathbf{r})}{\partial N^2}\right)_{N_S,v(\mathbf{r}),\mathbf{B}(\mathbf{r})} \tag{6.14}$$

$$\Delta f_{N_S,N_SN_S}(\mathbf{r}) = \left(\frac{\partial^2 \rho_S(\mathbf{r})}{\partial N_S^2}\right)_{N,v(\mathbf{r}),\mathbf{B}(\mathbf{r})} \tag{6.15}$$

$$\Delta f_{N,N_SN_S}(\mathbf{r}) = \left(\frac{\partial^2 \rho(\mathbf{r})}{\partial N_S^2}\right)_{N,v(\mathbf{r}),\mathbf{B}(\mathbf{r})} \tag{6.16}$$

$$\Delta f_{N_S,NN}(\mathbf{r}) = \left(\frac{\partial^2 \rho_S(\mathbf{r})}{\partial N^2}\right)_{N_S,v(\mathbf{r}),\mathbf{B}(\mathbf{r})} \tag{6.17}$$

$$\Delta f_{N,NN_S}(\mathbf{r}) = \left(\frac{\partial^2 \rho(\mathbf{r})}{\partial N \partial N_S}\right)_{v(\mathbf{r}),\mathbf{B}(\mathbf{r})} \tag{6.18}$$

$$\Delta f_{N_S,NN_S}(\mathbf{r}) = \left(\frac{\partial^2 \rho_S(\mathbf{r})}{\partial N \partial N_S}\right)_{v(\mathbf{r}),\mathbf{B}(\mathbf{r})} \tag{6.19}$$

These quantities were first proposed in 2008 by E. Chamorro and co-workers, and the chemical meaning of the four first ones examined [20].

The first quantity, $\Delta f_{N,NN}(\mathbf{r})$, is rather straightforward: it is the "usual" dual descriptor, that is, second-order response of the electron density to a change in the number of electrons, but *at fixed total spin*. As one may note, this constraint was not implemented in the previously mentioned approximations of Δf.

The fourth quantity, $\Delta f_{N_S,NN}(\mathbf{r})$, is also a descriptor associated to a response at fixed total spin, but this time of the spin density function. It is thus expected to provide information about the rearrangement of the unpaired electrons in space under a nucleophilic or electrophilic attack, but with a constant total spin.

The second and third quantities, $\Delta f_{N_S,N_SN_S}(\mathbf{r})$ and $\Delta f_{N,N_SN_S}(\mathbf{r})$, however, provide information about the reorganization of the spin-density and electron density under a change in the total spin of the molecule, at fixed total number of electrons. Said otherwise, they allow to map the reorganization of spin and electron density of a system when it changes its spin multiplicity. They were coined "twofold spin-polarization descriptors", and are expected to display negative values in regions were electron (for $\Delta f_{N,N_SN_S}(\mathbf{r})$) or spin (for $\Delta f_{N_S,N_SN_S}(\mathbf{r})$) density vary under a decrease in the spin multiplicity. Similarly, they should display positive values in regions where the electron or spin density vary under an increase in the spin multiplicity. Thus these quantities afford a description of the electron (or spin) density reshuffling under a change in the electronic state, as is for instance happening in singlet-triplet conversion.

6.2.2.7 Grand Canonical Ensemble Derivation

Up to now, all derivations we presented were conducted within the canonical statistical ensemble, that is, electronic configuration are determined by the total number of electrons N and external potential $v(\mathbf{r})$.

Because of this choice of statistical ensemble, limited transferability is granted to Δf: one cannot compare (quantitatively) systems with different N. In order to afford complete transferability, one should thus shift to a statistical ensemble in which N is allowed to fluctuate: the grand-canonical ensemble, defined by the external potential $v(\mathbf{r})$ and the chemical potential μ (conjugate of N).

In line with Eq. 6.4, we may express the second-order response of the electron density with respect to μ:

$$\left(\frac{\partial^2 \rho(\mathbf{r})}{\partial \mu^2}\right)_{v(\mathbf{r})} = \frac{\Delta f(\mathbf{r})}{\eta^2} + \frac{\gamma}{\eta^3} f(\mathbf{r}) \tag{6.20}$$

with γ the canonical hyperhardness (third-order response of energy to a change in N). Even though hyperhardness values are not necessarily negligible, one may expect that the ratio γ/η^3 will be quite small compared to $1/\eta^2$, and thus the second term in the previous equation is usually neglected. As a consequence, the grand-canonical equivalent of the dual descriptor can be directly approximated from the canonical function, by a simple weighting by the square of hardness. Hence the problem of transferability can be reduced to a simple scaling of Δf [16].

6.2.3 Real-Space Partitioning

As one may note, so far we only evaluated reactivity and selectivity through a local analysis of the dual descriptor (local meaning in that case "point by point in real space"). However, in chemistry one often reasons in a semi-local approach, comparing the reactivity of atoms or groups of atoms.

Conciliation of both approaches can be achieved by real-space partitioning, that is, division of whole space into non-overlapping volumes and integration within these volumes. For instance, one may obtain atomic dual descriptor values by integration of Δf into atomic basins, defined for instance in a Hirshfeld or AIM partitioning method. From Eq. 6.5, if we assume the atomic basins only marginally vary under the addition or subtraction of one electron to the whole molecule, we have

$$\Delta f(A) \approx -2q_N(A) + q_{N+1}(A) + q_{N-1}(A) \tag{6.21}$$

for a given atom A.

But one may also consider integrating Δf into any real-space function domains, as long as these domains span $\mathbb{R}^3$. This is noteworthy in the case of Δf, which may then be partitioned into domains of constant sign [21]. This affords in principle a partition into reactive domains (electrophilic versus nucleophilic). Nevertheless, examples have shown that often all nucleophilic (resp. electrophilic) basins merge into a single one, and thus very limited insight is gained. A possible solution is then to exclude regions in space exhibiting Δf values lower than a given cut-off. This indeed often allows reactive regions to be separated, but at the expense of an approximation in the calculated values. In this respect, the zero integration over all space (Eq. 6.2) affords a certain degree of control on the approximation. Indeed, one may check how much the chosen cut-off induces a strong error by comparing the sum of the integrated values to zero.

6.2.4 Dual Descriptor and Chemical Principles

Beyond its intrinsic value, the dual descriptor can also be used or retrieved in the development of other quantities of interest and principles.

6.2.4.1 Principle of Maximum Hardness

For instance, from Eq. 6.3, it appears that interaction between reagents is optimal when their hardness variation is also optimum. This is rather reminiscent of the Principle of Maximum Hardness (PMH) of Pearson, Parr, and Chattaraj [22, 23]. Actually, if the incoming reagent is an electrophile, one may expect $\delta v(\mathbf{r}) < 0$ (interaction with an electron-poor species). Thus in order to minimize $\Delta f(\mathbf{r})$, $\delta\eta$ must be positive: hardness is indeed increasing. Similarly, interaction with a nucleophile should result in $\delta v(\mathbf{r}) > 0$, and thus $\delta\eta > 0$ is expected if one wants to maximize $\Delta f(\mathbf{r})$. Thus it appears that optimum (local) interactions between reagents translates in a tendency of hardness to increase.

PMH can also be retrieved in the grand-canonical derivation in Eq. 6.20: high hardness values should indeed translate in low values for the grand-canonical dual descriptor, and thus low reactivity (which is expected for the end product of a spontaneous reaction in chemistry).

6.2.4.2 Local Hardness Descriptors

In the previous paragraph, we discussed the evolution of hardness along a chemical event. Hardness is a central quantity in reactivity, describing the tendency of a system to distort or not its electron density in consequence of a chemical event. Yet, this quantity is global, as it considers the system as a whole.

It then quickly appeared useful to dispose of a local counterpart of hardness, the so-called local hardness descriptor. Though local softness can be straightforwardly derived within the grand canonical ensemble (first-order response of electron density to a change in the chemical potential), there is unfortunately no unique definition of local hardness, and thus the concept was heavily debated over time.

Recently, Polanco-Ramirez and co-workers proposed a general strategy to derive local quantities from global descriptors, which they applied to hardness [24–26]. The central idea in this development is to use chain derivation, to rewrite global derivatives as composed functional derivatives of the electron density:

$$R = \left(\frac{\partial A}{\partial X}\right) = \int \frac{\delta A}{\delta\rho(\mathbf{r})}\left(\frac{\partial\rho(\mathbf{r})}{\partial X}\right)\delta\rho(\mathbf{r})d\mathbf{r}. \tag{6.22}$$

The integrand is then considered as the local counterpart of R, $R(\mathbf{r})$. Using this rule in the case of hardness, and by identification of classical C-DFT descriptors, it can ultimately be proposed that

$$\eta(\mathbf{r}) = \mu\Delta f(\mathbf{r}) + \eta f(\mathbf{r}) \tag{6.23}$$

represents a local hardness descriptor. Thus Δf is not only a valid reactivity descriptor, but it also allow to partially describe local hardness (and thus to convey local information on polarisability).

6.2.4.3 Local Electrophilicity and Nucleophilicity

In 1999, Parr, Von Szentpaly and Liu have shown that, at fixed geometry, a molecular system interacting with an electron reservoir can acquire a maximum number of $\Delta N = -\mu/\eta$ electrons, and undergo in this process a stabilization energy of $-\mu^2/2\eta$ [27]. In 2014, it could be shown that the same quantity can also be used to characterize nucleophilicity, since in the grand canonical ensemble, the number of electrons that a system may release when its chemical potential is raised from μ to zero, is equal to $\Delta N = \mu/\eta$ [28].

Additionally, the maximal variations of the electron number could be used to develop local electrophilicity and nucleophilicity descriptors. Indeed, expanding the electron density variation as a Taylor series on N, one gets at second order that

$$\Delta\rho(\mathbf{r}) = \left(\frac{\partial\rho(\mathbf{r})}{\partial N}\right)_{v(\mathbf{r})} \Delta N + \frac{1}{2}\left(\frac{\partial^2\rho(\mathbf{r})}{\partial N^2}\right)_{v(\mathbf{r})} \Delta N^2. \tag{6.24}$$

When interacting with a perfect electron reservoir at constant geometry, a molecule will then undergo an electron density variation such that:

$$\Delta\rho_{Elec}(\mathbf{r}) = -\frac{\mu}{\eta}f(\mathbf{r}) + \frac{1}{2}\left(\frac{\mu}{\eta}\right)^2 \Delta f(\mathbf{r}) \tag{6.25}$$

where we have replaced the previous derivatives by their usual notations (Fukui function and dual descriptor). Similarly, interaction of the molecule with a perfect electron attractor will lead to:

$$\Delta\rho_{Nuc}(\mathbf{r}) = \frac{\mu}{\eta}f(\mathbf{r}) + \frac{1}{2}\left(\frac{\mu}{\eta}\right)^2 \Delta f(\mathbf{r}). \tag{6.26}$$

As expected, the dual descriptor is involved in these two expressions as a second-order philicity descriptor.

6.2.4.4 Local Chemical Potential and Excited States Reactivity

Finally, it has been shown in 2009 that Δf can also be used to predict reactivity in excited states, and not only in the ground state [29]. The principle of the demonstration is the following: let one use an excited state electron density as trial density for the DFT functional. In the general case, the first-order response of energy to this density, which in the ground state is the chemical potential, is a non-constant function:

$$\left(\frac{\delta E}{\delta\rho_k(\mathbf{r})}\right)_N = \lambda_k(\mathbf{r}). \tag{6.27}$$

This local function translates the tendency of electron density to distort from the excited state to the ground state in order to minimize the energy [30]. Obviously this local chemical potential can be rewritten as the sum of the ground state (global) chemical potential and a local term

$$\lambda_k(\mathbf{r}) = \mu_0 + V_k(\mathbf{r}). \tag{6.28}$$

Expressing the excited state density as a distortion of the ground state density, and accordingly the response of energy to this excited state density as a modified response to the ground state density, it could then be shown that for the lowest energy excited states

$$V_k(\mathbf{r}) \approx \int \frac{\Delta f_k(\mathbf{r}')}{|\mathbf{r} - \mathbf{r}'|} d^3\mathbf{r}' \tag{6.29}$$

with Δf_k the k-th state-specific dual descriptor. The energy change associated to the electron density reshuffling $\delta\rho_k(\mathbf{r})$ towards the ground state configuration can then be expressed as

$$\delta E_k = \mu_0 dN + \int V_k(\mathbf{r})\delta\rho_k(\mathbf{r})d^3\mathbf{r}. \tag{6.30}$$

At constant number of electron ($dN = 0$), excited states will be stabilized by density variations such that

$$V_k(\mathbf{r})\delta\rho_k(\mathbf{r}) < 0. \tag{6.31}$$

This implies that the electron density will be decreasing in regions displaying positive values of Δf_k, and conversely the electron density will increase anywhere Δf_k is negative. Thus meanings of the sign of the state-specific dual descriptors are the opposite of the ones of the ground-state dual descriptor.

6.3 Illustrations

6.3.1 Woodward Hoffmann Rules in Diels-Alder Reactions

In paragraph 6.2.2.2, a clear link was established between Δf and the Frontier Molecular Orbital theory, through Eq. 6.6. Conclusions arising from FMO theory should thus be equally retrieved by a dual descriptor analysis.

This was nicely illustrated by a rederivation of the Woodward-Hoffmann rules [31, 32]. Let us for instance consider the classical Diels-Alder cycloaddition reaction. The HOMO and LUMO of ethylene and butadiene are schematically provided in Figure 6.1, and the expected shape of Δf in the FMO approximation is also depicted.

From these schemes, it is rather plain that suprafacial attack of ethylene on itself will result in a strong repulsion if both molecules are in their ground state, since sites with similar sign of the dual descriptors are placed in repulsive interaction.

Conversely, reaction of butadiene and ethylene, here also in their ground state, gives rise to overlap between dual descriptor domains with opposite sign, and should thus lead to a certain stabilization and make the [2 + 4] addition possible.

Using comparable arguments, it is then possible to show for all other polyenes that [n+m] suprafacial/suprafacial Diels-Alder reactions are thermally possible when n+m equals 4k+2 ($k \in \mathbb{N}$), in line with Woodward and Hoffmann's rule.

But furthermore, since we observed in paragraph 6.2.4.4 that the meaning of the sign of Δf is reversed in the excited states, the selectivity rule under photochemical conditions can also be retrieved. For instance, reaction of two ethylene molecules, one in its first excited state and the second in the ground state, leads in that case to a stabilizing interaction in the case of a suprafacial/suprafacial attack. Extending this argument to other polyenes, one eventually

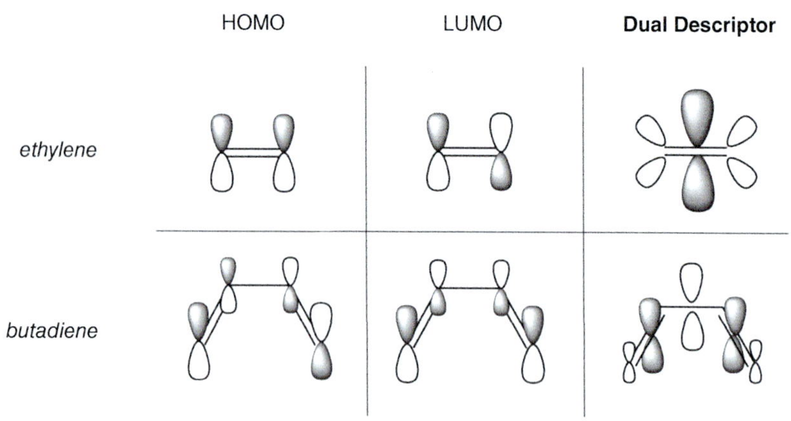

Figure 6.1 Schematic representation of the HOMO and LUMO of butadiene and ethylene as found in Hückel's theory, and FMO approximation of the dual descriptor.

retrieves Woodward and Hoffmann's rule: photochemical Diels-Alder reactions are possible if n+m equals 4k.

6.3.2 Perturbational MO Theory and Dual Descriptor

The dual descriptor is an interesting tool to explain the difference in reactivity between ethene and formaldehyde. It should be noted that the formaldehyde double bond reacts easily with nucleophiles whereas ethene does not.

This difference in reactivity can be explained by a detailled study of the dual descriptor shape. Indeed, in the case of ethylene, because of the molecular symmetry the two carbon atoms bear equivalent Δf contributions, with a marked nucleophilicity in the middle of the C=C double bond, while electrophilic areas are equally splitted at the molecule end sites. This is plain when deriving Δf in the frozen orbital scheme, as shown in Figure 6.2. Thus interaction with an electrophile is "geometrically" favored. On the other hand, in the case of formaldehyde a strong asymmetry is expected in Δf. Once again, this is plain when deriving the dual descriptor from the frontier orbitals densities, as shown on Figure 6.2. Electrophilic sites should indeed roughly correspond to the regions in which the LUMO develops most, and thus principally concentrate on the C atom. Because of the anti-bonding character of this orbital, one can furthermore state that the electrophilic basin on C will be pushed away from the C=O bond. Thus maximal interaction with a nucleophile is expected when the attack is performed at a nucleophile-C-O angle larger than 90°. Δf thus allows the long-known Dünitz-Burgi angle of attack of nucleophiles on carbonyls to be retrieved [33].

6.3.3 Markovnikov Rule

The regio- and stereoselectivity of the electrophilic additions to asymmetric alkenes are described following the well-known Markovnikov rule [34]. Therefore, the acidic proton addition to a double bond proceeds on the least substituted carbon atom. This rule is generalized for all addition reaction to alkenes. Thus, the most stable carbocation adduct is the most favorable electrophile addition product. Regioselectivity explanation based on the Frontier MOs is not simple. Starting from the

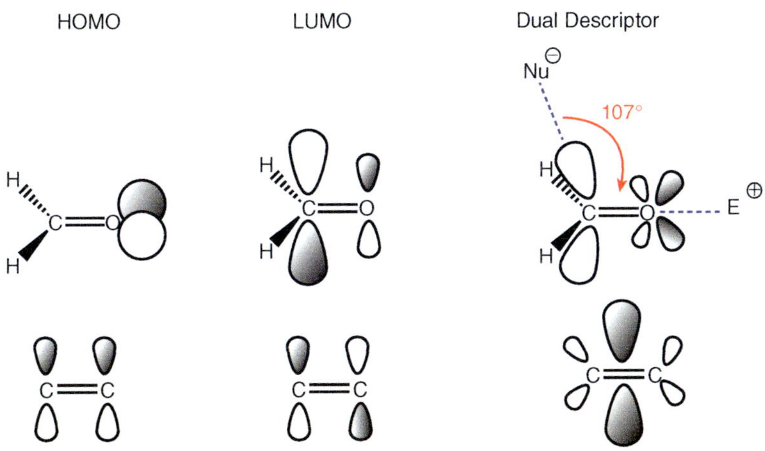

Figure 6.2 Schematic representation of the HOMO and LUMO of formaldehyde and ethylene and their dual descriptor with the grey lobes are negative region and white lobes are positive region.

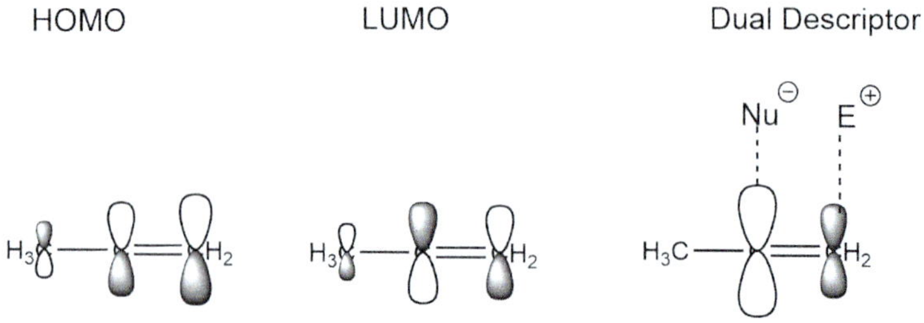

Figure 6.3 Schematic representation of Markovnikov description with dual descriptor, with the grey lobes are negative region and white lobes are positive region.

MO diagram of ethylene displayed in Figure 6.3, one would indeed expect that substitution at a C atom by an electron-donating group will end up in a distortion of the HOMO, the unsubstituted C now bearing the largest contribution and thus displaying the largest nucleophilicity. Yet, this distortion appears more than slight when calculating the MO diagram using classical methods (such as DFT); it is actually plain only within Hückel's theory. As a matter of fact, computed HOMOs and LUMOs (or equivalently Fukui functions) often suggest both C atoms of the double C=C bond may undergo electrophilic or nucleophilic attack. However, the dual descriptor successfully translates the expected selectivity, as asymmetric dual lobes are observed. The substituted carbon atom indeed bears a pronounced electrophilic character, whereas the unsubstituted carbon is an electrophile. The fact that the dual descriptor is successful when both Fukui functions or MOs are not can furthermore be simply accounted for. Indeed, while the asymmetry in the MOs is too slight to be noticed, even on the corresponding MO electron densities (which are approximations of the Fukui functions), by the subtraction in Eq. 6.1 one can expect trends to be amplified. For instance, on the substituted C atom, f^+ will be slighly larger than on the other C, while at the same time f^- will be smaller, and thus Δf will be more positive. Conversely, on the unsubstituted C atom, Δf is expected to be more negative. Overall, one retrieve the expected tendency: proton addition on the least substituted C (more nucleophilic), and electrophilic addition on the most substituted C.

6.4 Conclusions

In this chapter the last 15 years of development of the so-called dual descriptor were presented. The review started with the very first definition of the concept and its relations with early theories or concepts such as frontier molecular orbital or chemical hardness. Then, more recent developments pertaining to the dual descriptor were shown. Of particular interest was how molecular orbital degeneracy is tackled either through weighted combination of molecular orbital around the Fermi level or using electronic excited states densities. Definitions of dual descriptor counterparts within the grand canonical ensemble or spin polarized description were also recalled. Finally prototypical examples of chemical reactivity and selectivity rationalization were proposed in both ground and excited states. An alternative explanation of chemical rules was put forward using the dual descriptor as principal ingredient. Through this bird's-eye-view, it is plain to see that the dual descriptor is quite a powerful index for understanding chemical reactivity and selectivity. Over time it might replace with efficiency previous descriptors.

References

1 Gilles Klopman. Chemical reactivity and the concept of charge and frontier-controlled reactions. Journal of American Chemical Society, 90(2):223–234, 1968.

2 Lionel Salem. Intermolecular orbital theory of the interaction between conjugated systems. I. General theory. J. Am. Chem. Soc., 90(3):543–552, 1968.

3 Paul Geerlings, Frank De Proft, and Wilfried Langenaeker. Conceptual density functional theory. Chemical Reviews, 103:1793–1874, 2003.

4 Henry Chermette. Chemical reactivity indexes in density functional theory. Journal of Computational Chemistry, 20(1):129–154, 1999.

5 Robert G. Parr and Weitao Yang. Density-functional theory of atoms and molecules. International Series of Monographs on Chemistry. Oxford University Press, 1994.

6 José L. Gazquez. Perpectives on density functional theory of chemical reactivity. Journal of the Mexican Chemical Society, 52:3–10, 2008.

7 Shu-Bin Liu. Conceptual density functional theory and some recent developments. Acta Physico-Chimica Sinica, 25:590–600, 2009.

8 Paul Geerlings, Eduardo Chamorro, Pratim K Chattaraj, Frank De Proft, José-L Gazquez, Shu-Bin Liu, Christophe Morell, Alejandro Toro-Labbé, Alberto Vela, and Paul Woodson Ayers. Conceptual density functional theory: status, prospects, issues. Theoretical Chemistry Accounts, 193:8240–8247, 2020.

9 Pierre Hohenberg and Walter Kohn. Inhomogeneous electron gas. Physics Review, 136:864–871, 1964.

10 Christophe Morell, André Grand, and Alejandro Toro-Labbé. New dual descriptor for chemical reactivity. The Journal of Physical Chemistry A, 109(1):205–212, 2005. PMID: 16839107.

11 Weitao Yang, Robert Gormley Parr, and Robert Pucci. Electron density, Kohn-Sham frontier orbitals, and Fukui functions. Journal of Chemical Physics, 181:2862, 1984.

12 Kenichi Fukui. Role of frontier orbitals in chemical reactions. Science, 218(4574):747–754, 1982.

13 Kenichi Fukui and Hiroshi Fujimoto. An MO-theoretical interpretation of the nature of the chemical reaction. I. Partitioning analysis of the interaction energy. Bulletin of the Chemical Society of Japan, 41(9):1989–1987, 1968.

14 Vincent Tognetti, Christophe Morell, Paul W. Ayers, Laurent Joubert, and Henry Chermette. A proposal for an extended dual descriptor: a possible solution when frontier molecular orbital theory fails. Physical Chemistry Chemical Physics, 15:14465–14475, 2013.

15 Ralph G. Pearson. Electronic spectra and chemical reactivity. Journal of the American Chemical Society, 110:2092–2097, 1988.

16 Frédéric Guégan, Vincent Tognetti, Laurent Joubert, Henry Chermette, Dominique Luneau, and Christophe Morell. Towards the first theoretical scale of the trans effect in octahedral complexes. Physical Chemistry Chemical Physics, 18:982–990, 2016.

17 Jorge I. Martínez-Araya. A generalized operational formula based on total electronic densities to obtain 3d pictures of the dual descriptor to reveal nucleophilic and electrophilic sites accurately on closed-shell molecules. Journal of Computational Chemistry, 37(25):2279–2303, 2016.

18 Ricardo Pino-Rios, Diego Inostroza, Gloria Cárdenas-Jirón, and William Tiznado. Orbital-weighted dual descriptor for the study of local reactivity of systems with (quasi-) degenerate states. The Journal of Physical Chemistry A, 123(49):10556–10562, 2019. PMID: 31710492.

19 Patricia Perez, Eduardo Chamorro, and Paul Woodson Ayers. Universal mathematical identities in density functional theory: Results from three different spin-resolved representations. Journal of Chemical Physics, 128:204108–1–204108–21, 2008.

20 Eduardo Chamorro, Patricia Pérez, Mario Duque, Frank De Proft, and Paul Geerlings. Dual descriptors within the framework of spin-polarized density functional theory. The Journal of Chemical Physics, 129(6):064117, 2008.

21 Vincent Tognetti, Christophe Morell, and Laurent Joubert. Quantifying electro/nucleophilicity by partitioning the dual descriptor. Journal of Computational Chemistry, 36(9):649–659, 2015.

22 Ralph G. Pearson. The principle of maximum hardness. Accounts of Chemical Research, 26(5):250–255, 1993.

23 Robert G. Parr and Pratim K. Chattaraj. Principle of maximum hardness. J. Am. Chem. Soc., 113(5):1854–1855, 1991.

24 Carlos A. Polanco-Ramírez, Marco Franco-Pérez, Javier Carmona-Espíndola, José L. Gázquez, and Paul W. Ayers. Revisiting the definition of local hardness and hardness kernel. Physical Chemistry Chemical Physics, 19:12355–12364, 2017.

25 Frédéric Guégan, Walid Lamine, Henry Chermette, and Christophe Morell. Comment on "revisiting the definition of local hardness and hardness kernel" by Carlos A. Polanco-Ramírez, Marco Franco-Pérez, Javier Carmona Espindola, José L. Gázquez and Paul W. Ayers, Physical Chemistry Chemical Physics, 2017, 19, 12355. Physical Chemistry Chemical Physics, 20:9006–9010, 2018.

26 Marco Franco-Pérez, Carlos A. Polanco-Ramírez, José L. Gázquez, and Paul W. Ayers. Reply to the 'Comment on "revisiting the definition of local hardness and hardness kernel"' by Christophe Morell, Frédéric Guégan, Walid Lamine, and Henry Chermette, Physical Chemistry Chemical Physics, 2018, 20, doi: 10.1039/c7cp04100d. Physical Chemistry Chemical Physics, 20:9011–9014, 2018.

27 Robert G. Parr, László v. Szentpály, and Shubin Liu. Electrophilicity index. Journal of the American Chemical Society, 121(9):1922–1924, 1999.

28 Christophe Morell, José L. Gázquez, Alberto Vela, Frédéric Guégan, and Henry Chermette. Revisiting electroaccepting and electrodonating powers: proposals for local electrophilicity and local nucleophilicity descriptors. Physical Chemistry Chemical Physics, 16:26832–26842, 2014.

29 Christophe Morell, Vanessa Labet, André Grand, Paul W. Ayers, Frank De Proft, Paul Geerlings, and Henry Chermette. Characterization of the chemical behavior of the low excited states through a local chemical potential. Journal of Chemical Theory and Computation, 5(9):2274–2283, 2009.

30 Christophe Morell, Paul Woodson Ayers, André Grand, and HEnry Chermette. Application of electron density force to chemical reactivity. Physical Chemistry Chemical Physics, 13:9601–9608, 2011.

31 Paul Woodson Ayers, C. Morell, F. De Proft, and P. Geerlings. Understanding the Woodward-Hoffmann rules by using changes in electron density. Chemistry: A European Journal, 13:8240–8247, 2007.

32 Paul Geerlings, Paul Woodson Ayers, Alejandro Toro-Labbé, Pratim K Chattaraj, and Frank De Proft. The Woodward-Hoffmann rules reinterpreted by conceptual density functional theory. Chemistry: A European Journal, 13:8240–8247, 2007.

33 Hans-Beat Bürgi, Jack D. Dunitz, Jean-Marie Lehn, and Georges Wipff. Tetrahedron, 30:1563–1572, 1974.

34 Vladimir V. Markownikoff. I. Ueber die Abhängigkeit der Verschiedenen Vertretbarkeit des Radicalwasserstoffs in den isomeren Buttersäuren. Justus Liebigs Annalen der Chemie, 153(2):228–259, 1870.

7

Molecular Electrostatic Potentials: Significance and Applications

Peter Politzer and Jane S. Murray

Department of Chemistry, University of New Orleans, New Orleans, LA 70148, USA

7.1 A Quick Review of Some Classical Physics

Consider a point charge Q_A. It creates an electrical potential V_A in the surrounding space, the value of which at any distance r from Q_A is given by

$$V_A(r) = Q_A/r \tag{7.1}$$

in atomic units. This means that if a second point charge Q_B is placed at the distance r from Q_A, then the energy of interaction between the point charges will be,

$$\Delta E(r) = Q_B V_A(r) = Q_B Q_A/r \tag{7.2}$$

If Q_A and Q_B have the same sign, then $\Delta E > 0$ and the interaction is repulsive; if they have opposite signs, $\Delta E < 0$ and it is attractive.

From classical physics, the point charge Q_A also creates an electric field ε_A that is equal to the negative gradient of the electrical potential. Thus,

$$\varepsilon_A(r) = -d\left[V_A(r)\right]/dr = Q_A/r^2. \tag{7.3}$$

The electric field $\varepsilon_A(\mathbf{r})$ exerts a force $F_A(\mathbf{r})$ upon any point charge Q_B that is at the distance r from Q_A; this force is

$$F_A(r) = Q_B\varepsilon_B = Q_B Q_A/r^2 \tag{7.4}$$

This is simply Coulomb's Law. If Q_A and Q_B have the same sign, then the force is positive, i.e. in the direction of increasing r, and Q_A is repelling Q_B; if they have opposite signs, the force is negative, in the direction of decreasing r, and Q_A is attracting Q_B. These conclusions are consistent with the energy of interaction ΔE being repulsive in the former case, attractive in the latter.

Eqs. (7.1)–(7.4) are all forms of Coulomb's Law. They show that the electrical potential $V_A(\mathbf{r})$ governs the interaction energy of Q_A with Q_B, the electric field that Q_A creates and the force that it exerts upon Q_B.

7.2 Molecular Electrostatic Potentials

Our present focus is upon the electrical potential of a molecule. Since this is simply a collection of point charges (the nuclei and electrons), the electrical potential that it creates at any point $\mathbf{r}$ is

Chemical Reactivity in Confined Systems: Theory, Modelling and Applications,
First Edition. Edited by Pratim Kumar Chattaraj and Debdutta Chakraborty.

just the summation of Eq. (7.1) over all of them. By the Born-Oppenheimer approximation, the nuclei can be treated as having fixed positions but this does not apply to the electrons; for these, summation must be replaced by integration over the electronic density $\rho(\mathbf{r})$, the average electronic charge in every volume element $d\mathbf{r}$ in the surrounding space. The electrical potential produced by the molecule at r is then,

$$V(\mathbf{r}) = \sum_{A} \frac{Z_A}{|\mathbf{R}_A - \mathbf{r}|} - \int \frac{\rho(\mathbf{r}')\, d\mathbf{r}'}{|\mathbf{r}' - \mathbf{r}|} \tag{7.5}$$

where Z_A is the charge on nucleus A, located at $\mathbf{R}_A$. (Since a molecule lacks the perfect spherical symmetry of a point charge, it is now necessary to use vector notation for position, electric field and force.)

The electrical potential computed with Eq. (7.5) is for a fixed, static distribution of charge, i.e. the electronic density and nuclear positions that were used in calculating it. These usually correspond to the isolated free molecule. The resulting $V(\mathbf{r})$ is accordingly known as the *electrostatic* potential of the molecule.

7.3 The Electronic Density and the Electrostatic Potential

The electrostatic potential $V(\mathbf{r})$ and the electronic density $\rho(\mathbf{r})$ are rigorously related by both Eq. (7.5) and by Poisson's equation, Eq. (7.6):

$$\nabla^2 V(\mathbf{r}) = 4\pi\rho(\mathbf{r}) - 4\pi \sum_{A} Z_A \delta(\mathbf{r} - \mathbf{R}_A) \tag{7.6}$$

An important feature of both $V(\mathbf{r})$ and $\rho(\mathbf{r})$ is that they are real physical properties, observables. They can be determined experimentally, by diffraction methods [1–3], as well as computationally.

$V(\mathbf{r})$ and $\rho(\mathbf{r})$ are one-electron properties; they do not explicitly depend upon interactions between electrons. When computed at the Hartree-Fock level, Møller-Plesset perturbation theory shows them to be correct through first-order [4]; errors due to neglect of electronic correlation are no more than second-order effects, although this does not necessarily make them insignificant [5, 6].

A key difference between $V(\mathbf{r})$ and $\rho(\mathbf{r})$ is that while $\rho(\mathbf{r})$ explicitly represents only the amount of electronic charge at the point r, $V(\mathbf{r})$ is the resultant of the positive contributions of all of the nuclei and the negative ones of all of the electrons. This means that the sign and magnitude of $V(\mathbf{r})$ do not necessarily correlate with just the electronic density at $\mathbf{r}$ [7–10]. Thus, although an "electron-rich" portion of a molecule does often have a negative $V(\mathbf{r})$, this is not always the case. The effect of the nuclei needs to be taken into account. For instance, the internuclear regions of covalent bonds often have buildups of electronic density [3, 11] but frequently have positive electrostatic potentials because of the nearby nuclei [1, 7, 12].

The Hohenberg-Kohn theorem shows that the properties of a ground-state system of nuclei and electrons are completely determined by the electronic density [13]. For instance, the energy of the system is a functional of the density:

$$E = f[\rho(\mathbf{r})] \tag{7.7}$$

The theorem clearly establishes the fundamental role of $\rho(\mathbf{r})$. However, while numerous approximate forms of Eq. (7.7) have been developed and have revolutionized computational chemistry, the true form of the functional predicted by Eq. (7.7) has not yet been discovered. On the other hand, two years before Hohenberg and Kohn, Wilson derived an exact expression for molecular energies in terms of the electrostatic potentials at their nuclei [14].

As Galvez and Porras observed with respect to the electronic density of an atom [15], "It is hard to prove mathematical properties of this quantity in a rigorous way." We have pointed out earlier that the electrostatic potential is also a fundamental property, with which it may sometimes be easier to work [16].

Another example of this is the variation of $\rho(\mathbf{r})$ with radial distance from the nucleus of a ground-state neutral atom. It has been demonstrated empirically that $\rho(\mathbf{r})$ decreases monotonically [17–19] but efforts to obtain a proof have been unsuccessful. In contrast, it was shown fairly readily that $V(\mathbf{r})$ is positive everywhere in the space of the atom and decreases monotonically [20].

7.4 Characterization of Molecular Electrostatic Potentials

While the $V(\mathbf{r})$ of neutral ground-state atoms are positive everywhere [20], the atoms' interactions in forming molecules typically produce some regions of negative $V(\mathbf{r})$, often associated with lone pairs and π electrons. Each negative region must have one or more local minima, V_{min}, at which $V(\mathbf{r})$ attains its locally most negative values. It has been shown, however, that there are no local maxima, V_{max}, other than those associated with the positions of the nuclei [21], the magnitudes of which simply reflect the nuclear charges.

There are various ways in which molecular electrostatic potentials have been presented, e.g. at specific points within the space of a molecule [22], along certain axes [23], within planes through the molecule [12, 24], etc. Since about 1990, $V(\mathbf{r})$ has commonly been displayed on molecular "surfaces" [25–27], these often being defined by outer contours of the molecules' electronic densities as proposed by Bader et al. [28]. A surface chosen in this manner is specific to the molecule and reflects its particular features, such as lone pairs, π electrons and atomic anisotropy. We typically use the 0.001 au contour for this purpose, although other outer ones, e.g. the 0.002 au, would show the same trends [29].

When $V(\mathbf{r})$ is plotted on the 0.001 au molecular surface, it is labeled $V_S(\mathbf{r})$. Its locally most positive and most negative values, of which there may be several, are designated by $V_{S,max}$ and $V_{S,min}$, respectively. (There can be local maxima on the surface, even though spatial maxima are associated only with nuclei [21].) Note that the $V_{S,max}$ and $V_{S,min}$ are maxima and minima only on the 0.001 au surface, not spatially; e.g. the $V_{S,min}$ do not normally correspond to V_{min}.

As an example of the electrostatic potential on a molecular surface, Figure 7.1 displays $V_S(\mathbf{r})$ computed on the 0.001 au contour of 4-bromopyridine (Scheme 7.1). The calculations (and those to follow) were at the density functional B3PW91/6-31G(d,p) level [30] using the WFA-SAS code to obtain the electrostatic potential [31]. Comparative analyses confirm that Kohn-Sham hybrid density functional procedures within the generalized gradient approximation, which do include some correlation, are effective for computing $V(\mathbf{r})$ [32, 33], provided that polarization functions are included in the basis set. There will be differences in details depending upon the particular method but overall trends and relative values, which is usually what is of interest, are generally reliable if consistency is maintained with regard to method and basis set.

In principle, $V(\mathbf{r})$ should be given in units of potential, e.g. volts. In practice, however, it is presented most often in energy units, usually kcal/mol or kJ/mol. By Eq. (7.2), this would correspond to the interaction energy of the *static* charge distribution of the *isolated* molecule with a +1 point charge at r. This is an obvious contradiction, since the presence of the point charge means that the molecule is not isolated and its charge distribution would not remain static; it would be polarized

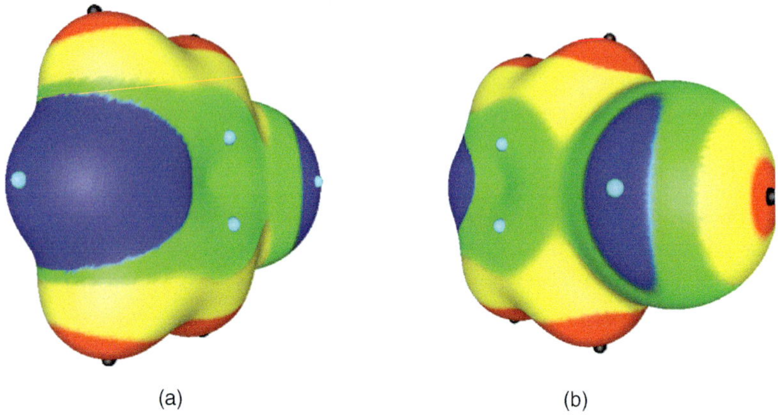

Figure 7.1 Computed electrostatic potential on the 0.001 au molecular surface of 1. In (a), the nitrogen is in the left foreground, the bromine in the right rear. In (b), the nitrogen is in the left rear, the bromine in the right foreground. Color ranges, in kcal/mol: Red, more positive than 12; yellow, between 12 and zero; green, between zero and −6; blue, more negative than −6. Black hemispheres indicate positions of most positive potentials, the $V_{S,max}$; blue hemispheres indicate positions of most negative potentials, the $V_{S,min}$.

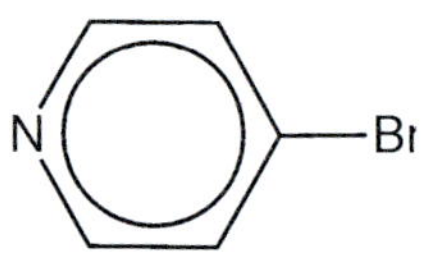

1

Scheme 7.1

by the electric field of the point charge (see section 7.5). It would accordingly be more meaningful to express V(**r**) in volts, since then no point charge is involved. For consistency, however, we shall follow the tradition of using energy units.

Figure 7.1 shows that 4-bromopyridine has a strongly-negative electrostatic potential associated with the ring nitrogen ($V_{S,min}$ = −34.1 kcal/mol) and weaker ones in the π regions above and below the ring ($V_{S,min}$ = −3.5 kcal/mol) and on the lateral sides of the bromine ($V_{S,min}$ = −7.7 kcal/mol). There are moderately strong positive potentials on the hydrogens ($V_{S,max}$ = 21.9 and 18.0 kcal/mol) and on the bromine ($V_{S,max}$ = 16.1 kcal/mol).

7.5 Molecular Reactivity

Probably the most extensive use of computed molecular electrostatic potentials has been in relation to reactive behavior. Such applications were one of the objectives of Scrocco, Tomasi, and their colleagues when they pioneered the analyses of molecular electrostatic potentials [24, 34, 35]. Tomasi et al. later gave an interesting detailed summary of this early work [36].

As pointed out above, if the electrostatic potential $V_A(\mathbf{r})$ of an isolated molecule A were assumed to be unaffected by the presence of a +1 point charge at r, then $V_A(\mathbf{r})$ would equal the protonation energy of the molecule at r. The locations and magnitudes of the $V_{S,min}$ would indicate and rank the energetically most favored protonation sites. This could readily be extended to point charges other than +1, by looking for the locations of the most negative values of the product $QV_A(\mathbf{r})$.

Within the approximation of static molecular charge distributions, the same approach could be applied to the interaction of molecule A with a second molecule B, treating B as a collection of point charges. This would involve summing the products of $V_A(\mathbf{r})$ and the nuclear charges and integrating the product $\rho_B(\mathbf{r})V_A(\mathbf{r})$, where $\rho_B(\mathbf{r})$ is the electronic density of B.

Thus, if the charge distributions of molecules A and B were truly static (fixed electronic densities and nuclear positions), then their electrostatic potentials would provide good guides to their interactive behavior. Electrophilic portions of each molecule would tend to interact favorably with the locally most negative regions of the other, especially the surface minima $V_{S,min}$. Nucleophilic portions of each would interact favorably with the regions of most positive electrostatic potential on the other, especially the surface local maxima $V_{S,max}$.

It might be protested that this would be treating the interactions as purely Coulombic. But the Hellmann-Feynman theorem shows rigorously that the forces felt by any nucleus in a system of nuclei and electrons are indeed entirely classical Coulombic [37, 38]. This is a direct consequence of the fact that all of the potential energy terms in the Schrödinger equation (which determine forces) are Coulombic.

In reality, however, the charge distributions of interacting molecules do not remain static. Even in the interaction of molecule A with point charge Q, the latter creates an electric field that exerts a force upon each nucleus and electron of A and causes some degree of distortion (polarization) of the charge distribution of A. This is not inconsistent with the Hellmann-Feynman theorem; electric fields and polarization are intrinsic parts of a Coulombic interaction, as shown by Eqs. (7.3) and (7.4). The result is that the product $QV_A(\mathbf{r})$, where $V_A(\mathbf{r})$ corresponds to the isolated molecule A, is at best an approximation to the interaction energy of A and Q.

What is rigorously correct, for the interaction of molecule A with a point charge Q at position r, is that the product $QV_A(\mathbf{r})$ is equal to the first-order term in a perturbation theory expansion of the A---Q interaction energy [12, 39]:

$$\Delta E\,(\text{A---Q}) = QV_A\,(\mathbf{r}) + Q^2P\,(\mathbf{r}) \tag{7.8}$$

The second-order term $P(\mathbf{r})$ essentially corrects for the polarization of A by the electric field of Q. Higher-order terms can usually be neglected [40, 41].

While Eq. (7.8) is written for interaction with a point charge Q, it has sometimes been extended to interactions with a second molecule B, using calculated at ? omic charges to approximate the charge distribution of B [36, 40]. Of course this encounters the problem that the charges assigned to the isolated molecule B would be affected by the polarizing electric field of A.

Including the second-order term can be expected to significantly change the magnitude of ΔE, but how much does it affect the qualitative predictions that would be made on the basis of $QV_A(\mathbf{r})$ alone? Several studies have addressed the use of perturbation theory to correct for polarization [12, 39–47].

Brinck has pointed out that the locations and ranking of the most favored interaction sites often remain essentially the same when the polarization correction is included as when predicted from just the $V_{S,max}$ and $V_{S,min}$ [39]. Long experience has indeed shown that molecular electrostatic potentials alone are frequently effective qualitative guides to reactive behavior [7, 12, 24, 25, 34–36, 39, 48, 49]. An early example was the protonation of a group of three-membered ring molecules [50].

On the other hand, the electrostatic potential sometimes fails to correctly predict and/or rank reaction sites, and polarization must be invoked [12, 39, 40, 42, 43, 46]. Consider the alkyl amines. The potentials of the nitrogen lone pairs become less negative in going from ammonia to trimethylamine, but the gas phase basicity *increases* because the amines are becoming more polarizable [51].

Electrophilic aromatic substitution provides more examples. The most negative $V_{S,min}$ of aniline and phenol are associated with the NH_2 and OH groups, respectively [52], yet electrophilic substitution is well-known to occur on the rings, at the *ortho* and *para* positions. This is because of the greater local polarizabilities at those positions (as measured by the local ionization energies [53]).

It follows therefore that relying upon the electrostatic potential alone to predict and interpret molecular reactive behavior is best limited to situations in which polarization can be anticipated to have only a minor effect, so that the approximation of static molecular charge distributions is at least somewhat well satisfied. Since the strength of an electric field (and the resulting polarization) diminishes with the square of distance, Eq. (7.3), this suggests that the focus be upon weak interactions in which the molecules are at relatively large separations. For instance, electrostatic potentials have had considerable success in describing noncovalent interactions; some examples will be given in section 7.6. Plotting $V(\mathbf{r})$ on 0.001 au molecular surfaces is appropriate in this context, since the 0.001 au contours are beyond typical van der Waals radii of the constituent atoms [23] and $V_S(\mathbf{r})$ is accordingly reasonably indicative of what an approaching reactant initially "sees" (however, note section 7.6.3).

It might be argued that atomic charges could also be used for these purposes, and are easier to calculate. However, the electrostatic potential is a real physical property, rigorously and uniquely defined and experimentally observable [1–3]. In contrast, atomic charges have no physical basis [54–56]. They are arbitrarily defined, and many procedures for doing so have been proposed, which sometimes seriously contradict each other. For instance, the carbon atom in CH_3NO_2 has been predicted to have charges ranging from −0.478 to +0.564 [57]; there is not even agreement as to whether it is positive or negative!

Bader's mathematically-elegant Quantum Theory of Atoms in Molecules (QTAIM) [58, 59] fares no better; the atomic charges that it produces are chemically unrealistic [60–64]. For example, carbon monoxide is described as ionic [58].

7.6 Some Applications of Electrostatic Potentials to Molecular Reactivity

7.6.1 σ-Hole and π-Hole Interactions

Two widely occurring types of noncovalent interactions that are currently receiving a great deal of attention involve what are known as σ-holes and π-holes. These terms refer to specific regions of lower electronic density in molecules, features of the anisotropies of covalently-bonded atoms [65–67]. σ-Holes are on the outer sides of covalently-bonded atoms of Groups IV–VII, on the extensions of the bonds. π-holes are above and below atoms of Groups III–VI in planar portions of molecules.

Due to their lower electronic densities, there are often (not always) positive electrostatic potentials associated with σ- and π-holes. An example is the positive region on the bromine in Figure 7.1(b), which is due to the σ-hole on the extension of the C-Br bond. In Figure 7.2, which shows $V_S(\mathbf{r})$ for the planar carbonyl (Scheme 7.2), the positive potentials above and below the carbon arise from the π-holes on the carbon. Their $V_{S,max}$ are 30.0 kcal/mol; there is also a $V_{S,max}$ of 19.5 kcal/mol associated with the chlorine σ-hole on the extension of the C-Cl bond. There is a $V_{S,min}$, −19.2 kcal/mol, on the fluorine.

Regions of positive electrostatic potential, such as those due to σ- and π-holes, can result in attractive noncovalent interactions with negative sites, e.g. lone pairs, π-electrons and anions [49, 68–71]. The interactions are directional; those involving σ-holes are approximately along the extensions of

Figure 7.2 Computed electrostatic potential on the 0.001 au molecular surface of 2. The chlorine is at the lower left, the fluorine at the lower right. Color ranges, in kcal/mol: red, more positive than 12; yellow, between 12 and zero; green, between zero and −6; blue, more negative than −6. Black hemispheres indicate positions of most positive potentials, the $V_{S,max}$; blue hemisphere indicates position of most negative potential, $V_{S,min}$.

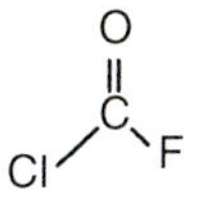

2

Scheme 7.2

the bonds that produced the σ-holes while the π-hole interactions are approximately perpendicular to the planar portions of the molecules.

These types of interactions had long been known experimentally, and in the case of halogen atoms were regarded as particularly enigmatic, since these are themselves expected to be negative in character. It was the analysis of molecular surface electrostatic potentials that provided the explanations, for σ-hole interactions in 1992 [72] and for π-hole interactions in 2012 [73].

For covalently-bonded atoms of Groups V–VII, $V_S(\mathbf{r})$ also reveals that the same atom can have regions of negative potential in addition to the positive ones due to σ-holes; note the bromine in Figure 7.1 and the chlorine in Figure 7.2. This means that such atoms can interact attractively with both negative and positive sites. Such dual reactivity has in fact been observed, both crystallographically [74–77] and computationally [78, 79].

Calculated atomic charges could not account either for σ-hole interactions or for dual reactivity, since each atom in a molecule is typically assigned a single charge. However, both are readily understood in terms of electrostatic potentials.

7.6.2 Hydrogen Bonding: A σ-Hole Interaction

Some of the earliest applications of electrostatic potentials were to hydrogen bonding [22, 35, 51, 80, 81], and it continues to be a prime focus of such studies [39, 48, 82–85]. For instance, Sandhu et al. recently used computed electrostatic potentials in analyzing co-crystal formation through hydrogen bonding [86].

Hydrogen bonding fits into the category of σ-hole interactions [48, 84, 87], although it is less directional [84, 88] because the positive potential associated with the σ-hole is more hemispherical and less focused than for other atoms. Compare, for example, the σ-hole positive potentials of the hydrogens and the bromine in Figure 7.1(b).

It has been demonstrated that the experimentally-determined hydrogen-bonding tendencies of a series of donors of different types correlate very well with the electrostatic potentials (the $V_{S,max}$) of the hydrogens; $R^2 = 0.982$ [89]. Furthermore, it was shown that the strengths of 24 hydrogen bonds between a variety of donors and acceptors could be expressed satisfactorily in terms of the product of the $V_{S,max}$ of the hydrogens and the $V_{S,min}$ of the negative sites [90]. The R^2 value was 0.908, with one outlier. It corresponded to the most strongly-bound complex, having an interaction energy of −13.1 kcal/mol, perhaps reflecting a higher degree of polarization. When this outlier was removed, R^2 increased to 0.931.

7.6.3 Interaction Energies

The interaction energy ΔE for molecules A and B forming a complex A---B is given in terms of the respective energies by,

$$\Delta E = E(A\text{---}B) - [E(A) + E(B)] \tag{7.9}$$

The more negative is ΔE, the stronger is the interaction.

It is sometimes possible to neglect polarization and to relate ΔE for a series of σ- or π-hole complexes to just the $V_{S,max}$ if the negative site stays the same, to just the $V_{S,min}$ if the σ- or π-hole molecule is kept the same, or to some combination of $V_{S,max}$ and $V_{S,min}$ if both participants vary [68, 91, 92]. An example is the relationship between ΔE and the product of $V_{S,max}$ and $V_{S,min}$ for hydrogen bonds [90], mentioned above.

Sometimes, especially for stronger interactions having ΔE more negative than about −9 kcal/mol, ΔE may not correlate with just $V_{S,max}$ and/or $V_{S,min}$, and it may be claimed that this shows the Coulombic interpretation to be inadequate [93, 94]. However, the problem in such cases is not the Coulombic interpretation, but rather the failure to take account of the polarization that is an intrinsic part of that interpretation.

We have already discussed perturbation theory as one way to include polarization. An overview of that and some other approaches for doing so has recently been presented [92].

Two different procedures have been applied, with an encouraging level of success, to σ- and π-hole interactions:

(a) For interaction with an anion as the negative site, in which the dominant polarization is anticipated to be that due to the anion's strong electric field, a negative point charge was used to model the polarizing effect [95–97].

(b) For interaction with a negative site in a neutral molecule, the electric field of the σ- or π-hole molecule and the polarizability of the negative site were explicitly included in the expression for ΔE [92, 98, 99]. The primary polarization is expected to be that of the negative site, since its usually high electronic density should increase its polarizability.

Since interactions with neutral molecules are probably more common, we will illustrate method (b) by means of a study in which it was applied to 54 σ-hole complexes using the regression relationship [99],

$$\Delta E = c_1\,[V(R)] + c_2[\varepsilon(R)]^2 + c_3\,[\alpha] + c_4\left[V_{S,min}\right] + c_5 \tag{7.10}$$

In Eq. (7.10), R is the equilibrium interaction separation, V(R) and ε(R) are the electrostatic potential and the electric field of the σ-hole molecule at the approximate position of the negative site, α is the polarizability of the negative site and $V_{S,min}$ is the most negative potential on the 0.001 au surface of the negative site.

For the 54 σ-hole complexes, the correlation between the ΔE predicted with Eq. (7.10) and those computed quantum chemically at the MP2/aug-cc-pVDZ level had $R^2 = 0.938$ and a root-mean-square error of 1.0 kcal/mol [99]. Similarly good results were obtained for 33 π-hole complexes and for all 87 together.

What is particularly notable is that Eq. (7.10) was able to represent interactions much stronger than what is normally expected for noncovalent bonding, having ΔE as negative as −42 kcal/mol. These were the interactions in which polarization made the greatest contributions [99] as measured by the polarization energies [100], $\Delta E(pol) = -0.5[\alpha][\varepsilon(R)]^2$.

It was suggested that these stronger interactions with higher levels of polarization might be described as having some "dative" or "coordinate covalent" character, corresponding to the polarization of some electronic density from the negative site toward the positive potential of the σ- or π-hole atom. Density difference plots confirm that such polarization does occur, along with electronic density of the σ- or π-hole atom being polarized away from the negative site [101–103].

7.6.4 Close Contacts and Interaction Sites

Attractive noncovalent interactions are commonly identified by looking for "close contacts" between atoms, by which is meant separations that are less than the sums of their van der Waals radii. This is often an effective and useful approach. However, certain limitations should be kept in mind.

The first relates to the very concept of van der Waals radii. They are not a rigorously defined property; various approaches have been used to assign their values. Pauling felt that they "are to be considered as reliable only to 0.05 or 0.10 Å" [104]. Bondi warned that his van der Waals radii "may not always be suitable for the calculation of contact distances in crystals" [105], which is exactly for what they are now widely used! As he pointed out, assuming covalently-bonded atoms to be spherical, which underlies the idea of van der Waals radii, is not valid; they are well known to be anisotropic [65–67]. For further discussion, see Dance [106] and Alvarez [107]. Our point is simply that van der Waals radii should be used cautiously, and not viewed as a rigid criterion for the existence of noncovalent interactions.

Another limitation upon using close contacts to identify noncovalent interactions is that it assumes interactions to always be between atoms. This is indeed often the case – but not always. A molecule's $V_{S,max}$ and $V_{S,min}$ are not invariably associated with specific atoms. The positive or negative potentials of atoms in close proximity frequently overlap and may result in just a single $V_{S,max}$ or $V_{S,min}$ at a point between the atoms [10, 49, 71, 108]. If a subsequent interaction is governed by this $V_{S,max}$ or $V_{S,min}$, then any atom-atom close contacts that may be present are not revealing the actual interaction.

Consider dimethyl urea (Scheme 7.3). Its crystal lattice is composed of chains of molecules in which each oxygen has two O---H-N close contacts [109]. This was interpreted as bifurcated hydrogen bonding (Scheme 7.4), which may seem reasonable on the basis of the atom-atom close contacts. However, the computed molecular electrostatic potential of Scheme 7.3 shows that the positive potentials of the urea hydrogens have coalesced into one long positive region having just a single $V_{S,max}$ at its center [108]. Furthermore, instead of the oxygen having a $V_{S,min}$ associated with each of its ostensible lone pairs, it has just one $V_{S,min}$ in the middle of its negative region. Accordingly, in terms of the electrostatic potentials and resulting electric fields that actually govern this interaction, the realistic way to describe it is by means of structure (Scheme 7.5). Analogous situations are encountered with other ureas [110].

Heterocyclic rings provide further examples of interactions that do not correspond to atom-atom close contacts [10, 49, 71, 108]. What is sometimes found is illustrated in Figure 7.3 for the heterocyclic molecule (Scheme 7.6). The carbons, sulfur and selenium are expected to have a σ-hole on

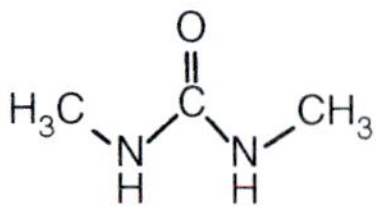

3

Scheme 7.3

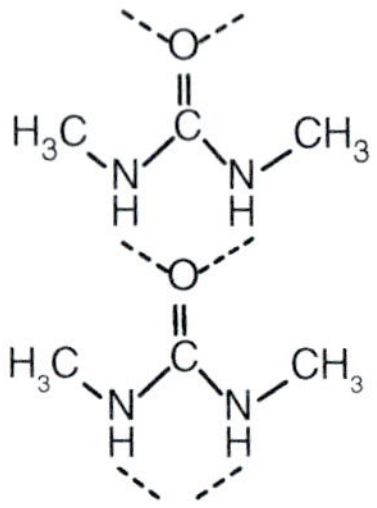

4

Scheme 7.4

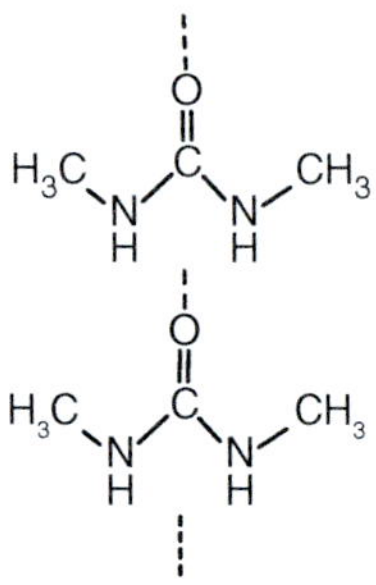

5

Scheme 7.5

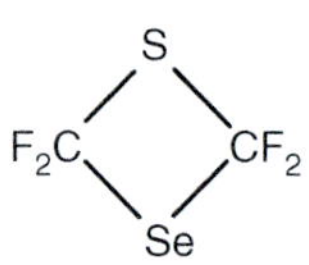

6

Scheme 7.6

the extension of each C-S and C-Se bond, producing two $V_{S,max}$ for each ring atom and a total of eight $V_{S,max}$ in the plane of the ring. However, the computed molecular surface electrostatic potential in Figure 7.3 shows only four $V_{S,max}$, of essentially 30 kcal/mol, and they are near the centers of the C-S and C-Se bonds, not associated specifically with any ring atoms. Apparently the positive potentials due to the ring atom σ-holes overlap and reinforce each other in the regions between the atoms, resulting in just a single $V_{S,max}$ located in the middle portion of each C-S and C-Se bond. It is with these middle $V_{S,max}$ that interactions with negative sites occur, not with the ring atoms [10, 49, 71, 108].

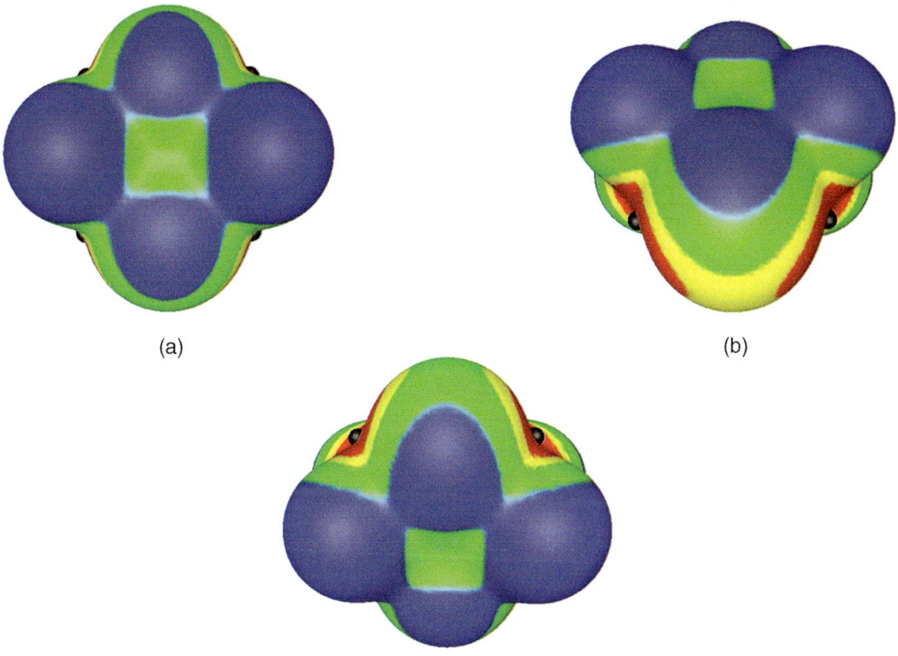

(a) (b) (c)

Figure 7.3 Computed electrostatic potential on the 0.001 au molecular surface of 6. In (a), the view is from above the molecular plane, the sulfur is at the top and the selenium at the bottom. In (b), the sulfur is at the top rear, the selenium at the bottom front. In (c), the sulfur is at the top front, the selenium at the bottom rear. Color ranges, in kcal/mol: red, more positive than 14; yellow, between 14 and 7; green, between 7 and −2; blue, more negative than −2. Black hemispheres indicate positions of most positive potentials, the $V_{S,max}$. Note that they are near the centers of the C-S and C-Se bonds.

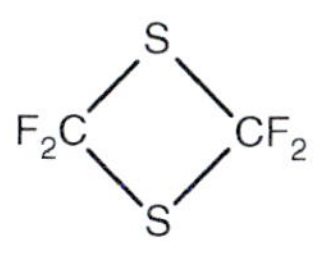

7

Scheme 7.7

In the crystal lattice of the dithietane (Scheme 7.7), for instance, intermolecular interactions may be anticipated to involve a fluorine on one molecule and the carbon or sulfur positive σ-hole potential on an adjacent molecule. Indeed the F---C and the F---S separations between neighboring molecules are less than the sums of the respective van der Waals radii, i.e. they are close contacts [111]. It could therefore be concluded that these are the attractive intermolecular interactions. However, the electrostatic potential of Scheme 7.7 shows $V_{S,max}$ near the centers of the C-S bonds [111], and in the crystal lattice of Scheme 7.7 the fluorines are positioned to interact with these $V_{S,max}$, not with the carbons or sulfurs. The attractive intermolecular interactions are actually F---$V_{S,max}$, not the close contacts.

Finally we will look at some cyclic polyketones, such as parabanic acid (Scheme 7.8) and alloxan (Scheme 7.9). In these molecules, there are not π-hole $V_{S,max}$ above and below each carbonyl carbon,

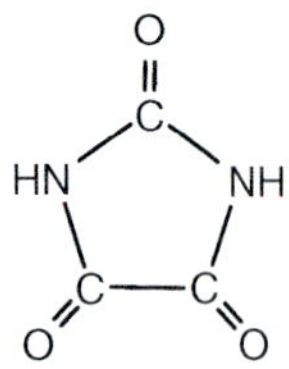

8

Scheme 7.8

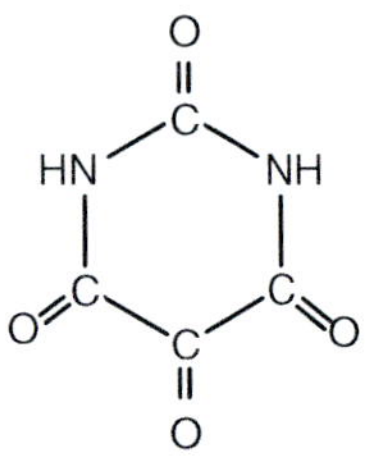

9

Scheme 7.9

as in Scheme 7.2 (Figure 7.2). Instead the positive potentials of the carbonyl carbons in Schemes 7.8 and 7.9 overlap, resulting in single $V_{S,max}$ above and below the approximate centers of the rings in both Scheme 7.8 [71] and Scheme 7.9 [112]. The crystal structures of Schemes 7.8 and 7.9 confirm that the dominant intermolecular interactions are between carbonyl oxygens and these central $V_{S,max}$ of neighboring molecules [113, 114].

Interactions are governed by the electric fields of molecules, and these are not necessarily linked to individual atoms; like the electrostatic potential, they reflect the whole molecule. Close contacts between atoms of two molecules do not always indicate the actual interactions [10, 49, 71, 108], which may involve $V_{S,max}$ or $V_{S,min}$ that are not associated with any particular atom.

7.6.5 Biological Recognition Interactions

Molecular electrostatic potentials have been extensively used in biochemistry and pharmacology to identify characteristic patterns of positive and negative potentials that either promote or inhibit particular types of biological activities [7, 12, 115–122]. Drug-receptor and enzyme-substrate interactions have been a common focus of attention.

To illustrate, we will consider the widely-varying toxicities of the halogenated derivatives of dibenzo-*p*-dioxin (Scheme 7.10), which range from virtually none for the parent compound (Scheme 7.10) to the high toxicity of the notorious 2,3,7,8-tetrachlorodibenzo-*p*-dioxin (Scheme 7.11) [123]. The problems associated with these compounds, to different extents, include carcinogenesis, hepatotoxicity, loss of lymphoid tissue, gastric lesions and urinary tract hyperplasia.

Comparisons of the computed electrostatic potentials of a series of variously halogenated dibenzo-*p*-dioxins plus some related molecules revealed a pattern [26, 124]. High levels of receptor binding and toxicity are linked to the presence of negative regions above and below the lateral positions (the chlorines in Scheme 7.11) and a large positive area above and below the three rings [26, 124]. This pattern apparently promotes attractive interaction with the biological receptor. The

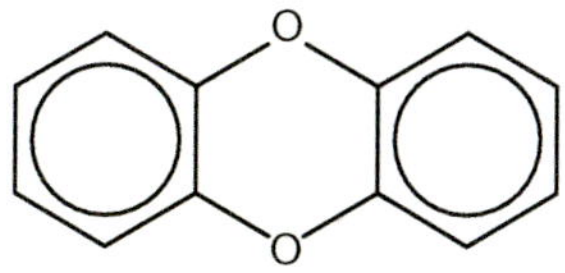

10

Scheme 7.10

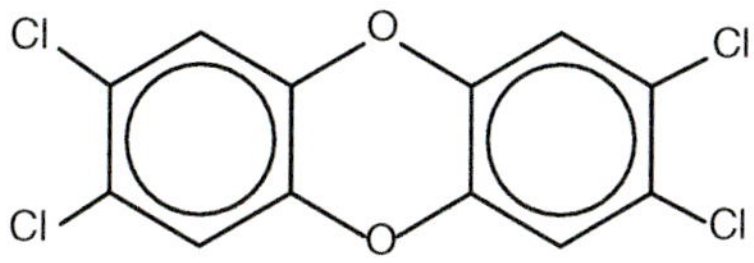

11

Scheme 7.11

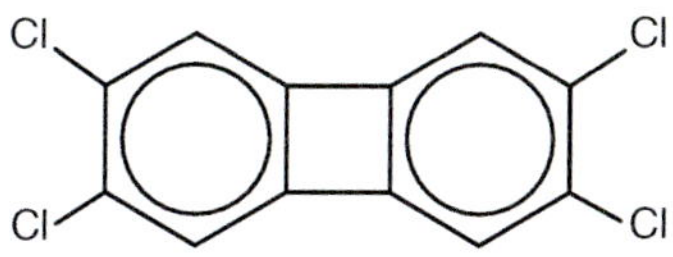

12

Scheme 7.12

oxygens of the middle ring appear to have no significant role since Scheme 7.12 has toxicity similar to that of Scheme 7.11. The negative potentials of the oxygens need to be small and weak, perhaps to avoid inhibiting repulsive interactions with nitrogen lone pairs of the receptor, which has been proposed to be porphine-like [125]. In Scheme 7.10, the oxygen negative potentials are relatively strong, since there are no electron-attracting chlorines, which may account for its being virtually non-toxic [123].

7.6.6 Statistical Properties of Molecular Surface Electrostatic Potentials

Applications of molecular surface electrostatic potentials, $V_S(\mathbf{r})$, have commonly focused upon the patterns of localized positive and negative regions and their $V_{S,max}$ and $V_{S,min}$. However, it can be quite fruitful to also analyze $V_S(\mathbf{r})$ in terms of other, statistically-defined features of the entire molecular surface, including:

- The average deviation of $V_S(\mathbf{r})$, a measure of the internal charge separation that is present even in molecules having zero dipole moments.
- The positive, negative, and total variance of $V_S(\mathbf{r})$, indicators of the variabilities (ranges) of the positive and negative surface potentials, and their sum.
- An electrostatic balance parameter, a measure of the degree of similarity between the positive and negative variances on a molecular surface, which indicates how well the molecule can interact through both its positive and negative surface potentials.

These statistical quantities are discussed in more detail, with formulas for their evaluation, in several reviews [39, 126, 127].

Good correlations have been found between subsets of these statistical quantities and a variety of condensed phase properties that depend upon noncovalent interactions [39, 126, 127]. These properties include heats of phase transitions, solubilities and solvation energies, boiling points and critical constants, partition coefficients, viscosities, surface tensions and diffusion constants. For example, the product of the total variance and the balance parameter is important in representing properties that involve interactions of molecules with their own kind, such as heats of phase transitions, boiling points and critical temperatures.

It is noteworthy that condensed phase properties can be expressed reasonably well in terms of features of electrostatic potentials computed for isolated, gas phase molecules. This follows from the fact that these condensed phase properties reflect noncovalent intermolecular interactions between individual molecules, and these interactions are Coulombic in nature, as has been discussed.

7.7 Electrostatic Potentials at Nuclei

The use of the electrostatic potential computed over a molecular surface to interpret and predict the molecule's reactive behavior is well known. What is less widely recognized is that the electrostatic potential can be related to the energy of an atom or molecule. This involves specifically the potential at each nucleus that is created by the electronic density and the other nuclei. These potentials, V(0) for an atom and $V(\mathbf{R}_A)$ for nucleus A in a molecule, are given by Eqs. (7.11) and (7.12); they follow directly from Eq. (7.5).

$$V(0) = -\int \frac{\rho(\mathbf{r})d\mathbf{r}}{\mathbf{r}} \tag{7.11}$$

$$V(\mathbf{R}_A) = \sum_{B \neq A} \frac{Z_B}{|\mathbf{R}_B - \mathbf{R}_A|} - \int \frac{\rho(\mathbf{r})d\mathbf{r}}{|\mathbf{r} - \mathbf{R}_A|} \tag{7.12}$$

By applying the Hellmann-Feynman theorem in its generalized form [37, 38, 128, 129], the energy of an atom or molecule can be formulated *exactly* in terms of V(0) for the atom and $V(R_A)$ for each nucleus in the molecule [14, 130–132]. What is particularly intriguing is that this allows electron-electron repulsion energies (two-electron properties) to be expressed rigorously in terms of electrostatic potentials at nuclei (one-electron properties). This can be viewed as demonstrating the Hohenberg-Kohn theorem (section 7.3).

The significance of the exact energy formulas has thus far, to our knowledge, been primarily conceptual. However, they have led to a variety of useful approximate relationships between atomic and molecular energies and electrostatic potentials at nuclei. These have been discussed in several reviews [133–137].

Experience has shown that electrostatic potentials at nuclei computed at the Hartree-Fock level are frequently more accurate than the associated Hartree-Fock energies [133–139]. Accordingly, sufficiently good formulas relating atomic and molecular energies to V(0) and $V(\mathbf{R}_A)$, used in conjunction with Hartree-Fock values of V(0) and $V_A(\mathbf{R}_A)$, might produce energies that are better than Hartree-Fock, i.e. include some electronic correlation. This has been confirmed [135, 138–140]. Particularly noteworthy was the derivation by Levy et al. of atomic energy expressions that yielded nearly all of the correlation energies of the atoms hydrogen through argon, using Hartree-Fock V(0) [135, 138], even though Hartree-Fock methodology does not include electronic correlation.

7.8 Discussion and Summary

The fundamental nature of atomic and molecular electrostatic potentials follows directly from the Schrödinger equation, which describes nuclear-nuclear, nuclear-electronic, and electronic-electronic interactions as purely Coulombic. The same is then true for molecular interactions.

This is not saying that they are entirely electrostatic. The only entirely electrostatic interactions are between point charges. The electrical potentials of the nuclei and electrons in an atom or molecule create an electric field that polarizes the charge distribution of any approaching atom or molecule. This polarization is an intrinsic part of the Coulombic interaction.

However, the polarization component of an interaction is sometimes relatively minor, especially for weak interactions at large separations, and the electrostatic potentials of the isolated free molecules or atoms can then provide good qualitative guides to reactivity. They have indeed been very effective in this role. Nevertheless, the possible effect of polarization should always be kept in mind.

A less known function of molecular electrostatic potentials is defining boundaries between covalently-bonded atoms. Since $V(\mathbf{r})$ has spatial maxima only at the positions of the nuclei [21], it must pass through a minimum at some point $\mathbf{r}_m$ along the internuclear axis between two bonded atoms. At $\mathbf{r}_m$, the gradient of $V(\mathbf{r})$ along that axis is therefore zero and so, by Eq. (7.3), an element of electronic charge at $\mathbf{r}_m$ feels no attractive force along the axis from either nucleus. Accordingly $\mathbf{r}_m$ defines a natural axial boundary between the two atoms. It has been confirmed [141, 142] that axial minima of $V(\mathbf{r})$ provide atomic radii that are in much better agreement with experimental observations and empirical covalent radii than are axial minima of $\rho(\mathbf{r})$, which have also been proposed as defining boundaries between atoms [58, 59].

We have already mentioned, in section 7.7, the interesting and conceptually significant relationships, both exact and approximate, between molecular energies and the electrostatic potentials at their nuclei. However, the importance of the $V(\mathbf{R}_A)$ is not limited to molecular energies. As was pointed out some time ago [134, 136], they are linked to both core-electron binding energies (and thus to electron spectroscopy) and to diamagnetic shielding (and thus to nuclear magnetic resonance). Electrostatic potentials at molecular nuclei clearly merit further investigation.

References

1 Stewart, R.F. (1979). On the mapping of electrostatic properties from Bragg diffraction data. Chemical Physics Letters 65: 335–342.

2 Politzer, P. and Truhlar, D.G. (eds.) (1981). Chemical Applications of Atomic and Molecular Electrostatic Potentials. New York: Plenum Press.

3 Klein, C.L. and Stevens, E.D. (1988). Experimental measurements of electron density distributions and electrostatic potentials. In: Structure and Reactivity (eds. J.F. Liebman and A. Greenberg), 25–64. New York: VCH Publishers.

4 Møller, C. and Plesset, M.S. (1934). Note on an approximate treatment for many-electron systems. Physical Review 46: 618–622.

5 Seeger, R. and Pople, J.A. (1975). Electron density in Møller-Plesset theory. Journal of Chemical Physics 62: 4566.

6 Köster, A.M., Lebouf, M., and Salahub, D.R. (1996). Molecular electrostatic potentials from density functional theory. In: Molecular Electrostatic Potentials: Concepts and Applications (eds. J.S. Murray and K. Sen), 105–142. Amsterdam: Elsevier.
7 Politzer, P. and Murray, J.S. (1991). Molecular electrostatic potentials and chemical reactivity. In: Reviews in Computational Chemistry, vol. 2 (eds. K.B. Lipkowitz and D.B. Boyd), 273–312. New York: VCH Publishers.
8 Wheeler, S.E. and Houk, K.N. (2009). Through-space effects of substituents dominate molecular electrostatic potentials of substituted arenes. Journal of Chemical Theory and Computation 5: 2301–2312.
9 Murray, J.S. and Politzer, P. (2017). Molecular electrostatic potentials and noncovalent interactions. WIREs Computational and Molecular Science 7: e1326.
10 Politzer, P., Murray, J.S. (2017) σ-Hole interactions: Perspectives and misconceptions, Crystals, 7, 212(1–14).
11 Coppens, P. (1977). Experimental electron densities and chemical bonding. Angewandte Chemie International Edition 16: 32–40.
12 Politzer, P. and Daiker, K.C. (1981). Models for chemical reactivity. In: The Force Concept in Chemistry (ed. B.M. Deb), 294–387. New York: Van Nostrand Reinhold.
13 Hohenberg, P. and Kohn, W. (1964). Inhomogeneous electron gas. Physical Review B 136: 864–871.
14 Wilson, E.B. Jr., (1962). Four-dimensional electron density. Journal of Chemical Physics 36: 2232–2233.
15 Galvez, F.J. and Porras, I. (1991). Atomic charge density at the nucleus and inequalities among radial expectation values. Physical Review A 44: 144–147.
16 Politzer, P. and Murray, J.S. (2002). The fundamental nature and role of the electrostatic potential in atoms and molecules. Theoretical Chemical Accounts 108: 134–142.
17 Sperber, G. (1971). Analysis of reduced density matrices in the coordinate representation. II. The structure of closed-shell atoms in the restricted Hartree-Fock approximation, International Journal of Quantum Chemistry 5: 189–214.
18 Weinstein, H., Politzer, P., and Srebrenik, S. (1975). A misconception concerning the electronic density of an atom. Theoretica Chimica Acta 38: 159–163.
19 Angulo, J.C., Schmider, H., Sagar, R.P., and Smith, V.H. Jr., (1994). Nonconvexity of the atomic charge density and shell structure. Physical Review A 49: 726–728.
20 Sen, K.D. and Politzer, P. (1989). Characteristic features of the electrostatic potentials of singly-negative monatomic ions. Journal of Chemical Physics 90: 4370–4372.
21 Pathak, R.K. and Gadre, S.R. (1990). Maximal and minimal characteristics of molecular electrostatic potentials. Journal of Chemical Physics 93: 1770–1773.
22 Kollman, P. (1977). A general analysis of noncovalent intermolecular interactions. Journal of the American Chemical Society 99: 4875–4894.
23 Murray, J.S. and Politzer, P. (2009). Molecular surfaces, van der Waals radii and electrostatic potentials in relation to noncovalent interactions. Croatica Chimica Acta 82: 267–275.
24 Bonaccorsi, R., Scrocco, E., and Tomasi, J. (1970). Molecular SCF calculations for the ground state of some three-membered ring molecules. Journal of Chemical Physics 52: 5270–5284.
25 Sjoberg, P. and Politzer, P. (1990). The use of the electrostatic potential at the molecular surface to interpret and predict nucleophilic processes. Journal of Physical Chemistry 94: 3959–3961.
26 Sjoberg, P., Murray, J.S., Brinck, T. et al. (1990). The use of the electrostatic potential at the molecular surface in recognition interactions: Dibenzo-p-dioxins and related systems. Journal of Molecular Graphics 8: 81–85.

27 Kendrick, J. and Fox, M. (1991). Calculation and display of electrostatic potentials. Journal of Molecular Graphics 9: 182–187.

28 Bader, R.F.W., Carroll, M.T., Cheeseman, J.R., and Chang, C. (1987). Properties of atoms in molecules. Atomic volumes, Journal of the American Chemical Society 109: 7968–7979.

29 Murray, J.S., Brinck, T., Grice, M.E., and Politzer, P. (1992). Correlations between molecular electrostatic potentials and some experimentally-based indices of reactivity. Journal of Molecular Structure (Theochem) 256: 29–45.

30 Frisch, M.J., Trucks, G.W., Schlegel, H.B. et al. Gaussian 09, revision A.1, Gaussian, Inc., Wallingford. CT.

31 Bulat, F.A., Toro-Labbé, A., Brinck, T. et al. (2010). Quantitative analysis of molecular surface properties: areas, volumes, electrostatic potentials and average local ionization energies. Journal of Molecular Modeling 16: 1679–1691.

32 Soliva, R., Orozco, M., and Luque, F.J. (1997). Suitability of density functional methods for calculation of electrostatic properties. Journal of Computational Chemistry 18: 980–991.

33 Riley, K.E., Tran, K.-A., Lane, P. et al. (2016). Comparative analysis of electrostatic potential maxima and minima on molecular surfaces, as determined by three methods and a variety of basis sets. Journal of Computational Science 17: 273–284.

34 Scrocco, E. and Tomasi, J. (1973). The electrostatic molecular potential as a tool for the interpretation of molecular properties. Topics in Current Chemistry 42: 95–170.

35 Scrocco, E. and Tomasi, J. (1978). Electronic molecular structure, reactivity and intermolecular forces: an euristic interpretation by means of electrostatic molecular potentials. Advances in Quantum Chemistry 11: 115–193.

36 Tomasi, J., Mennucci, B., and Cammi, R. (1996). A tool for interpretation and prediction. From molecular structure to solvation effects. In: Molecular Electrostatic Potentials: Concepts and Applications (eds. J.S. Murray and K. Sen), 1–103. Amsterdam: Elsevier.

37 Hellmann, H. (1937). Einführung in die Quantenchemie. Leipzig: Deuticke.

38 Feynman, R.P. (1939). Forces in molecules. Physical Review 56: 340–343.

39 Brinck, T. (1998). The use of the electrostatic potential for analysis and prediction of intermolecular interactions. In: Theoretical Organic Chemistry (ed. C. Parkányi), 51–93. Amsterdam: Elsevier.

40 Francl, M.M. (1985). Polarization corrections to electrostatic potentials. Journal of Physical Chemistry 89: 428–433.

41 Alkorta, I., Villar, H.O., and Perez, J.J. (1993). Effect of the basis set on the computation of molecular polarization. Journal of Physical Chemistry 97: 9112–9119.

42 Bartlett, R.J. and Weinstein, H. (1975). Theoretical treatment of multiple site reactivity in large molecules. Chemical Physics Letters 30: 441–447.

43 van der Neut, R.N. (1975). Approximate calculations on a protonation step in the reduction of propadienes. Tetrahedron 31: 2547–2549.

44 Weinstein, H., Chou, D., Johnson, C.L. et al. (1976). Tautomerism and the receptor action of histamine: A mechanistic model. Molecular Pharmacology 12: 738–745.

45 Haymet, A.D.J. and Oxtoby, D.W. (1982). Calculation of polarization potentials and three-body induction effects from perturbation theory. Journal of Chemical Physics 77: 2466–2474.

46 Dive, G. and Dehareng, D. (1993). Polarization correction of the electrostatic potential for aromatic compounds: Study of the nucleophilic attack. International Journal of Quantum Chemistry 46: 127–136.

47 Orozco, M. and Luque, F.J. (1996). Generalization of the molecular electrostatic potential for the study of noncovalent interactions. In: Molecular Electrostatic Potentials: Concepts and Applications (eds. J.S. Murray and K. Sen), 181–218. Amsterdam: Elsevier.

48 Politzer, P. and Murray, J.S. (2013). Molecular electrostatic potentials: Some observations. In: Concepts and Methods in Modern Theoretical Chemistry (eds. S.K. Ghosh and P.K. Chattaraj), 181–199. Boca Raton, FL: CRC Press.

49 Politzer, P., Murray, J.S., Clark, T., and Resnati, G. (2017). The σ-hole revisited. Physical Chemistry, Chemical Physics 19: 32166–32178.

50 Ghio, C. and Tomasi, J. (1973). The protonation of three-membered ring molecules. The ab initio SCF versus the electrostatic picture of the proton approach, Theoretica Chimica Acta 30: 151–158.

51 Kollman, P. and Rothenberg, S. (1977). Theoretical studies of basicity. Proton affinities, Li^+ affinities, and H-bond affinities of some simple bases. Journal of the American Chemical Society 49: 1333–1342.

52 Politzer, P., Murray, J.S., and Concha, M.C. (2002). The complementary roles of molecular surface electrostatic potentials and average local ionization energies with respect to electrophilic processes. International Journal of Quantum Chemistry 88: 19–27.

53 Politzer, P., Murray, J.S. (2018) An Occam's Razor approach to chemical hardness: lex parsimoniae, Journal of Molecular Modeling, 24, 332(1–10), and papers cited.

54 Stone, A.J. and Price, S.L. (1988). Some new ideas in the theory of intermolecular forces: anisotropic atom-atom potentials. Journal of Physical Chemistry 92: 3325–3335.

55 Bachrach, S.M. (1994) Population analysis and electron densities from quantum mechanics, in Reviews in Computational Chemistry, vol. 5 (eds. K.B. Lipkowitz and D.B. Boyd), VCH Publishers, New York, pp. 171–227.

56 Price, S.L. (1996). Applications of realistic electrostatic modelling to molecules in complexes, solids and proteins. Journal of the Chemical Society, Faraday Transactions 92: 2997–3008.

57 Wiberg, K.B. and Rablen, P.R. (1993). Comparison of atomic charges derived via different procedures. Journal of Computational Chemistry 14: 1504–1518.

58 Bader, R.F.W. (1990). Atoms in Molecules: A Quantum Theory. Oxford: Oxford University Press.

59 Bader, R.F.W. (1991). A quantum theory of molecular structure and its applications. Chemical Reviews 91: 893–928.

60 Perrin, C.L. (1991). Atomic size dependence of Bader electron populations: significance for questions of resonance stabilization. Journal of the American Chemical Society 113: 2865–2868.

61 Saethre, L.J., Siggel, M.R.F., and Thomas, T.D. (1991). Molecular charge distribution, core-ionization energies, and the point-charge approximation. Journal of the American Chemical Society 113: 5224–5230.

62 De Proft, F., Van Alsenoy, C., Peeters, A. et al. (2002). Atomic charges, dipole moments, and Fukui functions using the Hirshfeld partitioning of the electron density. Journal of Computational Chemistry 23: 1198–1209.

63 Guerra, C.F., Handgraaf, J.W., Baerends, E.J., and Bickelhaupt, F. (2004). Voronoi deformation density (VDD) charges. Assessment of the Mulliken, Bader, Hirshfeld, Weinhold and VDD methods for charge analysis, Journal of Computational Chemistry 25: 189–210.

64 Brom, J.M., Schmitz, B.J., Thompson, J.D. et al. (2003). A Class IV charge model for boron based on hybrid density functional theory. Journal of Physical Chemistry A 107: 6483–6488.

65 Nyburg, S.C. and Faerman, C.H. (1985). A revision of van der Waals atomic radii for molecular crystals: N. O, F, S, Cl, Se, Br and I bonded to carbon, Acta Crystallographica B 41: 274–279.

66 Ikuta, S. (1990). Anisotropy of electron density distribution around atoms in molecules: N, P, O and S atoms. Journal of Molecular Structure (Theochem) 205: 191–201.

67 Torii, H. (2003). The role of atomic quadrupoles in intermolecular electrostatic interactions of polar and nonpolar molecules. Journal of Chemical Physics 119: 2192–2198.

68 Politzer, P., Murray, J.S., and Clark, T. (2013). Halogen bonding and other σ-hole interactions: A perspective. Physical Chemistry Chemical Physics 15: 11178–11189.

69 Bauzá, A., Mooibroek, T.J., and Frontera, A. (2015). The bright future of unconventional σ/π-hole interactions. ChemPhysChem 16: 2496–2517.

70 Kolář, M.H. and Hobza, P. (2016). Computer modeling of halogen bonds and other σ-hole interactions. Chemical Reviews 116: 5155–5187.

71 Politzer, P., Murray, J.S. (2019) An overview of strengths and directionalities of noncovalent interactions: σ-Holes and π-holes, Crystals, 9, 165(1–15).

72 Brinck, T., Murray, J.S., and Politzer, P. (1992). Surface electrostatic potentials of halogenated methanes as indicators of directional intermolecular interactions. International Journal of Quantum Chemistry 44 (Suppl. 19): 57–64.

73 Murray, J.S., Lane, P., Clark, T. et al. (2012). σ-Holes, π-holes and electrostatically-driven interactions. Journal of Molecular Modeling 18: 541–548.

74 Rosenfeld, R.E. Jr., Parthasarathy, R., and Dunitz, J.D. (1977). Directional preferences of nonbonded atomic contacts with divalent sulfur. 1. Electrophiles and nucleophiles, Journal of the American Chemical Society 99: 4860–4862.

75 Guru Row, T.N. and Parthasarathy, R. (1981). Directional preferences of nonbonded atomic contacts with divalent sulfur in terms of its orbital orientations. 2. S---S interactions and nonspherical shape of sulfur in crystals, Journal of the American Chemical Society 103: 477–479.

76 Ramasubbu, N., Parthasarathy, R., and Murray-Rust, P. (1986). Angular preferences of intermolecular forces around halogen centers: Preferred directions of approach of electrophiles and nucleophiles around carbon-halogen bond. Journal of the American Chemical Society 108: 4308–4314.

77 Ramasubbu, N. and Parthasarathy, R. (1987). Stereochemistry of incipient electrophilic and nucleophilic reactions at divalent selenium center: Electrophile-nucleophile pairing and anisotropic shape of Se in Se---Se interactions. Phosphorus Sulfur 31: 221–229.

78 Politzer, P., Murray, J.S., and Concha, M.C. (2008). σ-Hole bonding between like atoms. A fallacy of atomic charges, Journal of Molecular Modeling 14: 659–665.

79 Politzer, P., Murray, J.S., and Clark, T. (2010). Halogen bonding: An electrostatically-driven highly directional noncovalent interaction. Physical Chemistry Chemical Physics 12: 7748–7757.

80 Kollman, P., McKelvey, J., Johansson, A., and Rothenberg, S. (1975). Theoretical studies of hydrogen-bonded dimers. Journal of the American Chemical Society 97: 955–965.

81 Morokuma, K. (1977). Why do molecules interact? The origins of electron donor-acceptor complexes, hydrogen bonding and proton affinity, Accounts of Chemical Research 10: 294–300.

82 Kenny, P.W. (1994). Prediction of hydrogen bond basicity from computed molecular electrostatic properties: Implications for comparative molecular field analysis. Journal of the Chemical Society, Perkin Transactions 2: 199–202.

83 Hunter, C.A. (2004). Quantifying intermolecular interactions: Guidelines for the molecular recognition toolbox. Angewandte Chemie International Edition 43: 5310–5324.

84 Shields, Z.P., Murray, J.S., and Politzer, P. (2010). Directional tendencies of halogen and hydrogen bonds. International Journal of Quantum Chemistry 110: 2823–2832.

85 Aakeröy, C.B., Wijethunga, T.K., and Desper, J. (2015). Molecular electrostatic potential dependent selectivity of hydrogen bonding. New Journal of Chemistry 39: 822–828.

86 Sandhu, B., McLean, A., Sinha, A.S. et al. (2018). Crystal Growth & Design 18: 466–478.

87 Politzer, P., Murray, J.S., and Clark, T. (2015). σ-Hole bonding: A physical interpretation. Topics in Current Chemistry 358: 19–42.

88 Legon, A.C. (1999). Prereactive complexes of dihalogens XY with Lewis bases B in the gas phase: A systematic case for the halogen analogue B---XY of the hydrogen bond B---HX. Angewandte Chemie International Edition 38: 2686–2714.

89 Hagelin, H., Brinck, T., Berthelot, M. et al. (1995). Family-independent relationships between computed molecular surface quantities and solute hydrogen bond acidity/basicity and solute-induced methanol O-H infrared frequency shifts. Canadian Journal of Chemistry 73: 483–488.

90 Murray, J.S. and Politzer, P. (2019). Hydrogen bonding: A Coulombic σ-hole interaction. Journal of the Indian Institute of Science 100: 21–30.

91 Wang, H., Wang, W., and Jin, W.H. (2016). σ-Hole bond vs π-hole bond: A comparison based on halogen bond. Chemical Reviews 116: 5072–5104.

92 Politzer, P., Murray, J.S. (2019) Electrostatics and polarization in σ- and π-hole noncovalent interactions: An overview, ChemPhysChem, DOI: 10.1002/cphc.201900968R1, and papers cited.

93 Ding, X., Tuikka, M., Haukka, M. (2012) Halogen bonding in crystal engineering, in Recent Advances in Crystallography (ed. J.B. Benedict), IntechOpen, DOI: https://doi.org/10.5772/48592.

94 Thirman, J., Engelage, E., Huber, S.M., and Head-Gordon, M. (2018). Characterizing the interplay of Pauli repulsion, electrostatics, dispersion and charge transfer in halogen bonding with energy decomposition analysis. Physical Chemistry Chemical Physics 20: 905–915.

95 Clark, T. and Hesselmann, A. (2018). The Coulomb σ-hole model describes bonding in CX_3I---Y^- complexes completely. Physical Chemistry Chemical Physics 20: 22849–22855.

96 Clark, T., Murray, J.S., and Politzer, P. (2018). The σ-hole Coulombic interpretation of trihalide anion formation. ChemPhysChem 19: 3044–3049.

97 Brinck, T., Borrfors, A.N. (2019) Electrostatics and polarization determine the strength of the halogen bond: A red card for charge transfer, Journal of Molecular Modeling, 25, 125(1–9).

98 Politzer, P., Murray, J.S., and Clark, T. (2019). Explicit inclusion of polarizing electric fields in σ- and π-hole interactions. Journal of Physical Chemistry A 123: 10123–10130.

99 Murray, J.S., Politzer, P. (2020) Interaction and polarization energy relationships in σ-hole and π-hole bonding, Crystals, 10, 76(1–18).

100 Hirschfelder, J.O., Curtiss, C.F., and Bird, R.B. (1954). Molecular Theory of Gases and Liquids, 852. New York: John Wiley & Sons.

101 Politzer, P., Riley, K.E., Bulat, F.A., and Murray, J.S. (2012). Perspectives on halogen bonding and other σ-hole interactions: lex parsimoniae (Occam's Razor). Computational & Theoretical Chemistry 998: 2–8.

102 Solimannejad, M., Malekani, M., and Alkorta, I. (2010). Cooperative and diminutive unusual weak bonding in F_3CX---HM_gH---Y and F_3CX---Y---HM_gH trimers (X = Cl, Br; Y = HCN and HNC). Journal of Physical Chemistry A 114: 12106–12111.

103 Clark, T. (2017). Polarization, donor-acceptor interactions, and covalent contributions in weak interactions: A clarification. Journal of Molecular Modeling 23: 297.

104 Pauling, L. (1960). The Nature of the Chemical Bond, 3rde. Ithaca, NY: Cornell University Press.

105 Bondi, A. (1964). van der Waals volumes and radii. Journal of Physical Chemistry 68: 441–451.

106 Dance, I. (2003). Distance criteria for crystal packing analysis of supramolecular motifs. New Journal of Chemistry 27: 22–27.

107 Alvarez, S. (2013). A cartography of the van der Waals territories. Dalton Transactions 42: 8617–8636.

108 Murray, J.S., Resnati, G., and Politzer, P. (2017). Close contacts and noncovalent interactions in crystals. Faraday Discussions 203: 113–130.

109 Näther, C., Döring, C., Jess, I. et al. (2013). Thermodynamic and structural relationships between the two polymorphs of 1,3-dimethylurea. Acta Crystallographica B 69: 70–76.

110 Politzer, P. and Murray, J.S. (2015). Quantitative analyses of molecular surface electrostatic potentials in relation to hydrogen bonding and co-crystallization. Crystal Growth & Design 15: 3767–3774.

111 Nayak, S.K., Kumar, V., Murray, J.S. et al. (2017). Fluorination promotes chalcogen bonding in crystalline solids. CrystEngComm 19: 4955–4959.

112 Politzer, P. and Murray, J.S. (2018). σ-Holes and π-holes: Similarities and differences. Journal of Computational Chemistry 39: 464–471.

113 Bent, H.A. (1968). Structural chemistry of donor-acceptor interactions. Chemical Reviews 68: 587–648.

114 Paulini, R., Müller, K., and Diederich, F. (2005). Orthogonal multipolar interactions in structural chemistry and biology. Angewandte Chemie International Edition 44: 1788–1805.

115 Naray-Szabó, G. (1979). Electrostatic isopotential maps for large biomolecules. International Journal of Quantum Chemistry 16: 265–272.

116 Weinstein, H., Osman, R., Topiol, S., and Green, J.P. (1981). Quantum chemical studies on molecular determinants for drug action. Annals of the New York Academy of Sciences 367: 434–451.

117 Weiner, P.K., Langridge, R., Blaney, J.M., Schaefer, R., Kollman, P.A. (1982) Electrostatic potential molecular surfaces, Proceedings of the National Academy of Sciences U.S.A., 79, 3754–3758.

118 Martin, M., Sanz, F., Campillo, M. et al. (1983). Quantum chemical study of the molecular patterns of MAO inhibitors and substrates. International Journal of Quantum Chemistry 23: 1627–1641.

119 Politzer, P., Laurence, P.R., and Jayasuriya, K. (1985). Molecular electrostatic potentials: An effective tool for the elucidation of biochemical phenomena. Environmental Health Perspectives 61: 191–202.

120 Beveridge, D.L. and Lavery, R. (eds.) (1991). Theoretical Biochemistry & Molecular Biophysics, Adenine Press. NY: Schenectady.

121 Platt, D.E. and Silverman, D. (1996). Registration, orientation and similarity of molecular electrostatic potentials through multipole matching. Journal of Computational Chemistry 17: 358–366.

122 Politzer, P., Murray, J.S., and Peralta-Inga, Z. (2001). Molecular surface electrostatic potentials in relation to noncovalent interactions in biological systems. International Journal of Quantum Chemistry 85: 676–684.

123 Poland, A. and Knutson, J.C. (1982). 2,3,7,8-Tetrachlorodibenzo-p-dioxin and related halogenated aromatic hydrocarbons: Examination of the mechanism of toxicity. Annual Reviews of Pharmacology and Toxicology 22: 517–554.

124 Murray, J.S. and Politzer, P. (1987). Electrostatic potentials of some dibenzo-p-dioxins in relation to their biological activities. Theoretica Chimica Acta 72: 507–517.

125 McKinney, J.D., Darden, T., Lyerly, J.A., and Pedersen, L.G. (1985). Dioxin and related compound binding to the Ah receptor(s). Theoretical model based on molecular parameters and molecular mechanics, Quantitative Structure-Activity Relationships 4: 166–172.

126 Murray, J.S. and Politzer (1998). Statistical analysis of the molecular surface electrostatic potential: An approach to describing noncovalent interactions in condensed phases. Journal of Molecular Structure (Theochem) 425: 107–114.

127 Politzer, P. and Murray, J.S. (2001). Computational prediction of condensed phase properties from statistical characterization of molecular surface electrostatic potentials. Fluid Phase Equilibria 185: 129–137.

128 Güttinger, P. (1932). Das Verhalten von Atomen in magnetischen Drehfeld. Zeitschrift für Physik 73: 169–184.

129 Hellmann, H. (1933). Zur Rolle der kinetischen Electronenenergie für die zwischenatomaren Kräfte. Zeitschrift für Physik 85: 180–190.

130 Foldy, L.L. (1951). A note on atomic binding energies. Physical Review 83: 397–399.

131 Politzer, P. and Parr, R.G. (1974). Some new energy formulas for atoms and molecules. J. Chem. Phys. 61: 4258–4262.

132 Politzer, P. (2004). Atomic and molecular energies as functionals of the electrostatic potential. Theoretical Chemical Accounts 111: 395–399.

133 Politzer, P. (1980). Observations on the significance of the electrostatic potentials at the nuclei of atoms and molecules. Israel Journal of Chemistry 19: 224–232.

134 Politzer, P. (1981). Relationships between the energies of atoms and molecules and the electrostatic potentials at their nuclei. In: Chemical Applications of Atomic and Molecular Electrostatic Potentials (eds. P. Politzer and D.G. Truhlar), 7–28. New York: Plenum Press.

135 Levy, M., Clement, S.C., and Tal, Y. (1981). Correlation energies from Hartree-Fock electrostatic potentials at nuclei and generation of electrostatic potentials from asymptotic and zero-order information. In: Chemical Applications of Atomic and Molecular Electrostatic Potentials (eds. P. Politzer and D.G. Truhlar), 29–50. New York: Plenum Press.

136 Politzer, P. (1987). Atomic and molecular energy and energy difference formulas based upon electrostatic potentials at nuclei. In: The Single-Particle Density in Physics and Chemistry (eds. N.H. March and B.M. Deb), 59–72. San Diego: Academic Press.

137 Politzer, P. and Murray, J.S. (2002). The fundamental nature and role of the electrostatic potential in atoms and molecules. Theoretical Chemical Accounts 108: 134–142.

138 Levy, M., Tal, Y. (1980) Atomic binding energies from fundamental theorems involving the electron density, $<r^{-1}>$, and the Z-1 perturbation expansion, Journal of Chemical Physics, 72, 3416–3417.

139 Levy, M. and Perdew, J.P. (1986). Success of quantum mechanical approximations for molecular geometries and electron-nuclear attraction expectation values: Gift of the Coulomb potential? Journal of Chemical Physics 84: 4519–4523.

140 Politzer, P. and Sjoberg, P. (1983). A formula for calculating molecular energy differences from electrostatic potentials at nuclei. Journal of Chemical Physics 78: 7008–7010.

141 Wiener, J.J.M., Grice, M.E., Murray, J.S., and Politzer, P. (1996). Molecular electrostatic potentials as indicators of covalent radii. Journal of Chemical Physics 104: 5109–5112.

142 Politzer, P., Murray, J.S., and Lane, P. (2003). Electrostatic potentials and covalent radii. Journal of Computational Chemistry 24: 505–511.

8

Chemical Reactivity Within the Spin-Polarized Framework of Density Functional Theory

E. Chamorro[1*] *and P. Pérez*[2]

Universidad Andres Bello, Facultad de Ciencias Exactas. Departamento de Ciencias Químicas, Avenida República 275, 8370146 Santiago, Chile

8.1 Introduction

Density functional theory (DFT) [1, 2] has proved to be an effective and fruitful framework for the detailed exploration of useful *concepts* and *reactivity principles* in chemistry [3–11]. Such a long-standing and continuously evolving effort [12–41] contributes to the aim of developing a "nonempirical, mathematically and physically sound, density-based, quantum-mechanical theory for interpreting and predicting chemical phenomena, especially chemical reactions" [42]. Such a framework of research, i.e., conceptual DFT (CDFT), enables the establishment of critical definitions and the implementation of computational approaches for quantities that help us to rationalize global stability, local selectivity, and non-local activation/deactivation relative responses of chemical species [3–11]. This perturbative approximation to reactivity has continued to develop successfully over the years [43–46], broadening the applications in several areas with significant impact on our understanding of the chemical phenomena [30, 36, 47–52]. Very recently, a brief perspective of the current status and prospects of CDFT have been presented [42].

The DFT framework offers a formal mathematical structure for the interpretation/prediction of experimental/theoretical *chemical reactivity* patterns on the basis of a series of responses of state functions (e.g., the electronic energy E for instance) to changes or perturbations in basic ground-state variables (e.g., the number of electrons N or the external potential, $v(\mathbf{r})$ [53–57]. These descriptors conform full hierarchies defined in terms of Taylor series expansions of the energy functional within Legendre-transformed ensembles representations (e.g., canonical, grand-canonical, isomorphic, and grand-isomorphic) and formulation of DFT [3, 43, 44, 58–61]. Within this perspective, global, local, and nonlocal descriptors are identified to the coefficients of the functional perturbative expansions [7, 53, 62, 63]. Those coefficients describing global responses against global perturbations, such as the electronic chemical potential [3, 64, 65] $\mu = [\partial E/\partial N]_{v(\mathbf{r})}$ (i.e., identified with the negative of electronegativity [66]), the chemical hardness, $\eta = [\partial \mu/\partial N]_{v(\mathbf{r})}$ [67, 68], and the softness $S = [\partial N/\partial \mu]_{v(\mathbf{r})}$ [69], are quantities characterizing the entire system as an entity. These indices are naturally associated to primary global energetic differences. Local descriptors (i.e., dependent on the $\mathbf{r}$ – spatial coordinate within the molecular framework) are associated to global or local responses of the system against local or global

[1] E-mail: echamorro@unab.cl; ORCID: https://orcid.org/0000-0002-9200-9859
[2] E-mail: p.perez@unab.cl; ORCID: https://orcid.org/0000-0002-6920-703X

Chemical Reactivity in Confined Systems: Theory, Modelling and Applications,
First Edition. Edited by Pratim Kumar Chattaraj and Debdutta Chakraborty.

perturbations, respectively. The electron density itself $\rho(\mathbf{r}) = [\partial E/\partial v(\mathbf{r})]_N$ [3], the Fukui function $f(\mathbf{r}) = [\partial \rho(\mathbf{r})/\partial \mathrm{N}]_{v(\mathbf{r})} = [\delta \mu/\delta v(\mathbf{r})]_N$ [70–73], and the local softness $s(\mathbf{r}) = [\partial \rho(\mathbf{r})/\partial \mu]_{v(\mathbf{r})}$ [70–72], are well know examples of local descriptors [74, 75]. These quantities are therefore suitable to describe the molecular selectivity, providing key information about the relative reactivity of different sites in a molecule, and playing a central role to make predictions about regioselectivity and kinetics. Both global and local electrophilicities, i.e., $\omega = \mu^2/2\eta$ and $\omega(\mathbf{r}) = \omega f(\mathbf{r})$ respectively [9, 76–78], are also key quantities entering the arsenal of reactivity tools, helping us to describe electronic reactivity in diverse chemical problems [79–81]. Dual descriptors [82] and reaction indicators [39, 83, 84] have been also proposed on the basis of CDFT. On the other hand, local responses to local perturbations correspond to the so-called non-local electronic responses [43, 85–88] (i.e., quantities depending on two or more spatial positions, $\mathbf{r}, \mathbf{r}'$,.. etc.). These, such as for instance the linear responses associated with density [30, 89] and Fukui function, $\chi(\mathbf{r}, \mathbf{r}_1) = [\delta \rho(\mathbf{r})/\delta v(\mathbf{r}')]_N$ and $f(\mathbf{r}, \mathbf{r}_1) = [\delta f(\mathbf{r})/\delta v(\mathbf{r}')]_N$, respectively, or the softness $s(\mathbf{r}, \mathbf{r}_1) = [\delta \rho(\mathbf{r})/\delta v(\mathbf{r}')]_\mu$ kernel [90], either measure a molecular polarization with respect to its environment or the change in polarization associated with density reorganization in electron transfer processes [45, 91, 92]. Conceptual and computational advances [16, 19, 27, 74, 93, 94] have been extensively discussed in this field in connection the electron descriptors [95–101]. Concepts founded in the DFT framework have, for instance, been found particularly successful in the analysis and rationalization of the electronegativity equalization principle (EEM) [102–106], the maximum hardness principle (MHP) [16, 23, 107–110], the hard and soft acids and bases principle (HSAB) [102, 109, 111–114], electrophilicity/nucleophilicity [21, 76, 91, 115, 116], frontier molecular orbital theory [3, 5, 8, 117–122], the quality of leaving groups [123–128], redox potentials [129–132], extensions to finite-temperature [21, 75, 114, 133], inclusion of relativistic effects [40], and the characterization of polar interactions driving the nucleophilic-electrophilic interplay in substitution/elimination and even pericyclic reactions of organic chemistry [80, 134–152]. It has, indeed, been argued that electron transfer and electrostatics play a key role within the aim of gaining a deeper understanding of chemical processes. All the above-mentioned global, local, and non-local DFT descriptors are related to the intrinsic electron density susceptibilities of a given electronic system at some specific nuclear configuration. The so-called nuclear responses have been explored through explicit considerations of changes in the external potential, i.e., $v(\mathbf{r}) \equiv \sum_\delta Z_\delta/|\mathbf{r} - \mathbf{R}_\delta|$ [7], or equivalently in terms of the origin of such changes, i.e., within a Born-Oppenheimer approximation, from the analysis of forces on nuclei $\mathbf{F}_\delta = Z_\delta[\int \rho(\mathbf{r})(\mathbf{r} - \mathbf{R}_\delta)/|\mathbf{r} - \mathbf{R}_\delta|^3 d\mathbf{r} - \sum_{\varsigma \neq \delta} Z_\varepsilon(\mathbf{R}_\delta - \mathbf{R}_\varsigma)/|\mathbf{R}_\delta - \mathbf{R}_\varsigma|^3]$, where Z_δ and $\mathbf{R}_\delta$ represent charge and position of nucleus δ, respectively [95–101, 153–156]. Such a framework of electronic and nuclear reactivity hierarchies illustrates the evident success of concepts developed within a spin-independent DFT framework.

It should, however, be emphasized [45, 91, 92, 129, 157–175] that the reactivity tools that were derived to model *electron transfer* are *not ideal* when both electron and spin (e.g., spin polarization) *transfers* are also important factors to be considered in the treatment of radical reactivity [176, 177]. It was just the need for a general framework [178–180] for discussing chemical reactivity including both *electron transfer* and *spin polarization* that motivated the development of the conceptual spin-dependent limit of the density functional theory (SP-DFT) [181–185]. As illustrated by the different interests devoted to the development and application of predictive tools based on DFT [38, 45, 91, 92, 150, 158–162, 173, 176, 177, 185–187, 189–191], main goals within such a field strive towards the further formal mathematical exploration and practical implementation of both electronic and nuclear reactivity indices and fundamental chemical principles at different levels of perturbation theory within a SP-DFT framework [38, 45, 91, 92, 150, 158–162, 184, 185, 187, 188, 190, 192]. As a challenging example, *free-radical* chemical reactivity remains an attractive open subject, still very elusive [193–195] to be fully rationalized by models developed within the conventional framework of DFT approaches [137, 138, 161, 196–205]. Changes in electron spin are usually coupled to electron transfer in most radical reactions, although in some rearrangements and termination processes the extent of such interplay is expected to be minimal. Furthermore, being free radical energetically unstable species, entropic factors are expected to influence the observed reactivity. As is well known [202–205], typical steps in complex radical mechanisms include ionization or excitation into a diradical state, transferring of electrons of a specific spin, molecular rearrangements of functional groups, and/or combinations of radicals to form closed-shell species. The generalized SP-DFT framework properly describe spin-dependent reactivity, as involved in free radical chemistry [45, 91, 92, 158–162, 184, 185]. It is also customary, however, to emphasize that that, by construction [183], our intended developments and applications are devised within a nonrelativistic spin-polarized limit of DFT [206]. Hence, the *spin* characteristics become considered only after the introduction of *spin orbitals*, and they are not explicitly derived from a proper relativistic treatment to the electronic structure problem. This clarification is important in order to limit the scope and applicability of the reactivity descriptors in the particular perturbative approximation to chemical reactivity on which we are focusing in this presentation [184].

8.2 The Spin-Polarized Density Functional Theory as a Suitable Framework to Describe Both Charge and Spin Transfer Processes

Let first start by recalling some formal aspects of the nonrelativistic spin-polarized DFT framework [183, 185, 207] within a canonical representation of an N-electron system in the field created by M nuclei at positions $\{\mathbf{R}_\delta\}$, i.e., $v(\mathbf{r}) = \sum_\delta^M Z_\delta/|\mathbf{r} - \mathbf{R}_\delta|$ and in the presence of an external magnetic field $\mathbf{B}(\mathbf{r})$, the electronic energy E can be expressed as

$$E_{V,\mathbf{B}}[\rho, \mathbf{m}] = F[\rho, \mathbf{m}] + \int \rho(\mathbf{r})v(\mathbf{r})\,dr - \int \mathbf{B}(\mathbf{r}) \cdot \mathbf{m}(\mathbf{r})\,dr \tag{8.1}$$

where $F[\rho, \mathbf{m}]$ is a universal functional only dependent of the electron density $\rho(\mathbf{r})$ and of the magnetization $\mathbf{m}(\mathbf{r})$ [183, 208–211]. $F[\rho, \mathbf{m}]$ refers specifically to the sum of the total kinetic electronic energy $T_e[\rho, \mathbf{m}]$ and the total electron-electron repulsion $V_{ee}[\rho, \mathbf{m}]$ [212–216]. In practical applications of SP-DFT, the magnetic field is either absent or constrained to be constant in the analysis of radical chemical reactions. It is clear that for a zero external magnetic field, the formalism of Eq. (8.1) is still valid for ground states having open shell electronic structures within a Kohn-Sham formulation of DFT that explicitly introduces spin orbitals. For the case of magnetic field $\mathbf{B}(\mathbf{r})$ and magnetization $\mathbf{m}(\mathbf{r})$ entirely collinear with a chosen spatial direction (e.g., z-component for instance), we can write that [183, 208–211],

$$m(\mathbf{r}) = -\mu_B \rho_S(\mathbf{r}), \quad \text{and} \quad \mathbf{B}(\mathbf{r}) = \mu_B^{-1} v_S(\mathbf{r}) \mathbf{1}_Z \tag{8.2}$$

where μ_B is the electron Bohr magneton, $\rho_S(\mathbf{r})$ is the spin density, and $v_S(\mathbf{r})$ is the external potential associated to the magnetic field directed along the direction of the unit vector. In this case, changes in the magnetic field $\mathbf{B}(\mathbf{r})$ become equivalent to changes in the specific potential $v_S(\mathbf{r})$. Hereafter, by introducing spin-up, $\rho_\alpha(\mathbf{r})$, and spin-down, $\rho_\beta(\mathbf{r})$, electron densities, and their associated *external* potentials, $v_\alpha(\mathbf{r})$ and $v_\beta(\mathbf{r})$, we assume for electron and spin densities and external potentials that,

$$\rho(\mathbf{r}) \equiv \rho_\alpha(\mathbf{r}) + \rho_\beta(\mathbf{r}), \quad \rho_S(\mathbf{r}) \equiv \rho_\alpha(\mathbf{r}) - \rho_\beta(\mathbf{r})$$

and

$$v(\mathbf{r}) \equiv v_\alpha(\mathbf{r}) + v_\beta(\mathbf{r}), \quad v_S(\mathbf{r}) \equiv v_\alpha(\mathbf{r}) - v_\beta(\mathbf{r}) \tag{8.3}$$

Normally, the external potentials for the $\alpha-$ or $\beta-$ spin electrons will be the same. This is not true, however, in the presence of a magnetic field [184, 206]. The simultaneous minimization of the energy functional in Eq. (8.1) with respect to $\rho(\mathbf{r})$ and $\rho_S(\mathbf{r})$ under the normalization constraints $\int \rho(\mathbf{r}) = N$ and $\int \rho_S(\mathbf{r}) = N_S$; or equivalently, with respect to $\rho_\alpha(\mathbf{r})$ and $\rho_\beta(\mathbf{r})$, subject to $\int \rho_\alpha(\mathbf{r}) = N_\alpha$ and $\int \rho_\beta(\mathbf{r}) = N_\beta$, yields the fundamental equations,

$$\mu_N = \left[\frac{\delta F}{\delta \rho}\right]_{\rho_S} + v(\mathbf{r}), \text{ and } \mu_S = \left[\frac{\delta F}{\delta \rho_S}\right]_{\rho} + v_S(\mathbf{r})$$

or

$$\mu_\alpha = \left[\frac{\delta F}{\delta \rho_\alpha}\right]_{\rho_\beta} + v_\alpha(\mathbf{r}), \text{ and } \mu_\beta = \left[\frac{\delta F}{\delta \rho_\beta}\right]_{\rho_\alpha} + v_\beta(\mathbf{r}) \tag{8.4}$$

that concurrently solved provide the ground state associated to the chosen values of N and spin multiplicity. In Kohn–Sham DFT, the M_S quantum number is chosen to have the maximum possible value, being the spin multiplicity of the system given by $|N_S| + 1|$ [184]. Hence, it becomes clear that the SP-DFT formalism can be chemically ("conceptually") explored using the $[\rho, \rho_S]$ or $[N, N_S]$ and $[\rho_\alpha, \rho_\beta]$ or $[N_\alpha, N_\beta]$ representations [45, 91, 92, 158–162, 185], or still more general, any of their associated Legendre-transformed resolutions [183, 185, 207].

Note that in SP-DFT the mapping between spin-densities and spin-potentials is not unique [208, 209, 212–216]. However, $F[\rho_\alpha, \rho_\beta]$ can still be properly defined and the Hohenberg-Kohn theorem holds for spin densities [63, 206, 217]. Even though the raise of renewed attention on *spin-resolved reactivity indicators*, almost all research has focused until now on the $[N, N_S]$ representation, instead of alternative representations as the $[N_\alpha, N_\beta]$ one [184]. The $[N, N_S]$ representation seems properly suited for cases where there is spin transfer without electron transfer or vice versa, as the case of molecular excitation/deexcitation between the lowest-energy states of different multiplicities [45, 91, 92, 129, 158–165, 167–169]. Studies on chemical processes related to *excitation to the lowest-energy state of a specified multiplicity* [91, 116, 157, 159, 218–220] *as well as in termination processes by radical recombination* [161, 168] shows the $[N, N_S]$ representation to be highly appealing in these cases. However, it should be emphasized [184] that in the $[N, N_S]$ representation, coupling between electron transfer and spin transfer become only incorporated through higher-order terms, most of them, difficult to be evaluated in practice [183, 185, 207].

The first general treatment of chemical reactivity comprising *global* and *local* (i.e., $\mathbf{r}-$ dependent) electronic chemical reactivity descriptors within SP-DFT were introduced by Galván and co-workers within a $[N, N_S]$ context [221], and by Ghanty and Ghosh within a $[N_\alpha, N_\beta]$ representation [222]. The Lagrange multipliers μ_N and μ_S of Eq. (8.4) were identified as the *constrained chemical* and *spin potentials*, respectively [223], with μ_N being analogous to the chemical potential of the spin-free DFT theory except that it should be also evaluated at constant spin number, and μ_S is related to the ability of such system to undergo spin polarization, at constant number of electrons, $\mu_N = [\partial E/\partial N]_{N_S, v(\mathbf{r}), v_S(\mathbf{r})}$, $\mu_S = [\partial E/\partial N_S]_{N, v(\mathbf{r}), v_S(\mathbf{r})}$ [181, 183, 185, 207]. These indices describe responses of the entire system against global perturbations. Other global quantities in such a representation, i.e., the constrained hardness $\eta_{NN} = [\partial \mu_N/\partial N]_{N_S, v(\mathbf{r}), v_S(\mathbf{r})}$, $\eta_{NS} = [\partial \mu_N/\partial N_S]_{N, v(\mathbf{r}), v_S(\mathbf{r})} \equiv \eta_{SN} = [\partial \mu_S/\partial N]_{N_S, v(\mathbf{r}), v_S(\mathbf{r})}$ and $\eta_{SS} = [\partial \mu_S/\partial N_S]_{N, v(\mathbf{r}), v_S(\mathbf{r})}$, were analyzed in terms of changes in μ_N and μ_S, with respect to variations in the number of electrons N, at constant N_S (i.e., constrained charge transfer), or in the spin number N_S at constant N (i.e., spin-polarization processes). Local generalized $[N, N_S]$ SP-DFT Fukui functions [181, 183, 185], $f_{NN}(\mathbf{r}) = [\partial \rho(\mathbf{r})/\partial N]_{N_S, v(\mathbf{r}), v_S(\mathbf{r})}$, $f_{NS}(\mathbf{r}) = [\partial \rho(\mathbf{r})/\partial N_S]_{N, v(\mathbf{r}), v_S(\mathbf{r})}$, $f_{SN}(\mathbf{r}) = [\partial \rho_S(\mathbf{r})/\partial N]_{N_S, v(\mathbf{r}), v_S(\mathbf{r})}$, and $f_{SS}(\mathbf{r}) = [\partial \rho_S(\mathbf{r})/\partial N_S]_{N, v(\mathbf{r}), v_S(\mathbf{r})}$, were identified as the derivatives of the generalized chemical μ_N

and μ_S potentials with respect to variations of $v(\mathbf{r})$ and $v_S(\mathbf{r})$ at N and/or N_s held constant. Maxwell relations in such representation give a direct connection of these responses with variations of electron density $\rho(\mathbf{r})$ and spin density $\rho_S(\mathbf{r})$ against changes in N and/or N_S. Hence, $f_{NN}(\mathbf{r})$ and $f_{SN}(\mathbf{r})$ provide responses in density and spin density, respectively, to *constrained* charge transfer occurring at constant N_S; while $f_{NS}(\mathbf{r})$ and $f_{SS}(\mathbf{r})$ correspond to density and spin density responses, respectively, against perturbations in the spin number, at constant, N [92]. The local regioselectivity can be explicitly investigated in terms of these reactivity descriptors in a SP-DFT framework [181, 183, 185]. Evaluation of the simplest condensed-to-atom SP-DFT representation of Fukui functions [92] is straightforwardly available from electronic structure calculations performed via commercial [224, 225] and/or shareware [226] software.

The usual methodological framework is based on the so-called "perturbative approximation to chemical reactivity" [6, 7, 46, 53, 62], which is explored within a conceptual spin-polarized density functional theory framework (SP-DFT). The global and local electronic SP-DFT hierarchies constitute the basic ingredients to discuss reactivity in processes that involve both *charge transfer* and/or *spin polarization* (i.e., *spin transfer*). In this context, a *matrix-vector notation* which enables the derivation of key identities linking the two basic $[N, N_S]$ and $[N_\alpha, N_\beta]$ SP-DFT representations has been discussed [183, 185, 207]. Furthermore, in terms of such notation and simple transformation rules *all* of the identities in conceptual DFT exhibit *the same* mathematical structure. Given the state functions representing the electronic energy E, the grand potential Ω, and the universal functional F, i.e., $\Omega = E - \vec{\mu} \cdot \vec{N}$, and $F = E - \int \overleftrightarrow{\rho}(\mathbf{r}) \cdot \vec{v}(\mathbf{r})d\mathbf{r}$ of a given system, Taylor series expansions are used to represent the way the ground state function responds to changes in proper variables defining a closed-system (i.e. E is the state function), an open-system (Ω is the state function), and/or a density (e.g. F is the state function) representation, respectively. Namely, we write for the sensitivity of a system to a perturbation that [184],

$$
\begin{aligned}
\Delta E &= \vec{\mu} \cdot \Delta\vec{N} + \int \vec{\rho}(\mathbf{r}) \cdot (\Delta\vec{v}(\mathbf{r}))d\mathbf{r} + \frac{1}{2!}[(\Delta\vec{N})^T \cdot \overleftrightarrow{\eta} \cdot (\Delta\vec{N})] \\
&\quad + \int (\Delta\vec{v}(\mathbf{r}))^T \cdot \overleftrightarrow{f}(\mathbf{r}) \cdot (\Delta\vec{N})d\mathbf{r} + \frac{1}{2!}\int\int (\Delta\vec{v}(\mathbf{r}))^T \cdot \overleftrightarrow{\chi}(\mathbf{r},\mathbf{r}') \cdot (\Delta\vec{v}(\mathbf{r}'))d\mathbf{r}\ d\mathbf{r}' + \ldots, \\
\Delta\Omega &= -\vec{N} \cdot \Delta\vec{\mu} + \int \vec{\rho}(\mathbf{r}) \cdot (\Delta\vec{v}(\mathbf{r}))d\mathbf{r} - \frac{1}{2!}[(\Delta\vec{\mu})^T \cdot \vec{S} \cdot (\Delta\vec{\mu})] \\
&\quad + \int (\Delta\vec{v}(\mathbf{r}))^T \cdot \overleftrightarrow{s}(\mathbf{r}) \cdot (\Delta\vec{\mu})d\mathbf{r} - \frac{1}{2!}\int\int (\Delta\vec{v}(\mathbf{r}))^T \cdot \overleftrightarrow{s}(\mathbf{r},\mathbf{r}') \cdot (\Delta\vec{v}(\mathbf{r}'))d\mathbf{r}\ d\mathbf{r}' + \ldots, \\
\Delta F &= -\int \vec{u}(\mathbf{r}) \cdot (\Delta\vec{\rho}(\mathbf{r}))d\mathbf{r} + \frac{1}{2!}\int\int (\Delta\vec{v}(\mathbf{r}))^T \cdot \overleftrightarrow{\eta}(\mathbf{r},\mathbf{r}') \cdot (\Delta\vec{\rho}(\mathbf{r}'))d\mathbf{r}\ d\mathbf{r}' + \ldots
\end{aligned}
\tag{8.5}
$$

These perturbative expansions are to be considered for chemical conceptual purposes under the following assumptions: (i) differentiability of the state function, non degeneracy of the ground state, (iii) equivalency of mixed derivatives, and (iv) the radius of convergence of the Taylor series is large enough to be useful. The supporting hypothesis of the research in this field is built on the context and scope of validity of these assumptions and within the (limited) scope of a nonrelativistic spin-polarized DFT described in the limit of a spin polarized Kohn-Sham DFT framework [184].

In the energy changes given in Eq. (8.5), corresponding to changes form a ground state to another one within a thermodynamic analogy, global quantities associated to the number of electrons $\vec{N}$, the chemical potential $\vec{\mu}$, softness $\overleftrightarrow{S}$, and hardness $\overleftrightarrow{\eta}$ matrix-vector quantities are defined as [184],

$$
\vec{N} = \begin{bmatrix} N_x \\ N_y \end{bmatrix}, \quad \vec{\mu} = \begin{bmatrix} \mu_x \\ \mu_y \end{bmatrix}, \quad \overleftrightarrow{S} = \begin{bmatrix} S_{xx} & S_{xy} \\ S_{yx} & S_{yy} \end{bmatrix}, \quad \overleftrightarrow{\eta} = \begin{bmatrix} \eta_{xx} & \eta_{xy} \\ \eta_{yx} & \eta_{yy} \end{bmatrix}
\tag{8.6}
$$

Similarly, local descriptors corresponding to the electron density $\vec{\rho}(\mathbf{r})$, the external potential $\vec{v}(\mathbf{r})$, the grand potential $\vec{u}(\mathbf{r})$, the softness $\overleftrightarrow{s}(\mathbf{r})$, and the Fukui function $\overleftrightarrow{f}(\mathbf{r})$, are written as [184],

$$\vec{\rho}(\mathbf{r}) = \begin{bmatrix} \rho_x(\mathbf{r}) \\ \rho_y(\mathbf{r}) \end{bmatrix}, \quad \vec{v}(\mathbf{r}) = \begin{bmatrix} v_x(\mathbf{r}) \\ v_y(\mathbf{r}) \end{bmatrix}, \quad \vec{u}(\mathbf{r}) = \begin{bmatrix} v_x(\mathbf{r}) - \mu_x \\ v_y(\mathbf{r}) - \mu_y \end{bmatrix}; \tag{8.7}$$

and non local descriptors such as the linear response of density function $\overleftrightarrow{\chi}(\mathbf{r}, \mathbf{r}')$, the softness $\overleftrightarrow{s}(\mathbf{r}, \mathbf{r}')$ and hardness $\overleftrightarrow{\eta}(\mathbf{r}, \mathbf{r}')$ kernels are simply [184],

$$\overleftrightarrow{\chi}(\mathbf{r}, \mathbf{r}') = \begin{bmatrix} \chi_{xx}(\mathbf{r}, \mathbf{r}') & \chi_{xy}(\mathbf{r}, \mathbf{r}') \\ \chi_{yx}(\mathbf{r}, \mathbf{r}') & \chi_{yy}(\mathbf{r}, \mathbf{r}') \end{bmatrix},$$

$$\overleftrightarrow{s}(\mathbf{r}, \mathbf{r}') = \begin{bmatrix} s_{xx}(\mathbf{r}, \mathbf{r}') & s_{xy}(\mathbf{r}, \mathbf{r}') \\ s_{yx}(\mathbf{r}, \mathbf{r}') & s_{yy}(\mathbf{r}, \mathbf{r}') \end{bmatrix},$$

and

$$\overleftrightarrow{\eta}(\mathbf{r}, \mathbf{r}') = \begin{bmatrix} \eta_{xx}(\mathbf{r}, \mathbf{r}') & \eta_{xy}(\mathbf{r}, \mathbf{r}') \\ \eta_{yx}(\mathbf{r}, \mathbf{r}') & \eta_{yy}(\mathbf{r}, \mathbf{r}') \end{bmatrix} \tag{8.8}$$

The derivatives entering the matrix-vector notation corresponds to global reactivity components [184],

$$N_q \equiv \left[\frac{\partial \Omega}{\partial \mu_q}\right]_{\mu_{\tau \neq q}, v_x(\mathbf{r}), v_y(\mathbf{r})}, \quad \mu_q \equiv \left[\frac{\partial E}{\partial N_q}\right]_{N_{\tau \neq q}, v_x(\mathbf{r}), v_y(\mathbf{r})},$$

$$S_{pp'} \equiv \left[\frac{\partial}{\partial \mu_{p'}}\left[\frac{\partial \Omega}{\partial \mu_p}\right]_{\mu_{\tau \neq p}, v_x(\mathbf{r}), v_y(\mathbf{r})}\right]_{\mu_{\tau' \neq p'}, v_x(\mathbf{r}), v_y(\mathbf{r})},$$

and

$$\eta_{pp'} \equiv \left[\frac{\partial}{\partial N_{p'}}\left[\frac{\partial E}{\partial N_p}\right]_{N_{\tau \neq p}, v_x(\mathbf{r}), v_y(\mathbf{r})}\right]_{N_{\tau' \neq p'}, v_x(\mathbf{r}), v_y(\mathbf{r})} \tag{8.9}$$

local reactivity descriptors [184],

$$\rho_p(\mathbf{r}) \equiv \left[\frac{\delta E}{\delta v_p(\mathbf{r})}\right]_{N_x, N_y, v_{\tau \neq p}(\mathbf{r})} = \left[\frac{\delta \Omega}{\delta v_p(\mathbf{r})}\right]_{\mu_x, \mu_y, v_{\tau \neq p}(\mathbf{r})}, \quad u_p(\mathbf{r}) \equiv -\left[\frac{\delta F}{\delta \rho_p(\mathbf{r})}\right]_{\rho_{\tau \neq p}(\mathbf{r})},$$

$$S_{pp'}(\mathbf{r}) \equiv \left[\frac{\partial}{\partial \mu_{p'}}\left[\frac{\partial \Omega}{\partial v_p(\mathbf{r})}\right]_{\mu_x, \mu_y, v_{\tau \neq p}(\mathbf{r})}\right]_{\mu_{\tau' \neq p'}, v_x(\mathbf{r}), v_y(\mathbf{r})},$$

and

$$f_{pp'}(\mathbf{r}) \equiv \left[\frac{\partial}{\partial N_{p'}}\left[\frac{\delta E}{\delta v_p(\mathbf{r})}\right]_{N_x, N_y, v_{\tau \neq p}(\mathbf{r})}\right]_{N_{\tau' \neq p'}, v_x(\mathbf{r}), v_y(\mathbf{r})} \tag{8.10}$$

and non-local indices [184]

$$\chi_{pp'}(\mathbf{r}) \equiv \left[\frac{\partial}{\partial v_{p'}(\mathbf{r}')}\left[\frac{\delta E}{\delta v_p(\mathbf{r})}\right]_{N_x, N_y, v_{\tau \neq p}(\mathbf{r})}\right]_{N_x, N_y, v_{\tau' \neq p'}(\mathbf{r})},$$

$$S_{pp'}(\mathbf{r}) \equiv \left[\frac{\delta}{\delta v_{p'}(\mathbf{r}')}\left[\frac{\delta \Omega}{\delta v_p(\mathbf{r})}\right]_{\mu_x, \mu_y, v_{\tau \neq p}(\mathbf{r})}\right]_{\mu_x, \mu_y, v_{\tau' \neq p'}(\mathbf{r})},$$

and

$$\eta_{\sigma\sigma'}(\mathbf{r}) \equiv \left[\frac{\delta}{\delta\rho_{\sigma'}(\mathbf{r}')} \left[\frac{\delta F}{\delta\rho_{\sigma}(\mathbf{r})} \right]_{\rho_{\tau\neq\sigma}(\mathbf{r})} \right]_{\rho_{\tau'\neq\sigma'}(\mathbf{r})} . \tag{8.11}$$

In the above derivatives the index q refers to either x or y, and in such notation x, x', y, y', τ, and τ' are indices that assume either of the spin values, α and β in the $[N_\alpha, N_\beta]$ representation, or either of the N and N_S values in the $[N, N_S]$ representation [184]. Further, it should be explicitly noted that $\overleftrightarrow{s}(\mathbf{r})$ and $\overleftrightarrow{f}(\mathbf{r})$ are not symmetric matrices and that in the case of local and non-local descriptors, the first index in the notation always refers to the spin character of the external potential derivative. Note also that the subscript N is redundant in the cases of N_N and $v_N(\mathbf{r})$ and it is dropped in the usual manipulation of these quantities. The universal character of well-known relationships linking global, local and nonlocal reactivity indicators have been emphasized using this notation as for instance: the inverse relationships between the global softness and global hardness [90, 227]; the inverse relationship between the hardness kernel and the softness kernel [43, 45, 90]; the Berkowitz-Parr identity [90, 227, 228],

$$\overleftrightarrow{\chi}(\mathbf{r}, \mathbf{r}') = \overleftrightarrow{f}(\mathbf{r}) \cdot \overleftrightarrow{S} \cdot (\overleftrightarrow{f}(\mathbf{r}'))^T - \overleftrightarrow{s}(\mathbf{r}, \mathbf{r}') \tag{8.12}$$

and the Harbola-Chattaraj-Cedillo-Parr identity [184, 229–232],

$$\overleftrightarrow{\eta} = \int \overleftrightarrow{\eta}(\mathbf{r}, \mathbf{r}') \cdot \overleftrightarrow{f}(\mathbf{r}')d\mathbf{r}' \tag{8.13}$$

Note that the above relevant results can be concisely stated using a new powerful matrix-vector notation, with which the fundamental relationships in conceptual-DFT assume indeed a *universal* form [184]. These equations are a sound basis for the exploration of the reactivity descriptor hierarchies within the SP-DFT framework and the link between the Fukui function and the local softness [90, 227, 228]. These facts constitute a sound basis for a deeper treatment for general electrophilicity/nucleophilicity interactions, and the analysis of chemical principles (e.g., MHP [107–110], HSAB [102, 109, 111–113]) in the context of chemical process involving both charge transfer and spin polarization effects. The universal matrix-vector notation for conceptual DFT enables easy transfer of results to *any* formulation of spin-resolved DFT and even to spin-free conceptual DFT, offering an unifying perspective on conceptual DFT as a whole [183–185, 207].

8.3 Practical Applications of SP-DFT Indicators

Early explorations of the SP-DFT framework focused on the exploration of the key global chemical potentials and local Fukui functions, in problems associated to atomic systems, chemical binding, charge redistribution between states of different multiplicities, and forbidden singlet-triplet transitions have been previously reviewed [183, 185, 207]. The SP-DFT Fukui indices have been found to provide valuable insights characterizing selectivity in a wide variety of applications related to the rational design of drugs and materials sciences related fields, including for instance selectivity characterization of derivatives based on quinoline [233–235], thiourea [236–240], benzoxazole [241–243], benzene [244], xylene [245], imidazole [246–250], oxadiazole [251, 252], triazole [253], thiazole [254], pyrazine [252, 255], semicarbazide [256], amide [257–263], acid [264, 265], pyrazole [266], chalcone [267], quinoline [268, 269], antidepressant-like [270], pyridoxylidene aminoguanidine [271], arylpiperazine-based [272], oxime [273], isoindoline [274], pirymidine [275, 276], pyrrolizidine alkaloids [277], and peptidomimetics [278]. Illustrative applications to radical chemical processes (e.g., the hydrogenation reaction of the succinimidyl radical and the

Bergman cyclization) were first given by Melin et al. [279]. It was found that the changes for the chemical ($\Delta\mu_N^0$) and the spin potential ($\Delta\mu_S^0$) were positive and negative respectively, indicating that charge transfer is an unfavorable process (i.e. accompanied by an energy increase) at the beginning of the reaction, whereas the spin polarization appears to be more favorable. It was thus concluded that the hydrogenation of the succinimidyl radicals can be attributed more to the change in the spin-polarization rather than to charge transfer. This conclusion was strengthened by the fact that the Fukui function $f_{NN}^{+}(\mathbf{r})$ is very small at the hydrogenation site of the radical, whereas $f_{NS}^{+}(\mathbf{r})$, accounting for the variation in the electron density with respect to changes in the spin number, is consistently the highest at this site. Conversely, in the study of the Bergman cyclization by the same authors, it was concluded that the intramolecular charge transfer seems to be the controlling factor. In this case, $f_{NS}^{+}(\mathbf{r})$ is now zero for all the sites of the reactant and the authors concluded that this reaction is an electrocyclic rather than a radical process. These predictions are in complete agreement with experimental available evidence [280] and demonstrate in the light of developments [184] the practical usefulness of global and local descriptors defined within the $[N, N_S]$ representation to provide a description of *spin polarization effects* in this type of radical processes. Ring-closure mechanisms, for instance, have proved to be well modeled [161] in terms of both charge transfer and spin polarization factors. In such a context, for instance, the regioselectivity of the set of ring-forming radical reactions was also investigated within the framework of spin-polarized conceptual density functional theory by Pinter et al. [161]. The different radical cyclization reactions investigated included alkyl and acyl-substituted hexenyl radicals and the complex cascade radical reaction of N-alkenyl-2-aziridinylmethyl radicals [161]. Again the results allowed us to conclude that the use of the non-spin polarized Fukui function $f^0(\mathbf{r})$ [73] does not lead to a satisfactory explanation of the regioselectivity in these intramolecular radical reactions. In fact, and according to a local Hard and Soft Acids and Bases principle, the smallest absolute value of the difference in local Fukui functions between the nitrogen or carbon radical on one hand and one of the carbon atoms of the double bond (C_{exo} or C_{endo}) on the other would correspond to the more susceptible atom for a radical attack. The condensed values of the conventional Fukui function $f^0(\mathbf{r})$ for the relevant atoms reveals that the *endo* carbon atom is predicted to be more susceptible to radical attack than the *exo* one. In the past, spin independent reactivity indices were applied to these species through $f^0(\mathbf{r})$ [73, 137, 138, 161, 195, 197–199, 201], although the results were not always satisfactory. A prediction that is not in agreement with the experimentally observed outcome that the hexenyl radical and its substituted analogues cyclize with high regioselectivity to give five-membered rings. For acyl-substituted radicals, where, experimentally, a clear preference for the 6-endo process is found, the calculated regioselectivity was obtained, however, it was in good agreement with the experiment. The calculated Fukui functions for the second and third step of radical cascade cyclizations of N-alkenyl-2-aziridinylmethyl radicals to pyrrolizidines and indolizidines show that in both steps, the Fukui function value of the C_{endo} is always higher than C_{exo}. It is thus expected that the radical attack will preferentially be at the endo carbon atom. Also these results are not in agreement with experimental data; because the observed regioselectivity and the calculated activation barriers show that the cyclization of the nitrogen radical (second step) and the carbon radical (third step) always prefer the attack on the C_{exo}, resulting in the formation of the smaller ring. The use of the non-spin polarized Fukui function $f^0(\mathbf{r})$ [73] is not recommended to lead to a satisfactory explanation of the regioselectivity in these intramolecular radical reactions. As in a bimolecular radical addition, one can expect little charge transfer between the two molecules [279]. When charge is transferred from the donor to the acceptor the spin state of both of the reactants is changing. Thus, considering the reactants separately, the non-spin-polarized Fukui function measures the response of the density for each reactant to a change in number of electrons at fixed external potential but variable spin number and can be expected to give the right regioselectivity. The key step in modeling these reactions is the simple assumption that in the intramolecular addition, during the charge transfer from the donor part to the acceptor part of the molecule, the spin state of the molecule remains constant. This basic assumption can be examined and further developed in the context of using a proper representation for both charge and spin transfer provided by a $[N_\alpha, N_\beta]$ SP-DFT framework [184]. The approaching radical (nucleophilic, electrophilic) adds to the π bond of the carbon-carbon double bond to form a closed ring doublet radical. In such a process, the number of electrons is changing from a local point of view (charge transfer occurs from one part of the molecule to another one) at constant global spin number N_S (both the reactant and the closed ring are in the doublet state). The generalized Fukui function $f_{NN}(\mathbf{r})$ was used as the key SP-DFT quantity in a $[N, N_S]$ representation [183, 185, 207], to investigate the regioselectivity in this type of reactions. Note, however, that the change in density is always *constrained* to be held a constant spin number in such simplifying model [279]. Also in this case, the smallest absolute value of the difference in local spin-polarized Fukui functions between the nitrogen or carbon radical on one hand and one of the carbon atoms of the double bond (C_{exo} or C_{endo}) on the other, should identify the atom more susceptible to radical attack in agreement with the local HSAB principle. In order to test these results under unconstrained conditions, we propose to introduce analog $[N_\alpha, N_\beta]$ Fukui descriptors in order to study the regioselectivity in different cases considered above, as this cyclization involves a *doublet* radical compound. The calculated values of the differences Δf_{NN} for the different radicals studied [279] reveals that it is always smaller between the radical and C_{exo} of the double bond than between the radical and the terminal carbon atom (C_{endo}), i.e. it shows a preference for the formation of 5-exo rather than the 6-endo products, in good agreement with the experimental results. In the case of acyl-substituted radicals, the 6-endo pathway is predicted to be generally favored; the only exception occurs when an alkyl substituent connects to the radical carbon atom. Although there are no available experimental data for these specific systems, theoretical calculations showed that the lowest activation barrier is found for the endo-cyclization. Values for the nitrogen radical and for the carbon 6-endo- and 5-exo-cyclizations of the cascade reaction also reveal a good agreement between our computational results and available experimental data. Experimentally both cyclization steps of the domino reaction were found to be a regioselective process as it leads, if it occurs, always to the smaller ring. The better matching always corresponds to the exo-ring closure. This application thus clearly highlights that the experimentally observed regioselectivity for these ring closure steps can be predicted using the spin-polarized Fukui functions for radical attack. The predictive nature of SP-DFT reactivity indices and their relationships reported in the $[N_\alpha, N_\beta]$ representation is straightforward [183–185, 207].

The conceptual SP-DFT framework have further incorporated discussions about nonlocal descriptors and hardness and softness kernels [45, 281, 282], nuclear reactivity descriptors [158, 164], generalized local Fukui functions in a condensed-to-atoms scheme [92], and generalized philicities [91]. A critical analysis about the use of the frozen-core and finite difference approximations in such context has been reported [283]. Further work incorporate the analysis of dual descriptors [227, 284, 285] intended to describe in full detail spin polarization in the $[N, N_S]$ representation [218], and extensions towards alternative representations [184]. Illustrative applications of the $[N, N_S]$ SP-DFT descriptors have been reviewed [185]. Additional examples include the characterization of global and local spin proclivities of phenylhalocarbenes [160] as well as highly reactive divalent and monovalent compounds from Groups 14 and 15. In particular, global- and local-spin reactivity indicators were applied to series of reactive intermediates such as carbenes, silylenes, germylenes, and stannylenes [116, 141, 157, 219, 220, 286], where the nature and interplay of the electronic ground and excited states (e.g., triplet or singlet) define the reactivity. The nature of local reactivity in nitrenes, phosphinidenes, and phenylhalocarbenes was similarly explored [160]. Additional applications of spin-polarized DFT based reactivity indicators in radical cyclization reactions have been performed [161]. Guerra et al. studied homolytic substitution reactions, in analogy with the conceptual DFT description of leaving group capabilities [287]. These authors have applied the electro and nucleofugalities descriptors [123, 288] introducing the concept of homofugality to probe the leaving group ability of radical fragments in these transformations [289, 290]. This index was defined as the regional spin-philicity of the leaving group in the substrate that undergoes the radical attack. Jaramillo has also reported on feasibility of intrinsic nucleofugality scales [124]. Further applications of the nucleofugality descriptor were given by Campodonico [127, 128, 291], and Geerlings [292]. Following an analogy with Sanderson's electronegativity equalization principle, a principle of *spin potential equalization* was proposed in the $E[N_\alpha, N_\beta]$ representation by Guerra et al. [163]. The model was computationally illustrated for a series of addition reactions of electrophilic, nucleophilic, and ambiphilic carbenes (i.e., exhibiting both electrophilic and nucleophilic character) to alkenes in their singlet and triplet multiplicities [119]. Other examples of application include the analysis of spectrochemical and nephelauxetic series of different octahedral Ru complexes which were explored using local spin-philicity and SP-DFT Fukui descriptors [129], and the study of regioselectivity and spin coupling in the first biradical formation step of the photochemical [2+2]-cycloaddition of α, β–unsaturated carbonyl compounds (enones) to substituted alkenes, investigated using the concepts of local spin-philicity and donicity [165, 168, 169]. The regioselectivity was interpreted as resulting from *the spin-coupling of the local site on the alkene with the highest spin-philicity* (i.e. the smallest destabilization upon a global increase of the spin number of the molecule) with the site of the highest change in spin number upon a decreasing global multiplicity of the enone. A comparable approach was used by some of the same authors in the study of the regioselectivity through the f_{NN} descriptor of the photochemical [2+2] cycloaddition of triplet carbonyl compounds with a series of ground state electron-rich and electron-poor alkenes, the Paterno-Buchi reaction [168], yielding oxetanes. The regioselectivity of the [2+2] photocycloaddition reaction between triplet $(\pi - \pi^*)$ acrolein and substituted olefins in their ground states was investigated in more detail using the reaction force concept and reactivity indices from conceptual spin-polarized density functional theory [169]. In addition to these exciting applications, there has been progress on the theoretical and the computational methods associated with spin-resolved conceptual DFT.

The *spin-philicity* and *spin-donicity* descriptors have been defined as analogues to the electrophilicity in conventional (spin-free) conceptual DFT [91, 115, 116, 141, 159, 186, 219, 220]. These indicators describe the energetic proclivity of a reagent to receive spin (increasing the spin-multiplicity of the reagent) and donate spin to another molecule (increasing the spin-multiplicity of reagent's reaction partner), respectively. These indicators were further rationalized as indicators of *philicity for spin polarization* [91]. In this context, we will use the now-standard terminology of *donation/acceptance of spin* when we wish to describe how a system responds to decreases/increases in spin-multiplicity at fixed electron number [184]. The spin-resolved analogs of other key spin-free reactivity indicators have also been defined. For example, there are spin-DFT extensions to nuclear reactivity indicators, to *local* reactivity indicators (for regioselectivity), and to nonlocal reactivity indicators (for molecular responses) [184]. On the computational side, a simple scheme, which is based on a frozen core approximation, has been developed for approximating the condensed Fukui functions within such SP-DFT framework [92]. There has also been interest in approximating reactivity indicators associated with SP-DFT using *ab initio* methods such as electron propagator theory and the extended Koopmans' theorem [293]. The nature of models based on finite-difference approximations for spin-reactivity indicators has been carefully discussed [186]. Concerning the response of nuclei in processes involving charge transfer and/or spin polarization upon excitation or de-excitation, as occurring both in chemical reactions (8.3) in spectroscopic experiments, we have also introduced nuclear reactivity indexes within such a framework, stressing its importance in the context of the Berlin theorem of chemical binding [158]. Dual descriptors intended to describe regions prone to undergo spin charge reorganization during spin polarization processes have been introduced within a $[N, N_S]$ representation [218]. Extensions regarding the proper treatment and application of nuclear indicators in the case of open shell systems will deserve further attention [25, 38, 155, 156, 173, 176, 177, 186, 190, 282, 294, 295].

The key SP-DFT quantities arise as derivatives of appropriate state functions, and different representations could be chosen for the chemical reactivity problem at hand: a *closed system* representation, where N_α, N_β, $v_\alpha(\mathbf{r})$, and $v_\beta(\mathbf{r})$ are the parameters defining the system, an *open system* representation, where the chemical potentials associated with electrons of each spin, μ_α, μ_β, $v_\alpha(\mathbf{r})$, and $v_\beta(\mathbf{r})$ are the parameters defining the system, and a *density* representation, where the electron spin-densities, $\rho_\alpha(\mathbf{r})$, and $\rho_\beta(\mathbf{r})$, are the parameters defining the system. All of these different choices of representation are equivalent, and the relationships between reactivity indicators in the different representations have been explicitly derived [184]. It was there advanced that using the matrix-vector formulation, every identity in the $[N_\alpha, N_\beta]$ family of representations can be trivially extended to the $[N, N_S]$ family of representations or even to the conventional spin-unresolved representations of conceptual DFT. That is, with this notation, the interpretation of the symbol changes between the $[N_\alpha, N_\beta]$ representation, the $[N, N_S]$ representation, and the spin-free case, but the form of the equation remains the same. The relationships between reactivity indicators in the spin-resolved conceptual DFT and the conventional or *density only* conceptual DFT have been also emphasized. With the theoretical foundations almost complete, results relating electronic and nuclear descriptors may be also be explored. The matrix-vector notation facilitates the derivation of electron-transfer processes in spin-resolved conceptual DFT [91]. In fact, a simple and compact expression for the total *spin-electrophilicity* showing that the total spin-electrophilicity is equal to the sum of the component spin-electrophilicities only if the off-diagonal elements of the hardness matrix are zero, has been presented [184]. As emphasized above, each of SP-DFT representations is associated with a characteristic set of reactivity indicators, and they become certainly equivalent under suitable conditions. Thus, in a closed-system representation $[N_\alpha, N_\beta]$ it is usually possible to assign to each molecule a precise number of electrons. Hence such representation is appropriate for molecules that interact weakly with their environment (e.g., molecules in the gas phase). For "no isolated" molecules, the number of electrons can be considered to fluctuate significantly and one should use an open-system picture, i.e., $[\mu_\alpha, \mu_\beta]$. Note that in these representations, in which the external potential is a variable, the electron density is assumed to change in response to changes in the nuclear positions (e.g., an electron-following picture). In a density representation $[\rho_\alpha, \rho_\beta]$, on the other hand, the electron density is the fundamental variable and the nuclei must adapt to changes in the electron density (e.g., the electron-preceding picture) [158, 296–300].

8.4 Concluding Remarks and Perspectives

The above presentation above emphasizes the usefulness of the spin-resolved DFT representation in terms of the number of electrons with each spin (N_α and N_β). The progress given in the framework of SP-DFT towards a characterization of exact relationships between electronic [45, 91, 92] and nuclear chemical descriptors [158, 165], are expected to provide a general reference framework for the exploration of chemical responses as intended within such a framework [184]. These responses are indeed assumed directly representable through changes in the external potential – or even in the effective Kohn-Sham potential – of a given system. Further exploration of plausible even more general scope for the hard and soft acids and bases (HSAB) principle should be expected within the $E[N_\alpha, N_\beta]$ representation [184]. Such generalizations would extend, in the light of the ideas behind the HSAB principle, a general treatment of both *electron transfer* and *spin polarization phenomena in key free radical* processes. Further work in this area could be related to the systematic computational probe of the key SP-DFT relationships along such a process.

One important quantity to be further explored is the role of electrophilicity and nucleophilicity indices within a SP-DFT perspective and its impact on the rationalization of free radical reactivity. Most free-radical chemical reactions feature electron transfer. Given that the transfer of electron spin is coupled to the electron itself, the $[N_\alpha, N_\beta]$ representation seems the most appropriate to use in these cases [184]. The Parr's electrophilicity index [184] becomes written simply as

$$\omega = \frac{1}{2}(\vec{\mu})^T \cdot \overleftrightarrow{S} \cdot \vec{\mu} = \frac{1}{2}(\vec{\mu})^T \cdot (\overleftrightarrow{\eta})^{-1} \cdot \vec{\mu}, \tag{8.14}$$

which certainly open new possibilities within the goals being here proposed to further advance these concepts for SP-DFT framework. For instance, if the hardness matrix is diagonal, then one can decompose the electrophilicity as a sum of two contributions. These contributions define the popular spin-electrophilicities [184], e.g., using the above subscripts, $\omega = \omega_x + \omega_y$. These findings constitute a methodological basis for the further exploration of a generalized nucleophilicity matrix formulation, for instance. By starting from the $[N_\alpha, N_\beta, v_\alpha(\mathbf{r}), v_\beta(\mathbf{r})]$, further work is also desired in order to advance the state of the SP-DFT for arbitrarily Legendre-transformed representations. It seems also reasonable to look for new relationships that explicitly incorporate nucleophilicities and the changes in the nuclei positions using as a basis the previous development on nuclear reactivities given in the $[N, N_S]$ representation [158, 184]. The number of studies concerning the application of conceptual-based SP-DFT descriptors to chemical problems of general interest (e.g., Fukui indicators as tools for selectivity in drug discovery and material sciences) is certainly increasing as simple implementations are available [301]. The partition density functional theory (PDFT) and embedding/partitioning related approaches [302, 303] offers also an interesting framework to further exploration of conceptual spin-polarized density functional theory. Applications regarding characterization of magnetic materials [304, 305] and/or spintronic applications [306–309] constitute research fields which can provide an attractive field for exploring conceptual SP-DFT developments, both providing a basis for the definition of new concepts, but also within considerations of computational implementation of reactivity models. The above developments contribute to enable further growing on both development and interesting fundamental applications given in the context free radical reactivity within a conceptual DFT based-approach [138, 161, 172, 176, 177, 310]. In particular, and related to a possible *nucleophilicity* indicator defined within SP-DFT, further exploration could be concerned to electron transfer energy between two molecules in the absence of constraints on the total spin-multiplicity [184]. The above methodologies can be straightforwardly extended for instance to the analysis of the (2-ethynylphenyl) triazene derivatives cyclization, where the electronic nature of the ring closure certainly remains open to further debate [311, 312]. A related example is the analysis of [3+2] cycloaddition reactions [313–317], which in fact appear to have a more unpredictable behavior than the Diels-Alder cycloadditions [150, 318–320]. In such a context, the local SP-DFT Fukui functions have probed usefulness characterizing the most reactive sites along polar organic chemical reactions [146, 149, 150, 190, 310]. We finally recall that conventional DFT-based reactivity indicators are built entirely on electronic energetic criterion. The consideration of entropic contributions or even a description based on the full free energy to the SP-DFT state functions is a subject still open to further research [184, 193, 321, 322]. Nuclear reactivity descriptors can be also incorporated through the explicit consideration of Taylor expansions, the total energy of the system (i.e., adding the nucleus-nucleus interaction). In such context the treatment of nuclei changes has been explored through explicit considerations, within a Born-Oppenheimer approximation, of the changes in the forces on nuclei. The general formulation has been already advanced for the $[N, N_S]$ and spin-free representations on the basis of derivatives of forces on nuclei, e.g., the so-called nuclear Fukui functions [158, 165], in connection with the electron descriptors [25, 185, 186]. This fact certainly offers, for instance, an opportunity to the generation of new insights and further advancement of a spin-polarized DFT framework to include both electronic and nuclear chemical reactivity coupled descriptors and temperature-dependent formulations [21, 114, 133], broadening, moreover, the scope of chemical principles within the context of a chemical reactivity theory rooted in DFT framework [42].

Acknowledgements

Authors acknowledge to Fondecyt Chile, Grant Nos. 1180348 and 1181582 for continuous support.

References

1 Hohenberg, P. and Kohn, W. (1964). Inhomogeneous electron gas. Physical Review B 136: 864.

2 Kohn, W. and Sham, L.J. (1965). Self-consistent equations including exchange and correlation effects. Physical Review 140A: 1133.

3 Parr, R.G. and Yang, W. (1989). Density Functional Theory of Atoms and Molecules. Oxford: Oxford University Press.

4 Cohen, M.H. (1996). Strengthening the foundations of chemical reactivity theory. Topics in Current Chemistry 183 (Density Functional Theory IV): 143–173.

5 Chermette, H. (1999). Chemical reactivity indexes in density functional theory. Journal of Computational Chemistry 20 (1): 129–154.

6 Ayers, P.W. and Parr, R.G. (2000). Variational principles for describing chemical reactions: The Fukui function and chemical hardness revisited. Journal of the American Chemical Society 122 (9): 2010–2018.

7 Ayers, P.W. and Parr, R.G. (2001). Variational principles for describing chemical reactions. reactivity indices based on the external potential. Journal of the American Chemical Society 123 (9): 2007–2017.

8 Geerlings, P., De Proft, F., and Langenaeker, W. (2003). Conceptual density functional theory. Chemical Reviews 103 (5): 1793–1873.

9 Chattaraj, P.K., Sarkar, U., and Roy, D.R. (2006). Electrophilicity index. Chemical Reviews 106 (6): 2065–2091.

10 Geerlings, P. and De Proft, F. (2008). Conceptual DFT: the chemical relevance of higher response functions. Physical Chemistry Chemical Physics 10 (21): 3028–3042.

11 Liu, S.B. (2009). Conceptual density functional theory and some recent developments. Acta Physico-Chimica Sinica 25 (3): 590–600.

12 von Szentpaly, L. (2020). Theorems and rules connecting bond energy and bond order with electronegativity equalization and hardness maximization. Theoretical Chemistry Accounts 139 (3): 9.

13 Guegan, F., Pigeon, T., De Proft, F. et al. (2020). Understanding chemical selectivity through well selected excited states. Journal of Physical Chemistry A 124 (4): 633–641.

14 Sanchez-Marquez, J., Garcia, V., Zorrilla, D., and Fernandez, M. (2019). New insights in conceptual DFT: New model for the calculation of local reactivity indices based on the Sanderson's principle. International Journal of Quantum Chemistry 119 (7): 13.

15 Sanchez-Marquez, J. (2019). New advances in conceptual-DFT: an alternative way to calculate the Fukui function and dual descriptor. Journal of Molecular Modeling 25 (5): 7.

16 Miranda-Quintana, R.A. and Ayers, P.W. (2019). The "|| big is good" rule, the maximum hardness, and minimum electrophilicity principles. Theoretical Chemistry Accounts 138 (3): 6.

17 Figueredo, S., Paez, M., and Torres, F. (2019). The electrophilic descriptor. Computational and Theoretical Chemistry 1157: 34–39.

18 Liu, S.B. and Zhang, X.J. (2018). Chemical concepts from density functional theory. Acta Physico-Chimica Sinica 34 (6): 563–566.

19 Anderson, J. and Ayers, P. (2018). Conceptual DFT formulated general purpose reactivity indicator for predicting and classifying the reactivity of sites within a molecule. Abstracts of Papers of the American Chemical Society 255: 1.

20 Zhou, P.P., Liu, S.B., Ayers, P.W., and Zhang, R.Q. (2017). Bonding reactivity descriptor from conceptual density functional theory and its applications to elucidate bonding formation. Journal of Chemical Physics 147 (13): 7.

21 Miranda-Quintana, R.A., Chattaraj, P.K., and Ayers, P.W. (2017). Finite temperature grand canonical ensemble study of the minimum electrophilicity principle. Journal of Chemical Physics 147 (12): 4.

22 Miranda-Quintana, R.A. and Ayers, P.W. (2017). Interpolating Hamiltonians in chemical compound space. International Journal of Quantum Chemistry 117 (14): 4.

23 Miranda-Quintana, R.A. (2017). Perturbed reactivity descriptors: the chemical hardness. Theoretical Chemistry Accounts 136 (7): 8.

24 Miranda-Quintana, R.A. (2017). Thermodynamic electrophilicity. Journal of Chemical Physics 146 (21): 8.

25 Heidar-Zadeh, F., Richer, M., Fias, S. et al. (2016). An explicit approach to conceptual density functional theory descriptors of arbitrary order. Chemical Physics Letters 660: 307–312.

26 Geerlings, P., Boisdenghien, Z., De Proft, F., and Fias, S. (2016). The $E = E$ N, v functional and the linear response function: a conceptual DFT viewpoint. Theoretical Chemistry Accounts 135 (9): 8.

27 Franco-Perez, M., Ayers, P.W., and Gazquez, J.L. (2016). Average electronic energy is the central quantity in conceptual chemical reactivity theory. Theoretical Chemistry Accounts 135 (8): 8.

28 Skara, G., Pinter, B., Top, J. et al. (2015). Conceptual quantum chemical analysis of bonding and noncovalent interactions in the formation of frustrated Lewis pairs. Chemistry-a European Journal 21 (14): 5510–5519.

29 Perez, P., Yepes, D., Jaque, P. et al. (2015). A computational and conceptual DFT study on the mechanism of hydrogen activation by novel frustrated Lewis pairs. Physical Chemistry Chemical Physics 17 (16): 10715–10725.

30 Geerlings, P., Fias, S., Boisdenghien, Z., and De Proft, F. (2014). Conceptual DFT: chemistry from the linear response function. Chemical Society Reviews 43 (14): 4989–5008.

31 Bultinck, P., Cardenas, C., Fuentealba, P. et al. (2014). How to compute the Fukui matrix and function for systems with (quasi-)degenerate states. Journal of Chemical Theory and Computation 10 (1): 202–210.

32 Tognetti, V., Morell, C., Ayers, P.W. et al. (2013). A proposal for an extended dual descriptor: a possible solution when Frontier Molecular Orbital Theory fails. Physical Chemistry Chemical Physics 15 (34): 14465–14475.

33 Putz, M.V. and Chattaraj, P.K. (2013). Electrophilicity kernel and its hierarchy through softness in conceptual density functional theory. International Journal of Quantum Chemistry 113 (18): 2163–2171.

34 Yang, W.T., Cohen, A.J., De Proft, F., and Geerlings, P. (2012). Analytical evaluation of Fukui functions and real-space linear response function. Journal of Chemical Physics 136 (14): 9.

35 Sablon, N., De Proft, F., Sola, M., and Geerlings, P. (2012). The linear response kernel of conceptual DFT as a measure of aromaticity. Physical Chemistry Chemical Physics 14 (11): 3960–3967.

36 Geerlings, P., Ayers, P.W., Toro-Labbe, A. et al. (2012). The Woodward-Hoffmann rules reinterpreted by conceptual density functional theory. Accounts of Chemical Research 45 (5): 683–695.

37 Matito, E. and Putz, M.V. (2011). New link between conceptual density functional theory and electron delocalization. Journal of Physical Chemistry A 115 (45): 12459–12462.

38 Araya, J.I.M. (2011). The dual descriptor: Working equations applied on electronic open-shell molecular systems. Chemical Physics Letters 506 (1–3): 104–111.

39 Ceron, M.L., Echegaray, E., Gutierrez-Oliva, S. et al. (2011). The reaction electronic flux in chemical reactions. Science China. Chemistry 54 (12): 1982–1988.

40 Sablon, N., Mastalerz, R., De Proft, F. et al. (2010). Relativistic effects on the Fukui function. Theoretical Chemistry Accounts 127 (3): 195–202.

41 Sablon, N., De Proft, F., and Geerlings, P. (2010). The linear response kernel of conceptual DFT as a measure of electron delocalisation. Chemical Physics Letters 498 (1–3): 192–197.

42 Geerlings, P., Chamorro, E., Chattaraj, P.K. et al. (2020). Conceptual density functional theory: status, prospects. issues. Theoretical Chemistry Accounts 139 (2): 18.

43 Chamorro, E., Contreras, R., and Fuentealba, P. (2000). Some relationships within the nonlocal (pair-site) chemical reactivity formalism of density functional theory. Journal of Chemical Physics 113 (24): 10861–10866.

44 Chamorro, E., Fuentealba, P., and Contreras, R. (2001). Higher order derivatives for nuclear indexes in the framework of density functional theory. Journal of Chemical Physics 115 (15): 6822–6826.

45 Chamorro, E., De Proft, F., and Geerlings, P. (2005). Hardness and softness reactivity kernels within the spin-polarized density-functional theory. Journal of Chemical Physics 123 (15): 154104.

46 Ayers, P.W., Fias, S., and Heidar-Zadeh, F. (2018). The axiomatic approach to chemical concepts. Computational and Theoretical Chemistry 1142: 83–87.

47 Domingo, L.R., Rios-Gutierrez, M., and Perez, P. (2016). Applications of the conceptual density functional theory indices to organic chemistry reactivity. Molecules 21 (6): 22.

48 Li, H.P., Zhu, W.S., Zhu, S.W. et al. (2016). The selectivity for sulfur removal from oils: an insight from conceptual density functional theory. Aiche Journal 62 (6): 2087–2100.

49 Pinter, B., Nagels, N., Herrebout, W.A., De Proft, F. (2013) Halogen bonding from a hard and soft acids and bases perspective: investigation by using density functional theory reactivity indices. Chemistry-a European Journal, 19(2), 518–529.

50 Chattaraj, P.K., Bandaru, S., and Mondal, S. (2011). Hydrogen storage in clathrate hydrates. Journal of Physical Chemistry A 115 (2): 187–193.
51 Gurkan, Y.Y., Turkten, N., Hatipoglu, A., and Cinar, Z. (2012). Photocatalytic degradation of cefazolin over N-doped TiO_2 under UV and sunlight irradiation: Prediction of the reaction paths via conceptual DFT. Chemical Engineering Journal 184: 113–124.
52 Feixas, F., Matito, E., Poater, J., and Sola, M. (2015). Quantifying aromaticity with electron delocalisation measures. Chemical Society Reviews 44 (18): 6434–6451.
53 Ayers, P.W., Anderson, J.S.M., and Bartolotti, L.J. (2005). Perturbative perspectives on the chemical reaction prediction problem. International Journal of Quantum Chemistry 101 (5): 520–534.
54 Parr, R.G., Ayers, P.W., and Nalewajski, R.F. (2005). What is an atom in a molecule? Journal of Physical Chemistry A 109 (17): 3957–3959.
55 Cohen, M.H. and Wasserman, A. (2003). Revisiting N-continuous density-functional theory: Chemical reactivity and “atoms” in “molecules”. Israel Journal of Chemistry 43 (3–4): 219–227.
56 Cohen, M.H. and Wasserman, A. (2007). On the foundations of chemical reactivity theory. Journal of Physical Chemistry A 111 (11): 2229–2242.
57 Cohen, M.H., Wasserman, A., and Burke, K. (2007). Partition theory: A very simple illustration. Journal of Physical Chemistry A 111 (49): 12447–12453.
58 Nalewajski, R.F. and Parr, R.G. (1982). Legendre transforms and Maxwell relations in density functional theory. Journal of Chemical Physics 77 (1): 399–407.
59 Nalewajski, R.F. (1983). Reduction of derivatives and simple applications of the Legendre transformed density functional theory. Journal of Chemical Physics 78 (10): 6112–6120.
60 Cedillo, A. (1994). A new representation for ground states and its Legendre transforms. International Journal of Quantum Chemistry, Quantum Chemistry Symposium 28: 231–240.
61 Baekelandt, B.G., Cedillo, A., and Parr, R.G. (1995). Reactivity indexes and fluctuation formulas in density functional theory: isomorphic ensembles and a new measure of local hardness. Journal of Chemical Physics 103 (19): 8548–8556.
62 Ayers, P.W., Morrison, R.C., and Roy, R.K. (2002). Variational principles for describing chemical reactions: Condensed reactivity indices. Journal of Chemical Physics 116 (20): 8731–8744.
63 Ayers, P.W., Golden, S., and Levy, M. (2006). Generalizations of the Hohenberg-Kohn theorem: I. Legendre transform constructions of variational principles for density matrices and electron distribution functions. Journal of Chemical Physics 124 (5): 054101.
64 Gyftopoulos, E.P. and Hatsopoulos, G.N. (1965). Quantum-thermodynamic definition of electronegativity. Proceedings of the National Academy of Sciences of the United States of America 60: 786.
65 Perdew, J.P., Parr, R.G., Levy, M., and Balduz, J.L. Jr., (1982). Density-functional theory for fractional particle number: derivative discontinuities of the energy. Physical Review Letters 49 (23): 1691–1694.
66 Parr, R.G., Donnelly, R.A., Levy, M., and Palke, W.E. (1978). Electronegativity: The density functional viewpoint. Journal of Chemical Physics 68 (8): 3801–3807.
67 Parr, R.G. and Pearson, R.G. (1983). Absolute hardness: companion parameter to absolute electronegativity. Journal of the American Chemical Society 105 (26): 7512–7516.
68 Berkowitz, M., Ghosh, S.K., and Parr, R.G. (1985). On the concept of local hardness in chemistry. Journal of the American Chemical Society 107 (24): 6811–6814.
69 Yang, W. and Parr, R.G. (1985). Hardness, softness, and the Fukui function in the electronic theory of metals and catalysis. Proceedings of the National Academy of Sciences of the United States of America 82 (20): 6723–6726.

70 Parr, R.G. and Yang, W. (1984). Density functional approach to the frontier-electron theory of chemical reactivity. Journal of the American Chemical Society 106: 4049.

71 Ayers, P.W., Levy, M. (2000) Perspective on "Density functional approach to the frontier-electron theory of chemical reactivity" - Parr RG, Yang W (1984) J Am Chem Soc 106: 4049–4050. Theoretical Chemistry Accounts, 103(3–4), 353–360.

72 Fuentealba, P. and Contreras, R. (2002). Fukui function in chemistry. Reviews of Modern Quantum Chemistry 2: 1013–1052.

73 Yang, W., Parr, R.G., and Pucci, R. (1984). Electron density, Kohn-Sham frontier orbitals. and Fukui functions. Journal of Chemical Physics 81 (6): 2862–2863.

74 Franco-Perez, M., Polanco-Ramirez, C.A., Gazquez, J.L., and Ayers, P.W. (2018). Local and nonlocal counterparts of global descriptors: the cases of chemical softness and hardness. Journal of Molecular Modeling 24 (10): 8.

75 Franco-Perez, M., Ayers, P.W., Gazquez, J.L., and Vela, A. (2015). Local and linear chemical reactivity response functions at finite temperature in density functional theory. Journal of Chemical Physics 143 (24): 9.

76 Parr, R.G., Szentpaly, L.V., and Liu, S. (1999). Electrophilicity index. Journal of the American Chemical Society 121 (9): 1922–1924.

77 Chattaraj, P.K. and Roy, D.R. (2007). Update 1 of: Electrophilicity index. Chemical Reviews, 107(9). In: PR46-PR74.

78 Chattaraj, P.K., Giri, S., and Duley, S. (2011). Update 2 of: Electrophilicity index. Chemical Reviews, 111(2). In: PR43-PR75.

79 Perez, P., Domingo, L.R., Duque-Norena, M., and Chamorro, E. (2009). A condensed-to-atom nucleophilicity index. An application to the director effects on the electrophilic aromatic substitutions. Journal of Molecular Structure-Theochem 895 (1–3): 86–91.

80 Chamorro, E., Duque-Norena, M., and Perez, P. (2009). A comparison between theoretical and experimental models of electrophilicity and nucleophilicity. Journal of Molecular Structure-Theochem 896 (1–3): 73–79.

81 Gonzalez, M.M., Cardenas, C., Rodriguez, J.I. et al. (2018). Quantitative Electrophilicity Measures. Acta Physico-Chimica Sinica 34 (6): 662–674.

82 Franco-Perez, M., Gazquez, J.L., Ayers, P.W., and Vela, A. (2018). Thermodynamic dual descriptor. Acta Physico-Chimica Sinica 34 (6): 683–691.

83 Matute, R.A., Perez, P., Chamorro, E. et al. (2018). Reaction electronic flux perspective on the mechanism of the Zimmerman Di-pi-methane rearrangement. Journal of Organic Chemistry 83 (11): 5969–5974.

84 Echegaray, E. and Toro-Labbe, A. (2008). Reaction electronic flux: a new concept to get insights into reaction mechanisms. study of model symmetric nucleophilic substitutions. Journal of Physical Chemistry A 112 (46): 11801–11807.

85 Senet, P. (1996). Nonlinear electronic responses, Fukui functions and hardnesses as functionals of the ground-state electronic density. Journal of Chemical Physics 105 (15): 6471–6489.

86 Senet, P. (1997). Kohn-Sham orbital formulation of the chemical electronic responses. including the hardness. Journal of Chemical Physics 107 (7): 2516–2524.

87 Contreras, R., Andres, J., Perez, P. et al. (1998). Theory of non-local (pair site) reactivity from model static-density response functions. Theoretical Chemistry Accounts 99 (3): 183–191.

88 Contreras, R., Domingo, L.R., Andres, J. et al. (1999). Nonlocal (pair site) reactivity from second-order static density response function: gas- and solution-phase reactivity of the acetaldehyde enolate as a test case. Journal of Physical Chemistry A 103 (10): 1367–1375.

89 Sablon, N., De Proft, F., and Geerlings, P. (2010). The linear response kernel: inductive and resonance effects quantified. Journal of Physical Chemistry Letters 1 (8): 1228–1234.

90 Berkowitz, M. and Parr, R.G. (1988). Molecular hardness and softness, local hardness and softness, hardness and softness kernels. and relations among these quantities. Journal of Chemical Physics 88 (4): 2554–2557.
91 Chamorro, E., Perez, P., De Proft, F., and Geerlings, P. (2006). Philicity indices within the spin-polarized density-functional theory framework. Journal of Chemical Physics 124 (4): 044105.
92 Chamorro, E. and Perez, P. (2005). Condensed-to-atoms electronic Fukui functions within the framework of spin-polarized density-functional theory. Journal of Chemical Physics 123 (11): 114107.
93 Chamorro, E., De Proft, F., and Geerlings, P. (2005). Hardness and softness reactivity kernels within the spin-polarized density-functional theory. Journal of Chemical Physics 123 (15).
94 Heidar-Zadeh, F., Fias, S., Vohringer-Martinez, E. et al. (2016). The local response of global descriptors. Theoretical Chemistry Accounts 136 (1): 9.
95 Cohen, M.H. and Ganduglia-Pirovano, M.V. (1994). Electronic and nuclear chemical reactivity. Journal of Chemical Physics 101 (10): 8988–8997.
96 Cohen, M.H., Ganduglia-Pirovano, M.V., and Kudrnovsky, J. (1995). Reactivity kernels, the normal modes of chemical reactivity, and the hardness and softness spectra. Journal of Chemical Physics 103 (9): 3543–3551.
97 Baekelandt, B.G. (1996). The nuclear Fukui function and Berlin's binding function in density functional theory. Journal of Chemical Physics 105 (11): 4664–4667.
98 Ordon, P. and Komorowski, L. (1998). Nuclear reactivity and nuclear stiffness in density functional theory. Chemical Physics Letters 292 (1,2): 22–27.
99 De Proft, F., Liu, S.B., and Geerlings, P. (1998). Calculation of the nuclear Fukui function and new relations for nuclear softness and hardness kernels. Journal of Chemical Physics 108 (18): 7549–7554.
100 Balawender, R., De Proft, F., and Geerlings, P. (2001). Nuclear Fukui function and Berlin's binding function: Prediction of the Jahn-Teller distortion. Journal of Chemical Physics 114 (10): 4441–4449.
101 Balawender, R. and Geerlings, P. (2001). Nuclear Fukui function from coupled perturbed Hartree-Fock equations. Journal of Chemical Physics 114 (2): 682–691.
102 Ayers, P.W. (2005). An elementary derivation of the hard/soft-acid/base principle. Journal of Chemical Physics 122 (14): 141102.
103 Parr, R.G., Donnelly, R.A., Levy, M., and Palke, W.E. (1978). Electronegativity. Journal of Chemical Physics 68: 3801.
104 Mortier, W.J., Ghosh, S.K., and Shankar, S. (1986). Electronegativity-equalization method for the calculation of atomic charges in molecules. Journal of the American Chemical Society 108 (15): 4315–4320.
105 Bultinck, P., Langenaeker, W., Lahorte, P. et al. (2002). The electronegativity equalization method I: Parametrization and validation for atomic charge calculations. Journal of Physical Chemistry A 106 (34): 7887–7894.
106 Bultinck, P., Langenaeker, W., Lahorte, P. et al. (2002). The electronegativity equalization method II: Applicability of different atomic charge schemes. Journal of Physical Chemistry A 106 (34): 7895–7901.
107 Pearson, R.G. (1997). Chemical Hardness. Oxford: Wiley.
108 Pearson, R.G. (1993). The Principle of Maximum Hardness. Accounts of Chemical Research 26 (5): 250–255.

109 Pearson, R.G. (1990). Hard and soft acids and bases-the evolution of a chemical concept. Coordination Chemistry Reviews 100: 403–425.

110 Pearson, R.G. (2005). Chemical hardness and density functional theory. Journal of Chemical Sciences 117 (5): 369–377.

111 Ayers, P.W. (2007). The physical basis of the hard/soft acid/base principle. Faraday Discussions 135: 161–190.

112 Chattaraj, P.K. and Ayers, P.W. (2005). The maximum hardness principle implies the hard/soft acid/base rule. Journal of Chemical Physics 123 (8): 086101.

113 Pearson, R.G. (1995). The HSAB Principle - more quantitative aspects. Inorganica Chimica Acta 240 (1–2): 93–98.

114 Miranda-Quintana, R.A., Kim, T.D., Cardenas, C., and Ayers, P.W. (2017). The HSAB principle from a finite-temperature grand-canonical perspective. Theoretical Chemistry Accounts 136 (12): 6.

115 Perez, P., Andres, J., Safont, V.S. et al. (2002). Spin-philicity and spin-donicity as auxiliary concepts to quantify spin-catalysis phenomena. Journal of Physical Chemistry A 106 (21): 5353–5357.

116 Olah, J., De Proft, F., Veszpremi, T., and Geerlings, P. (2004). Spin-Philicity and Spin-Donicity of Substituted Carbenes, Silylenes, Germylenes, and Stannylenes. Journal of Physical Chemistry A 108 (3): 490–499.

117 Berkowitz, M. (1987). Density functional approach to frontier controlled reactions. Journal of the American Chemical Society 109 (16): 4823–4825.

118 Fukui, K. (1982). Role of frontier orbitals in chemical reactions. Science 218 (4574): 747–754.

119 Ayers, P.W., Morell, C., De Proft, F., and Geerlings, P. (2007). Understanding the Woodward-Hoffmann rules by using changes in electron density. Chemistry - A European Journal 13: 8240–8247.

120 De Proft, F., Ayers, P.W., Fias, S., and Geerlings, P. (2006). Woodward-Hoffmann rules in density functional theory: Initial hardness response. Journal of Chemical Physics 125 (21): 214101.

121 De Proft, F., Chattaraj, P.K., Ayers, P.W. et al. (2008). Initial hardness response and hardness profiles in the study of Woodward-Hoffmann rules for electrocyclizations. Journal of Chemical Theory and Computation 4 (4): 595–602.

122 Dreizler, R.M. and Gross, E.K.U. (1990). Density Functional Theory. Berlin: Springer Verlag.

123 Ayers, P.W., Anderson, J.S.M., Rodriguez, J.I., and Jawed, Z. (2005). Indices for predicting the quality of leaving groups. Physical Chemistry Chemical Physics 7 (9): 1918–1925.

124 Jaramillo, P., Domingo, L.R., and Perez, P. (2006). Towards an intrinsic nucleofugality scale: The leaving group (LG) ability in CH(3)LG model system. Chemical Physics Letters 420 (1–3): 95–99.

125 Campodonico, P.R., Aizman, A., and Contreras, R. (2006). Empirical scale of nucleophilicity for substituted pyridines. Chemical Physics Letters 422 (1–3): 204–209.

126 Campodonico, P.R., Aizman, A., Contreras, R. (2009) Electrophilicity of quinones and its relationship with hydride affinity. Chemical Physics Letters, 471(1–3), 168–173.

127 Campodonico, P.R., Andres, J., Aizman, A., and Contreras, R. (2007). Nucleofugality index in alpha-elimination reactions. Chemical Physics Letters 439 (1–3): 177–182.

128 Campodonico, P.R., Perez, C., Aliaga, M. et al. (2007). Electrofugality index for benzhydryl derivatives. Chemical Physics Letters 447 (4–6): 375–378.

129 Moens, J., Jaque, P., De Proft, F., and Geerlings, P. (2009). A new view on the spectrochemical and nephelauxetic series on the basis of spin-polarized conceptual DFT. Chemphyschem 10 (5): 847–854.

130 Moens, J., Jaque, P., De Proft, F., and Geerlings, P. (2008). The study of redox reactions on the basis of conceptual DFT principles: EEM and vertical quantities. Journal of Physical Chemistry A 112 (26): 6023–6031.

131 Moens, J., Roos, G., Jaque, P. et al. (2007). Can electrophilicity act as a measure of the redox potential of first-row transition metal ions? Chemistry - A European Journal 13 (33): 9331–9343.

132 Jaque, P., Marenich, A.V., Cramer, C.J., and Truhlar, D.G. (2007). Computational electrochemistry: The aqueous $Ru^{3+}|Ru^{2+}$ reduction potential. Journal of Physical Chemistry C 111 (15): 5783–5799.

133 Gazquez, J.L., Franco-Perez, M., Ayers, P., and Vela, A. (2019). Temperature-dependent approach to chemical reactivity concepts in density functional theory. International Journal of Quantum Chemistry 119 (2): e25797.

134 Chamorro, E., Duque-Norena, M., and Perez, P. (2009). Further relationships between theoretical and experimental models of electrophilicity and nucleophilicity. Journal of Molecular Structure-Theochem 901 (1–3): 145–152.

135 Cedillo, A., Contreras, R., Galvan, M. et al. (2007). Nucleophilicity index from perturbed electrostatic potentials. Journal of Physical Chemistry A 111 (12): 2442–2447.

136 Campodonico, P., Santos, J.G., Andres, J., and Contreras, R. (2004). Relationship between nucleophilicity/electrophilicity indices and reaction mechanisms for the nucleophilic substitution reactions of carbonyl compounds. Journal of Physical Organic Chemistry 17 (4): 273–281.

137 De Vleeschouwer, F., Van Speybroeck, V., Waroquier, M. et al. (2007). Electrophilicity and nucleophilicity index for radicals. Organic Letters 9 (14): 2721–2724.

138 De Vleeschouwer, F., Van Speybroeck, V., Waroquier, M. et al. (2008). An intrinsic radical stability scale from the perspective of bond dissociation enthalpies: a companion to radical electrophilicities. Journal of Organic Chemistry 73 (22): 9109–9120.

139 Jaramillo, P., Fuentealba, P., and Perez, P. (2006). Nucleophilicity scale for n- and pi-nucleophiles. Chemical Physics Letters 427 (4–6): 421–425.

140 Jaramillo, P., Perez, P., Contreras, R. et al. (2006). Definition of a nucleophilicity scale. Journal of Physical Chemistry A 110 (26): 8181–8187.

141 Olah, J., De Proft, F., Veszpremi, T., Geerlings, P. (2006) Relationship between electrophilicity and spin-philicity of divalent and monovalent species of group 14 and 15 elements. Journal of Molecular Structure-Theochem, 771(1–3), 135–140.

142 Padmanabhan, J., Parthasarathi, R., Elango, M. et al. (2007). Multiphilic descriptor for chemical reactivity and selectivity. Journal of Physical Chemistry A 111 (37): 9130–9138.

143 Miranda-Quintana, R.A., Gonzalez, M.M., Hernandez-Castillo, D. et al. (2017). Conceptual DFT analysis of the regioselectivity of 1,3-dipolar cycloadditions: nitrones as a case of study. Journal of Molecular Modeling 23 (8): 15.

144 Domingo, L.R., Rios-Gutierrez, M., Chamorro, E., and Perez, P. (2019). Are one-step aromatic nucleophilic substitutions of non-activated benzenes concerted processes? Organic & Biomolecular Chemistry 17 (35): 8185–8193.

145 Domingo, L.R., Rios-Gutierrez, M., Perez, P., Chamorro, E. (2016) Understanding the 2n+2n Reaction Mechanism between a Carbenoid Intermediate and CO_2. Molecular Physics, 114(7–8), 1374–1391.

146 Domingo, L.R., Ríos-Gutiérrez, M., and Pérez, P. (2016). Applications of the conceptual density functional indices to organic chemistry reactivity. Molecules 2: 748.

147 Domingo, L.R., Rios-Gutierrez, M., Duque-Norena, M. et al. (2016). Understanding the carbenoid-type reactivity of nitrile ylides in 3+2 cycloaddition reactions towards electron-deficient ethylenes: a molecular electron density theory study. Theoretical Chemistry Accounts 135 (7): 12.

148 Domingo, L.R., Rios-Gutierrez, M., Chamorro, E., and Perez, P. (2016). Electrophilic activation of CO_2 in cycloaddition reactions towards a nucleophilic carbenoid intermediate: new defying insights from the molecular electron density theory. Theoretical Chemistry Accounts 136 (1): 13.

149 Domingo, L.R., Pérez, P., and Sáez, J.A. (2013). Understanding the local reactivity in polar organic reactions through electrophilic and nucleophilic Parr functions. RSC Advances 3: 1486–1494.

150 Chamorro, E., Perez, P., and Domingo, L.R. (2013). On the nature of Parr functions to predict the most reactive sites along organic polar reactions. Chemical Physics Letters 582: 141–143.

151 Domingo, L.R., Chamorro, E., and Perez, P. (2008). Understanding the reactivity of captodative ethylenes in polar cycloaddition reactions. a theoretical study. Journal of Organic Chemistry 73 (12): 4615–4624.

152 Domingo, L.R., Chamorro, E., and Perez, P. (2008). An Understanding of the electrophilic/nucleophilic behavior of electro-deficient 2,3-disubstituted 1,3-butadienes in polar Diels-Alder reactions. A density functional theory study. Journal of Physical Chemistry A 112 (17): 4046–4053.

153 Ordon, P. and Tachibana, A. (2005). Nuclear reactivity indices within regional density functional theory. Journal of Molecular Modeling 11 (4–5): 312–316.

154 Ordona, P. and Tachibana, A. (2007). Use of nuclear stiffness in search for a maximum hardness principle and for the softest states along the chemical reaction path: A new formula for the energy third derivative gamma. Journal of Chemical Physics 126 (23): 234115.

155 Giannerini, T. and Borges, I. (2015). Molecular electronic topology and fragmentation onset via charge partition methods and nuclear Fukui functions: 1,1-diamino-2,2-dinitroethylene. Journal of the Brazilian Chemical Society 26 (5): 851–859.

156 Ordon, P., Komorowski, L., and Jeedrzejewski, M. (2017). Conceptual DFT analysis of the fragility spectra of atoms along the minimum energy reaction coordinate. Journal of Chemical Physics 147 (13): 10.

157 Rincon, E., Perez, P., and Chamorro, E. (2007). Global and local reactivity of simple substituted nitrenes and phosphinidenes within the spin-polarized density functional theory framework. Chemical Physics Letters 448: 273–279.

158 Chamorro, E., De Proft, F., Geerlings, P. (2005) Generalized nuclear Fukui functions in the framework of spin-polarized density-functional theory. Journal of Chemical Physics, 123(8), 084104.

159 Chamorro, E., Santos, J.C., Escobar, C.A., and Perez, P. (2006). Electrophilicity and spin polarization of simple substituted silylenes. Chemical Physics Letters 431 (1–3): 210–215.

160 Guerra, D., Andres, J., Chamorro, E., and Perez, P. (2007). Understanding the chemical reactivity of phenylhalocarbene systems: an analysis based on the spin-polarized density functional theory. Theoretical Chemistry Accounts 118 (2): 325–335.

161 Pinter, B., De Proft, F., Van Speybroeck, V. et al. (2007). Spin-polarized conceptual density functional theory study of the regioselectivity in ring closures of radicals. Journal of Organic Chemistry 72 (2): 348–356.

162 Chamorro, E., Perez, P., Duque, M. et al. (2008). Dual descriptors within the framework of spin-polarized density functional theory. Journal of Chemical Physics 129 (064117).

163 Guerra, D., Contreras, R., Cedillo, A. et al. (2009). On the principle of spin potential equalization. Journal of Physical Chemistry A 113 (7): 1390–1396.

164 Cardenas, C., Lamsabhi, A.M., and Fuentealba, P. (2006). Nuclear reactivity indices in the context of spin polarized density functional theory. Chemical Physics 322 (3): 303–310.

165 De Proft, F., Geerlings, P., Chamorro, E. (2005) The nuclear Fukui function: generalization within spin-polarized conceptual density functional theory. Advances in Computational Methods in Sciences and Engineering 2005, Vols 4 A & 4 B, 4A-4B, 1240–1243.

166 De Proft, F., Fias, S., Van Alsenoy, C., and Geerlings, P. (2005). Spin-polarized conceptual density functional theory study of the regioselectivity in the 2+2 photocycloaddition of enones to substituted alkenes. Journal of Physical Chemistry A 109 (28): 6335–6343.

167 Pinter, B., De Proft, F., Veszpremi, T., Geerlings, P. (2008) Theoretical study of the orientation rules in photonucleophilic aromatic substitutions. Journal of Organic Chemistry, 73(4), 1243–1252.

168 Pinter, B., De Proft, F., Veszpremi, T., and Geerlings, P. (2005). Regioselectivity in the 2+2 cyclo-addition reaction of triplet carbonyl compounds to substituted alkenes (Paterno-Buchi reaction): A spin-polarized conceptual DFT approach. Journal of Chemical Sciences 117 (5): 561–571.

169 Jaque, P., Toro-Labbe, A., Geerlings, P., De Proft, F. (2009) Theoretical study of the regioselectivity of 2+2 photocycloaddition reactions of acrolein with olefins. Journal of Physical Chemistry A, 113(1), 332–344.

170 Escano, M.C.S., Nakanishi, H., and Kasai, H. (2009). Spin-polarized density functional theory study of reactivity of diatomic molecule on bimetallic system: the case of O_2 dissociative adsorption on Pt monolayer on Fe(001). Journal of Physical Chemistry A 113 (52): 14302–14307.

171 Gal, T., Ayers, P.W., De Proft, F., Geerlings, P. (2009) Nonuniqueness of magnetic fields and energy derivatives in spin-polarized density functional theory. Journal of Chemical Physics, 131(15), 7.

172 Ugur, I., De Vleeschouwer, F., Tuzun, N. et al. (2009). Cyclopolymerization reactions of diallyl monomers: exploring electronic and steric effects using DFT reactivity indices. Journal of Physical Chemistry A 113 (30): 8704–8711.

173 Feng, X.T., Yu, J.G., Liu, R.Z. et al. (2010). Why iron? A spin-polarized conceptual density functional theory study on metal-binding specificity of porphyrin. Journal of Physical Chemistry A 114 (21): 6342–6349.

174 Perez, P. and Chamorro, E. (2010). Global and local reactivity of N-heterocyclic carbenes with boron and phosphorus atoms: An analysis based on spin polarized density functional framework. Journal of Molecular Structure-Theochem 943 (1–3): 110–114.

175 Lozano-Espinosa, M., Guevara-Garcia, A., and Galvan, M. (2016). Sensing the active site properties of enzymes as a function of the size of an effective peptidic environment using DFT reactivity parameters. Theoretical Chemistry Accounts 135 (9): 13.

176 De Vleeschouwer, F., Jaque, P., Geerlings, P. et al. (2010). Regioselectivity of radical additions to substituted alkenes: insight from conceptual density functional theory. Journal of Organic Chemistry 75 (15): 4964–4974.

177 De Vleeschouwer, F., De Proft, F., and Geerlings, P. (2010). Conceptual density functional theory based intrinsic radical stabilities: Application to substituted silylenes and p-benzynes. Journal of Molecular Structure-Theochem 943 (1–3): 94–102.

178 Ziane, M., Amitouche, F., Bouarab, S., and Vega, A. (2017). Density functional study of the structural, electronic, and magnetic properties of Mo-n and MonS (n=1–10) clusters. Journal of Nanoparticle Research 19 (12): 15.

179 Tazibt, S., Chilthaoui, A., Bouarab, S., and Vega, A. (2017). Structural, electronic, and magnetic properties of iron disulfide $Fe_nS_2^{0/\pm}$ (n=1–6) Clusters. Journal of Physical Chemistry A 121 (19): 3768–3780.

180 Navarro-Santos, P., Rodriguez-Olalde, N.E., Gallo, M. et al. (2019). On the initial stages of lignin polymerization through spin-polarized density functional theory. Chemical Physics Letters 730: 289–296.

181 Vargas, R., Cedillo, A., Garza, J., and Galvan, M. (2002). Reactivity criteria in spin-polarized density functional theory. Reviews of Modern Quantum Chemistry 2: 936–965.

182 Galvan, M. and Vargas, R. (1992). Spin Potential in Kohn Sham Theory. Journal of Physical Chemistry 96 (4): 1625–1630.

183 Vargas, R. and Galván, M. (2009). Chapter 10. Spin-polarized density functional theory: chemical reactivity. In: Chemical Reactivity Theory : A Density Functional View (eds. P.K. Chattaraj and editor), 137–153. Boca Raton: CRC Press Taylor & Francis Group.

184 Perez, P., Chamorro, E., and Ayers, P.W. (2008). Universal mathematical identities in density functional theory: Results from three different spin-resolved representations. Journal of Chemical Physics 128 (20): 204108.

185 De Proft, F., Chamorro, E., Pérez, P. et al. (2009). Spin-polarized reactivity indices from density functional theory: theory and applications. In: Specialist Periodical Reports: Chemical Modeling: Applications and Theory (ed. M. eds Springborg), 63–111. Cambridge: The Royal Society of Chemistry (RSC) Publishing.

186 Miranda-Quintana, R.A. and Ayers, P.W. (2016). Systematic treatment of spin-reactivity indicators in conceptual density functional theory. Theoretical Chemistry Accounts 135 (10): 1–18.

187 Fias, S., Boisdenghien, Z., De Proft, F., Geerlings, P. (2014) The spin polarized linear response from density functional theory: Theory and application to atoms. Journal of Chemical Physics, 141(18), 12.

188 Boisdenghien, Z., Fias, S., Van Alsenoy, C. et al. (2014). Evaluating and interpreting the chemical relevance of the linear response kernel for atoms II: open shell. Physical Chemistry Chemical Physics 16 (28): 14614–14624.

189 Reyes, A., Cuervo, P.A., Orozco, F. et al. (2013). Theoretical investigation of the selectivity in intramolecular cyclizations of some 2′-aminochalcones to dihydroquinolin-8-ones and indolin-3-ones. Journal of Molecular Modeling 19 (9): 3611–3618.

190 Alcoba, D.R., Lain, L., Torre, A. et al. (2013). Fukui and dual-descriptor matrices within the framework of spin-polarized density functional theory. Physical Chemistry Chemical Physics 15 (24): 9594–9604.

191 Chan, G.K.L. (1999). A fresh look at ensembles: Derivative discontinuities in density functional theory. Journal of Chemical Physics 110 (10): 4710–4723.

192 Gal, T. and Geerlings, P. (2010). Derivative of the Lieb definition for the energy functional of density-functional theory with respect to the particle number and the spin number. Physical Review A 81 (3): 8.

193 Chandra, A.K. and Nguyen, M.T. (2007). Use of DFT-based reactivity descriptors for rationalizing radical addition reactions: applicability and difficulties. Faraday Discussions 135: 191–201.

194 Hemelsoet, K., Van Speybroeck, V., and Waroquier, M. (2007). How useful are reactivity indicators for the description of hydrogen abstraction reactions on polycyclic aromatic hydrocarbons? Chemical Physics Letters 444 (1–3): 17–22.

195 Hemelsoet, K., Van Speybroeck, V., Marin, G.B. et al. (2004). Reactivity indices for radical reactions involving polyaromatics. Journal of Physical Chemistry A 108 (35): 7281–7290.

196 De Proft, F., Pauwels, E., Lahorte, P. et al. (2004). Density functional theory as a tool for the structure determination of radiation-induced bioradicals. Magnetic Resonance in Chemistry 42: S3–S19.

197 Pauwels, E., Lahorte, P., Vanhaelewyn, G. et al. (2002). Tentative structures for the radiation-induced radicals in crystalline beta-D-fructose using density functional theory. Journal of Physical Chemistry A 106 (51): 12340–12348.

198 Vanhaelewyn, G., Lahorte, P., De Proft, F. et al. (2001). Electron magnetic resonance study of stable radicals in irradiated D-fructose single crystals. Physical Chemistry Chemical Physics 3 (9): 1729–1735.

199 Lahorte, P., De Proft, F., Vanhaelewyn, G. et al. (1999). Density functional calculations of hyperfine coupling constants in alanine-derived radicals. Journal of Physical Chemistry A 103 (33): 6650–6657.

200 Lahorte, P., De Proft, F., Callens, F. et al. (1999). A density functional study of hyperfine coupling constants in steroid radicals. Journal of Physical Chemistry A 103 (50): 11130–11135.

201 DeProft, F. and Geerlings, P. (1996). Ab initio and density functional study of the Jahn-Teller distortion in the silane radical cation. Chemical Physics Letters 262 (6): 782–788.

202 Togo, H. (2004). Advanced Free Radical Reactions for Organic Synthesis. Elsevier Science.

203 Carey, F.A. and Sundberg, R.J. (2008). Advanced Organic Chemistry: Part A: Structure and Mechanisms/Part B: Reaction and Synthesis. Springer.

204 Smith, M.B. and March, J. (2007). Advanced Organic Chemistry: Reactions, Mechanisms, and Structure. Wiley-Interscience.

205 Fleming, I. (1976). Frontier Orbitals and Organic Chemical Reactions. Wiley.

206 Ayers, P.W. and Yang, W.T. (2006). Legendre-transform functionals for spin-density-functional theory. Journal of Chemical Physics 124 (22): 224108.

207 Vargas, R., Cedillo, A., Garza, J., and Galván, M. (2002). Reviews of modern quantum chemistry: a celebration of the contributions of Robert, G. Parr. In: Reviews of Modern Quantum Chemistry (eds. R.G. Parr, K.D. Sen and editors), 936–965. Singapore: World Scientific Publishing Company.

208 Barth, U.V. and Hedin, L. (1972). A local exchange-correlation potential for the spin polarized case. I. J Phys C 5: 1629.

209 Eschrig, H. and Pickett, W.E. (2001). Density functional theory of magnetic systems revisited. Solid State Commun 118: 123.

210 Ullrich, C.A. (2005). Nonuniqueness in spin-density-functional theory on lattices. Phys Rev B 72 073102.

211 Holas, A. and Balawender, R. (2006). Comment on "Legendre-transform functionals for spin-density-functional theory". Journal of Chemical Physics 125: 247101.

212 Capelle, K. and Vignale, G. (2001). Nonuniqueness of the potentials of spin-density-functional theory. Physical Review Letters 86 (24): 5546–5549.

213 Capelle, K. and Vignale, G. (2002). Nonuniqueness and derivative discontinuities in density-functional theories for current-carrying and superconducting systems. Physical Review B 65 (11).

214 Capelle, K. and Libero, V.L. (2005). Spin-density functional theory: Some open problems and application to inhomogeneous Heisenberg models. International Journal of Quantum Chemistry 105 (6): 679–686.

215 Gritsenko, O.V. and Baerends, E.J. (2004). The spin-unrestricted molecular Kohn-Sham solution and the analogue of Koopman's theorem for open-shell molecules. Journal of Chemical Physics 120 (18): 8364–8372.

216 Gritsenko, O.V. and Baerends, E.J. (2002). The analog of Koopman's theorem in spin-density functional theory. Journal of Chemical Physics 117 (20): 9154–9159.

217 Ayers, P.W. (2006). Axiomatic formulations of the Hohenberg-Kohn functional. Physical Review A 73 (1).

218 Chamorro, E., Perez, P., Duque, M. et al. (2008). Dual descriptors within the framework of spin-polarized density functional theory. Journal of Chemical Physics 129 (6): 064117.

219 Olah, J., Veszpremi, T., and Nguyen, M.T. (2005). Spin-philicity and spin-donicity of simple nitrenes and phosphinidenes. Chemical Physics Letters 401 (4–6): 337–341.

220 Olah, J., Veszpremi, T., De Proft, F., and Geerlings, P. (2007). Silylenes: A unified picture of their stability, acid-base and spin properties, nucleophilicity, and electrophilicity via computational and conceptual density functional theory. Journal of Physical Chemistry A 111 (42): 10815–10823.

221 Galvan, M., Vela, A., and Gazquez, J.L. (1988). Chemical reactivity in spin-polarized density functional theory. Journal of Physical Chemistry 92 (22): 6470–6474.

222 Ghanty, T.K. and Ghosh, S.K. (1994). Spin-polarized generalization of the concepts of electronegativity and hardness and the description of chemical binding. Journal of the American Chemical Society 116 (9): 3943–3948.

223 Galvan, M. and Vargas, R. (1992). Spin-potential in Kohn-Sham theory. Journal of Physical Chemistry 96 (4): 1625–1630.

224 Bochevarov, A.D., Harder, E., Hughes, T.F. et al. (2013). Jaguar: A high-performance quantum chemistry software program with strengths in life and materials sciences. International Journal of Quantum Chemistry 113 (18): 2110–2142.

225 Frisch, M.J., Trucks, G.W., Schlegel, H.B. et al. (2017). Gaussian 16, Revision. In: B, vol. 01. Wallingford CT.: Gaussian, Inc.

226 Lu, T. and Chen, F. (2012). Multiwfn: A multifunctional wavefunction analyzer. Journal of Computational Chemistry 33 (5): 580–592.

227 Morell, C., Ayers, P.W., Grand, A. et al. (2008). Rationalization of Diels-Alder reactions through the use of the dual reactivity descriptor Delta f(r). Physical Chemistry Chemical Physics 10 (48): 7239–7246.

228 Ayers, P.W. and Parr, R.G. (2008). Beyond electronegativity and local hardness: Higher-order equalization criteria for determination of a ground-state electron density. Journal of Chemical Physics 129 (5) 054111.

229 Harbola, M.K., Chattaraj, P.K., and Parr, R.G. (1991). Aspects of the softness and hardness concepts of density-functional theory. Israel Journal of Chemistry 31 (4): 395–402.

230 Harbola, M.K., Parr, R.G., and Lee, C.T. (1991). Hardnesses from Electrostatic Potentials. Journal of Chemical Physics 94 (9): 6055–6056.

231 Chattaraj, P.K., Cedillo, A., and Parr, R.G. (1995). Variational method for determining the Fukui function and chemical hardness of an electronic system. Journal of Chemical Physics 103 (17): 7645–7646.

232 Ghosh, S.K. (1990). Energy derivatives in density-functional theory. Chemical Physics Letters 172 (1): 77–82.

233 Sureshkumar, B., Mary, Y.S., Panicker, C.Y. et al. (2020). Quinoline derivatives as possible lead compounds for anti-malarial drugs: Spectroscopic, DFT and MD study. Arabian Journal of Chemistry 13 (1): 632–648.

234 Klisuric, O.R., Armakovic, S.J., Armakovic, S. et al. (2020). Structural, biological and in-silico study of quinoline-based chalcogensemicarbazones. Journal of Molecular Structure 1203: 12.

235 Sureshkumar, B., Mary, Y.S., Suma, S. et al. (2018). Spectroscopic characterization of 8-hydroxy-5-nitroquinoline and 5-chloro-8-hydroxy quinoline and investigation of its reactive properties by DFT calculations and molecular dynamics simulations. Journal of Molecular Structure 1164: 525–538.

236 Bielenica, A., Beegum, S., Mary, Y.S. et al. (2020). Experimental and computational analysis of 1-(4-chloro-3-nitrophenyl)-3-(3,4-dichlorophenyl)thiourea. Journal of Molecular Structure 1205: 10.

237 Menon, V.V., Mary, Y.S., Mary, Y.S. et al. (1155). (2018) Combined spectroscopic, 184–195. DFr: TD-DFT and MD study of newly synthesized thiourea derivative. Journal of Molecular Structure.

238 War, J.A., Jalaja, K., Mary, Y.S. et al. (2017). Spectroscopic characterization of 1- 3-(1H-imidazol-1-yl)propyl -3-phenylthiourea and assessment of reactive and optoelectronic properties employing DFT calculations and molecular dynamics simulations. Journal of Molecular Structure 1129: 72–85.

239 Aswathy, V.V., Mary, Y.S., Jojo, P.J. et al. (2017). Investigation of spectroscopic, reactive, transport and docking properties of 1-(3,4-dichloropheny1)-3 3-(trifluoromethyl)phenyl thiourea (ANF-6): Combined experimental and computational study. Journal of Molecular Structure 1134: 668–680.

240 Mary, Y.S., Aswathy, V.V., Panicker, C.Y. et al. (2016). Spectroscopic, single crystal XRD structure, DFT and molecular dynamics investigation of 1-(3-chloro-4-fluorophenyl)-3-3-(trifluoromethyl)phenyl thiourea. RSC Advances 6 (113): 111997–112015.

241 Mary, Y.S., Al-Shehri, M.M., Jalaja, K. et al. (2017). Synthesis, vibrational spectroscopic investigations, molecular docking, antibacterial studies and molecular dynamics study of 5- (4-nitrophenyl)acetamido -2-(4-tert-butylphenyl)benzoxazole. Journal of Molecular Structure 1133: 557–573.

242 Aswathy, V.V., Alper-Hayta, S., Yalcin, G. et al. (2017). Modification of benzoxazole derivative by bromine-spectroscopic, antibacterial and reactivity study using experimental and theoretical procedures. Journal of Molecular Structure 1141: 495–511.

243 Beegum, S., Mary, Y.S., Panicker, Y. et al. (2019). Spectroscopic, antimicrobial and computational study of novel benzoxazole derivative. Journal of Molecular Structure 1176: 881–894.

244 Vennila, P., Govindaraju, M., Venkatesh, G. et al. (2018). A complete computational and spectroscopic study of 2-bromo-1, 4-dichlorobenzene – A frequently used benzene derivative. Journal of Molecular Structure 1151: 245–255.

245 Venkatesh, G., Kamal, C., Vennila, P. et al. (1171). (2018) Molecular dynamic simulations, 253–267. ALIE surface: Fukui functions geometrical, molecular docking and vibrational spectra studies of tetra chloro p and m-xylene. Journal of Molecular Structure.

246 Crisan, L., Borota, A., Bora, A., and Pacureanu, L. (2019). Diarylthiazole and diarylimidazole selective COX-1 inhibitor analysis through pharmacophore modeling, virtual screening, and DFT-based approaches. Structural Chemistry 30 (6): 2311–2326.

247 Thomas, R., Hossain, M., Mary, Y.S. et al. (2018). Spectroscopic analysis and molecular docking of imidazole derivatives and investigation of its reactive properties by DFT and molecular dynamics simulations. Journal of Molecular Structure 1158: 156–175.

248 Hossain, M., Thomas, R., Mary, Y.S. et al. (2018). Understanding reactivity of two newly synthetized imidazole derivatives by spectroscopic characterization and computational study. Journal of Molecular Structure 1158: 176–196.

249 Benzon, K.B., Sheena, M.Y., Panicker, C.Y. et al. (2017). Studies on the synthesis, spectroscopic analysis, molecular docking and DET calculations on 1-hydroxy-2-(4-hydroxyphenyl)-4,5-dimethyl-imidazol 3-oxide. Journal of Molecular Structure 1130: 644–658.

250 Benzon, K.B., Mary, Y.S., Varghese, H.T. et al. (2017). Spectroscopic, DFT, molecular dynamics and molecular docking study of 1-butyl-2-(4-hydroxypheny1)-4,5-dimethyl-imidazole 3-oxide. Journal of Molecular Structure 1134: 330–344.

251 Mary, Y.S., Miniyar, P.B., Mary, Y.S. et al. (2018). Synthesis and spectroscopic study of three new oxadiazole derivatives with detailed computational evaluation of their reactivity and pharmaceutical potential. Journal of Molecular Structure 1173: 469–480.

252 Al-Tamimi, A.M.S., Mary, Y.S., Miniyar, P.B. et al. (2018). Synthesis, spectroscopic analyses, chemical reactivity and molecular docking study and anti-tubercular activity of pyrazine and condensed oxadiazole derivatives. Journal of Molecular Structure 1164: 459–469.

253 Mary, Y.S., Al-Omary, F.A.M., Mostafa, G.A.E. et al. (2017). Insight into the reactive properties of newly synthesized 1,2,4-triazole derivative by combined experimental (FT-IR and FR-Raman) and theoretical (DFT and MD) study. Journal of Molecular Structure 1141: 542–550.

254 Devi, A.S., Aswathy, V.V., Mary, Y.S. et al. (2017). Synthesis, XRD single crystal structure analysis, vibrational spectral analysis, molecular dynamics and molecular docking studies of 2-(3-methoxy-4-hydroxyphenyl) benzothiazole. Journal of Molecular Structure 1148: 282–292.

255 El-Azab, A.S., Mary, Y.S., Abdel-Aziz, A.A.M. et al. (2018). Synthesis, spectroscopic analyses (FT-IR and NMR), vibrational study, chemical reactivity and molecular docking study and anti-tubercular activity of condensed oxadiazole and pyrazine derivatives. Journal of Molecular Structure 1156: 657–674.

256 Muthukkumar, M., Bhuvaneswari, T., Venkatesh, G. et al. (2018). Synthesis, characterization and computational studies of semicarbazide derivative. Journal of Molecular Liquids 272: 481–495.

257 Murthy, P.K., Suneetha, V., Armakovic, S. et al. (2018). Synthesis, characterization and computational study of the newly synthetized sulfonamide molecule. Journal of Molecular Structure 1153: 212–229.

258 Beegum, S., Mary, Y.S., Mary, Y.S. et al. (2020). Exploring the detailed spectroscopic characteristics, chemical and biological activity of two cyanopyrazine-2-carboxamide derivatives using experimental and theoretical tools. Spectrochimica Acta Part a-Molecular and Biomolecular Spectroscopy 224: 13.

259 Mary, S.Y., Al-Abdullah, E.S., Aljohar, H.I. et al. (2018). 4- (4-Acetylphenyl)amino -2-methylidene-4-oxobutanoic acid, a newly synthesized amide with hydrophilic and hydrophobic segments: Spectroscopic characterization and investigation of its reactive properties. Journal of the Serbian Chemical Society 83 (1): 1–18.

260 Ranjith, P.K., Al-Abdullah, E.S., Al-Omary, F.A.M. et al. (2017). FT-IR and FT-Raman characterization and investigation of reactive properties of N-(3-iodo-4-methylphenyl)pyrazine-2-carboxamide by molecular dynamics simulations and DFT calculations. Journal of Molecular Structure 1136: 14–24.

261 Murthy, P.K., Mary, Y.S., Mary, Y.S. et al. (2017). Synthesis, crystal structure analysis, spectral investigations, DFT computations and molecular dynamics and docking study of 4-benzyl-5-oxomorpholine-3-carbamide, a potential bioactive agent. Journal of Molecular Structure 1134: 25–39.

262 Bonsaii, M., Gholivand, K., Khosravi, M., and Abdi, K. (2017). Negative hyperconjugation effect on the reactivity of phosphoramide mustard derivatives as a DNA alkylating agent: theoretical and experimental insights. New Journal of Chemistry 41 (19): 11036–11052.

263 Beegum, S., Mary, Y.S., Varghese, H.T. et al. (2017) Vibrational spectroscopic analysis of cyanopyrazine-2-carboxamide derivatives and investigation of their reactive properties by DFT calculations and molecular dynamics simulations. Journal of Molecular Structure, 1131, 1–15.

264 Murthy, P.K., Krishnaswamy, G., Armakovic, S. et al. (2018). Structural and spectroscopic characterization, reactivity study and charge transfer analysis of the newly synthetized 2-(6-hydroxy-1-benzofuran-3-yl) acetic acid. Journal of Molecular Structure 1162: 81–95.

265 Mary, Y.S., Mary, Y.S., Panicker, C.Y. et al. (1148). (2017) Investigation of reactive and spectroscopic properties of oxobutanoic acid derivative: Combined spectroscopic, 266–275. DFT: MD and docking study. Journal of Molecular Structure.

266 Nayak, N., Prasad, K.S., Pillai, R.R. et al. (2018). Remarkable colorimetric sensing behavior of pyrazole-based chemosensor towards Cu(II) ion detection: synthesis, characterization and theoretical investigations. RSC Advances 8 (32): 18023–18029.

267 Zainuri, D.A., Arshad, S., Khalib, N.C. et al. (2017). Synthesis, XRD crystal structure, spectroscopic characterization (FT-IR, H-1 and C-13 NMR), DFT studies, chemical reactivity and bond dissociation energy studies using molecular dynamics simulations and evaluation of antimicrobial and antioxidant activities of a novel chalcone derivative, (E)-1-(4-bromopheny1)-3-(4-iodophenyl)prop-2-en-l-one. Journal of Molecular Structure 1128: 520–533.

268 Sureshkumar, B., Mary, Y.S., Panicker, C.Y. et al. (2017). Spectroscopic analysis of 8-hydroxyquinoline-5-sulphonic acid and investigation of its reactive properties by DFT and molecular dynamics simulations. Journal of Molecular Structure 1150: 540–552.

269 Ranjith, P.K., Mary, Y.S., Panicker, C.Y. et al. (2017). New quinolone derivative: Spectroscopic characterization and reactivity study by DFT and MD approaches. Journal of Molecular Structure 1135: 1–14.

270 Sultan, M.A., Almansour, A.I., Pillai, R.R. et al. (2017). Synthesis, theoretical studies and molecular docking of a novel chlorinated tetracyclic: (Z/E)-3-(1,8-dichloro-9,10-dihydro-9,10-ethanoanthracen-11-yl)acrylaldehyde. Journal of Molecular Structure 1150: 358–365.

271 Radanovic, M.M., Rodic, M.V., Armakovic, S. et al. (2017). Pyridoxylidene aminoguanidine and its copper(II) complexes - Syntheses, structure, and DFT calculations. Journal of Coordination Chemistry 70 (16): 2870–2887.

272 Onawole, A.T., Al-Ahmadi, A.F., Mary, Y.S. et al. (2017). Conformational, vibrational and DFT studies of a newly synthesized arylpiperazine-based drug and evaluation of its reactivity towards the human GABA receptor. Journal of Molecular Structure 1147: 266–280.

273 Jalaja, K., Mary, Y.S., Panicker, C.Y. et al. (2017). Spectroscopic characterization of 4 2-(5-Ethylpyridin-2-yl)ethoxy benzaldehyde oxime and investigation of its reactive properties by DFT calculations and molecular dynamics simulations. Journal of Molecular Structure 1128: 245–256.

274 El-Azab, A.S., Mary, Y.S., Mary, Y.S. et al. (2017). Newly synthesized dihydroquinazoline derivative from the aspect of combined spectroscopic and computational study. Journal of Molecular Structure 1134: 814–827.

275 Murthy, P.K., Valverde, C., Suneetha, V. et al. (2019). An analysis of structural and spectroscopic signatures, the reactivity study of synthetized 4,6-dichloro-2-(methylsulfonyl)pyrimidine: A potential third-order nonlinear optical material. Journal of Molecular Structure 1186: 263–275.

276 Al-Omary, F.A.M., Mary, Y.S., Beegum, S. et al. (2017). Molecular conformational analysis, reactivity, vibrational spectral analysis and molecular dynamics and docking studies of

6-chloro-5-isopropylpyrimidine-2,4(1H,3H)-dione, a potential precursor to bioactive agent. Journal of Molecular Structure 1127: 427–436.

277 Fashe, M.M., Juvonen, R.O., Petsalo, A. et al. (2015). In silico prediction of the site of oxidation by cytochrome P450 3A4 that leads to the formation of the toxic metabolites of pyrrolizidine alkaloids. Chemical Research in Toxicology 28 (4): 702–710.

278 El-Labbad, E.M., Ismail, M.A.H., Abou El Ella, D.A. et al. (2015). Discovery of novel peptidomimetics as irreversible CHIKV NsP2 protease inhibitors using quantum mechanical-based ligand descriptors. Chemical Biology & Drug Design 86 (6): 1518–1527.

279 Melin, J., Aparicio, F., Galvan, M. et al. (2003). Chemical reactivity in the {N, N-S, nu(r)} space. Journal of Physical Chemistry A 107 (19): 3831–3835.

280 Kraka, E. and Cremer, D. (1994). CCSD(T) Investigation of the Bergman cyclization of enediyne. Relative stability of o-, m-, and p-didehydrobenzene. Journal of the American Chemical Society 116 (11): 4929–4936.

281 Guerra, D., Contreras, R., Perez, P., and Fuentealba, P. (2006). Hardness and softness kernels, and related indices in the spin polarized version of density functional theory. Chemical Physics Letters 419 (1–3): 37–43.

282 Peng, D.G. and Yang, W.T. (2013). Fukui function and response function for nonlocal and fractional systems. Journal of Chemical Physics 138 (18): 15.

283 Vargas, R., Garza, J., and Cedillo, A. (2005). Koopmans-like approximation in the Kohn-Sham method and the impact of the frozen core approximation on the computation of the reactivity parameters of the density functional theory. Journal of Physical Chemistry A 109 (39): 8880–8892.

284 Morell, C., Grand, A., and Toro-Labbe, A. (2005). New dual descriptor for chemical reactivity. Journal of Physical Chemistry A 109 (1): 205–212.

285 Franco-Perez, M., Polanco-Ramirez, C.A., Ayers, P.W. et al. (2017). New Fukui, dual and hyper-dual kernels as bond reactivity descriptors. Physical Chemistry Chemical Physics 19 (24): 16095–16104.

286 Perez, P. (2003). Theoretical evaluation of the global and local electrophilicity patterns of singlet carbenes. Journal of Physical Chemistry A 107 (4): 522–525.

287 Guerra, D., Castillo, R., Andres, J. et al. (2006). Homofugality: A new reactivity index describing the leaving group ability in homolytic substitution reactions. Chemical Physics Letters 424 (4–6): 437–442.

288 Anderson, J.S.M. and Ayers, P.W. (2007). Predicting the reactivity of ambidentate nucleophiles and electrophiles using a single, general-purpose, reactivity indicator. Physical Chemistry Chemical Physics 9 (19): 2371–2378.

289 Contreras, R., Galvan, M., Oliva, M. et al. (2008). Two state reactivity mechanism for the rearrangement of hydrogen peroxynitrite to nitric acid. Chemical Physics Letters 457 (1–3): 216–221.

290 Guerra, D., Fuentealba, P., Aizman, A., and Contreras, R. (2007). beta-Scission of thioimidoyl radicals (R-1-N-C-center dot=S-R-2): A theoretical scale of radical leaving group ability. Chemical Physics Letters 443 (4–6): 383–388.

291 Campodonico, P.R., Aizman, A., and Contreras, R. (2006). Group electrophilicity as a model of nucleofugality in nucleophilic substitution reactions. Chemical Physics Letters 422 (4–6): 340–344.

292 Broeckaert, L., Moens, J., Roos, G. et al. (2008). Intrinsic nucleofugality scale within the framework of density functional reactivity theory. Journal of Physical Chemistry A 112 (47): 12164–12171.

293 Melin, J., Ayers, P.W., and Ortiz, J.V. (2005). The electron-propagator approach to conceptual density-functional theory. Journal of Chemical Sciences 117 (5): 387–400.

294 Martinez-Araya, J.I., Yepes, D., and Jaque, P. (2018). A 3D visualization of the substituent effect A brief analysis of two components of the operational formula of dual descriptor for open-shell systems. Journal of Molecular Modeling 24 (1): 8.

295 Martinez-Araya, J.I., Grand, A., and Glossman-Mitnik, D. (2015). Towards the rationalization of catalytic activity values by means of local hyper-softness on the catalytic site: a criticism about the use of net electric charges. Physical Chemistry Chemical Physics 17 (44): 29764–29775.

296 Nakatsuji, H. (1974). Electron-cloud following and preceding and the shapes of molecules. Journal of the American Chemical Society 96 (1): 30–37.

297 Nakatsuji, H. (1974). Common nature of the electron cloud of a system undergoing change in nuclear configuration. Journal of the American Chemical Society 96 (1): 24–30.

298 Nakatsuji, H. (1973). Electrostatic force theory for a molecule and interacting molecules. I. Concept and illustrative applications. Journal of the American Chemical Society 95 (2): 345–354.

299 Nakatsuji, H., Kanda, K., and Yonezawa, T. (1980). Force in SCF Theories. Chemical Physics Letters 75 (2): 340–346.

300 Cardenas, C., Chamorro, E., Galvan, M., and Fuentealba, P. (2007). Nuclear Fukui functions from nonintegral electron number calculations. International Journal of Quantum Chemistry 107 (4): 807–815.

301 Bochevarov, A.D., Harder, E., Hughes, T.F. et al. (2013). Jaguar: A high-performance quantum chemistry software program with strengths in life and materials sciences. International Journal of Quantum Chemistry 113 (18): 2110–2142.

302 Elliott, P., Burke, K., Cohen, M.H., and Wasserman, A. (2010). Partition density-functional theory. Physical Review A 82 (2): 4.

303 Mosquera, M.A. and Wasserman, A. (2013). Partition density functional theory and its extension to the spin-polarized case. Molecular Physics 111 (4): 505–515.

304 Otrokov, M.M., Klimovskikh, I.I., Bentmann, H. et al. (2019). Prediction and observation of an antiferromagnetic topological insulator. Nature 576 (7787): 416–422.

305 Deng, Y.J., Yu, Y.J., Song, Y.C. et al. (2018). Gate-tunable room-temperature ferromagnetism in two-dimensional Fe_3GeTe_2. Nature 563 (7729): 94–99.

306 Wolf, S.A., Awschalom, D.D., Buhrman, R.A. et al. (2001). Spintronics: A spin-based electronics vision for the future. Science 294 (5546): 1488–1495.

307 Zutic, I., Fabian, J., and Das Sarma, S. (2004). Spintronics: Fundamentals and applications. Reviews of Modern Physics 76 (2): 323–410.

308 Coronado, E. (2020). Molecular magnetism: from chemical design to spin control in molecules, materials and devices. Nature Reviews Materials 5 (2): 87–104.

309 Smidstrup, S., Markussen, T., Vancraeyveld, P. et al. (2020). QuantumATK: an integrated platform of electronic and atomic-scale modelling tools. Journal of Physics-Condensed Matter 32 (1): 36.

310 Domingo, L.R. and Pérez, P. (2013). Global and local reactivity indices for electrophilic/nucleophilic free radicals. Organic & Biomolecular Chemistry 11 (26): 4350–4358.

311 Cardenas, C., Chamorro, E., and Notario, R. (2005). Nature of bonding in the cyclization reactions of (2-ethynylphenyl)triazene and 2-ethynylstyrene. Journal of Physical Chemistry A 109 (19): 4352–4358.

312 Kimball, D.B., Weakley, T.J.R., Herges, R., and Haley, M.M. (2002). Deciphering the mechanistic dichotomy in the cyclization of 1-(2-ethynylphenyl)-3,3-dialkyltriazenes: Competition between pericyclic and pseudocoarctate pathways. Journal of the American Chemical Society 124 (45): 13463–13473.
313 Nakayama, J., Furuya, T., Suzuki, Y. et al. (2009). face-selective 1,3-dipolar cycloadditions of 3,4-di-tert-butylthiophene 1-oxide with 1,3-dipoles. Phosphorus Sulfur and Silicon and the Related Elements 184 (5): 1175–1183.
314 Ess, D.H. and Houk, K.N. (2008). Theory of 1,3-dipolar cycloadditions: Distortion/interaction and frontier molecular orbital models. Journal of the American Chemical Society 130 (31): 10187–10198.
315 Ess, D.H. and Houk, K.N. (2007). Distortion/Interaction energy control of 1,3-dipolar cycloaddition reactivity. Journal of the American Chemical Society 129 (35): 10646–10647.
316 Ibata, T. and Singh, G.S. (1994). Formation and reaction of azomethine ylide by the reaction of Cu(Acac)2-ketocarbenoids with 1,1-diphenylmethanimine. Chemistry Letters 1: 81–84.
317 Padwa, A. (1984). 1,3-Dipolar Cycloaddition Chemistry. New York: Wiley-Interscience.
318 Domingo, L.R., Chamorro, E., and Perez, P. (2009). An analysis of the regioselectivity of 1,3-dipolar cycloaddition reactions of benzonitrile N-oxides based on global and local electrophilicity and nucleophilicity indices. European Journal of Organic Chemistry 2009 (18): 3036–3044.
319 Domingo, L.R. (1999). Theoretical study of the 1,3-dipolar cycloaddition reactions of azomethine ylides. A DFT study of reaction between trifluoromethyl thiomethyl azomethine ylide and acronitrile. Journal of Organic Chemistry 64 (11): 3922–3929.
320 Domingo, L.R., Chamorro, E., and Perez, P. (2010). Understanding the high reactivity of the azomethine ylides in [3 + 2] cycloaddition reactions. Letters in Organic Chemistry 7 (6).
321 Franco-Perez, M., Ayers, P.W., Gazquez, J.L., and Vela, A. (2017). Thermodynamic responses of electronic systems. Journal of Chemical Physics 147 (9): 11.
322 Franco-Perez, M., Gazquez, J.L., Ayers, P.W., and Vela, A. (2017). Thermodynamic hardness and the maximum hardness principle. Journal of Chemical Physics 147 (7): 7.

9

Chemical Binding and Reactivity Parameters: A Unified Coarse Grained Density Functional View

Swapan K. Ghosh

UM-DAE-Centre for Excellence in Basic Sciences, University of Mumbai, Kalina, Santacruz (East), Mumbai 400098, India

9.1 Introduction

This chapter is about portraying of a physical picture of chemical binding and chemical reactivity within a single theoretical framework based on Density Functional Theory (DFT) [1–3]. Chemical binding is essentially a collective effect comprised of atomic confinement, interatomic charge and spin transfer, density reorganization, and response properties. Thus, in this report, the emphasis has been on the conceptual aspects rather than accurate calculation by solving some variant of the Schrodinger equation in a more elaborate manner.

When atoms form molecules, their freedom gets restricted, due to spatial confinement, perturbation by other participating atoms and related consequences, although there are situations where electrons acquire more freedom due to delocalization effects. The atoms, however, may enjoy the warmth of interatomic interaction producing molecules which are at the root of whole of chemistry, physics, biology, and other sciences, a truly remarkable union. The attempt to have a deeper understanding of the molecule formation has thus been an old pursuit and in the process a wide range of chemical concepts have crept in from time to time as a stream of novelty, richness, and unity in diversity. Since the driving force behind the birth of these concepts was mainly to explain and rationalize experimental facts, sometimes many of them might have faced criticism for lack of rigor, but due to their versatility and usefulness, they have survived the test of time. Of course, attempt has always been made to provide physical foundation to these chemical concepts, particularly whenever a new theoretical machinery has been available. Since molecule formation or chemical binding is essentially a quantum mechanical effect, a purely classical picture may not be possible. Although many quantum mechanical approaches in the form of valence bond or molecular orbital theory have been quite successful, a true physical picture of the chemical bond would be highly desirable. Since it is a consequence of electron density reorganization, DFT, being a theory of electron density, will be the most appropriate tool for discussing the theoretical approaches to the chemical concepts. The field dealing with conceptual aspects within the framework of DFT is often called conceptual DFT [4].

Molecules or materials can be viewed as a collection of atoms bonded together by the electron glue, and different systems correspond to different selection and arrangements of the atoms. Different phenomena analogously arise from change of one configuration of atoms to another.

*Email: swapankghosh@gmail.com, swapan.ghosh@cbs.ac.in

Chemical Reactivity in Confined Systems: Theory, Modelling and Applications,
First Edition. Edited by Pratim Kumar Chattaraj and Debdutta Chakraborty.

Since the binding energy of molecules or cohesive energy of solids is much smaller in comparison to the electronic energies of individual atoms, it is quite appropriate to model chemical binding as a consequence of interatomic perturbation of the electron cloud. The electron density rearrangement can be described in a coarse-grained prescription in terms of charge transfer from one atom to another. The introduction of the concepts of electronegativity [5] (a measure of the chemical potential of the electron cloud) and chemical hardness [6] of the atoms, which dictate the direction and extent of charge flow respectively, has been quite successful. This has analogy to the temperature and heat capacity which play the role of controlling the direction and extent of heat flow from hot body to cold body, or the concept of electric potential and capacitance determining the direction and extent of electricity flow. These classical-like concepts of electronegativity and hardness are, however, suitable for describing only ionic binding, but chemistry is rich in examples of covalent binding, which was thus initially not amenable to this electronegativity-based picture, since, for example, in a homonuclear diatomic molecule the two atoms are identical and hence there is no interatomic charge transfer. However, two important aspects have to be taken into account in this context. It is known that there is a charge build-up in the bond midpoint, which can be handled by introducing the concept of bond electronegativity and bond hardness [7–10]. Analogously for covalent binding there needs to be unpaired electrons in the atoms, which can be treated using spin-polarized DFT [11–12]. Thus, the spin up and spin down electrons will be different in number in each of the participating atoms. Thus, the up spin electronegativity and down spin electronegativity will be different. Thus between two atoms there can be up spin electron transfer from one atom to the other, while down spin electron transfer may be in the opposite direction. There may not be any resultant charge transfer but there is a net spin transfer, which can be responsible for covalent binding.

In the process of these charge transfer phenomena, the atoms in the molecule acquire partial charges [13], which can be used to investigate the reactivity of particular reactive sites in the molecule. Many reactivity indices [14] have been introduced from time to time, but charge-related ones are linked with the binding aspects directly [11, 15, 16]. Some of the reactivity parameters, such as electrophilicity [17, 18], nucleophilicity, frontier orbital charges [3, 19], response kernels [20, 21], local and global hardness and softness [22–24], etc. have gained much importance and popularity.

All the binding aspects considered above are, however, at the equilibrium geometry of the molecular species. Since the electronegativity difference drives the charge transfer, even at large distance there will be an interatomic charge transfer which is unphysical. This can be taken care of by demanding that there needs to be chemical contact between the atoms for binding to occur, just as there should be thermal contact for heat transfer and a mere presence of temperature difference may not be sufficient. This has been taken into account by several conceptual developments. For example, the bond hardness may be modelled to increase at larger separation. Or equivalently one can assume a probability of charge transfer which will decay exponentially or as per other functional form, with an increase in interatomic separation. This will lead to the correct dissociation limit of the molecule.

The objective of this work is to provide an overview of the developments in all these areas of research. No exhaustive treatment is, however, aimed at and only glimpses of the main features will be highlighted. In what follows, we first introduce the concepts of electronegativity and hardness and their role in describing the partial atomic charges and ionic binding in diatomic as well as polyatomic molecular systems. The modifications within the basic framework leading to other variants in this approach are then highlighted.

9.2 Theory

9.2.1 Concept of Electronegativity, Chemical Hardness, and Chemical Binding

9.2.1.1 Electronegativity and Hardness

The introduction of the concept of electronegativity has been linked with the rationalization and prediction of the polar nature of the molecules, as indicated by the dipole moment. It is also close to the heart of prediction of ionic contribution to chemical binding. For an N-electron system, the presently accepted definition of electronegativity parameter χ is given by

$$\chi = -(\partial E/\partial N)_{v(r)} \tag{9.1}$$

while the hardness parameter η is defined as

$$\eta = (1/2)\left(\partial^2 E/\partial N^2\right)_{v(r)} \tag{9.2}$$

where v(r) is the external potential characterizing the quantum system. Within a finite difference approximation, the two derivatives can be written as $\chi=(I + A)/2$ and $\eta=(I - A)/2$ in terms of the ionization potential (I) and electron affinity (A) of the system. A direct quantum mechanical calculation is possible using DFT, where the energy is expressed as a functional $E_v[\rho]$ of the electron density $\rho(r)$ given by

$$E_v[\rho] = \int v(r)\rho(r)\,d\mathbf{r} + F[\rho], \tag{9.3}$$

with $F[\rho]$ representing a universal functional of density. For fixed v(r), the energy functional obeys the variational (minimum) principle, leading to the result

$$\mu = \left(\delta E_v[\rho]/\delta\rho(r)\right) = v(r) + \left(\delta F[\rho]/\delta\rho(r)\right), \tag{9.4}$$

where μ, the chemical potential of the electron cloud, appears in the theory as a Lagrange multiplier for imposition of the normalization condition $\int \rho(r)dr = N$ and can be identified as $\mu = -\chi$. The fact is that although μ is equal to sum of position-dependent terms, it is quite clear that the value of μ is the same at every point. This aspect serves as a basis for electronegativity equalization, that has been widely used for the calculation of partial atomic charges as well as the ionic binding energy in a molecular system

9.2.1.2 Interatomic Charge Transfer in Molecular Systems

When two or more dissimilar atoms (n) are brought close to each other, due to difference in electronegativity, there is interatomic charge transfer and the electronegativity of the charged atoms, will change till they all become the same and the charge transfer stops bringing the system to equilibrium. The electronegativity of a charged atom (i-th) is given by

$$\chi_i^{eff} = \chi_i^0 + 2\eta_i^0 q_i + \Sigma_j \eta_{ij}^0 q_j \text{ for } i = 1, 2, \ldots n \tag{9.5}$$

where $q_i = -\Delta N_i$ and charge conservation $\Sigma\, q_i = 0$ holds good to maintain overall neutrality. Here, the additional contribution (the last term on the right) arises from the influence of other atoms and the cross or mutual hardness parameter is defined as the cross derivative $\eta_{ij} = (\partial^2 E/\partial N_i \partial N_j)_{v(r)}$ whose diagonal element is the self hardness $\eta_i = (1/2)\eta_{ii}$.

The energy change ΔE of the system can be expressed as a Taylor expansion in terms of the change in the number of electrons in each atom, and is given by

$$\Delta E = \Sigma_i \Delta E_i = -\Sigma_i \chi_i^0 \Delta N_i + \Sigma_i \Sigma_j \eta_{ij}^0 \Delta N_i \Delta N_j \tag{9.6a}$$

$$= -\Sigma_i \chi_i^0 \Delta N_i + \Sigma_i \eta_{ii}^0 (\Delta N_i)^2 + \Sigma_i \Sigma_{j\neq i} \eta_{ij}^0 \Delta N_i \Delta N_j \tag{9.6b}$$

which on minimization with respect to ΔN_i (or equivalently q_i) produces the same set of linear equations as obtained by equating the effective electronegativities given by Eq. (9.5). These (n−1) linear equations together with the charge conservation condition result in the n linear equations for n unknown charges, which can be obtained by solving the matrix equations of the form

$$\mathbf{Aq} = \mathbf{b} \tag{9.7}$$

where the vector **q** stands for the atomic charges (q_1, q_2, q_3, … q_n), vector **b** is contributed by the electronegativity of various atoms, and the matrix **A** consists of combination of the hardness parameters. Thus, the partial atomic charges as well as the binding energy of the molecular system can be obtained from the solution of this set of equations. The parametrization of the cross or mutual hardness parameter can be carried out by any of the several routes that have been proposed for this purpose or by introducing any other new prescription.

One can have, for example, the simple model based on the average of the softness (inverse of the hardness, i.e. $S = 1/\eta$) of the two atoms, viz.

$$\eta_{ij} = 2/\left(\eta_{ii}^{-1} + \eta_{jj}^{-1}\right) \tag{9.8}$$

Or in terms of the interatomic distance R_{ij} and the average hardness, viz. $a_{ij} = 2/(\eta_{ii} + \eta_{jj})$, viz.

$$\eta_{ij} = 1/\left(a_{ij} + R_{ij}\right) \tag{9.9}$$

in the spirit of Mataga-Nishimoto type empirical formula of the electronic structure theory [11].

Eq. (9.7) will yield the charges on each atom, which can be substituted in the energy expression given by Eq. (9.6) to obtain the ionic contribution to the binding energy. For heteronuclear diatomic molecule, methods have often been proposed to add to it the covalent contribution given by the geometric mean or arithmetic mean of the binding energies of the corresponding homonuclear diatomic counterparts, viz. $(\Delta E_{AB})_{cov} = (\Delta E_{AA}\ \Delta E_{BB})^{1/2}$ or $(\Delta E_{AB})_{cov} = (\Delta E_{AA} + \Delta E_{BB})/2$. This approach, however, is suitable for the equilibrium geometry only, and at large interatomic distance it predicts some spurious results for charge transfer and binding energy.

Now, this approach suffers from two major drawbacks, viz. (i) it cannot account for the effect of covalent binding, and (ii) it cannot predict the correct dissociation limit.

9.2.1.3 Concept of Chemical Potential and Hardness for the Bond Region

For describing the covalent binding within an electronegativity-based picture, mention may be made of two major approaches. The first one is inspired by the charge accumulation at the bond region in covalent binding, which can be investigated by introducing an electronegativity and hardness parameter for the bond region. This is quite appropriate in view of Eq. (9.4), where the chemical potential is defined at all points in space. This leads to a lattice model for molecules where, in addition to the atomic sites, each bonded atom pair is also associated with the bond (say, middle of the bond approximately) region, which is associated with the two values μ^{bond} and η^{bond}. The energy expression of Eq. (9.6) can now be written as (retaining terms up to second order).

$$\begin{aligned}\Delta E &= \Sigma_i \Delta E_i^{atom} + \Sigma_{ij} \Delta E_{ij}^{bond} + \text{cross terms}\\ &= -\Sigma_i \chi_i^0 q_i + \Sigma_i \Sigma_j \eta_{ij}^0 q_i q_j + \Sigma_{ij} \chi_{ij}^{bond} q_{ij}^{bond} + \Sigma_{ij} \Sigma_{kl} \eta_{ij,kl}^{bond} q_{ij}^{bond} q_{kl}^{bond}\\ &\quad + \Sigma_i \Sigma_{kl} \eta_{i,kl}^{atom/bond} q_i q_{kl}^{bond} \qquad (9.10a)\\ &= -\Sigma_i \chi_i^0 q_i + \Sigma_i \eta_{ii}^0 q_i^2 + \Sigma_i \Sigma_{j \neq i} \eta_{ij}^0 q_i q_j\\ &\quad + \Sigma_{ij} \chi_{ij}^{bond} q_{ij}^{bond} + \Sigma_{ij} \eta_{ij}^{bond} \left(q_{ij}^{bond}\right)^2 + \Sigma_{ij} \Sigma_{kl \neq ij} \eta_{ij,kl}^{bond} q_{ij}^{bond} q_{kl}^{bond}\\ &\quad + \Sigma_i \Sigma_{kl} \eta_{i,kl}^{atom/bond} q_i q_{kl}^{bond} \qquad (9.10b)\end{aligned}$$

where the bond electronegativity χ_{ij}^{bond} and bond hardness η_{ij}^{bond} correspond to the bond region for the bond between the i-th atom and j-th atom. The mutual bond-bond hardness $\eta_{ij,kl}^{bond}$ correspond to two bonds: one between the i-th atom and j-th atom while the other is between the k-th atom and *l*-th atom. The atom-bond hardness $\eta_{i,kl}^{atom/bond}$ similarly refers to i-th atom and bond between the k-th atom and *l*-th atom. This includes the case of i-th atom and bond of the i-th atom with other atoms as well. Eq. (9.10b) has been written from Eq. (9.10a) by separating out the diagonal and off-diagonal terms in the quadratic terms involving atom-atom or bond-bond situations.

As far as modeling of the bond electronegativity is concerned, since the chemical potential of an atom is approximately the electrostatic potential at the atomic radius site, one can assume the bond electronegativity to be proportional to the sum of the chemical potentials of the two concerned atoms which form the bond, i.e. $\chi_{ij}^{bond} = K(\chi_i^0 + \chi_j^0)$, with K as an empirical constant. The various bond hardness parameters have to be modeled along the lines of the same for mutual hardness parameters for atom-atom cases.

Here, the charges on the atoms q_i, q_j etc. and the bond charge q_{ij} are independent variables and hence the energy expression of Eq. (9.10) can be minimized with respect to both q_i and q_{ij}, and the resulting values can be used to calculate the effective partial charge. For example, the full charge Q_i can be calculated as $Q_i = q_i + (1/2)\, q_{ij}$, etc. Alternatively, one can take the derivative of the energy expression with respect to q_i and define an effective atomic electronegativity and then equalize them for all the atoms, resulting in a set of linear equations, which can be solved along with the same obtained by evaluating first an effective bond electronegativity followed by their equalization, to calculate the atomic charges and the bond charges and hence the binding energy. For diatomic molecules the equations are few and hence the calculation becomes very simple.

9.2.1.4 Spin-Polarized Generalization of Chemical Potential and Hardness

In covalent binding, besides accumulation of density at the bond region, another important issue is the presence of unpaired spin in the reactant atoms. Thus pairing of spin is an important aspect in covalent binding. Therefore, spin polarized generalization of the concepts of chemical potential and hardness is essential. Considering a system of N electrons, of which N_α and N_β are the numbers of spin up and spin down electron numbers, one can define the up spin and down spin electronegativity and hardness as a generalization of Eqs (9.1) and (9.2), viz.

$$\chi_\alpha = -\left(\partial E/\partial N_\alpha\right)_{v(r)} \quad \chi_\beta = -\left(\partial E/\partial N_\beta\right)_{v(r)} \tag{9.11}$$

while the hardness parameter η_α and η_β are defined as

$$\eta_\alpha = (1/2)\left(\partial^2 E/\partial N_\alpha^2\right)_{v(r)} \qquad \eta_\beta = (1/2)\left(\partial^2 E/\partial N_\beta^2\right)_{v(r)} \tag{9.12}$$

The spin polarized cross hardness parameter $\eta_{\alpha\beta}$ is defined as

$$\eta_{\alpha\beta} = \left(\partial^2 E/\partial N_\alpha \partial N_\beta\right)_{v(r)} \tag{9.13}$$

Considering the energy as $E(N_\alpha, N_\beta)$ the energy change is given by

$$\begin{aligned}\Delta E = &-\Sigma_i \chi_{\alpha,i}{}^0 \Delta N_{\alpha,i} - \Sigma_i \chi_{\beta,i}{}^0 \Delta N_{\beta,i} + \Sigma_i \Sigma_j \eta_{\alpha\alpha,ij}{}^0 \Delta N_{\alpha,i} \Delta N_{\alpha,j} \\ &+ \Sigma_i \Sigma_j \eta_{\beta\beta,ij}{}^0 \Delta N_{\beta,i} \Delta N_{\beta,j} + \Sigma_i \Sigma_j \eta_{\alpha\beta,ij}{}^0 \Delta N_{\alpha,i} \Delta N_{\beta,j}\end{aligned} \tag{9.14}$$

One can minimize this expression and find out optimum spin dependent charge transfer and calculate the energy change involved, which includes both ionic and covalent contributions to chemical binding.

Alternatively, one can arrive at expressions for the effective spin dependent chemical potential of the atoms from Eq. (9.14), viz.

$$\chi_{\alpha,i}\,(\text{eff}) = \chi_{\alpha,i}{}^{0} - 2\,\eta_{\alpha\alpha,ii}{}^{0}\,\Delta N_{\alpha,i} - \Sigma_j\,\eta_{\alpha\alpha,ij}{}^{0}\,\Delta N_{\alpha,j} - \Sigma_j\,\eta_{\alpha\beta,ij}{}^{0}\,\Delta N_{\beta,j} \tag{9.15a}$$

$$\chi_{\beta,i}\,(\text{eff}) = \chi_{\beta,i}{}^{0} - 2\,\eta_{\beta\beta,ii}{}^{0}\,\Delta N_{\beta,i}{}^{0} - \Sigma_j\,\eta_{\beta\beta,ij}{}^{0}\,\Delta N_{\beta,j} - \Sigma_j\,\eta_{\beta\alpha,ij}{}^{0}\,\Delta N_{\beta,i} \tag{9.15b}$$

and then equate the up-spin and down-spin chemical potentials separately for all the atoms to obtain a set of linear equations which can be solved to obtain the quantities $\Delta N_{\alpha,\,i}$ and $\Delta N_{\beta,\,i}$ for the each (i-th) atom, and thereby obtain the binding energy.

The net charge transfer between any two bonded pair can be thus calculated by summing the two spin contributions and the net spin transfer from their difference. For homonuclear molecules, there is no charge transfer and it is the spin transfer that accounts for the binding energy. Modeling the other spin-dependent parameters can be done along the same lines as done for spin-averaged parameters.

9.2.1.5 Charge Equilibriation Methods: Split Charge Models and Models with Correct Dissociation Limits

The fact is that the models described so far all deal with situations near equilibrium bond length and fail miserably in the dissociation limit, where an unphysical charge transfer is predicted. Several approaches with the general name of split charge model [25–29] are suggested, where charge transfer variables q_{ij} rather than q_i or q_j are used for the Taylor expansion. These q_{ij} variables denoting charge transfer from i-th to j-th atom span the bond space in the spirit of bond charge introduced earlier. They obey the relation $q_{ji} = -q_{ij}$ and define the atomic charge $q_i = \Sigma_j\, q_{ij}$. With this new variables, the energy expression is written as

$$\Delta E = \Sigma_{ij}\,\chi_{ij}{}^{0}\,q_{ij} + \Sigma_{ij}\,\Sigma_{kl}\,\eta_{ij,kl}\,q_{ij}\,q_{kl} \tag{9.16}$$

using parameters analogous to those used in bond space electronegativity and hardnesses.

Use of some linear transformation, however, can lead to reformulation in terms of atomic charge variables, given by

$$\Delta E = \Sigma_i\,q_i\,\left(\Sigma_j\,\left(\chi_{ji}^{0} - \chi_{ij}^{0}\right)/n\right) + \Sigma_{ij}\,\eta_{ij}\,q_i\,q_j \tag{9.17}$$

The bracketed quantity in the first term is shown to be some sort of effective electronegativity of the i-th atom. It is also shown that this quantity can be expressed as

$$\chi_i^{\text{eff}} = (1/n)\,\Sigma_j\,k_{ij}\,\left(\chi_i^{0} - \chi_j^{0}\right)S_{ij} \tag{9.18}$$

with S_{ij} being the overlap integral between two orbitals located at i-th and j-th atoms. With similar modeling of k_{ij}, the final expression is given by

$$\chi_i^{\text{eff}} = \Sigma_j\,\left(\chi_i^{0} - \chi_j^{0}\right)S_{ij}\,(R_{ij})\,/S_{ij}\,\left(R_{ij}^{0}\right) \tag{9.19}$$

Alternative expressions have also been written where the hardness term is expressed to have a quadratic charge difference dependence. This one-to-one correspondence between a quadratic charge model in bond space and an equivalent quadratic charge model in atom space, although quite interesting, is indicative of the nature of the transformation relations.

Chen et al. [25] have examined the long range dissociation behavior of the prediction of binding in molecules and shown that essentially the hardness parameter shows a quadratic dependence on the charge transfer. This shows that charge transfer will not take place when the distance is large.

The same conclusion has also been reached by Mathieu [28], who showed a quadratic behavior of the hardness kernel. The work of Valone [30] in this connection also reports very interesting conclusions in this context.

9.2.1.6 Density Functional Perturbation Approach: A Coarse Graining Procedure

We have so far considered the number of electrons as the variables. However, it is possible to arrive at this coarse-grained approach from a more detailed density-based approach. Consider an N-electron molecular system, being formed from n atoms. If the atoms are placed in a certain geometrical arrangement $\{R_\alpha\}$, their unperturbed density (let it be $\rho(r)$) will undergo change ($\delta\rho(r)$) due to mutual interatomic interaction, which may change the original potential v(r) to v(r)+ δv(r). The corresponding energy change can be expressed as a functional Taylor expansion given by (up to second order)

$$\begin{aligned}
\Delta E &= E_{v(r)+\delta v(r)}\left[\rho(r)+\delta\rho(r)\right] - E_{v(r)}\left[\rho(r)\right] \\
&= \int \mathbf{dr}\, \delta\rho(r)\left[(\delta E/\delta\rho(r))\right] + \int \mathbf{dr}\, \delta v(r)\rho(r) + \int \mathbf{dr}\int \mathbf{dr}'\, \delta\rho(r)\delta v(r')\left[(\delta^2 E/\delta\rho(r)\delta v(r'))\right] \\
&\quad + (1/2)\int \mathbf{dr}\int \mathbf{dr}'\, \delta\rho(r)\delta\rho(r')\left[(\delta^2 E/\delta\rho(r)\delta\rho(r'))\right] \\
&\quad + (1/2)\int \mathbf{dr}\int \mathbf{dr}'\, \delta v(r)\delta v(r')\left[(\delta^2 E/\delta v(r)\delta v(r'))\right] \\
&= \int \mathbf{dr}\, \delta\rho(r)\mu + \int \mathbf{dr}\, \delta v(r)\rho(r) + (1/2)\int \mathbf{dr}\int \mathbf{dr}'\, \delta\rho(r)\delta\rho(r')\,\eta(r,r') \\
&\quad + (1/2)\int \mathbf{dr}\int \mathbf{dr}'\, \delta v(r)\delta v(r')\,\chi(r,r') + \int \mathbf{dr}\int \mathbf{dr}'\, \delta\rho(r)\delta v(r')\left[(\delta^2 E/\delta\rho(r)\delta v(r'))\right]
\end{aligned} \tag{9.20}$$

where we have defined the response function $\chi(r,r') = [(\delta^2E/\delta v(r)\delta v(r')]$ and the hardness kernel $\eta(r,r') = [(\delta^2F/\delta\rho(r)\delta\rho(r')]$, in terms of the functional derivatives of the energy quantities.

Now, for simplification, we express the density change as sum of density changes at each atom, i.e. $\delta\rho(r) = \Sigma_\alpha\, \delta\rho_\alpha(r)$, and rewrite Eq. (9.20) as

$$\begin{aligned}
\Delta E &= \Sigma_\alpha \int \mathbf{dr}_\alpha \delta\rho_\alpha(r_\alpha)\,\mu_\alpha + \Sigma_\alpha\Sigma_\beta \int \mathbf{dr}_\alpha \delta v_\beta(r_\alpha)\,\rho_\alpha(r_\alpha) \\
&\quad + (1/2)\Sigma_\alpha\Sigma_\beta \int \mathbf{dr}_\alpha \int \mathbf{dr}_\beta{}'\, \delta\rho_\alpha(r_\alpha)\,\delta\rho_\beta(r_\beta{}')\,\eta(r_\alpha, r_\beta{}') \\
&\quad + (1/2)\Sigma_\alpha\Sigma_\beta \int \mathbf{dr}_\alpha \int \mathbf{dr}_\beta{}'\, \delta v(r_\alpha)\,\delta v(r_\beta{}')\,\chi(r_\alpha, r_\beta{}') \\
&\quad + \Sigma_\alpha\Sigma_\beta \int \mathbf{dr}_\alpha \int \mathbf{dr}_\beta{}'\, \delta\rho_\alpha(r_\alpha)\,\delta v(r_\beta{}')\left[(\delta^2 E/\delta\rho_\alpha(r_\alpha)\,\delta v(r_\beta{}'))\right]
\end{aligned} \tag{9.21}$$

The corresponding expression for the chemical potential is given by

$$\begin{aligned}
\mu_\alpha(r_\alpha) &= \mu_\alpha^0 + \delta v(r_\alpha) + \Sigma_\beta \int \mathbf{dr}_\beta{}'\, \delta\rho_\beta(r_\beta{}')\,\eta(r_\alpha, r_\beta{}') \\
&\quad + \Sigma_\beta \int \mathbf{dr}_\beta{}'\, \delta v(r_\beta{}')\left[(\delta\mu_\alpha/\delta v(r_\beta{}'))\right]
\end{aligned} \tag{9.22}$$

The expression has been simplified due to near-sightedness of the quantities [31]. More detailed expressions using the bond charges etc. can be found elsewhere [16]. Instead of considering the density change written as sum of the atomic site densities, one can consider also the bond regions. This will generate the formalism with the additional quantities, viz. bond charge, bond electronegativity, and bond hardness.

Now, we employ further coarse graining and consider Taylor series expansion of the potential change and the hardness kernel, around the atomic sites (the positions of the nuclei). Thus, we have

$$\delta v_\alpha(r_\alpha) = \delta v_\alpha(\alpha) + \mathbf{r}_\alpha.\nabla_\alpha \delta v_\alpha(\alpha) + \ldots\ldots \tag{9.23a}$$

$$\eta(r_\alpha, r_\beta) = \eta(\alpha, \beta) + \mathbf{r}_\alpha.\nabla_\alpha \eta(\alpha, \beta) + \mathbf{r}_\beta.\nabla_\beta \eta(\alpha, \beta) + \ldots\ldots. \tag{9.23b}$$

which, on substitution in Eqs (9.21) and (9.22), leads to the result for the energy change as

$$\begin{aligned}\Delta E = {} & \Sigma_\alpha \int d\mathbf{r}_\alpha\, \delta\rho_\alpha(r_\alpha)\, \mu_\alpha + (1/2)\Sigma_\alpha\Sigma_\beta \int d\mathbf{r}_\alpha \int d\mathbf{r}_\beta{}'\, \delta\rho_\alpha(r_\alpha)\, \delta\rho_\beta(r_\beta{}')\, \eta(\alpha, \beta) \\ & + (1/2)\Sigma_\alpha\Sigma_\beta \int d\mathbf{r}_\alpha \int d\mathbf{r}_\beta\, \delta\rho_\alpha(r_\alpha)\, \delta\rho_\beta(r_\beta)\, \mathbf{r}_\alpha.\nabla_\alpha\eta(\alpha, \beta) \\ & + (1/2)\Sigma_\alpha\Sigma_\beta \int d\mathbf{r}_\alpha \int d\mathbf{r}_\beta{}'\, \delta\rho_\alpha(r_\alpha)\, \delta\rho_\beta(r_\beta{}')\, \mathbf{r}_\beta.\nabla_\beta\eta(\alpha, \beta)\end{aligned} \tag{9.24}$$

The corresponding expression for the chemical potential simplifies to

$$\begin{aligned}\mu_\alpha(r_\alpha) = {} & \mu_\alpha^0 + \delta v(\alpha) + \mathbf{r}_\alpha.\nabla_\alpha\delta v_\alpha(\alpha) + \Sigma_\beta\eta(\alpha, \beta) \int d\mathbf{r}_\beta\, \delta\rho_\beta(r_\beta) + r_\alpha.\Sigma_\beta\nabla_\alpha\eta(\alpha, \beta) \int d\mathbf{r}_\beta{}'\, \delta\rho_\beta(r_\beta{}') \\ & + \Sigma_\beta \int d\mathbf{r}_\beta\, \delta\rho_\beta(r_\beta)\, \mathbf{r}_\beta.\nabla_\beta\eta(\alpha, \beta) + \ldots..\end{aligned} \tag{9.25}$$

Using the definitions of the atomic charges and atomic dipoles, these expressions can be written as

$$\begin{aligned}\Delta E = {} & \Sigma_\alpha q_\alpha \mu_\alpha + (1/2)\Sigma_\alpha \Sigma_\beta q_\alpha q_\beta \eta(\alpha, \beta) \\ & + (1/2)\Sigma_\alpha \Sigma_\beta q_\beta \mathbf{p}_\alpha.\nabla_\alpha\eta(\alpha, \beta) + (1/2)\Sigma_\alpha \Sigma_\beta q_\alpha \mathbf{p}_\beta.\nabla_\beta\eta(\alpha, \beta)\end{aligned} \tag{9.26}$$

where the dipole-dipole interaction term is dropped for simplicity.

$$\begin{aligned}\mu_\alpha(r_\alpha) = {} & \mu_\alpha^0 + \delta v(\alpha) + \mathbf{r}_\alpha.\nabla_\alpha\delta v_\alpha(\alpha) + \Sigma_\beta\eta(\alpha, \beta)\, q_\beta \\ & + \mathbf{r}_\alpha.\, \Sigma_\beta q_\beta \nabla_\alpha\eta(\alpha, \beta) + \Sigma_\beta \mathbf{p}_\beta.\nabla_\beta\eta(\alpha, \beta) + \ldots..\end{aligned} \tag{9.27}$$

Eq. (9.27) can easily lead to the result

$$\mu_\alpha(r_\alpha) = \mu_\alpha^0 + \delta v(\alpha) + \Sigma_\beta\eta(\alpha, \beta)\, q_\beta + \Sigma_\beta \mathbf{p}_\beta.\nabla_\beta\eta(\alpha, \beta) + \tag{9.28}$$

where the effect due to atomic charges and atomic dipoles of other sites is quite clear. One can perform the equalization of the effective chemical potentials and obtain equations to determine the atomic charges and dipoles, and hence also the binding energy.

9.2.1.7 Atomic Charge Dipole Model for Interatomic Perturbation and Response Properties

The density-based approach has generated what is known as atomic charge dipole model for interatomic perturbation, which has been investigated earlier [15, 16] and used for calculation of response properties like polarizability and other related quantities.

9.2.1.8 Force Field Generation in Molecular Dynamics Simulation

The formalism can easily be used to determine on the fly the atomic charges and dipoles and thereby obtain the forces in molecular dynamics simulation [32]. Several groups are active in this area of research. Some of these approaches are even coupled with ab-initio MD simulation, with these variables as the dynamical variables instead of the full electron density.

9.3 Perspective on Model Building for Chemical Binding and Reactivity

Modeling chemical binding and reactivity has played a significant and important role in the development of chemical concepts, which is not only highly useful in rationalizing and systemizing the experimental results but is also associated with tremendous ability to predict the unknown. This is particularly relevant in the era of machine learning based approaches which seem to become a very powerful tool in science in the years to come. In this work we have explicitly considered one particular reactivity index, viz. the atomic charge and discuss about its implications and usefulness in the context of chemical binding. Work in the time-dependent domain [33, 34] might also be rewarding.

9.4 Concluding Remarks

This work has provided an overall picture of coarse-grained density-based view of how atom- (or bond-) centered variables can predict reasonable quantitative estimates of the reactivity and molecular binding effects. In particular, the chemical potential of the electron cloud which was identified with the electronegativity concept in chemistry almost 40 years ago in the laboratory of Professor Robert Parr, has now generated many new ideas, giving birth to an altogether new field of research, the so-called "Conceptual density functional theory". The report is limited by the understanding of the author. Much more broader aspects can be discussed.

Acknowledgements

The work and views reported here have been possible due to enlightening discussions I had with many teachers, collaborators, and friends. In particular, I express my gratefulness to Prof B.M. Deb and Prof Robert G. Parr, from whom I have learnt a lot over the years. I also thank my colleagues Dr Alok Samanta, Dr Tapan Ghanty, Dr K.R.S. Chandrakumar, Dr Amita Wadehra, and many others for helpful discussions and collaborations. Finally I am thankful to Prof Pratim Chattaraj, who is always available for help, in all possible ways.

References

1 Hohenberg, P. and Kohn, W. (1964). Inhomogeneous electron gas. Phys. Rev. 136: B 864.

2 Kohn, W. and Sham, L.J. (1965). Self-consistent equations including exchange and correlation effects. Physical Review 140: A1133.

3 Parr, R.G., Yang, W. (1989) Density-Functional Theory of Atoms and Molecules. Oxford University Press, New York, and Clarendon Press, Oxford.

4 Geerlings, P., Chamorro, E., Chattaraj, P.K. et al. (2020). Conceptual density functional theory: status, prospects, issues. Theoretical Chemistry Accounts 139 (2): 36. https://doi.org/10.1007/s00214-020-2546-7.

5 Parr, R.G., Donnelly, R.A., Levy, M., and Palke, W.E. (1978). Electronegativity: the density functional viewpoint. The Journal of Chemical Physics 68 (8): 3801–3807.

6 Parr, R.G. and Pearson, R.G. (1983). Absolute hardness: companion parameter to absolute electronegativity. Journal of the American Chemical Society 105 (26): 7512–7516.
7 Ghosh, S.K. and Parr, R.G. (1987). Toward a semiempirical density functional theory of chemical binding. Theoretica Chimica Acta 72: 379–391.
8 Ghanty, T.K. and Ghosh, S.K. (1991). Electronegativity and covalent binding in homonuclear diatomic molecules. The Journal of Physical Chemistry 95 (17): 6512–6514.
9 Ghanty, T.K. and Ghosh, S.K. (1992). Electronegativity, hardness, and chemical binding in simple molecular systems. Inorganic Chemistry 31 (10): 1951–1955.
10 Ghanty, T.K. and Ghosh, S.K. (1994). A new simple density functional approach to chemical binding. The Journal of Physical Chemistry 98 (7): 1840–1843.
11 Ghosh, S.K. (1994). Electronegativity, hardness, and a semiempirical density functional theory of chemical binding. International Journal of Quantum Chemistry 49: 239–251.
12 Ghanty, T.K. and Ghosh, S.K. (1994). Spin-polarized generalization of the concepts of electronegativity and hardness and the description of chemical binding. Journal of the American Chemical Society 116 (9): 3943–3948.
13 Mortier, W.J., Ghosh, S.K., and Shankar, S. (1986). Electronegativity-equalization method for the calculation of atomic charges in molecules. Journal of the American Chemical Society 108 (15): 4315–4320.
14 Chattaraj, P.K. (ed.) (2009). Chemical Reactivity Theory: A Density Functional View. Boca Raton, USA: CRC Press.
15 Wadehra, A. and Ghosh, S.K. (2005). A density functional theory-based chemical potential equalisation approach to molecular polarizability. Journal of Chemical Sciences 117: 401–409.
16 Ghosh, S.K. (2010). A coarse grained density functional theory, chemical potential equalization and electric response in molecular systems. Journal of Molecular Structure: THEOCHEM 943: 178–182.
17 Parr, R.G., Szentpály, L.V., and Liu, S. (1999). Electrophilicity index. Journal of the American Chemical Society 121 (9): 1922–1924.
18 Chattaraj, P.K., Sarkar, U., and Roy, D.R. (2006). Electrophilicity index. Chemical Reviews 106 (6): 2065–2091.
19 Parr, R.G. and Yang, W. (1984). J Am Chem Soc 106: 4049–4050.
20 Ghosh, S.K. (1990). Energy derivatives in density-functional theory. Chemical Physics Letters 172 (1): 77–82.
21 Sablon, N., De Proft, F., and Geerlings, P. (2010). The linear response kernel: Inductive and resonance effects quantified. The Journal of Physical Chemistry Letters 1 (8): 1228–1234.
22 Ghosh, S.K. and Berkowitz, M. (1985). A classical fluid-like approach to the density-functional formalism of many-electron systems. The Journal of Chemical Physics 83 (6): 2976–2983.
23 Berkowitz, M., Ghosh, S.K., and Parr, R.G. (1986). On the concept of local hardness in chemistry. Journal of the American Chemical Society 107 (24): 6811–6814.
24 Berkowitz, M. and Parr, R.G. (1988). Molecular hardness and softness, local hardness and softness, hardness and
softness kernels, and relations among these quantities. J. Chem. Phys. 88: 2554–2557.
25 Chen, J. and Martinez, T.J. (2007). QTPIE: Chare transfer with polarization current equalization: A fluctuating charge model with correct asymptotics. Chemical Physics Letters 438: 313–320.
26 Chen, J., Hundertmark, D., and Martínez, T.J. (2008). A unified theoretical framework for fluctuating-charge models in atom-space and in bond-space. The Journal of Chemical Physics 129: 214113.

27 Mathieu, D. and Lucas, A. (2007). Computational approaches to the dynamics of ions and electrons in materials under extreme conditions. Computational Materials Science 38: 514.

28 Mathieu, D. (2007). Split charge equilibration method with correct dissociation limits. The Journal of Chemical Physics 127: 224103.

29 Chen, J., Martínez, T.J. (2009). The dissociation catastrophe in fluctuating-charge models and its implications for the concept of atomic electronegativity. In: Advances in the Theory of Atomic and Molecular Systems, Progress in Theoretical Chemistry and Physics (ed. P. Piecuch et al), Springer, 397–415

30 Valone, S.M. and Atlas, S.R. (2004). An empirical charge transfer potential with correct dissociation limits. Journal of Chemical Physics 120: 7262–7273.

31 Fias, S., Heldar-Zadeh, F., Geerlings, P., and Ayers, P.W. (2017). Chemical transferability of functional groups follows from the nearsightedness of electronic matter. Proceedings National Academy of Sciences USA 114: 11633–11638.

32 York, D.M. and Yang, W. (1996). A chemical potential equalization method for molecular simulations. The Journal of Chemical Physics 104 (1): 159–172.

33 Ghosh, S.K. and Dhara, A.K. (1988). Density-functional theory of many-electron systems subjected to time-dependent electric and magnetic fields. Physical Review A 38: 1149.

34 Chattaraj, P. and Maiti, B. (2001). Reactivity dynamics in atom– field interactions: a quantum fluid density functional study. The Journal of Physical Chemistry A 105 (1): 169–183.

10

Softness Kernel and Nonlinear Electronic Responses

Patrick Senet

Laboratoire Interdisciplinaire Carnot de Bourgogne UMR 6303 CNRS-Université de Bourgogne Franche-Comté 9 Avenue Alain Savary, B. P. 47870, F-21078 Dijon Cedex, France

10.1 Introduction

Conceptual Density Functional Theory (CDFT) [1] is developed in the Born-Oppenheimer approximation and brings working empirical reactivity concepts in a formal and rigorous framework. In CDFT, chemical reactivity is embedded in formal derivatives of the electronic density, the central function in DFT. CDFT is mainly a linear response theory, the nonlinear chemical reactivity responses were introduced and formal relations were established between them, but they remain largely unexplored [2–4]. CDFT has been developed for more than 30 years and particularly active research groups who contribute to strengthening CDFT are the teams of Paul Ayers, Carlos Cardenas, Pratim Chattaraj, Patricio Fuentealba, Paul Geerlings, Roman Nalewasjki, Weitao Yang…among others. The literature of CDFT is huge and impossible to summarize in the present chapter. We invite the interested reader to consult the following excellent reviews of CDFT [5–9].

They are two common approaches to chemical reactivity in CDFT [1].

In the first approach, reactivity of a chemical moeity is described as its response to a change of its global number of electrons, N. In this first approach, the positions of the atom nuclei are fixed and the perturbation of the chemical moeity is embedded in a scalar number, ΔN, representing an electron transfer from or towards the molecule studied. As in thermodynamics, the environment of the molecule of interest is simplified. The electronic responses are defined as N-derivatives of the electron density and of the energy. Because N is an integer number, the derivatives of the electronic density relative to N are either defined as finite difference derivatives for an isolated system or defined in the Grand Canonical Ensemble.

In the second approach, reactivity of a chemical moitie is described as a change of the external potential, v_{ext}, i.e. a change of the electrostatic potential of the nuclei. In this case, reactivity is described by either a change of the positions of the nuclei of the molecule studied or by an external electrostatic (point charge) potential, representing a perturbative reactivity probe. The electronic responses are defined as functional derivatives of the electronic density and of the energy relative to the external potential, at constant N.

In both approaches, one aims to extract the reactive regions of an isolated molecule from its molecular responses, at the lowest perturbation order. Of course, the predictions of CDFT cannot be quantitative, as only one reagent is described at a time. The most quantitative description of reactivity remains to simulate the chemical reaction itself. However, numerical computation of the

Chemical Reactivity in Confined Systems: Theory, Modelling and Applications,
First Edition. Edited by Pratim Kumar Chattaraj and Debdutta Chakraborty.

solutions of the Schrödinger equation do not necessarily provide a straightforward understanding of molecular reactivity. The purpose of CDFT is thus also to provide an *understanding* of the experimental and simulated outcomes of a reaction by introducing molecular responses to a variation of N and/or v_{ext}.

The present chapter focuses exclusively on the second approach of CDFT: the response of a molecule to a change to its external potential, Δv_{ext}. In the linear response theory, the four key quantities which are involved are: the global softness S (and its inverse, the hardness η), the Fukui function $f(\mathbf{r})$, the hardness kernel $h(\mathbf{r}_1, \mathbf{r}_2)$ (and its inverse, the softness kernel $h^{-1}(\mathbf{r}_1, \mathbf{r}_2)$) and finally the density polarizability kernel $\chi_1(\mathbf{r}_1, \mathbf{r}_2)$ [1]. The hardness kernel is the second functional derivative of the energy at constant external potential [1]. Although that the Fukui function is usually defined as derivatives of the electronic density relative to N, it can be properly defined in terms of the softness kernel, $h^{-1}(\mathbf{r}_1, \mathbf{r}_2)$ [10]. Then it describes, as shown in [11, 12], the electron reservoir of the molecule to build the polarization charge induced by an external potential. The Fukui function can be also derived from a variational problem involving the hardness kernel [13].

The density polarizability function $\chi_1(\mathbf{r}_1, \mathbf{r}_2)$ describes the change of the electronic density at a point $\mathbf{r}_1$ if a localized perturbative potential is applied at a point $\mathbf{r}_2$ [8]. The molecular multipolar polarizabilities can be computed by quadrature from $\chi_1(\mathbf{r}_1, \mathbf{r}_2)$.

In CDFT, the softness kernel,$h^{-1}(\mathbf{r}_1, \mathbf{r}_2)$, is intimately related to $\chi_1(\mathbf{r}_1, \mathbf{r}_2)$ through the well-known Berkowitz-Parr relation [10]. This means that all linear responses can be formulated entirely in terms of the softness kernel only. Moreover, the important message of the present chapter is that the central quantity of CDFT is indeed the softness kernel because all nonlinear electronic responses can be formulated as the *linear response* to an effective potential [3]. The importance of $h^{-1}(\mathbf{r}_1, \mathbf{r}_2)$ to describe chemical reactivity has been underestimated so far. Focusing on the response of $\chi_1(\mathbf{r}_1, \mathbf{r}_2)$ instead of $h^{-1}(\mathbf{r}_1, \mathbf{r}_2)$ to describe reactivity is an equivalent fruitful alternative in which Geerling's group has been engaged for several years [8, 14]. This polarizability density approach avoids the problem of defining responses as derivatives of the energy relative to a number of electrons and is worth pursuing.

Owing to the fundamental importance of the softness kernel, it is interesting to investigate its spectral decomposition. The kernel being symmetrical, it can be decomposed in eigenmodes [15–17]. We demonstrated recently that these modes are solutions of a variational problem and that modes of χ_1 can be formulated in terms of the modes of h^{-1} [17]. In the present chapter, we discuss analytical expressions of these modes for a non-interacting quantum gas confined in a one-dimensional box of length L using the von Weizacker kinetic energy functional [18]. More details can be found in [17].

The purpose of the present chapter is to provide a short introduction to nonlinear electronic responses and to emphasize the central role of the softness kernel h^{-1} in CDFT. We hope that this formal introduction will stimulate the reader to investigate numerical applications of these concepts. The chapter is organized as follows. We review the linear response theory first. Then we describe how the nonlinear electronic responses can be formulated in terms of the softness kernel and polarizability density kernels. Explicit expressions are discussed at the second-order of perturbation. Finally, we derive the analytical expressions of the eigenmodes of the softness kernel and provide a numerical application to a non-interacting quantum gas confined in a one-dimensional box.

10.2 Linear and Nonlinear Electronic Responses

10.2.1 Linear Response Theory

10.2.1.1 Ground-State

We briefly summarize the linear response theory [1] and introduce notation. One considers an *isolated* molecule. In the Born-Oppenheimer approximation, the system is described by an electrostatic Hamiltonian $\hat{H}$ defined by the number of electrons N and the one-electron Coulomb potential due to the nuclei:

$$\hat{H} = \hat{F} + \sum_{i=1}^{N} \upsilon_{ext}(\mathbf{r}_i), \tag{10.1}$$

where $\hat{F}$ is independent of the nuclei positions,

$$\hat{F} \equiv \sum_{i=1}^{N} \frac{\hbar^2}{2m}\nabla^2 + \frac{1}{2}\sum_{i=1}^{N}\sum_{j=1, j\neq i}^{N} \frac{e^2}{\left|\mathbf{r}_i - \mathbf{r}_j\right|}, \tag{10.2}$$

and

$$\upsilon_{ext}(\mathbf{r}) \equiv \sum_{i=1}^{N}\sum_{J=1}^{M} \frac{e^2 Z_J}{\left|\mathbf{r}_i - \mathbf{R}_J\right|}. \tag{10.3}$$

In Eq. (10.3), M is the total number of nuclei, Z_J and $\mathbf{R}_J$ are respectively the nuclear charge and position of the Jth atom.

According to the fundamental theorems of DFT, the lowest eigenvalue E_0 of $\hat{H}$, corresponding to the normalized ground-state wave function $|\Psi_0\rangle$, is a functional $E[\rho]$ of the electron density $\rho(\mathbf{r})$ [19–21]. The so-called universal functional $F[\rho]$ at the solution point is [1]:

$$F[\rho_0] = \langle\Psi_0|\,\hat{F}\,|\Psi_0\rangle = E_0 - \int d\mathbf{r}\ \rho_0(\mathbf{r})\upsilon_{ext}(\mathbf{r}), \tag{10.4}$$

where $\rho_0(\mathbf{r})$ is the ground-state density. Several functionals $F[\rho]$ [19–21] can be constructed which all obey the following relations:

$$E_{\upsilon_{ext}}\left[\rho_0(\mathbf{r})\right] = E_0 = \langle\Psi_0|\,\hat{H}\,|\Psi_0\rangle, \tag{10.5}$$

and

$$E_{\upsilon_{ext}}\left[\rho(\mathbf{r})\right] > E_0. \tag{10.6}$$

10.2.1.2 Linear Responses [1]

One considers a variation of the external potential, $\Delta\upsilon_{ext}(\mathbf{r})$. The second-order variation of the energy relative to the ground-state value is given by

$$\delta E^{(2)}_{\upsilon_{ext}} = \int d\mathbf{r}_1\ \rho_0(\mathbf{r}_1)\Delta\upsilon_{ext}(\mathbf{r}_1) + \frac{1}{2}\int d\mathbf{r}_1\ \delta\rho_1(\mathbf{r}_1)\Delta\upsilon_{ext}(\mathbf{r}_1), \tag{10.7}$$

where the first-order variation of the electronic density is related to the perturbation by

$$\delta\rho_1(\mathbf{r}_1) = \int d\mathbf{r}_2\ \chi_1(\mathbf{r}_1, \mathbf{r}_2)\Delta\upsilon_{ext}(\mathbf{r}_2), \tag{10.8}$$

in which χ_1 is the density polarizability kernel [8]. The second-order variation of the energy can be alternatively formulated in terms of the hardness kernel $h(\mathbf{r}_1, \mathbf{r}_2)$ [1]:

$$\delta E^{(2)}_{v_{ext}} = \int d\mathbf{r}_1\ \rho_0(\mathbf{r}_1)\Delta v_{ext}(\mathbf{r}_1) + \frac{1}{2}\int d\mathbf{r}_1 \int d\mathbf{r}_2\ h(\mathbf{r}_1, \mathbf{r}_2)\delta\rho_1(\mathbf{r}_1)\delta\rho_1(\mathbf{r}_2), \tag{10.9}$$

with by definition,

$$\left[\frac{\delta^2 E}{\delta\rho(\mathbf{r}_1)\delta\rho(\mathbf{r}_2)}\right]_{0,v_{ext}} = \left[\frac{\delta^2 F}{\delta\rho(\mathbf{r}_1)\delta\rho(\mathbf{r}_2)}\right]_0 = h(\mathbf{r}_1, \mathbf{r}_2). \tag{10.10}$$

Comparing Eq. (10.7) with Eq. (10.9), one deduces the first-order perturbation equation [1]:

$$\delta\mu_1 - \Delta v_{ext}(\mathbf{r}_1) = \int d\mathbf{r}_2\ h(\mathbf{r}_1, \mathbf{r}_2)\delta\rho_1(\mathbf{r}_2), \tag{10.11}$$

where $\delta\mu_1$ is a constant because the perturbed density integrates to zero and the external potential can be defined only within a constant. It can be interpreted as the first-order change of the chemical potential (properly defined for model functionals or in the Grand Canonical Ensemble at low T). Here, we consider only variations of μ induced by a potential at constant N. Using the softness kernel,

$$\int d\mathbf{r}_2\ h^{-1}(\mathbf{r}_1, \mathbf{r}_2)h(\mathbf{r}_2, \mathbf{r}_3) = \delta(\mathbf{r}_1 - \mathbf{r}_3), \tag{10.12}$$

the solution of Eq. (10.11) is found:

$$\delta\rho_1(\mathbf{r}_1) = \int d\mathbf{r}_2\ h^{-1}(\mathbf{r}_1, \mathbf{r}_2)\left[\delta\mu_1 - \Delta v_{ext}(\mathbf{r}_2)\right]. \tag{10.13}$$

Integrating Eq. (10.13), one gets

$$\delta\mu_1 = \int d\mathbf{r}_1\ f(\mathbf{r}_1)\Delta v_{ext}(\mathbf{r}_1), \tag{10.14}$$

in which we have introduced the Fukui function:

$$f(\mathbf{r}_1) = \frac{\int d\mathbf{r}_2\ h^{-1}(\mathbf{r}_1, \mathbf{r}_2)}{S}, \tag{10.15}$$

and the chemical softness:

$$S = \int d\mathbf{r}_1 \int d\mathbf{r}_2\ h^{-1}(\mathbf{r}_1, \mathbf{r}_2). \tag{10.16}$$

Finally, it comes

$$\chi_1(\mathbf{r}_1, \mathbf{r}_2) = -h^{-1}(\mathbf{r}_1, \mathbf{r}_2) + S\,f(\mathbf{r}_1)f(\mathbf{r}_2), \tag{10.17}$$

which is the well-known Berkowitz-Parr relation [10].

10.2.2 Nonlinear Responses and the Softness Kernel

The total energy variation $\Delta E_{v_{ext}}$ induced by a change of external potential Δv_{ext} can be found by applying the Hellman-Feynman theorem [3]:

$$\Delta E_{v_{ext}} = \int d\mathbf{r}_1\ \rho_0(\mathbf{r}_1)\Delta v_{ext}(\mathbf{r}_1) + \sum_{n=1}^{\infty}\frac{1}{n+1}\int d\mathbf{r}_1\ \delta\rho_n(\mathbf{r}_1)\Delta v_{ext}(\mathbf{r}_1), \tag{10.18}$$

where $\delta\rho_n(\mathbf{r}_1)$ is the variation of the electronic density at nth-order perturbation, i.e.

$$\begin{aligned}\delta\rho_n(\mathbf{r}_1) = \tfrac{1}{n!}\int d\mathbf{r}_2 \int d\mathbf{r}_3 \ldots \int d\mathbf{r}_n\ \chi_n(\mathbf{r}_1, \mathbf{r}_2, \ldots, \mathbf{r}_{n+1})\\ \times\ \Delta v_{ext}(\mathbf{r}_2)\Delta v_{ext}(\mathbf{r}_3)\ldots\Delta v_{ext}(\mathbf{r}_{n+1}),\end{aligned} \tag{10.19}$$

where the polarizability density kernels at any order n are defined by [3]

$$\chi_n(\mathbf{r}_1, \mathbf{r}_2, ..., \mathbf{r}_{n+1}) = \left[\frac{\delta^n \rho(\mathbf{r}_1)}{\delta v_{ext}(\mathbf{r}_2)\delta v_{ext}(\mathbf{r}_3)...\delta v_{ext}(\mathbf{r}_{n+1})}\right]_0. \tag{10.20}$$

The total variation of the density is given by

$$\Delta\rho(\mathbf{r}_1) = \sum_{i=1}^{\infty} \delta\rho_n(\mathbf{r}_1). \tag{10.21}$$

Using the Taylor theorem for functionals [1], the total energy variation at constant potential is

$$\Delta E_{v_{ext\,fixed}} = \sum_{n=2}^{\infty} \frac{1}{n!} \int d\mathbf{r}_1 \int d\mathbf{r}_2 ... \int d\mathbf{r}_n\, h(\mathbf{r}_1, \mathbf{r}_2, ..., \mathbf{r}_n)\Delta\rho(\mathbf{r}_1)\Delta\rho(\mathbf{r}_2)...\Delta\rho(\mathbf{r}_n), \tag{10.22}$$

where the term $n = 1$ is zero because of the variational principle of DFT. The quantities h are generalizations of the hardness kernel [3]:

$$\left[\frac{\delta^n E}{\delta\rho(\mathbf{r}_1)\delta\rho(\mathbf{r}_2)...\delta\rho(\mathbf{r}_n)}\right]_{0,v_{ext}} = \left[\frac{\delta^n F}{\delta\rho(\mathbf{r}_1)\delta\rho(\mathbf{r}_2)...\delta\rho(\mathbf{r}_n)}\right]_0 = h(\mathbf{r}_1, \mathbf{r}_2 ... \mathbf{r}_n). \tag{10.23}$$

The nonlinear perturbation equations ($n > 1$) are deduced from the comparison of Eqs (10.18) and (10.22). For the first nonlinear orders, we find explicitly [3]

$$\begin{aligned}\delta\mu_2 &= \int d\mathbf{r}_1\, h(\mathbf{r}_1, \mathbf{r}_2)\delta\rho_2(\mathbf{r}_2) \\ &+ \frac{1}{2}\int d\mathbf{r}_2 \int d\mathbf{r}_3\, h(\mathbf{r}_1, \mathbf{r}_2, \mathbf{r}_3)\delta\rho_1(\mathbf{r}_2)\delta\rho_1(\mathbf{r}_3),\end{aligned} \tag{10.24}$$

and

$$\begin{aligned}\delta\mu_3 &= \int d\mathbf{r}_2\, h(\mathbf{r}_1, \mathbf{r}_2)\delta\rho_3(\mathbf{r}_2) \\ &+ \int d\mathbf{r}_2 \int d\mathbf{r}_3\, h(\mathbf{r}_1, \mathbf{r}_2, \mathbf{r}_3)\delta\rho_1(\mathbf{r}_2)\delta\rho_2(\mathbf{r}_3) \\ &+ \frac{1}{3!}\int d\mathbf{r}_2 \int d\mathbf{r}_3 \int d\mathbf{r}_4\, h(\mathbf{r}_1, \mathbf{r}_2, \mathbf{r}_3, \mathbf{r}_4)\delta\rho_1(\mathbf{r}_2)\delta\rho_1(\mathbf{r}_3)\delta\rho_1(\mathbf{r}_4).\end{aligned} \tag{10.25}$$

An important finding is that nonlinear perturbation equations can be formulated as a linear perturbation equation with an effective perturbation potential [3]:

$$\delta\mu_n - \Delta v_n(\mathbf{r}_1) = \int d\mathbf{r}_2\, h(\mathbf{r}_1, \mathbf{r}_2)\delta\rho_n(\mathbf{r}_2), \tag{10.26}$$

where the effective potential $\Delta v_1 = \Delta v_{ext}$ and $\Delta v_{n>1}$ is an effective potential depending on all the density variations $\delta\rho_n$ at order smaller than n. For example at the second order, we have

$$\Delta v_2(\mathbf{r}_1) = \frac{1}{2}\int d\mathbf{r}_2 \int d\mathbf{r}_3\, h(\mathbf{r}_1, \mathbf{r}_2, \mathbf{r}_3)\delta\rho_1(\mathbf{r}_2)\delta\rho_1(\mathbf{r}_3). \tag{10.27}$$

It is interesting to give an example of effective potential for a simple model functional. Let consider the Thomas-Fermi-Dirac LDA functional:

$$F(\rho) = C_F \int d\mathbf{r}_1\, \rho(\mathbf{r}_1)^{5/3} - \alpha\, C_x \int d\mathbf{r}_1\, \rho(\mathbf{r}_1)^{4/3} + J(\rho), \tag{10.28}$$

with

$$J(\rho) = \frac{e^2}{2}\int d\mathbf{r}_1 \int d\mathbf{r}_2 \frac{\rho(\mathbf{r}_1)\rho(\mathbf{r}_2)}{|\mathbf{r}_1 - \mathbf{r}_2|}, \tag{10.29}$$

and $C_F = 3\hbar^2/10m(3\pi^2)^{2/3}$ and $C_x = 3e^2\hbar^2/16\epsilon_0 m(3/\pi)^{1/3}$.

Taking the functional derivatives of Eq. (10.28), one gets for the first nonlinear hardness kernel $h(\mathbf{r}_1, \mathbf{r}_2, \mathbf{r}_3)$:

$$h(\mathbf{r}_1, \mathbf{r}_2, \mathbf{r}_3) = \delta(\mathbf{r}_1 - \mathbf{r}_2)\delta(\mathbf{r}_1 - \mathbf{r}_3)\left[\frac{8\alpha C_x}{27}\frac{1}{\rho(\mathbf{r}_1)^{5/3}} - \frac{10C_F}{27}\frac{1}{\rho(\mathbf{r}_1)^{4/3}}\right]. \tag{10.30}$$

It is worth noting that any functional derivative of the energy at an order larger than 2 does not involve the Coulomb interaction because the electron electrostatic repulsive energy, J, is a quadratic functional of the electron density. Therefore, the nonlinear responses will be more sensitive to kinetic and exchange-correlation functionals derivatives than the linear response, for which the Coulomb interaction is a quantitatively important contribution. From Eqs (10.27) and (10.30), one finds

$$'\Delta v_2(\mathbf{r}_1) = \frac{\delta(\rho_1(\mathbf{r}_1))^2}{\rho_1(\mathbf{r}_1)}\left[\frac{4\alpha C_x}{27}\frac{1}{\rho(\mathbf{r}_1)^{2/3}} - \frac{5C_F}{27}\frac{1}{\rho(\mathbf{r}_1)^{1/3}}\right]. \tag{10.31}$$

Interestingly, the nonlinear effective potential, at least for a LDA functional as the TFD model, has large values in regions of low densities.

The formal solution of the nonlinear perturbation equations, Eq. (10.26), involved the softness kernel. The formal solution at any order n is indeed

$$\delta\rho_n(\mathbf{r}_1) = \int d\mathbf{r}_2\, h^{-1}(\mathbf{r}_1, \mathbf{r}_2)\left[\delta\mu_n - \Delta v_n(\mathbf{r}_2)\right], \tag{10.32}$$

from which we deduced

$$\delta\rho_n(\mathbf{r}_1) = \int d\mathbf{r}_2\, \chi_1(\mathbf{r}_1, \mathbf{r}_2)\Delta v_n(\mathbf{r}_2), \tag{10.33}$$

$$\delta\mu_n = \int d\mathbf{r}_1\, f(\mathbf{r}_1)\Delta v_n(\mathbf{r}_1). \tag{10.34}$$

where both χ_1 and f can be built from the softness kernel (see Eqs (10.15) and (10.17)).

10.2.3 Eigenmodes of Reactivity

As the softness kernel h^{-1} and the density polarizability kernel χ_1 are the central quantities in the nonlinear perturbation theory, it is interesting to examine their spectral representation. In the present chapter, we focus on the eigenmodes of the softness/hardness kernel. Modes of the density polarizability kernel are discussed in our recent paper [17].

As the hardness kernel is symmetric, it can be expanded in its orthonormalized eigenmodes [22]:

$$h(\mathbf{r}_1, \mathbf{r}_2) = \sum_{n=1}^{\infty} \beta_n\, \Delta\rho_n(\mathbf{r}_1)\Delta\rho_n(\mathbf{r}_2), \tag{10.35}$$

solutions of

$$\int d\mathbf{r}_2 h(\mathbf{r}_1, \mathbf{r}_2)\Delta\rho_n(\mathbf{r}_2) = \beta_n \Delta\rho_n(\mathbf{r}_1). \tag{10.36}$$

All eigenvalues are positive and the kernel is positive-define (the smallest eigenvalue, β_1, is non-zero for a stable ground-state system and we sort the values as $\beta_1 < \beta_2 < \beta_3, \ldots$). The physical unit of the eigenvalues is EV where E is the energy unit and V is the volume unit whereas the unit of the eigenvectors is $1/\sqrt{V}$ (as the wave-function).

The softness kernel is [22]:

$$h^{-1}(\mathbf{r}_1, \mathbf{r}_2) = \sum_{i=1}^{\infty} \frac{1}{\beta_n}\, \Delta\rho_n(\mathbf{r}_1)\Delta\rho_n(\mathbf{r}_2). \tag{10.37}$$

The mode 1 contributes the most to the inverse kernel as $\beta_1 < \beta_n$ for $n > 1$.

The eigenmodes of the hardness kernel were introduced for the first time by Nalewajski in the context of a empirical discrete model, the Charge Sensivity Analysis [15], and were discussed by Cohen et al. [16].

The integral of the eigenvector, i.e.

$$\int d\mathbf{r}_1 \; \Delta\rho_n(\mathbf{r}_1) = \Delta_n N, \tag{10.38}$$

permits separation of the modes as the *polarization modes* for which $\Delta N_n = 0$, and the others we named *charging modes* where $\Delta N_n \neq 0$ [17]. ΔN_n can be interpreted as a virtual charge transfer (electron flow within the isolated molecule). However, it is worth noting that ΔN_n is not dimensionless, its unit is $\sqrt{V}$ and this quantity is proportional to a charge transfer [17].

The polarization modes have the nice property to be orthogonal to the Fukui function [17]. Indeed, multiplying Eq. (10.15) by $\Delta\rho_n$ and using Eq. (10.37), we find

$$\int d\mathbf{r}_1 \; \Delta\rho_n(\mathbf{r}_1) f(\mathbf{r}_1) = \frac{\Delta_n N}{S\beta_n}. \tag{10.39}$$

The Fukui function is built only from the charging modes. According to Eqs (10.15) and (10.37) [17],

$$f(\mathbf{r}_1) = \sum_{\text{charging modes}}^{\infty} \frac{\Delta_n N}{S\beta_n} \Delta\rho_n(\mathbf{r}_1). \tag{10.40}$$

Finally, the chemical softness (Eq. (10.16)) reads [17]

$$S = \sum_{\text{charging modes}}^{\infty} \frac{(\Delta_n N)^2}{\beta_n}. \tag{10.41}$$

10.3 One-Dimensional Confined Quantum Gas: Analytical Results from a Model Functional

In the hope to gain some insights, we examine the eigenmodes of the softness kernel for a one-dimensional noninteracting quantum gas confined in a box using a model functional. This problem was examined in our recent paper [17]. We summarize here the results.

We assume the von Weizacker kinetic energy functional [18, 23–26] (in one dimension)

$$T_W = \frac{\hbar^2}{8m} \int dx_1 \frac{\left[\frac{\partial\rho(x_1)}{\partial x_1}\right]^2}{\rho(x_1)}. \tag{10.42}$$

The eigenmode equation associated with T_W (Eq. (10.36)) is similar to the first-order equation of perturbation [1] (see Equation (77) in [3]) and is given by [17]:

$$-\frac{d}{dx}\left[\rho_0(x)\frac{dg_n(x)}{dx}\right] = \frac{4m}{\hbar^2}\rho_0^2(x)\beta_n g_n(x), \tag{10.43}$$

where we introduced the auxiliary functions g_n by $\Delta\rho_n(x) = \rho_0(x) g_n(x)$. An approximate solution of the Eq. (10.43) can be found by replacing $\rho_0(x)$ by an average value $\langle \rho_0(x) \rangle \equiv \overline{\rho}_0$. We find,

$$\begin{aligned} \Delta\rho_n(x) &= A\overline{\rho}_0 \sin(k_n x) \\ \beta_n &= \frac{\hbar^2 k_n^2}{4m\overline{\rho}_0}, \end{aligned} \tag{10.44}$$

where k_n is defined by the boundary conditions and A is a normalization constant. Assuming an quantum gas confined in a box of length L, the boundary conditions are $k_n = n\pi/L, n = 1, 2, ..$ and the normalization constant is $A = \frac{\sqrt{2L}}{N}$ where $N = \overline{\rho}_0 L$ is the number of electrons. Interestingly, the modes corresponding to odd n numbers are charging modes whereas the modes with even n numbers are polarization modes [17]:

$$\int_0^L dx\, sin(\frac{n\pi}{L}x) = \frac{2L}{n\pi}\, \delta(n, odd), \tag{10.45}$$

with $\Delta N_n = 2\sqrt{2L}/n\pi$ for the charging modes (n odd).

Using Eq. (10.44) in Eq. (10.37), the softness kernel reads:

$$h^{-1}(x, x') = N\left(\frac{32m}{h^2}\right)\left[\sum_{n=1}^{\infty}\frac{sin(\frac{n\pi}{L}x)sin(\frac{n\pi}{L}x')}{n^2}\right], \tag{10.46}$$

where $N = L\overline{\rho}_0$ is the number of electrons.

The Fukui function is found from (Eqs (10.15) using Eq. (10.16):

$$f(x) = \frac{48}{L\pi^3}\left[\sum_{n=1}^{\infty}\frac{sin(\frac{n\pi}{L}x)}{(2n-1)^3}\right] \tag{10.47}$$

The Fukui function has nodes at the box boundaries and is maximum in the middle of the box. One can easily check that the Fukui function is orthogonal to the polarization modes.

According to the Berkowitz-Parr relation (Eq. (10.17)) and the first-order perturbation equation (Eq. (10.11)), the softness kernel represents the density polarizability kernel χ_1 evaluated at constant chemical potential (for $\delta\mu = 0$), i.e. the response χ_1 of a large macromolecule for example, for which μ should not vary too much as a macromolecule is a large reservoir of electrons.

One important question is the convergence of the softness kernel as function n. Are only a few modes needed to represent the kernel? In particular the mode with the smallest eigenvalue ($n = 1$ in Eq. (10.37)), which contributes the most to the softness kernel, is it representative of the full kernel? As shown elsewhere the mode $n = 1$ can be evaluated from a variational principle, which opens a practical computational method to approximate h^{-1} in an actual molecule [17].

To examine the contributions of the different modes to the softness kernel in the present simple model of a confined gas, we computed $h^{-1}(x_1, x_2)$ for $x_1 = x_2$ in Figure 10.1. We recall that $-h^{-1}(x_1, x_2)$ represents the first-order variation of the density at x_1 if a localized perturbative potential is applied at x_2 for a molecule at constant chemical potential (either in contact with an electron

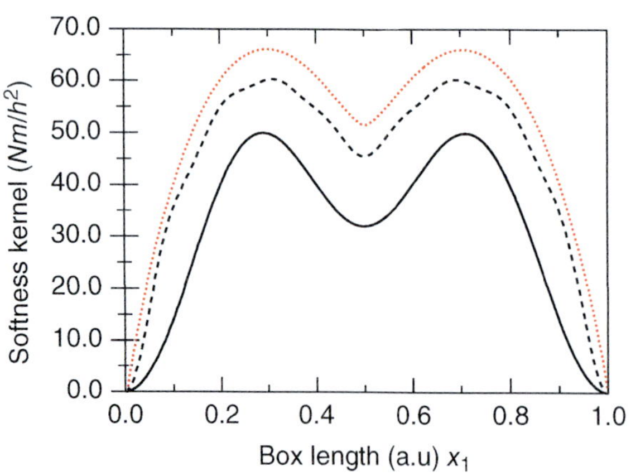

Figure 10.1 Diagonal elements of the softness kernel of a quantum gas in a box of length $L = 1$ as a function of the position $x_1 = x_2$. The contribution of the first eigenmode to the softness kernel is represented by a full line. The sum of the contributions of the modes $n = 1$ to 4 is represented by a dashed line. The full kernel is shown by a red dotted line.

reservoir or a very large macromolecule). In Fig. (10.1), the contribution of the mode with the lowest eigenvalue has maxima at the same positions than the softness kernel. One can see that the contributions of the sum of the four first modes, $n = 1$ to 4, is very close to the softness kernel. This example shows, although in a simple model, that a few reactivity modes is enough to represent the softness kernel. Although that the density is averaged, the softness kernel has a well-defined structure closely related to the constraints imposed by the boundary conditions, i.e. by the confinement of the quantum gas.

To test the convergence of the nonlocality of the response, we repeated the calculation of the softness kernel for $x_2 = L/2$ (Figure 10.2) and $x_2 = L/4$ (Figure 10.3). Again, one observes that the mode $n = 1$ compare qualitatively well with the softness kernel but his maxima is shifted for $x_2 = L/4$.

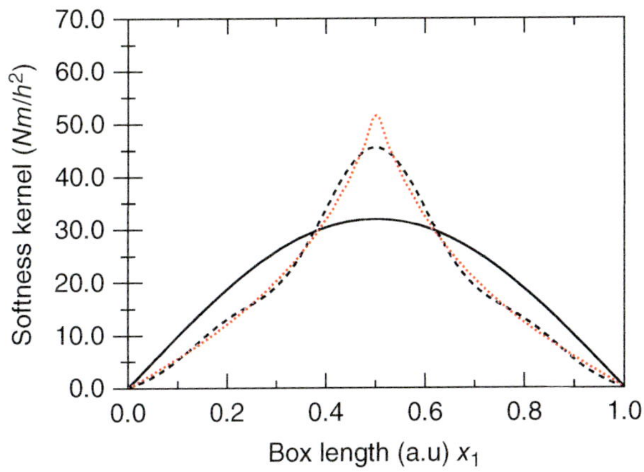

Figure 10.2 Non diagonal elements of the softness kernel of a quantum gas in a box of length $L = 1$ as a function of the position x_1 for $x_2 = L/2$. The contribution of the first eigenmode to the softness kernel is represented by a full line. The sum of the contributions of the modes $n = 1$ to 4 is represented by a dashed line. The full kernel is shown by a red dotted line.

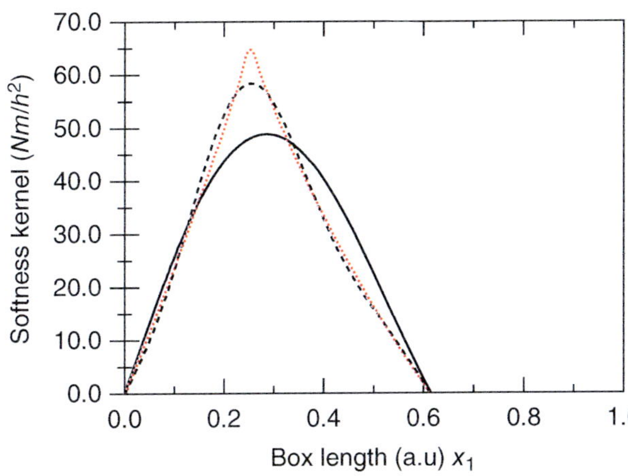

Figure 10.3 Non diagonal elements of the softness kernel of a quantum gas in a box of length $L = 1$ as a function of the position x_1 for $x_2 = L/4$. The contribution of the first eigenmode to the softness kernel is represented by a full line. The sum of the contributions of the modes $n = 1$ to 4 is represented by a dashed line. The full kernel is shown by a red dotted line.

The sum of the four modes fits very well the softness kernel. Again for a confined quantum gas, only a few modes are needed to reproduce the linear electronic response. As expected, one observes that the deformation of the density (value of h^{-1}) is maximum at the position of the localized perturbation ($x_1 = x_2$) and decreases with the distance to it as a piece linear function. For $x_2 = L/2$, the decrease has a change of slope at $L/4$.

We recall that the numerical results presented were obtained for the von Weizacker kinetic energy functional, which is exact for a one-electron system and for a noninteracting boson gas. There is no exact kinetic energy functional which includes the Pauli principle exactly. In specific examples, in one dimension, it was possible to derive an explicit kinetic energy functional for a Fermion gas, and in this case the kinetic energy functional appears highly nonlocal [25]. Probably the best approach to include the Pauli principle will remain to use the Kohn-Sham formalism, for which a practical calculation of the softness kernel is possible by using density perturbations [17], following the method developed to compute the density polarizability kernel [8].

10.4 Conclusion

We have provided a short introduction to nonlinear response functions in the framework of CDFT. We emphasized the central role of the softness kernel, h^{-1}, from which the linear Fukui function $f(r_1)$ and the density polarizability kernel can be built. As all nonlinear responses can be formulated in terms of the linear one by quadrature, an analysis and computation of h^{-1} is worth pursuing. For model functionals, we give a simple analytical example for a quantum gas confined in a box, for which we examined the eigenmodes of reactivity. In this example, only a few modes are necessary to reproduce the full diagonal and nondiagonal elements of h^{-1}. As the Fukui function and the chemical hardness can be built from h^{-1}, we hope that the present formal work will stimulate the computation of eigenmodes of reactivity for actual molecules. The computational method developed by the Geerlings's group to compute the density polarizability kernel from the Kohn-Sham orbitals [8] could be applied to evaluate the softness kernel and the eigenmodes of reactivity for molecular systems [17].

References

1 Robert G. Parr and Yang Weitao. Density-Functional Theory of Atoms and Molecules. Oxford University Press, May 1994.

2 Patricio Fuentealba and Robert G. Parr. Higher-order derivatives in density-functional theory, especially the hardness derivative $\partial\eta/\partial$N. The Journal of Chemical Physics, 94(8):5559–5564, 1991.

3 P. Senet. Nonlinear electronic responses, Fukui functions and hardnesses as functionals of the ground-state electronic density. The Journal of Chemical Physics, 105(15):6471–6489, 1996.

4 P. Senet. Kohn-Sham orbital formulation of the chemical electronic responses, including the hardness. The Journal of Chemical Physics, 107(7):2516–2524, 1997.

5 H. Chermette. Chemical reactivity indexes in density functional theory. Journal of Computational Chemistry, 20(1):129–154, 1999.

6 Frank De Proft and Paul Geerlings. Conceptual and computational DFT in the study of aromaticity. Chemical Reviews, 101(5):1451–1464, 2001.

7 P. Geerlings, F. De Proft, and W. Langenaeker. Conceptual density functional theory. Chemical Reviews, 103(5):1793–1874, 2003.

8 Paul Geerlings, Stijn Fias, Zino Boisdenghien, and Frank De Proft. Conceptual DFT: chemistry from the linear response function. Chemical Society Reviews, 43(14):4989–5008, 2014.

9 Patricio Fuentealba and Carlos Cárdenas. Density functional theory of chemical reactivity. In *Chemical Modelling*, pages 151–174. 2014.

10 Max Berkowitz and Robert G. Parr. Molecular hardness and softness, local hardness and softness, hardness and softness kernels, and relations among these quantities. The Journal of Chemical Physics, 88(4):2554–2557, 1988.

11 P. Senet and M. Yang. Relation between the Fukui function and the Coulomb hole. Journal of Chemical Sciences, 117(5):411–418, 2005.

12 Patrice Delarue and Patrick Senet. Polarization, reactivity and quantum molecular capacitance: From electrostatics to density functional theory. Indian Journal of Chemistry, 53A:1052–1057, 2014.

13 Pratim K. Chattaraj, Andrés Cedillo, and Robert G. Parr. Variational method for determining the Fukui function and chemical hardness of an electronic system. The Journal of Chemical Physics, 103(17):7645–7646, 1995.

14 Paul Geerlings, Stijn Fias, Thijs Stuyver, Paul Ayers, Robert Balawender, and Frank De Proft. New Insights and Horizons from the Linear Response Function in Conceptual DFT. Density Functional Theory, 2018.

15 Roman F. Nalewajski. Chemical reactivity concepts in charge sensitivity analysis. International Journal of Quantum Chemistry, 56(5):453–476, 1995.

16 Morrel H. Cohen, M. V. Ganduglia-Pirovano, and J. Kudrnovský. Reactivity kernels, the normal modes of chemical reactivity, and the hardness and softness spectra. The Journal of Chemical Physics, 103(9):3543–3551, September 1995.

17 Patrick Senet. Varitaional principle for eigenmodes of reactivity in conceptual density functional theory. Submitted, 2020.

18 C. F. v. Weizsäcker. Zur Theorie der Kernmassen. Zeitschrift für Physik, 96(7):431–458, 1935.

19 P. Hohenberg and W. Kohn. Inhomogeneous electron gas. Physical Review,136(3B):B864–B871, 1964.

20 W. Kohn and L. J. Sham. Self-consistent equations including exchange and correlation effects. Physical Review, 140(4A):A1133–A1138, 1965.

21 Mel Levy. Universal variational functionals of electron densities, first-order density matrices, and natural spin-orbitals and solution of the v-representability problem. Proceedings of the National Academy of Sciences, 76(12):6062–6065, 1979.

22 Francesco Giacomo Tricomi. Integral Equations. Courier Corporation, March 1985.

23 R. P. Feynman, N. Metropolis, and E. Teller. Equations of state of elements based on the generalized fermi-thomas theory. Physical Review, 75(10):1561–1573, 1949.

24 Conyers Herring. Explicit estimation of ground-state kinetic energies from electron densities. Physical Review A, 34(4):2614–2631, 1986.

25 N. H. March, P. Senet, and V. E. Van Doren. Non-local kinetic energy functional for an arbitrary number of Fermions moving independently in one-dimensional harmonic oscillator potential. Physics Letters A, 270(1):88–92, 2000.

26 Junchao Xia, Chen Huang, Ilgyou Shin, and Emily A. Carter. Can orbital-free density functional theory simulate molecules? The Journal of Chemical Physics, 136(8):084102, 2012.

11

Conceptual Density Functional Theory in the Grand Canonical Ensemble

José L. Gázquez[1,*], *Marco Franco-Pérez*[2], *Paul W. Ayers*[3], *and Alberto Vela*[4]

[1] *Departamento de Química, Universidad Autónoma Metropolitana-Iztapalapa, Av. San Rafael Atlixco 186, Ciudad de México 09340, México*
[2] *Facultad de Química, Universidad Nacional Autónoma de México, Cd. Universitaria Ciudad de México 04510, México*
[3] *Department of Chemistry, McMaster University, Hamilton, Ontario L8S 4M1, Canada*
[4] *Departamento de Química, Centro de Investigación y de Estudios Avanzados Av. Instituto Politécnico Nacional 2508, Ciudad de México 07360, México*

11.1 Introduction

The study of chemical reactivity in the framework of density functional theory (DFT) has shown to be a useful approach to explain a wide variety of interactions through a chemically meaningful language [1–11]. The starting point is based on the total differential of the energy as a function of the number of electrons, N, and the external potential, $v(\mathbf{r})$, that introduces the first-order response functions, from which one can also derive second and higher order response functions, to integrate a set of indicators that allow one to characterize the intrinsic reactivity of a chemical species from the electronic structure it has when it is isolated. The response functions are given by derivatives of the energy with respect to N, to $v(\mathbf{r})$ or to both. Thus, through these response functions one may infer the behavior of a given molecule when it interacts with different families of reactants. A relevant aspect regarding this approach is that the framework provided by density functional theory allows one to identify these derivatives [12–18] with intuitive chemical concepts [19–26] that are related with general, and also intuitive, reactivity principles [23–29]. This way, it has been possible to get a better understanding of the conditions required by the interacting species to fulfill these reactivity principles [30–69]. In addition, through this framework it has been possible to generalize these concepts in terms of a broader perspective conformed by global, local and non-local response functions [70–92]. In summary, this approach, better known as Conceptual Density Functional Theory (C-DFT), has become an important tool for the examination of chemical interactions in terms of the electronic structure of the participating species.

Since the study of chemical reactivity in the framework of C-DFT is based on the analysis of the pertinent global, local and non-local response functions determined for a specific species under specific conditions, it represents an attractive route to understand some aspects of the behavior of confined systems [93]. Among other possibilities, for example, one may simply compare the differences between the response functions of the free and the confined molecule, to get information about the changes induced in the electronic structure that result from the confinement. The benefit of this exercise lies in the fact that the response functions translate the information contained in

*Corresponding author. Email: jlgm@xanum.uam.mx

Chemical Reactivity in Confined Systems: Theory, Modelling and Applications,
First Edition. Edited by Pratim Kumar Chattaraj and Debdutta Chakraborty.

the electronic structure into quantities that are directly related to the chemical reactivity, so that one can immediately observe the changes induced by confinement.

In this context, the object of the present chapter is to give a brief perspective of the basic aspects of conceptual density functional theory. However, we will adopt a more general approach than the one normally followed, by making use of the grand canonical ensemble to incorporate the temperature from the beginning, not only to establish its effects on the reactivity indicators, but also to discuss through a more complete vision the fundamental issues that characterize this theoretical framework.

It is important to mention that some aspects of the temperature effects in C-DFT have been considered in the past by other authors [94–103]. However, the material presented in this chapter is based on the work we have developed [104–116].

11.2 Fundamental Equations for Chemical Reactivity

The consideration of the energy as a function of the number of electrons and the external potential, as the starting point to describe the intrinsic chemical reactivity of a molecule is based on the idea that the number of electrons N and the external potential $v(\mathbf{r})$ generated by the nuclei located at the ground state geometry will change, in the presence of another species to $N + \Delta N$, as a result of charge transfer, and to $v(\mathbf{r}) + \delta v(\mathbf{r})$ as a result of the presence of the electrons and nuclei of the other species. This way, the derivatives of the energy with respect to N, to $v(\mathbf{r})$ or to both, evaluated at the values of the isolated molecule, will reflect the way in which this one will respond in the early stages of the interaction. Since in this approach one needs to treat the number of electrons as a continuous variable, one must consider the molecule as an open system that can exchange electrons with the reservoir (bath) in which it is immersed (a molecule in a highly diluted solution), so that the number of electrons will fluctuate.

This situation implies that the appropriate framework in this case corresponds to the grand canonical ensemble, in which the natural variables are the chemical potential of the reservoir μ_{Bath}, the external potential $v(\mathbf{r})$, and the temperature T. In this ensemble the partition function is defined as

$$\Xi(\mu_{Bath}, T)[v(\mathbf{r})] = \sum_N \sum_i e^{-\beta(E_{N,i} - \mu_{Bath} N)}, \tag{11.1}$$

where $E_{N,i}$ is the energy of the ith N- electron excited state ($i = 0$ corresponds to the ground state) and $\beta = 1/k_B T$, where k_B is Boltzmann's constant. The grand canonical potential is given by

$$\Omega(\mu_{Bath}, T)[v(\mathbf{r})] = -kT \ln\{\Xi(\mu_{Bath}, T)[v(\mathbf{r})]\} = \langle E \rangle - \mu_{Bath} \langle N \rangle - T \langle S_T \rangle, \tag{11.2}$$

where S_T is the electronic entropy. This quantity together with E and N are written in between angular brackets to indicate that they correspond to the average value of the ensemble, that is

$$\langle E \rangle = \sum_N \sum_i w_{N,i} E_{N,i}, \tag{11.3}$$

$$\langle N \rangle = \sum_N \sum_i w_{N,i} N \tag{11.4}$$

and

$$\langle S_T \rangle = -k_B \sum_N \sum_i w_{N,i} \ln w_{N,i}. \tag{11.5}$$

Similarly, the average electronic density is expressed as

$$\langle \rho(\mathbf{r})\rangle = \sum_N \sum_i w_{N,i}\rho_{N,i}(\mathbf{r}). \tag{11.6}$$

The weighting factor that appears in these expressions is written in terms of the grand partition function in the form

$$w_{N,i} = \frac{1}{\Xi} e^{-\beta(E_{N,i}-\mu_{Bath}N)}. \tag{11.7}$$

As previously mentioned, the natural variables for the usual development of chemical reactivity in the framework of DFT at *zero temperature* are N and $v(\mathbf{r})$, and the fundamental equation to construct the set of reactivity indicators is the total differential of the energy in terms of these two variables [1, 2, 12, 14, 117]. However, the natural variables for the grand canonical ensemble are μ_{Bath}, $v(\mathbf{r})$ and T, so that the total differential of $\langle E\rangle$ is

$$d\langle E\rangle = \left(\frac{\partial \langle E\rangle}{\partial \mu_{Bath}}\right)_{T,v(\mathbf{r})} d\mu_{Bath} + \int \left(\frac{\delta \langle E\rangle}{\delta v(\mathbf{r})}\right)_{T,\mu_{Bath}} \delta v(\mathbf{r})d\mathbf{r} + \left(\frac{\partial \langle E\rangle}{\partial T}\right)_{\mu_{Bath},v(\mathbf{r})} dT \tag{11.8}$$

One can show through a series of algebraic manipulations that this differential can be transformed to a differential in terms of the variables $\langle N\rangle$, $v(\mathbf{r})$ and T of the form [107],

$$d\langle E\rangle = \left(\frac{\partial \langle E\rangle}{\partial \langle N\rangle}\right)_{T,v(\mathbf{r})} d\langle N\rangle + \int \left(\frac{\delta \langle E\rangle}{\delta v(\mathbf{r})}\right)_{T,\langle N\rangle} \delta v(\mathbf{r})d\mathbf{r} + \left(\frac{\partial \langle E\rangle}{\partial T}\right)_{\langle N\rangle,v(\mathbf{r})} dT. \tag{11.9}$$

The derivative in the first term of the right-hand side of this expression corresponds to the extension to any temperature of the usual C-DFT definition at zero temperature of the electronic chemical potential [12, 105],

$$\left(\frac{\partial \langle E\rangle}{\partial \langle N\rangle}\right)_{T,v(\mathbf{r})} = \mu_e, \tag{11.10}$$

(e stands for electronic). The electronic chemical potential is equal to the chemical potential of the reservoir plus an entropic contribution that vanishes at zero temperature,

$$\mu_e = \mu_{Bath} + T\left(\frac{\partial \langle S_T\rangle}{\partial \langle N\rangle}\right)_{T,v(\mathbf{r})} \tag{11.11}$$

The derivative in the second term of the right-hand side of Eq. (11.9) is equal to the average electron density, which would be the equivalent of the usual C-DFT [1], but there is also an entropy contribution that vanishes at $T = 0$,

$$\left(\frac{\delta \langle E\rangle}{\delta v(\mathbf{r})}\right)_{T,\langle N\rangle} = \langle \rho(\mathbf{r})\rangle + T\left(\frac{\delta \langle S_T\rangle}{\delta v(\mathbf{r})}\right)_{T,\langle N\rangle}, \tag{11.12}$$

The derivative in the third term of the right-hand side of Eq. (11.9), which evidently in a temperature independent derivation is not present, can be identified as the electronic heat capacity [107],

$$\left(\frac{\partial \langle E\rangle}{\partial T}\right)_{\langle N\rangle,v(\mathbf{r})} = C_{v(\mathbf{r})} \tag{11.13}$$

Now, if one considers the ensemble composed by the systems with $N_0 - 1$, N_0 and $N_0 + 1$ electrons in their ground state, one can show that the entropy contribution in Eq. (11.12), which disappears at $T = 0$, is negligible up to temperatures around 10^4 K. Since this ensemble has been shown to include all the basic contributions required to evaluate the chemical reactivity indexes, one can see that for temperatures of chemical interest Eq. (11.9) is the finite temperature version of the fundamental equation of chemical reactivity theory [1, 2, 12, 14, 117]. A relevant aspect of this expression is that it has the same form (neglecting the second term in the right-hand side of Eq. (11.12)) as the one obtained and used in C-DFT at $T = 0$, and therefore it indicates that one can derive the temperature-dependent expressions of the natural set of response functions through the derivatives of $\langle E\rangle$ and $\langle\rho(\mathbf{r})\rangle$, Eqs (11.3) and (11.6), with respect to $\langle N\rangle$ (Eq. (11.4)) and $v(\mathbf{r})$, and, in addition, one can also derive the response functions with respect to changes in the temperature, for example Eq. (11.13). In this general procedure, one can recover the usual definitions by evaluating the expressions at $T = 0$.

The total differential of the average energy provides the first-order response functions, so that for the second-order response functions one needs to consider the total differentials of the first-order responses [1, 14] in terms of the variables $\langle N\rangle$, $v(\mathbf{r})$ and T, in the finite temperature framework. Thus, for the electronic chemical potential one has that

$$d\mu_e = \left(\frac{\partial\mu_e}{\partial\langle N\rangle}\right)_{T,v(\mathbf{r})} d\langle N\rangle + \int\left(\frac{\delta\mu_e}{\partial v(\mathbf{r})}\right)_{T,\langle N\rangle}\delta v(\mathbf{r})\,d\mathbf{r} + \left(\frac{\partial\mu_e}{\partial T}\right)_{\langle N\rangle,v(\mathbf{r})} dT \tag{11.14}$$

In this case, the first derivative in this expression corresponds to the temperature-dependent definition of the electronic hardness [13, 105],

$$\left(\frac{\partial\mu_e}{\partial\langle N\rangle}\right)_{T,v(\mathbf{r})} = \left(\frac{\partial^2\langle E\rangle}{\partial\langle N\rangle^2}\right)_{T,v(\mathbf{r})} = \eta_e, \tag{11.15}$$

where Eq. (11.10) has been used to obtain the first equality. For the second derivative in Eq. (11.14) one finds that

$$\left(\frac{\delta\mu_e}{\delta v(\mathbf{r})}\right)_{T,\langle N\rangle} = f_e(\mathbf{r}) + T\left[\frac{\partial}{\partial\langle N\rangle}\left(\frac{\delta\langle S_T\rangle}{\delta v(\mathbf{r})}\right)_{T,\langle N\rangle}\right]_{T,v(\mathbf{r})} \tag{11.16}$$

where the equality comes from a Maxwell relation for the differential of the energy as a function of $\langle N\rangle$, $v(\mathbf{r})$ and T. The first term in the right-hand side of this expression corresponds to the extension to any temperature of the usual C-DFT definition at zero temperature of the Fukui function [14–16, 106],

$$f_e(\mathbf{r}) = \left(\frac{\partial\langle\rho(\mathbf{r})\rangle}{\partial\langle N\rangle}\right)_{T,v(\mathbf{r})}, \tag{11.17}$$

while the entropy contribution that appears in the second term can also be neglected at temperatures of chemical interest. Finally, the third derivative in Eq. (11.14) is given by

$$\left(\frac{\partial\mu_e}{\partial T}\right)_{\langle N\rangle,v(\mathbf{r})} = \left(\frac{\partial C_{v(\mathbf{r})}}{\partial\langle N\rangle}\right)_{T,v(\mathbf{r})}, \tag{11.18}$$

where the second equality corresponds also to a Maxwell relation for the differential of the energy as a function of $\langle N\rangle$, $v(\mathbf{r})$ and T.

In the case of the average electronic density one has that [14, 106, 107, 112]

$$d\langle\rho(\mathbf{r})\rangle = \left(\frac{\partial\langle\rho(\mathbf{r})\rangle}{\partial\langle N\rangle}\right)_{T,v(\mathbf{r})} d\langle N\rangle + \int\left(\frac{\delta\langle\rho(\mathbf{r})\rangle}{\delta v(\mathbf{r}')}\right)_{T,\langle N\rangle}\delta v(\mathbf{r}')\,d\mathbf{r}' + \left(\frac{\partial\langle\rho(\mathbf{r})\rangle}{\partial T}\right)_{\langle N\rangle,v(\mathbf{r})} dT. \tag{11.19}$$

where the first derivative has already been identified with the Fukui function, Eq. (11.17), the second derivative corresponds to the electronic linear density response function at finite temperature,

$$\left(\frac{\delta\langle\rho(\mathbf{r})\rangle}{\delta v(\mathbf{r}')}\right)_{T,\langle N\rangle} = \chi_e(\mathbf{r},\mathbf{r}'), \tag{11.20}$$

and the third derivative may be identified with a local heat capacity [107, 112],

$$\left(\frac{\partial\langle\rho(\mathbf{r})\rangle}{\partial T}\right)_{\langle N\rangle,v(\mathbf{r})} = C_{v(\mathbf{r})}(\mathbf{r}). \tag{11.21}$$

In a trivial extension one could also consider the changes in the electronic hardness through the total differential. In this case one obtains the temperature-dependent expressions for the first hyperhardness [118], γ_e as the third derivative of the $\langle E\rangle$ with respect to $\langle N\rangle$ and the dual descriptor [17, 18], $\Delta f_e(\mathbf{r})$, as the second derivative of $\langle\rho(\mathbf{r})\rangle$ with respect to $\langle N\rangle$. These latter third-order response functions are also important for chemical reactivity. Additionally one also obtains a third reactivity indicator that corresponds to the second derivative of the electronic heat capacity with respect to $\langle N\rangle$.

Thus, we have seen that starting from the total differential of the average energy in the grand canonical ensemble in terms of its natural variables, μ_{Bath}, $v(\mathbf{r})$, and T, one can obtain the total differential for the average energy in terms of the variables $\langle N\rangle$, $v(\mathbf{r})$, and T, and that this latter expression is basically equal in form to the zero temperature counterpart, except for an entropy contribution that is negligible at temperatures of chemical interest. From this identification, following the same approach used in the usual zero temperature formulation of C-DFT, we have found that the finite temperature definition of the response coefficients associated with changes in $\langle N\rangle$ and in $v(\mathbf{r})$ is the same, except that in the grand canonical ensemble one needs to consider the average ensemble values of the energy, the number of electrons, and the electronic density.

11.3 Temperature-Dependent Response Functions

According to the response functions described previously, in this section we proceed to evaluate the derivatives of $\langle E\rangle$ and $\langle\rho(\mathbf{r})\rangle$ with respect to $\langle N\rangle$, through the grand canonical ensemble expressions given in Eqs (11.3), (11.6), and (11.4), respectively, and to analyze their relevance to describe chemical reactivity.

By making use of the three ground states ensemble model previously described, one finds that the fractional part of the number of electrons is given by [105]

$$\omega = \langle N\rangle - N_0 = \frac{\exp[\beta(EA+\mu_{Bath})] - \exp[-\beta(IP+\mu_{Bath})]}{1+\exp[\beta(EA+\mu_{Bath})] + \exp[-\beta(IP+\mu_{Bath})]}, \tag{11.22}$$

where IP and EA correspond to the vertical first ionization potential and electron affinity, respectively, of the reference system with N_0 electrons. In the case of the difference between the average energy and the ground state energy of the N_0 electron system, $E_{N_0,0}$, one finds that

$$\Delta\langle E\rangle = \langle E\rangle - E_{N_0,0} = \frac{IP\exp[-\beta(IP+\mu_{Bath})] - EA\exp[\beta(EA+\mu_{Bath})]}{1+\exp[\beta(EA+\mu_{Bath})] + \exp[-\beta(IP+\mu_{Bath})]}. \tag{11.23}$$

These two expressions may be combined to find an expression for the energy difference in terms of the fractional charge, namely [114]

$$\Delta\langle E\rangle = -\tfrac{1}{2}(IP+EA)\omega + \tfrac{1}{2}(IP-EA)\frac{\alpha+\omega^2}{1+\alpha} \tag{11.24}$$

with

$$\alpha = (\omega^2 + 4(1-\omega^2)\exp[-\beta(IP-EA)])^{1/2} \tag{11.25}$$

The corresponding expressions for the difference between the average electronic density and the ground state electronic density, $\rho_{N_0,0}(\mathbf{r})$, are

$$\Delta\langle\rho(\mathbf{r})\rangle = \langle\rho(\mathbf{r})\rangle - \rho_{N_0,0}(\mathbf{r}) = \frac{f_e^+(\mathbf{r})\mathrm{e}^{\beta(EA+\mu_{Bath})} - f_e^-(\mathbf{r})\mathrm{e}^{-\beta(IP+\mu_{Bath})}}{1+\mathrm{e}^{\beta(EA+\mu_{Bath})}+\mathrm{e}^{-\beta(IP+\mu_{Bath})}}, \tag{11.26}$$

and [113]

$$\Delta\langle\rho(\mathbf{r})\rangle = \tfrac{1}{2}[f_e^-(\mathbf{r})+f_e^+(\mathbf{r})]\omega + \tfrac{1}{2}[f_e^+(\mathbf{r})-f_e^-(\mathbf{r})]\frac{\alpha+\omega^2}{1+\alpha}, \tag{11.27}$$

where

$$f_e^-(\mathbf{r}) = \rho_{N_0,0}(\mathbf{r}) - \rho_{N_0-1,0}(\mathbf{r}) \tag{11.28}$$

and

$$f_e^+(\mathbf{r}) = \rho_{N_0+1,0}(\mathbf{r}) - \rho_{N_0,0}(\mathbf{r}) \tag{11.29}$$

Using Eqs (11.24) and (11.27) one can obtain the finite temperature expressions for the first and second derivatives of $\langle E\rangle$ and $\langle\rho(\mathbf{r})\rangle$ with respect to $\langle N\rangle$. The resulting equations together with the limit when T goes to zero are reported in Table 11.1. It is important to mention that the expressions in this limit are the ones that can be used to calculate these quantities for temperatures of chemical interest, because the effects on their values require temperatures of at least around 10^4 K.

A very important aspect of the present development is related with the differentiability of the energy as a function of the number of electrons. One can confirm, from Eqs (11.22), (11.24) and

Table 11.1 Temperature-dependent expressions for the response functions (second row in each key is the frontier orbital approximation of the quantity above)[a].

Indicator	Expressions at temperatures of chemical interest		
	$\omega<0$	$\omega>0$	$\omega=0$
$\mu_e = \left(\frac{\partial\langle E\rangle}{\partial\langle N\rangle}\right)_{T,v(\mathbf{r})} = -\frac{1}{2}(IP+EA)+\frac{1}{2}(IP-EA)\frac{\omega}{\alpha}$	$-IP$	$-EA$	$-(IP+EA)/2$
	ε_H	ε_L	$(\varepsilon_H+\varepsilon_L)/2$
$\eta_e = \left(\frac{\partial^2\langle E\rangle}{\partial\langle N\rangle^2}\right)_{T,v(\mathbf{r})} = (IP-EA)P(\omega)$[b]	0	0	$(IP-EA)\delta(\omega)$
	0	0	$(\varepsilon_L-\varepsilon_H)\delta(\omega)$
$f_e(\mathbf{r}) = \left(\frac{\partial\langle\rho(\mathbf{r})\rangle}{\partial\langle N\rangle}\right)_{T,v(\mathbf{r})} = \frac{1}{2}[f_e^-(\mathbf{r})+f_e^+(\mathbf{r})]+\frac{1}{2}[f_e^+(\mathbf{r})-f_e^-(\mathbf{r})]\frac{\omega}{\alpha}$	$f^-(\mathbf{r})$	$f^+(\mathbf{r})$	$[f^-(\mathbf{r})+f^+(\mathbf{r})]/2$
	$\rho_H(\mathbf{r})$	$\rho_L(\mathbf{r})$	$[\rho_H(\mathbf{r})+\rho_L(\mathbf{r})]/2$
$\Delta f_e(\mathbf{r}) = \left(\frac{\partial^2\langle\rho(\mathbf{r})\rangle}{\partial\langle N\rangle^2}\right)_{T,v(\mathbf{r})} = [f_e^+(\mathbf{r})-f_e^-(\mathbf{r})]P(\omega)$[b]	0	0	$[f^+(\mathbf{r})-f^-(\mathbf{r})]\delta(\omega)$
	0	0	$[\rho_L(\mathbf{r}) - {}_H(\mathbf{r})]\delta(\omega)$
$\eta_T = \frac{\beta}{2}C_T(IP-EA)$[c]	0	0	$\eta_T = \beta(IP-EA)/2$[d]
	0	0	$\eta_T = \beta(\varepsilon_L-\varepsilon_H)/2$
$\Delta f_T(\mathbf{r}) = \frac{\beta}{2}C_T[f_e^+(\mathbf{r})-f_e^-(\mathbf{r})]$[c]	0	0	$\Delta f_T(\mathbf{r}) = \beta[f^+(\mathbf{r})-f^-(\mathbf{r})]/2$
	0	0	$\Delta f_T(\mathbf{r}) = \beta[\rho_L(\mathbf{r})-\rho_H(\mathbf{r})]/2$

a) ε_H and ε_L are the eigenvalues of the highest occupied and lowest unoccupied molecular orbitals, while $\rho_H(\mathbf{r})$ and $\rho_L(\mathbf{r})$ are the electron densities of these orbitals.

b) The function $P(\omega) = 2\,e^{-\beta(IP-EA)}/\alpha^3$ reduces to the Dirac delta function $\delta(\omega)$ in the zero temperature limit.

c) $C_T = \left[\mathrm{sech}^2[\beta\Delta\mu] - \mathrm{sech}[\beta\Delta\mu]e^{-\frac{\beta}{2}(IP-EA)}\right]\left[1+2\,\mathrm{sech}[\beta\Delta\mu]e^{-\frac{\beta}{2}(IP-EA)}\left(1+4e^{-\frac{\beta}{2}(IP-EA)}\mathrm{sech}[\beta\Delta\mu]\right)\right]^{-1}$ with $\Delta\mu = \mu_{Bath}+(IP+EA)/2$.

d) for $\mu_{Bath} = -(IP+EA)/2 \quad (\omega=0)$ and for $\mu_{Bath} \neq -(IP+EA)/2 \quad (\omega\neq 0)$.

(11.25) that in the limit of zero temperature one recovers the straight lines connecting the integer values of $\langle N \rangle$ [119–121], which leads to a discontinuous first derivative and an "ill defined" second derivative. However, at $T \neq 0$ one can see that the derivatives of any order of $\langle E \rangle$ with respect to $\langle N \rangle$ exist and can be determined analytically [105]. Therefore, the temperature-dependent approach to C-DFT provides support to one of the crucial and debated issues of chemical reactivity, namely, the differentiability with respect to the number of electrons.

The connection between the response functions considered in the previous section and DFT was established by Parr and collaborators [12], who recognized that the Lagrange multiplier associated with the minimization of the energy with respect to the electronic density, subject to the condition that the latter must integrate to the number of electrons, is equal to the derivative of the minimum energy with respect to the number of electrons [122], that is, this Lagrange multiplier is equal to the electronic chemical potential μ_e, an therefore it is also equal to the negative of electronegativity χ_e as defined by Iczkowski and Margrave [22]. Thus, one can see that a fundamental variable of DFT, the Lagrange multiplier of the variational equation, is identified as a chemical potential, which is then associated with the intuitive concept of electronegativity proposed by Pauling [19, 21], who defined it as "the power of an atom in a molecule to attract electrons to itself". An important support to this identification is the expression for μ_e at $T = 0$ for the reference system given in Table 11.1. This relationship is equal to the negative of the absolute electronegativity defined by Mulliken [20] ($\chi_e = (IP + EA)/2$). When one determines the values of χ_e using the experimental values for IP and EA, one finds that they follow approximately the same trends of the scale established by Pauling. Another relevant aspect of this identification comes from the fact that the principle of electronegativity equalization, proposed by Sanderson [27, 28] in terms of the charge transfer between interacting species, can be understood as the consequence of a chemical potential equalization process.

The identification of the second derivative of the average energy with respect to the number of electrons with the intuitive concept of hardness was the next step [13]. This one is related with the hard and soft acids and bases (HSAB) principle [23, 24] that establishes that hard acids prefer to bind to hard bases and soft acids prefer to bind to soft bases. However, note that the expression for η_e, in the limit of zero temperature, given in Table 11.1, is equal to $(IP - EA)$, but multiplied by the Dirac delta function. To avoid the latter, Parr and Pearson replaced the $\langle E \rangle$ versus $\langle N \rangle$ straight lines behavior at $T = 0$ by a smooth quadratic interpolation between the energy of the systems with $N_0 - 1$, N_0 and $N_0 + 1$ electrons, namely

$$\Delta E = \mu(N - N_0) + \frac{1}{2}\eta(N - N_0)^2, \tag{11.30}$$

with $\mu = -(IP + EA)/2$ and $\eta = IP - EA$. As in the case of electronegativity, if one determines the values of η using the experimental values for IP and EA, one finds that they follow approximately the same trends of the qualitative scale established by Pearson, providing a strong support to the identification given in Eq. (11.15). Additionally, the study of many different aspects involved in charge transfer processes through Eq. (11.30) [53–55, 58, 65–69, 123–139], because of its simplicity and clear chemical meaning, also constitutes further evidence for the connection between these intuitive concepts and the first and second derivatives of $\langle E \rangle$ with respect to $\langle N \rangle$.

It is important to mention that the softness, companion concept of hardness, was originally defined in the grand canonical ensemble as [71]

$$S = \left(\frac{\partial \langle N \rangle}{\partial \mu_{Bath}} \right)_{T,v(\mathbf{r})}. \tag{11.31}$$

Through the present approach one can see, by taking the derivative of Eq. (11.11) with respect to $\langle N \rangle$, that [105]

$$\eta_e = \frac{1}{S} + T\left(\frac{\partial^2 \langle S_T \rangle}{\partial \langle N \rangle^2}\right)_{T,v(\mathbf{r})}, \tag{11.32}$$

which shows that at $T = 0$ softness is the inverse hardness. However, when the temperature is different from zero this no longer holds because there is an entropic contribution.

This definition of hardness, Eq. (11.15), together with the chemical potential, Eq. (11.10), have been used to get a better understanding of the HSAB and the maximum hardness principles and the conditions that are required for their validity [30–69]. The latter was proposed by Pearson [29], who stated that "there seems to be a rule of nature that molecules arrange themselves so as to be as hard as possible".

Now, while the chemical potential and the hardness are global reactivity indexes that characterize the molecule as a whole, the Fukui function is a local response function that, in principle, allows one to distinguish the reactivity of different sites within a molecule. It can be shown that the Fukui function is the one that minimizes the energy change that results from adding or subtracting an electron [47]. This way, it indicates [14] through the values of $f_e^+(\mathbf{r})$, according to Eq. (11.29), which are the places in a molecule more prone to receive charge (nucleophilic attack), and through $f_e^-(\mathbf{r})$, Eq. (11.28), those more likely to donate charge (electrophilic attack). The average of these two quantities characterizes the places for radical attack.

From Eqs (11.22), (11.25), and (11.27) one can find that, as in the case of the energy, the electronic density in the limit of zero temperature is given by a set of straight lines connecting the integer values of $\langle N \rangle$ [119–121], that leads to a discontinuous first derivative and an "ill defined" second derivative. By smoothening this behavior through a procedure analogous to the one done for the energy, one finds that the dual descriptor is given by $\Delta f_e(\mathbf{r}) = f_e^+(\mathbf{r}) - f_e^-(\mathbf{r})$, which corresponds to the original expression proposed for this descriptor by Morell and collaborators [17, 18].

It is important to note that the response functions arising from the derivatives with respect to $\langle N \rangle$ reduce, under certain approximations, to properties that are normally used in frontier orbital theory, as can be seen in Table 11.1, where we present the expressions for the electronic chemical potential, hardness, Fukui function and dual descriptor, in terms of the eigenvalues and the electronic distributions of the highest occupied and lowest unoccupied molecular orbitals [140].

In the original work regarding the Fukui function, Parr and Yang [14], based on the analysis of the total differential of the electronic chemical potential, Eq. (11.14), postulated that among several possible sites on which a reagent may interact with a molecule, the one preferred is the one for which the change in the chemical potential is the largest. This statement indicates that the most reactive sites of a molecule can be associated with the ones where the Fukui function has large values. Due to the close relation between the Fukui function and the frontier orbitals, Parr and Yang conjectured that frontier orbital theory is equivalent to the assumption of the rule "$|\Delta\mu|$ big is good". This last statement was recently proven and shown to imply the HSAB and maximum hardness principles [66, 69].

So far we have considered the response functions that are related with the derivatives of $\langle E \rangle$ and $\langle \rho(\mathbf{r}) \rangle$ with respect to $\langle N \rangle$, and we have found that the second derivatives are "ill defined". However, another advantage of the framework provided by the grand canonical ensemble is that one can also consider the response functions with respect to changes in the chemical potential of the reservoir, μ_{Bath}. This is important because when $v(\mathbf{r})$ and T are kept constant, a change in μ_{Bath} implies that $\langle N \rangle$ will change, so that the derivative with respect to μ_{Bath} leads to a similar response. Thus, since the second derivatives with respect to $\langle N\rangle$ are "ill defined", one can consider as an equivalent alternative, the responses of the electronic chemical potential and Fukui function to changes in μ_{Bath}. This procedure leads to the definition of what we call the thermodynamic hardness [110], $\eta_T = (\partial\mu_e/\partial\mu_{Bath})_{T,v(\mathbf{r})}$, and the thermodynamic dual descriptor [113], $\Delta f_T(\mathbf{r}) = (\partial f_e(\mathbf{r})/\partial\mu_{Bath})_{T,v(\mathbf{r})}$. The expressions for these two quantities are also given in Table 11.1. One can see that although they tend to infinity as T approaches zero, they do not show the Dirac delta function behavior at temperatures of chemical interest, which is present in the hardness and the dual descriptor. However, it is important to note that for a temperature value different from zero η_T is proportional to $(IP - EA)$ and $\Delta f_T(\mathbf{r})$ is proportional to $(f^+(\mathbf{r}) - f^-(\mathbf{r}))$, and therefore the differences in these two quantities between different species, at the same temperature, are equal to the one obtained for these two properties using the well established expressions obtained through the smooth quadratic interpolation mentioned before.

In the case of the chemical potential and the hardness, it has also been proposed that at finite temperature one could consider the Helmholtz potential $\langle A\rangle = \langle E\rangle - T\langle S\rangle$, instead of $\langle E\rangle$, leading to a Helmholtz [100] chemical potential, μ_A, and hardness, η_A, given in terms of the first and second derivatives of $\langle A\rangle$ with respect to $\langle N\rangle$. This definition of hardness retains the behavior induced by the presence of Dirac delta function in the limit of zero temperature. However, an important aspect of the three temperature-dependent definitions of hardness, namely η_e, η_T and η_A, is that they all imply the HSAB and the maximum hardness principles [102, 103, 141].

Finally, we have seen that there are three new response functions related with changes in the temperature [107], the electronic heat capacity (Eq. (11.13)), the local heat capacity (Eq. (11.21)), and Eq. (11.18). The expressions obtained for the three ground states ensemble model in the low temperature limit are zero for $\omega < 0$ and $\omega > 0$ in the three cases, while for $\omega = 0$,

$$C_{v_r} = \left(\frac{\partial\langle E\rangle}{\partial T}\right)_{\langle N\rangle,v(\mathbf{r})} = \frac{e^{-\frac{\beta}{2}(I-A)}}{2k_BT^2}(I-A)^2 \quad \text{for} \quad \mu_{Bath} = \mu_0, \tag{11.33}$$

$$\left(\frac{\partial\mu_e}{\partial T}\right)_{\langle N\rangle,v(\mathbf{r})} = \left(\frac{\partial C_{v(\mathbf{r})}}{\partial\langle N\rangle}\right)_{T,v(\mathbf{r})} = -\frac{e^{-\frac{\beta}{2}(I-A)}}{4k_BT^2}(I-A)^2 \quad \text{for} \quad \mu_{Bath} = \mu_0, \tag{11.34}$$

and

$$C_{v_r}(\mathbf{r}) = \left(\frac{\partial\langle\rho(\mathbf{r})\rangle}{\partial T}\right)_{\langle N\rangle,v(\mathbf{r})} = \frac{e^{-\frac{\beta}{2}(I-A)}}{2k_BT^2}(I-A)[f_e^+(\mathbf{r}) - f_e^-(\mathbf{r})] \quad \text{for} \quad \mu_{Bath} = \mu_0. \tag{11.35}$$

Note that at $T = 0$ the three terms are equal to zero, so that, as expected, in this limit there is no contribution of the response functions associated to changes in the temperature.

However, it is also relevant to see that when $\langle N\rangle$ is an integer, at temperatures of chemical interest, which are the ones that correspond to the expressions given in Eqs (11.33–11.35), these response functions are different from zero. This result indicates that in the early stages of an interaction between two molecules, in which there has not been yet charge transfer among them and both species have an integer $\langle N\rangle$, the changes in temperature give rise to changes in the energy and in the chemical potential that induce the charge transfer process. Once $\langle N\rangle$ becomes a fractional number, the changes in temperature give rise directly to electron transfer. Thus one can see that the rule "$|\Delta\mu|$ big is good", based on Eq. (11.14), may be considered to have two components, one at the onset of an interaction induced by temperature changes, and another one once the charge transfer process begins. Through Eqs (11.35) and (11.14) one may conclude that the regioselectivity features of the Fukui function also have the same two components.

11.4 Local Counterpart of a Global Descriptor and Non-Local Counterpart of a Local Descriptor

A global descriptor characterizes a molecule as a whole. However, one may try to associate a local function to it, to obtain information about its spatial distribution in order to identify which are the sites or regions that make the largest contributions to the global value. Certainly, a desirable constraint on this local function would be to define it so that its integral over the whole space leads to the value of the global descriptor. In the case of a local descriptor one could proceed in a similar way, through a non-local function that identifies which are the sites or regions that make the largest contributions to the value of the property at a given point.

In this section we will analyze the cases of the electronic chemical potential [109, 116] and Fukui function [111, 116], which correspond to the global and local type reactivity descriptors, respectively, through the grand canonical ensemble formalism. The starting point takes into consideration that the variational constraint

$$\langle N \rangle = \int \langle \rho(\mathbf{r}) \rangle \, d\mathbf{r}, \tag{11.36}$$

suggests that the average electron density can be considered to be the local counterpart of the average number of electrons. Thus, since μ_e may be expressed as the ratio of the thermal fluctuations between $\langle N \rangle$ and $\langle E \rangle$, $\sigma_{NE} = \beta[\langle NE \rangle - \langle N \rangle\langle E \rangle]$, and the thermal fluctuations of $\langle N \rangle$, $\sigma_{NN} = \beta[\langle NN \rangle - \langle N \rangle\langle N \rangle]$, that is

$$\mu_e = \frac{\sigma_{NE}}{\sigma_{NN}} = \frac{\langle NE \rangle - \langle N \rangle \langle E \rangle}{\langle NN \rangle - \langle N \rangle \langle N \rangle}, \tag{11.37}$$

one can associate in the numerator of this expression the local counterpart of $\langle N \rangle$ with $\langle \rho(\mathbf{r}) \rangle$, so that one can express the local counterpart of μ_e as

$$\mu_e(\mathbf{r}) = \frac{\sigma_{\rho(\mathbf{r})E}}{\sigma_{NN}} = \frac{\langle \rho(\mathbf{r})E \rangle - \langle \rho(\mathbf{r}) \rangle \langle E \rangle}{\langle NN \rangle - \langle N \rangle \langle N \rangle}, \tag{11.38}$$

that, when integrated over the hole space [81, 84, 90, 142], leads to the global μ_e, that is

$$\mu_e = \int \mu_e(\mathbf{r}) \, d\mathbf{r}. \tag{11.39}$$

In Table 11.2, one can see the finite temperature expressions for the three ground states ensemble model, together with the expressions for the zero temperature limit, that, as in the previous cases, remain valid up to temperatures of chemical interest. The latter show that for $\omega < 0$ the local chemical potential becomes proportional to $IP \ f_e^-(\mathbf{r})$, a quantity that is closely related to the local ionization potential [143], which has been used to identify the sites in a molecule prone to receive an electrophilic attack [144–148], while for $\omega > 0$ it becomes proportional to $EAf_e^+(\mathbf{r})$, which has been used to identify the sites that will receive a nucleophilic attack [149–151]. For $\omega = 0$ one finds that this quantity becomes proportional $(IPf_e^-(\mathbf{r}) + EAf_e^+(\mathbf{r}))/2$, the local version of Mulliken's electronegativity.

A relevant aspect of the present approach is that taking the derivative with respect to $\langle N \rangle$ in Eq. (11.39) leads to a different expression for the local hardness than the one originally proposed in C-DFT, namely [90, 91, 109]

$$\eta_\tau(\mathbf{r}) = \left(\frac{\partial \mu_e(\mathbf{r})}{\partial \langle N \rangle} \right)_{T, v(\mathbf{r})}, \tag{11.40}$$

Table 11.2 Temperature-dependent expressions for the local and nonlocal counterparts of global and local reactivity indicators (second row in each key is the frontier orbital approximation of the quantity above).

Indicator	Expressions at temperatures of chemical interest		
	$\omega < 0$	$\omega > 0$	$\omega = 0$
$\mu_e(\mathbf{r}) = -\frac{1}{2}[IPf_e^-(\mathbf{r}) + EAf_e^+(\mathbf{r})] + \frac{1}{2}[IPf_e^-(\mathbf{r}) - EAf_e^+(\mathbf{r})]\frac{\omega}{\alpha} - [f_e^+(\mathbf{r}) - f_e^-(\mathbf{r})](IP - EA)\lambda^{-1}$	$-IPf_e^-(\mathbf{r})$ $\varepsilon_H\rho_H(\mathbf{r})$	$-EAf_e^+(\mathbf{r})$ $\varepsilon_L\rho_L(\mathbf{r})$	$-[IPf_e^-(\mathbf{r}) + EAf_e^+(\mathbf{r})]/2$ $[\varepsilon_H\rho_H(\mathbf{r}) + \varepsilon_L\rho_L(\mathbf{r})]/2$
$\eta_\tau(\mathbf{r}) = IPf_e^-(\mathbf{r}) - EAf_e^+(\mathbf{r}) + \frac{\omega}{2}(IP - EA)[f_e^+(\mathbf{r}) - f_e^-(\mathbf{r})]$	$IPf_e^-(\mathbf{r}) - EAf_e^+(\mathbf{r}) - (IP - EA)[f_e^+(\mathbf{r}) - f_e^-(\mathbf{r})]/2$ $\varepsilon_L\rho_L(\mathbf{r}) - \varepsilon_H\rho_H(\mathbf{r}) - (\varepsilon_L - \varepsilon_H)[\rho_L(\mathbf{r}) - \rho_H(\mathbf{r})]/2$	$IPf_e^-(\mathbf{r}) - EAf_e^+(\mathbf{r}) + (IP - EA)[f_e^+(\mathbf{r}) - f_e^-(\mathbf{r})]/2$ $\varepsilon_L\rho_L(\mathbf{r}) - \varepsilon_H\rho_H(\mathbf{r}) + (\varepsilon_L - \varepsilon_H)[\rho_L(\mathbf{r}) - \rho_H(\mathbf{r})]/2$	$IPf_e^-(\mathbf{r}) - EAf_e^+(\mathbf{r})$ $\varepsilon_L\rho_L(\mathbf{r}) - \varepsilon_H\rho_H(\mathbf{r})$
$f_e(\mathbf{r},\mathbf{r}') = \frac{1}{2}[f_e^-(\mathbf{r})f_e^-(\mathbf{r}') + f_e^+(\mathbf{r})f_e^+(\mathbf{r}')] + \frac{1}{2}[f_e^+(\mathbf{r})f_e^+(\mathbf{r}') - f_e^-(\mathbf{r})f_e^-(\mathbf{r}')]_\alpha^\omega + [f_e^+(\mathbf{r}) - f_e^-(\mathbf{r}')][f_e^-(\mathbf{r}) - f_e^+(\mathbf{r}')]\lambda^{-1}$	$f_e^-(\mathbf{r})f_e^-(\mathbf{r}')$ $\rho_H(\mathbf{r})\rho_H(\mathbf{r}')$	$f_e^+(\mathbf{r})f_e^+(\mathbf{r}')$ $\rho_L(\mathbf{r})\rho_L(\mathbf{r}')$	$[f_e^-(\mathbf{r})f_e^-(\mathbf{r}') + f_e^+(\mathbf{r})f_e^+(\mathbf{r}')]/2$ $[\rho_H(\mathbf{r})\rho_H(\mathbf{r}') + \rho_L(\mathbf{r})\rho_L(\mathbf{r}')]/2$
$\Delta f_\tau(\mathbf{r},\mathbf{r}') = f_e^+(\mathbf{r})f_e^+(\mathbf{r}') - f_e^-(\mathbf{r})f_e^-(\mathbf{r}') + \frac{\omega}{2}\Delta f_e(\mathbf{r})\Delta f_e(\mathbf{r}')$	$f_e^+(\mathbf{r})f_e^+(\mathbf{r}') - f_e^-(\mathbf{r})f_e^-(\mathbf{r}') - \frac{1}{2}[f^+(\mathbf{r}) - f^+(\mathbf{r})][f^+(\mathbf{r}') - f^+(\mathbf{r}')]$ $\rho_L(\mathbf{r})\rho_L(\mathbf{r}') - \rho_H(\mathbf{r})\rho_H(\mathbf{r}') - \frac{1}{2}[\rho_L(\mathbf{r}) - \rho_H(\mathbf{r})][\rho_L(\mathbf{r}') - \rho_H(\mathbf{r}')]$	$f_e^+(\mathbf{r})f_e^+(\mathbf{r}') - f_e^-(\mathbf{r})f_e^-(\mathbf{r}') + \frac{1}{2}[f^+(\mathbf{r}) - f^+(\mathbf{r})][f^+(\mathbf{r}') - f^+(\mathbf{r}')]$ $\rho_L(\mathbf{r})\rho_L(\mathbf{r}') - \rho_H(\mathbf{r})\rho_H(\mathbf{r}') + \frac{1}{2}[\rho_L(\mathbf{r}) - \rho_H(\mathbf{r})][\rho_L(\mathbf{r}') - \rho_H(\mathbf{r}')]$	$f_e^+(\mathbf{r})f_e^+(\mathbf{r}') - f_e^-(\mathbf{r})f_e^-(\mathbf{r}')$ $\rho_L(\mathbf{r})\rho_L(\mathbf{r}') - \rho_H(\mathbf{r})\rho_H(\mathbf{r}')$

a) $\lambda = 4 + \exp(\beta(IP + \mu_{Bath})) + \exp(-\beta(EA + \mu_{Bath}))$

whose temperature-dependent expression is given in Table 11.2. If one makes use of the expressions for the zero temperature limit given in Table 11.2, one can see that the integral over the whole space of $\eta_\tau(\mathbf{r})$, which can be considered as the global hardness η_τ associated to this local hardness, is equal to $(IP - EA)$, for the three values of the fractional occupation. Thus, in comparison with the expression for η_e when $\omega = 0$, reported in Table 11.1, the absence of the Dirac delta function in η_τ implies that the present approach leads to a well behaved global hardness. It is important to mention that the expression for the local hardness $IPf_e^-(\mathbf{r}) - EAf_e^+(\mathbf{r})$, for $\omega = 0$, was first proposed by Meneses et al. [77, 79] intuitively, and was later derived, together with the local chemical potential, at zero temperature through an approach based on the chain rule for functional derivatives [90, 91]. The expressions given in Table 11.2 for the local hardness at temperatures of chemical interest have been used to analyze some reactivity trends like the electrophilic attacks on substituted pyridines, substituted benzenes, and substituted ethenes.

It is important to note that the expressions obtained for the local chemical potential and the local hardness in the zero temperature limit are given in terms of the product of a global property and a local property. This form implies that a similar site in different molecules characterized for having similar values of the local part will be modified for each molecule by a different factor through the global part, allowing this way the analysis of the behavior of a similar site in different molecules, which imply different chemical environments.

In the case of the Fukui function, which is a local indicator, the ratio of thermal fluctuations is given by

$$f_e(\mathbf{r}) = \frac{\sigma_{\rho(\mathbf{r})N}}{\sigma_{NN}} = \frac{\langle \rho(\mathbf{r})N\rangle - \langle \rho(\mathbf{r})\rangle \langle N\rangle}{\langle N^2\rangle - \langle N\rangle \langle N\rangle}, \tag{11.41}$$

in which, as in the previous case, one can associate the local counterpart of $\langle N\rangle$ with $\langle \rho(\mathbf{r})\rangle$ in the numerator, so that one can express the non-local counterpart of $f_e(\mathbf{r})$, the Fukui kernel, as [111, 116]

$$f_e(\mathbf{r}, \mathbf{r}') = \frac{\sigma_{\rho(\mathbf{r})\rho(\mathbf{r}')}}{\sigma_{NN}} = \frac{\langle \rho(\mathbf{r})\rho(\mathbf{r}')\rangle - \langle \rho(\mathbf{r})\rangle \langle \rho(\mathbf{r}')\rangle}{\langle N^2\rangle - \langle N\rangle \langle N\rangle} \tag{11.42}$$

which satisfies the condition

$$f_e(\mathbf{r}) = \int f_e(\mathbf{r}, \mathbf{r}')\, d\mathbf{r}'. \tag{11.43}$$

On the other hand, if one considers the derivative of Eq. (11.43) with respect to $\langle N\rangle$, one can define the dual descriptor kernel through the integrand as

$$\Delta f_\tau(\mathbf{r}, \mathbf{r}') = \left(\frac{\partial f_e(\mathbf{r}, \mathbf{r}')}{\partial \langle N\rangle}\right)_{T, v(\mathbf{r})}. \tag{11.44}$$

In Table 11.2 we present the finite temperature expressions for these kernels that result from the three ground states ensemble model, together with their zero temperature limits.

These two indexes have been shown to be related with bond reactivity, due to their nonlocal nature. This can be seen more clearly by making use of the condensed to atom Fukui function values in the $T = 0$ expressions of Table 2, since one may consider that the condensation on atom k comes from the values of $f_e(\mathbf{r})$ around and at the position of k, while the condensation on atom k' comes from the values of $f_e(\mathbf{r}')$ around and at the position of k', which implies that the Fukui and dual descriptor kernels can be interpreted as reactivity descriptors of the bonding region between the two atoms, that will allow the characteristics of the bonds in the molecule to be established with respect to charge transfer processes.

11.5 Concluding Remarks

Conceptual density functional theory has become a very useful tool to analyze many aspects of the reactivity of a rather large variety of chemical interactions by transforming the complex information contained in the molecular wavefunction, into quantities that facilitate our understanding of some of the key features that characterize the reactivity of a given species, through a simple but meaningful language.

In this chapter we have presented the fundamental basis of C-DFT, starting from the more general perspective of the grand canonical ensemble, in order to treat the molecule as an open system that may exchange electrons with a reservoir, and to be able to analyze the effects of temperature.

Based on the fact that the three ground states ensemble has been shown to provide a satisfactory description, for temperatures of chemical interest, we have seen through Eqs (11.24) and (11.27) that at finite temperature, $\langle E \rangle$ and $\langle \rho(\mathbf{r}) \rangle$ become smooth functions of $\langle N \rangle$, indicating that one can obtain the analytical derivatives of any order of the average energy, and the average density with respect to the average number of electrons. This way, in Tables 11.1 and 11.2 we have presented the finite temperature equations of all the response functions considered, together with the expressions for temperatures of chemical interest. One can see that since the latter correspond to the zero temperature limit, the first derivatives, the electronic chemical potential, and the Fukui function retain the characteristics associated with the discontinuities that arise from the piecewise continuous nature of $\langle E \rangle$ and $\langle \rho(\mathbf{r}) \rangle$, and the second derivatives, the electronic hardness, and the dual descriptor show as a consequence the presence of the Dirac delta function. The analysis of these equations shows that the effect of temperature on the values of these reactivity indexes is negligible, indicating that their evaluation at temperatures of chemical interest can be done using the zero temperature limit expressions.

However, the grand canonical ensemble allows one to introduce alternative definitions to the hardness and the dual descriptor through the variations of the chemical potential of the reservoir, μ_{Bath}, that retain the same physical meaning, but at the same time eliminate the presence of the Dirac delta function. Thus, to distinguish them from the traditional electronic responses, η_e and Δf_e, we have named them the thermodynamic hardness and the thermodynamic dual descriptor.

Additionally, through the consideration of the average electronic density as the local counterpart of the average number of electrons, one may derive new response functions that may be identified as the local counterpart of a global descriptor, and the non-local counterpart of a local descriptor. The new local indexes provide complementary information to other local response functions, particularly for the comparison of a given site in different chemical environments, while the non-local indexes can be used to describe reactivity aspects of the bond region.

We have also shown that through the analysis of the fundamental equations for reactivity in the temperature-dependent framework, one is led to three additional response functions that contain important information about the effects of thermal changes at the onset of a chemical interaction.

Conceptual density functional theory is nowadays an important tool for the analysis of chemical reactivity. The general perspective presented in this chapter within the framework of the grand canonical ensemble shows that through this approach one recovers all the fundamental features of the original development, but complemented with the consequences of temperature in the formal aspects and in the analysis of the reactivity principles.

Acknowledgements

PWA and MFP thank NSERC, the Canada Research Chairs, Compute Canada, and Canarie for support. MFP also thanks Universidad Autónoma Metropolitana-Iztapalapa for a visiting professor invitation. JLG and AV thank Conacyt for grants 237045 and Fronteras-867, respectively.

References

1 Parr, R.G., Yang, W.T. (1989) Density-Functional Theory of Atoms and Molecules. Oxford University Press, New York.
2 Parr, R.G., Yang, W.T. (1995) Density-functional theory of the electronic-structure of molecules. Annual Review of Physical Chemistry, 46, 701–728.
3 Chermette, H. (1999) Chemical reactivity indexes in density functional theory. Journal of Computational Chemistry, 20, 129–154.
4 Geerlings, P., De Proft, F., Langenaeker, W. (2003) Conceptual density functional theory. Chemical Reviews, 103, 1793–1873.
5 Ayers, P.W., Anderson, J.S.M., Bartolotti, L.J. (2005) Perturbative perspectives on the chemical reaction prediction problem. International Journal of Quantum Chemistry, 101, 520–534.
6 Gázquez, J.L. (2008) Perspectives on the density functional theory of chemical reactivity. Journal of the Mexican Chemical Society, 52, 3–10.
7 Chattaraj, P.K. (ed) (2009) Chemical Reactivity Theory: A Density Functional View. CRC Press, Boca Raton.
8 Liu, S.B. (2009) Conceptual density functional theory and some recent developments. Acta Physico-Chimica Sinica, 25, 590–600.
9 Johnson, P.A., Bartolotti, L., Ayers, P.W., Fievez, T., Geerlings, P. (2012) Charge density and chemical reactions: a unified view from conceptual DFT. In: Modern Charge-Density Analysis (eds Gatti, C., Macchi, P.). Springer, Dordrecht, pp. 715–764.
10 Fuentealba, P., Cardenas, C. (2015) Density functional theory of chemical reactivity. In: Chemical Modelling: A Specialist Periodical Report (eds Joswig, J.O., Springborg, M.), vol. 11. Royal Society of Chemistry, pp. 151–174.
11 Geerlings, P., Chamorro, E., Chattaraj, P.K., De Proft, F., Gázquez, J.L., Liu, S.B., Morell, C., Toro-Labbe, A., Vela, A., Ayers, P. (2020) Conceptual density functional theory: status, prospects, issues. Theoretical Chemistry Accounts, 139, 36.
12 Parr, R.G., Donnelly, R.A., Levy, M., Palke, W.E. (1978) Electronegativity – density functional viewpoint. Journal of Chemical Physics, 68, 3801–3807.
13 Parr, R.G., Pearson, R.G. (1983) Absolute hardness – companion parameter to absolute electronegativity. Journal of the American Chemical Society, 105, 7512–7516.
14 Parr, R.G., Yang, W.T. (1984) Density functional approach to the frontier-electron theory of chemical reactivity. Journal of the American Chemical Society, 106, 4049–4050.
15 Yang, W.T., Parr, R.G., Pucci, R. (1984) Electron-density, Kohn-Sham frontier orbitals, and Fukui functions. Journal of Chemical Physics, 81, 2862–2863.
16 Ayers, P.W., Levy, M. (2000) Perspective on "Density functional approach to the frontier-electron theory of chemical reactivity" by Parr RG, Yang W (1984). Theoretical Chemistry Accounts, 103, 353–360.
17 Morell, C., Grand, A., Toro-Labbé, A. (2005) New dual descriptor for chemical reactivity. Journal of Physical Chemistry A, 109, 205–212.

18 Morell, C., Grand, A., Toro-Labbé, A. (2006) Theoretical support for using the Delta f(r) descriptor. Chemical Physics Letters, 425, 342–346.

19 Pauling, L. (1932) The nature of the chemical bond IV: The energy of single bonds and the relative electronegativity of atoms. Journal of the American Chemical Society, 54, 3570–3582.

20 Mulliken, R.S. (1934) A new electroaffinity scale; together with data on valence states and on valence ionization potentials and electron affinities. Journal of Chemical Physics, 2, 782–793.

21 Pauling, L. (1960) The Nature of the Chemical Bond. 3rd ed. Cornell University Press, New York.

22 Iczkowski, R., Margrave, J.L. (1961) Electronegativity. Journal of the American Chemical Society, 83, 3547–3551.

23 Pearson, R.G. (1963) Hard and soft acids and bases. Journal of the American Chemical Society, 85, 3533–3539.

24 Pearson, R.G. (1966) Acids and bases. Science, 151, 172–177.

25 Fukui, K., Yonezawa, T., Shingu, H. (1952) A molecular orbital theory of reactivity in aromatic hydrocarbons. Journal of Chemical Physics, 20, 722–725.

26 Fukui, K. (1982) Role of frontier orbitals in chemical-reactions. Science, 218, 747–754.

27 Sanderson, R.T. (1951) An interpretation of bond lengths and a classification of bonds. Science, 114, 670–672.

28 Sanderson, R.T. (1971) Chemical Bonds and Bond Energy. Academic Press, New York.

29 Pearson, R.G. (1987) Recent advances in the concept of hard and soft acids and bases. Journal of Chemical Education, 64, 561–567.

30 Zhou, Z.X., Parr, R.G. (1990) Activation hardness: New index for describing the orientation of electrophilic aromatic substitution. Journal of the American Chemical Society, 112, 5720–5724.

31 Chattaraj, P.K., Lee, H., Parr, R.G. (1991) HSAB Principle. Journal of the American Chemical Society, 113, 1855–1856.

32 Parr, R.G., Chattaraj, P.K. (1991) Principle of maximum hardness. Journal of the American Chemical Society, 113, 1854–1855.

33 Pearson, R.G., Palke, W.E. (1992) Support for a principle of maximum hardness. Journal of Physical Chemistry, 96, 3283–3285.

34 Gázquez, J.L. (1993) Hardness and softness in density functional theory. Structure and Bonding, 80, 27–43.

35 Gázquez, J.L., Martínez, A., Méndez, F. (1993) Relationship between energy and hardness differences. Journal of Physical Chemistry, 97, 4059–4063.

36 Ghanty, T.K., Ghosh, S.K. (1993) Correlation between hardness, polarizability, and size of atoms, molecules, and clusters. Journal of Physical Chemistry, 97, 4951–4953.

37 Gázquez, J.L., Méndez, F. (1994) The hard and soft acids and bases principle: An atoms in molecules viewpoint. Journal of Physical Chemistry, 98, 4591–4593.

38 Ghanty, T.K., Ghosh, S.K. (1994) A frontier orbital density functional approach to polarizability, hardness, electronegativity, and covalent radius of atomic systems. Journal of the American Chemical Society, 116, 8801–8802.

39 Ghanty, T.K., Ghosh, S.K. (1994) Simple density functional approach to polarizability, hardness, and covalent radius of atomic systems. Journal of Physical Chemistry, 98, 9197–9201.

40 Méndez, F., Gázquez, J.L. (1994) Chemical reactivity of enolate ions: The local hard and soft acids and bases principle viewpoint. Journal of the American Chemical Society, 116, 9298–9301.

41 Chattaraj, P.K., Cedillo, A., Parr, R.G. (1995) Variational method for determining the Fukui function and chemical hardness of an electronic system. Journal of Chemical Physics, 103, 7645–7646.

42 Chattaraj, P.K., Liu, G.H., Parr, R.G. (1995) The maximum hardness principle in the Gyftopoulos-Hatsopoulos three-level model for an atomic or molecular species and its positive and negative ions. Chemical Physics Letters, 237, 171–176.

43 Gázquez, J.L. (1997) The hard and soft acids and bases principle. Journal of Physical Chemistry A, 101, 4657–4659.

44 Gázquez, J.L. (1997) Bond energies and hardness differences. Journal of Physical Chemistry A, 101, 9464–9469.

45 Gázquez, J.L. (1997) Activation energies and softness additivity. Journal of Physical Chemistry A, 101, 8967–8969.

46 Chattaraj, P.K., Fuentealba, P., Jaque, P., Toro-Labbé, A. (1999) Validity of the minimum polarizability principle in molecular vibrations and internal rotations: An ab initio SCF study. Journal of Physical Chemistry A, 103, 9307–9312.

47 Ayers, P.W., Parr, R.G. (2000) Variational principles for describing chemical reactions: The Fukui function and chemical hardness revisited. Journal of the American Chemical Society, 122, 2010–2018.

48 Cedillo, A., Chattaraj, P.K., Parr, R.G. (2000) Atoms-in-molecules partitioning of a molecular density. International Journal of Quantum Chemistry, 77, 403–407.

49 Torrent-Sucarrat, M., Luis, J.M., Durán, M., Solá, M. (2001) On the validity of the maximum hardness and minimum polarizability principles for nontotally symmetric vibrations. Journal of the American Chemical Society, 123, 7951–7952.

50 Torrent-Sucarrat, M., Luis, J.M., Durán, M., Solá, M. (2002) Are the maximum hardness and minimum polarizability principles always obeyed in nontotally symmetric vibrations? Journal of Chemical Physics, 117, 10561–10570.

51 Gómez, B., Fuentealba, P., Contreras, R. (2003) The maximum hardness and minimum polarizability principles as the basis for the study of reaction profiles. Theoretical Chemistry Accounts, 110, 421–427.

52 Torrent-Sucarrat, M., Luis, J.M., Durán, M., Solá, M. (2004) The hardness profile as a tool to detect spurious stationary points in the potential energy surface. Journal of Chemical Physics, 120, 10914–10924.

53 Ayers, P.W. (2005) An elementary derivation of the hard/soft-acid/base principle. Journal of Chemical Physics, 122, 141102.

54 Chattaraj, P.K., Ayers, P.W. (2005) The maximum hardness principle implies the hard/soft acid/base rule. Journal of Chemical Physics, 123, 086101.

55 Ayers, P.W., Parr, R.G., Pearson, R.G. (2006) Elucidating the hard/soft acid/base principle: A perspective based on half-reactions. Journal of Chemical Physics, 124, 194107.

56 Ayers, P.W. (2007) On the electronegativity nonlocality paradox. Theoretical Chemistry Accounts, 118, 371–381.

57 Ayers, P.W. (2007) The physical basis of the hard/soft acid/base principle. Faraday Discussions, 135, 161–190.

58 Chattaraj, P.K., Ayers, P.W., Melin, J. (2007) Further links between the maximum hardness principle and the hard/soft acid/base principle: insights from hard/soft exchange reactions. Physical Chemistry Chemical Physics, 9, 3853–3856.

59 Ayers, P.W., Parr, R.G. (2008) Local hardness equalization: Exploiting the ambiguity. Journal of Chemical Physics, 128, 184108.

60 Ayers, P.W., Parr, R.G. (2008) Beyond electronegativity and local hardness: Higher-order equalization criteria for determination of a ground-state electron density. Journal of Chemical Physics, 129, 054111.

61 Chattaraj, P.K., Giri, S., Duley, S. (2010) Electrophilicity equalization principle. Journal of Physical Chemistry Letters, 1, 1064–1067.

62 Ayers, P.W., Cárdenas, C. (2013) Communication: A case where the hard/soft acid/base principle holds regardless of acid/base strength. Journal of Chemical Physics, 138, 181106.

63 Cardenas, C., Ayers, P.W. (2013) How reliable is the hard-soft acid-base principle? An assessment from numerical simulations of electron transfer energies. Physical Chemistry Chemical Physics, 15, 13959–13968.

64 Gázquez, J.L., Vela, A., Chattaraj, P.K. (2013) Local hardness equalization and the principle of maximum hardness. Journal of Chemical Physics, 138, 214103.

65 Miranda-Quintana, R.A. (2017) Note: The minimum electrophilicity and the hard/soft acid/base principles. Journal of Chemical Physics, 146, 046101.

66 Miranda-Quintana, R.A., Heidar-Zadeh, F., Ayers, P.W. (2018) Elementary derivation of the "$|\Delta\mu|$ big is good" Rule. Journal of Physical Chemistry Letters, 9, 4344–4348.

67 Miranda-Quintana, R.A., Ayers, P.W. (2018) Dipolar cycloadditions and the "$|\Delta\mu|$ big is good" rule: a computational study. Theoretical Chemistry Accounts, 137, 177.

68 Miranda-Quintana, R.A., Ayers, P.W. (2018) Note: Maximum hardness and minimum electrophilicity principles. Journal of Chemical Physics, 148, 196101.

69 Miranda-Quintana, R.A., Ayers, P.W. (2019) The "$|\Delta\mu|$ big is good" rule, the maximum hardness, and minimum electrophilicity principles. Theoretical Chemistry Accounts, 138, 44.

70 Berkowitz, M., Ghosh, S.K., Parr, R.G. (1985) On the concept of local hardness in chemistry. Journal of the American Chemical Society, 107, 6811–6814.

71 Yang, W.T., Parr, R.G. (1985) Hardness, softness, and the Fukui function in the electronic theory of metals and catalysis. Proceedings of the National Academy of Sciences of the United States of America, 82, 6723–6726.

72 Berkowitz, M., Parr, R.G. (1988) Molecular hardness and softness, local hardness and softness, hardness and softness kernels, and relations among these quantities. Journal of Chemical Physics, 88, 2554–2557.

73 Ghosh, S.K. (1990) Energy derivatives in density-functional theory. Chemical Physics Letters, 172, 77–82.

74 Vela, A., Gázquez, J.L. (1990) A relationship between the static dipole polarizability, the global softness and the Fukui function. Journal of the American Chemical Society, 112, 1490–1492.

75 Cohen, M.H., Ganduglia-Pirovano, M.V., Kudrnovsky, J. (1995) Reactivity kernels, the normal modes of chemical reactivity, and the hardness and softness spectra. Journal of Chemical Physics, 103, 3543–3551.

76 Fuentealba, P. (1995) A local model for the hardness kernel and related reactivity parameters in density functional theory. Journal of Chemical Physics, 103, 6571–6575.

77 Meneses, L., Tiznado, W., Contreras, R., Fuentealba, P. (2004) A proposal for a new local hardness as selectivity index. Chemical Physics Letters, 383, 181–187.

78 Chattaraj, P.K., Roy, D.R., Geerlings, P., Torrent-Sucarrat, M. (2007) Local hardness: a critical account. Theoretical Chemistry Accounts, 118, 923–930.

79 Meneses, L., Araya, A., Pilaquinga, F., Contreras, R., Fuentealba, P. (2007) Local hardness: An application to electrophilic additions. Chemical Physics Letters, 446, 170–175.

80 Torrent-Sucarrat, M., Salvador, P., Geerlings, P., Solá, M. (2007) On the quality of the hardness kernel and the Fukui function to evaluate the global hardness. Journal of Computational Chemistry, 28, 574–583.

81 Torrent-Sucarrat, M., De Proft, F., Geerlings, P., Ayers, P.W. (2008) Do the local softness and hardness indicate the softest and hardest regions of a molecule? Chemistry- A European Journal, 14, 8652–8660.

82 Torrent-Sucarrat, M., Salvador, P., Solá, M., Geerlings, P. (2008) The hardness kernel as the basis for global and local reactivity indices. Journal of Computational Chemistry, 29, 1064–1072.

83 Sablon, N., De Proft, F., Geerlings, P. (2010) The linear response kernel: inductive and resonance effects quantified. Journal of Physical Chemistry Letters, 1, 1228–1234.

84 Torrent-Sucarrat, M., De Proft, F., Ayers, P.W., Geerlings, P. (2010) On the applicability of local softness and hardness. Physical Chemistry Chemical Physics, 12, 1072–1080.

85 Gál, T. (2012) Why the traditional concept of local hardness does not work. Theoretical Chemistry Accounts, 131, 1223.

86 Fias, S., Geerlings, P., Ayers, P., De Proft, F. (2013) Sigma, pi aromaticity and anti-aromaticity as retrieved by the linear response kernel. Physical Chemistry Chemical Physics, 15, 2882–2889.

87 Geerlings, P., Fias, S., Boisdenghien, Z., De Proft, F. (2014) Conceptual DFT: chemistry from the linear response function. Chemical Society Reviews, 43, 4989–5008.

88 Zadeh, F.H., Fuentealba, P., Cardenas, C., Ayers, P.W. (2014) An information-theoretic resolution of the ambiguity in the local hardness. Physical Chemistry Chemical Physics, 16, 6019–6026.

89 Heidar-Zadeh, F., Fias, S., Vohringer-Martinez, E., Verstraelen, T., Ayers, P.W. (2016) The local response of global descriptors. Theoretical Chemistry Accounts, 136, 19.

90 Polanco-Ramírez, C.A., Franco-Pérez, M., Carmona-Espíndola, J., Gázquez, J.L., Ayers, P.W. (2017) Revisiting the definition of local hardness and hardness kernel. Physical Chemistry Chemical Physics, 19, 12355–12364.

91 Franco-Pérez, M., Polanco-Ramírez, C.A., Gázquez, J.L., Ayers, P.W. (2018) Local and nonlocal counterparts of global descriptors: the cases of chemical softness and hardness. Journal of Molecular Modeling, 24, 285.

92 Robles, A., Franco-Pérez, M., Gázquez, J.L., Cardenas, C., Fuentealba, P. (2018) Local electrophilicity. Journal of Molecular Modeling, 24, 245.

93 Chakraborty, D., Chattararaj, P.K. (2019) Bonding, Reactivity, and Dynamics in Confined Systems. Journal of Physical Chemistry A, 123, 4513–4531.

94 Gyftopoulos, E.P., Hatsopoulos, G.N. (1968) Quantum-thermodynamic definition of electronegativity. Proceedings of the National Academy of Sciences of the United States of America, 60, 786–793.

95 Sebastian, K.L. (1994) On the proof of the principle of maximum hardness. Chemical Physics Letters, 231, 40–42.

96 Baekelandt, B.G., Cedillo, A., Parr, R.G. (1995) Reactivity indexes and fluctuation formulas in density-functional theory – isomorphic ensembles and a new measure of local hardness. Journal of Chemical Physics, 103, 8548–8556.

97 Chattaraj, P.K., Cedillo, A., Parr, R.G. (1996) Chemical softness in model electronic systems: dependence on temperature and chemical potential. Chemical Physics, 204, 429–437.

98 Malek, A., Balawender, R. (2015) Revisiting the chemical reactivity indices as the state function derivatives. The role of classical chemical hardness. Journal of Chemical Physics, 142, 054104.

99 Miranda-Quintana, R.A., Ayers, P.W. (2016) Interpolation of property-values between electron numbers is inconsistent with ensemble averaging. Journal of Chemical Physics, 144, 244112.

100 Miranda-Quintana, R.A., Ayers, P.W. (2016) Fractional electron number, temperature, and perturbations in chemical reactions. Physical Chemistry Chemical Physics, 18, 15070–15080.

101 Miranda-Quintana, R.A. (2017) Thermodynamic electrophilicity. Journal of Chemical Physics, 146, 214113.

102 Miranda-Quintana, R.A., Chattaraj, P.K., Ayers, P.W. (2017) Finite temperature grand canonical ensemble study of the minimum electrophilicity principle. Journal of Chemical Physics, 147, 124103.

103 Miranda-Quintana, R.A., Kim, T.D., Cárdenas, C., Ayers, P.W. (2017) The HSAB principle from a finite-temperature grand-canonical perspective. Theoretical Chemistry Accounts, 136, 135.

104 Franco-Pérez, M., Gázquez, J.L., Vela, A. (2015) Electronic chemical response indexes at finite temperature in the canonical ensemble. Journal of Chemical Physics, 143, 024112.

105 Franco-Pérez, M., Gázquez, J.L., Ayers, P.W., Vela, A. (2015) Revisiting the definition of the electronic chemical potential, chemical hardness, and softness at finite temperatures. Journal of Chemical Physics, 143, 154103.

106 Franco-Pérez, M., Ayers, P.W., Gázquez, J.L., Vela, A. (2015) Local and linear chemical reactivity response functions at finite temperature in density functional theory. Journal of Chemical Physics, 143, 244117.

107 Franco-Pérez, M., Ayers, P.W., Gázquez, J.L. (2016) Average electronic energy is the central quantity in conceptual chemical reactivity theory. Theoretical Chemistry Accounts, 135, 199.

108 Franco-Pérez, M., Heidar-Zadeh, F., Ayers, P.W., Gázquez, J.L., Vela, A. (2017) Going beyond the three-state ensemble model: the electronic chemical potential and Fukui function for the general case. Physical Chemistry Chemical Physics, 19, 11588–11602.

109 Franco-Pérez, M., Ayers, P.W., Gázquez, J.L., Vela, A. (2017) Local chemical potential, local hardness, and dual descriptors in temperature dependent chemical reactivity theory. Physical Chemistry Chemical Physics, 19, 13687–13695.

110 Franco-Pérez, M., Gázquez, J.L., Ayers, P.W., Vela, A. (2017) Thermodynamic hardness and the maximum hardness principle. Journal of Chemical Physics, 147, 074113.

111 Franco-Pérez, M., Polanco-Ramirez, C.A., Ayers, P.W., Gázquez, J.L., Vela, A. (2017) New Fukui, dual and hyper-dual kernels as bond reactivity descriptors. Physical Chemistry Chemical Physics, 19, 16095–16104.

112 Franco-Pérez, M., Ayers, P.W., Gázquez, J.L., Vela, A. (2017) Thermodynamic responses of electronic systems. Journal of Chemical Physics, 147, 094105.

113 Franco-Pérez, M., Gázquez, J.L., Ayers, P.W., Vela, A. (2018) Thermodynamic dual descriptor. Acta Physico-Chimica Sinica, 34, 683–691.

114 Franco-Pérez, M., Gázquez, J.L., Ayers, P.W., Vela, A. (2018) Thermodynamic justification for the parabolic model for reactivity indicators with respect to electron number and a rigorous definition for the electrophilicity: the essential role played by the electronic entropy. Journal of Chemical Theory and Computation, 14, 597–606.

115 Gázquez, J.L., Franco-Pérez, M., Ayers, P.W., Vela, A. (2019) Temperature-dependent approach to chemical reactivity concepts in density functional theory. International Journal of Quantum Chemistry, 119, e25797.

116 Franco-Pérez, M., Polanco-Ramírez, C.A., Gázquez, J.L., Ayers, P.W., Vela, A. (2020) Study of organic reactions using chemical reactivity descriptors derived through a temperature-dependent approach. Theoretical Chemistry Accounts, 139, 44.

117 Nalewajski, R.F., Parr, R.G. (1982) Legendre transforms and Maxwell relations in density functional theory. Journal of Chemical Physics, 77, 399–407.

118 Fuentealba, P., Parr, R.G. (1991) Higher-order derivatives in density-functional theory, especially the hardness derivative delta-eta/delta-n. Journal of Chemical Physics, 94, 5559–5564.

119 Perdew, J.P., Parr, R.G., Levy, M., Balduz, J.L. (1982) Density-functional theory for fractional particle number – derivative discontinuities of the energy. Physical Review Letters, 49, 1691–1694.

120 Yang, W.T., Zhang, Y.K., Ayers, P.W. (2000) Degenerate ground states and a fractional number of electrons in density and reduced density matrix functional theory. Physical Review Letters, 84, 5172–5175.

121 Ayers, P.W. (2008) The dependence on and continuity of the energy and other molecular properties with respect to the number of electrons. Journal of Mathematical Chemistry, 43, 285–303.

122 Einhorn, M.B., Blankenbecler, R. (1971) Bounds on scattering amplitudes. Annals of Physics, 67, 480–517.

123 Pearson, R.G. (1988) Absolute electronegativity and hardness: Application to inorganic chemistry. Inorganic Chemistry, 27, 734–740.

124 Pearson, R.G. (1989) Absolute electronegativity and hardness: Application to organic chemistry. Journal of Organic Chemistry, 54, 1423–1430.

125 Parr, R.G., Von Szentpaly, L., Liu, S.B. (1999) Electrophilicity index. Journal of the American Chemical Society, 121, 1922–1924.

126 Ayers, P.W., Anderson, J.S.M., Rodríguez, J.I., Jawed, Z. (2005) Indices for predicting the quality of leaving groups. Physical Chemistry Chemical Physics, 7, 1918–1925.

127 Chattaraj, P.K., Sarkar, U., Roy, D.R. (2006) Electrophilicity index. Chemical Reviews, 106, 2065–2091.

128 Chattaraj, P.K., Roy, D.R. (2007) Update 1 of: Electrophilicity Index. Chemical Reviews, 107, PR46-PR74.

129 Gázquez, J.L., Cedillo, A., Vela, A. (2007) Electrodonating and electroaccepting powers. Journal of Physical Chemistry A, 111, 1966–1970.

130 Bagaria, P., Saha, S., Murru, S., Kavala, V., Patel, B.K., Roy, R.K. (2009) A comprehensive decomposition analysis of stabilization energy (CDASE) and its application in locating the rate-determining step of multi-step reactions. Physical Chemistry Chemical Physics, 11, 8306–8315.

131 Saha, S., Roy, R.K., Pal, S. (2010) CDASE-A reliable scheme to explain the reactivity sequence between Diels-Alder pairs. Physical Chemistry Chemical Physics, 12, 9328–9338.

132 Chattaraj, P.K., Giri, S., Duley, S. (2011) Update 2 of: Electrophilicity Index. Chemical Reviews, 111, PR43-PR75.

133 Sarmah, A., Roy, R.K. (2013) Understanding the interaction of nucleobases with chiral semiconducting single-walled carbon nanotubes: an alternative theoretical approach based on density functional reactivity theory. Journal of Physical Chemistry C, 117, 21539–21550.

134 Sarmah, A., Roy, R.K. (2015) Interaction between small gold clusters and nucleobases: a density functional reactivity theory based study. Journal of Physical Chemistry C, 119, 17940–17953.

135 Hamid, A., Anand, A., Roy, R.K. (2017) The charge transfer limit of a chemical adduct: the role of perturbation on external potential. Physical Chemistry Chemical Physics, 19, 10905–10912.

136 Orozco-Valencia, A.U., Gázquez, J.L., Vela, A. (2017) Global and local partitioning of the charge transferred in the Parr-Pearson model. Journal of Physical Chemistry A, 121, 4019–4029.

137 Orozco-Valencia, U., Gázquez, J.L., Vela, A. (2017) Donation and back-donation analyzed through a charge transfer model based on density functional theory. Journal of Molecular Modeling, 23, 207.

138 Orozco-Valencia, U., Gázquez, J.L., Vela, A. (2018) Role of reaction conditions in the global and local two parabolas charge transfer model. Journal of Physical Chemistry A, 122, 1796–1806.

139 Orozco-Valencia, U., Gázquez, J.L., Vela, A. (2018) Reactivity of indoles through the eyes of a charge-transfer partitioning analysis. Acta Physico-Chimica Sinica, 34, 692–698.

140 Gázquez, J.L. (2009) Chemical reactivity concepts in density functional theory. In: Chemical Reactivity Theory: A Density Functional View (ed. Chattaraj, P.K.), CRC Press, Boca Raton, pp. 7–22.

141 Miranda-Quintana, R.A., Franco-Pérez, M., Gázquez, J.L., Ayers, P.W., Vela, A. (2018) Chemical hardness: Temperature dependent definitions and reactivity principles. Journal of Chemical Physics, 149, 124110.

142 Gál, T., Geerlings, P., De Proft, F., Torrent-Sucarrat, M. (2011) A new approach to local hardness. Physical Chemistry Chemical Physics, 13, 15003–15015.

143 Toro-Labbé, A., Jaque, P., Murray, J.S., Politzer, P. (2005) Connection between the average local ionization energy and the Fukui function. Chemical Physics Letters, 407, 143–146.

144 Sjoberg, P., Murray, J.S., Brinck, T., Politzer, P. (1990) Average local ionization energies on the molecular surfaces of aromatic systems as guides to chemical reactivity. Canadian Journal of Chemistry-Revue Canadienne De Chimie, 68, 1440–1443.

145 Murray, J.S., Politzer, P. (1998) Average local ionization energies: significance and applications. In: Theoretical Organic Chemistry (ed. Parkanyi, C.), Elsevier, Amsterdam, pp. 189–202.

146 Murray, J.S., Peralta-Inga, Z., Politzer, P., Ekanayake, K., Lebreton, P. (2001) Computational characterization of nucleotide bases: Molecular surface electrostatic potentials and local ionization energies, and local polarization energies. International Journal of Quantum Chemistry, 83, 245–254.

147 Politzer, P., Murray, J.S., Concha, M.C. (2002) The complementary roles of molecular surface electrostatic potentials and average local ionization energies with respect to electrophilic processes. International Journal of Quantum Chemistry, 88, 19–27.

148 Chamorro, E., Duque-Norena, M. (2015) Understanding the highly varying pK(a) of arylamines. A perspective from the average local ionization condensed-to-atom framework. Journal of Physical Chemistry A, 119, 8156–8162.

149 Clark, T. (2010) The local electron affinity for non-minimal basis sets. Journal of Molecular Modeling, 16, 1231–1238.

150 Brinck, T., Carlqvist, P., Stenlid, J.H. (2016) Local electron attachment energy and its use for predicting nucleophilic reactions and halogen bonding. Journal of Physical Chemistry A, 120, 10023–10032.

151 Stenlid, J.H., Brinck, T. (2017) Nucleophilic aromatic substitution reactions described by the local electron attachment energy. Journal of Organic Chemistry, 82, 3072–3083.

12

Effect of Confinement on the Optical Response Properties of Molecules

Wojciech Bartkowiak, Marta Chołujv, and Justyna Kozłowska*

Department of Physical and Quantum Chemistry, Wroclaw University of Science and Technology, 50-370, Wroclaw, Wybrzeze Wyspianskiego 27, Poland

12.1 Introduction

The spatial confinement phenomenon occurs commonly in nature and its classic examples are, *inter alia*, (i) atoms and molecules enclosed in chemical cages, such as zeolites, mesoporous silica/organosilica, metal-organic frameworks (MOFs), or nanotubes and fullerenes, (ii) matter under high pressure, (iii) particles in a solvent environment, and (iv) semiconductor structures, such as wells, wires, and quantum dots [1, 2]. It is well known that the spatial restriction can significantly modify various chemical and physical properties. In the context of confined species, it is possible to suggest new paradigms for many areas of chemistry, physics, and material science including molecular nonlinear optics [3]. Thus, the problem of spatial confinement has become pivotal for scientists working in different fields.

Over the years, various approaches have been developed to describe the effect of spatial confinement on the theoretical basis [1, 4–6]. Among them, one of the most popular uses model external repulsive potentials. First reports in this regard date back to 1937 when Michels et al., in their pioneering work, enclosed the hydrogen atom within a hard sphere cavity to analyze the pressure effects on the atomic energy levels and polarizability [7]. Since then box models of quantum confinement (penetrable and impenetrable spherical or ellipsoidal boxes and other analytical model potentials) have been widely applied [1, 2, 8–10]. In this chapter we will summarize the findings of theoretical investigations devoted to the impact of spatial confinement, represented by the harmonic oscillator potential, on the linear and nonlinear electric properties of atoms, ions, molecules, and more complex molecular systems. The discussion will be preceded by a brief description of the basic concepts of molecular nonlinear optics as well as some formal aspects of the modelling of spatial confinement with the aid of analytical potentials.

The optical nonlinearity is observed at very high light intensities, such as those provided by pulsed lasers. At the molecular level, the relation between induced dipole moment ($\mu_i(F)$) and very strong external electric field (F) can be expressed in terms of the Taylor series expansion:

$$\mu_i(F) = \mu_i(0) + \sum_j \alpha_{ij} F_j + \frac{1}{2!}\sum_{j,k} \beta_{ijk} F_j F_k + \frac{1}{3!}\sum_{j,k,l} \gamma_{ijkl} F_j F_k F_l + \ldots, \quad (12.1)$$

*Corresponding Author: Wojciech Bartkowiak; wojciech.bartkowiak@pwr.edu.pl

Chemical Reactivity in Confined Systems: Theory, Modelling and Applications,
First Edition. Edited by Pratim Kumar Chattaraj and Debdutta Chakraborty.

where $\mu_i(0)$ is an i-th component of the permanent dipole moment, $i, j, k, l \in \{x, y, z\}$. Through this expression the components of polarizability as well as first and second hyperpolarizability tensors (α_{ij}, β_{ijk}, γ_{ijkl}, respectively) are defined. These quantities constitute the measure of a response of chemical object to the external electric fields (in general static and frequency dependent). Polarizability is closely related to some macroscopic physical quantities such as dielectric constant and refractive index, whereas first and second hyperpolarizabilities describe second (e.g. second harmonic generation) and third (e.g. two-photon absorption) order nonlinear processes, respectively. Investigations of the linear and nonlinear electric properties still remain a challenge for quantum chemical methods (especially if large molecular systems are considered). It is connected with the fact that the correlation effects for hyperpolarizabilities are usually large and therefore high-level methods and extended basis sets have to be used to obtain reliable results. The same applies to the calculations performed for spatially limited systems.

Theoretical evaluation of confinement effects is a complex problem, which requires a careful consideration of many aspects connected with the intermolecular interactions in rather unusual conditions (strong orbital overlap) [11–17]. Therefore, the simplified approach based on the model external potentials that act repulsively upon electrons is very important tool in this area of research. Although being quite drastic approximation of the real confining environments, it was proven to be useful for understanding various processes occurring in dense matter (under high pressure) and in tight spaces.

The effect of spatial confinement can be modeled by including the analytical external potential in the Hamiltonian of an isolated N-electron system ($\hat{H}^{(0)}$):

$$\hat{H} = \hat{H}^{(0)} + \hat{V}_{conf}(\vec{r}). \tag{12.2}$$

In the above expression, $\hat{V}_{conf}$ denotes the confining potential, being the sum of one-electron operators:

$$\hat{V}_{conf}(\vec{r}) = \sum_{i=1}^{N} \hat{V}_{conf}(\vec{r}_i), \tag{12.3}$$

where $\vec{r} = (\vec{r}_1, \vec{r}_2, ..., \vec{r}_N)$ describes the position of all electrons and $\vec{r}_i = (x_i, y_i, z_i)$ denotes the position of i-th electron. Assuming that only electron motion is considered, the problem comes down to solving the electronic Schrödinger equation in the form:

$$\left[\hat{H}^{(0)} + \sum_{i=1}^{N} \hat{V}_{conf}(\vec{r}_i)\right] \Psi_{el} = E_{el}\Psi_{el}. \tag{12.4}$$

The repulsive analytical potentials act directly only on electrons and allow to represent very important aspects of the confining environments, which are connected with the valence repulsion (i.e. the orbital compression effect). This way of modeling the spatial confinement corresponds to the case in which the studied object interacts with chemically and electronically inert environment [1]. One of the most commonly used potentials is the harmonic oscillator potential of spherical

$$V_{conf}(\vec{r}_i) = \frac{1}{2}\omega^2(x_i^2 + y_i^2 + z_i^2) \tag{12.5}$$

or cylindrical

$$V_{conf}(\vec{r}_i) = \frac{1}{2}\omega^2(x_i^2 + y_i^2) \tag{12.6}$$

symmetry – particularly suitable for modelling fullerene-like or nanotube-like chemical cages. The ω parameter in the above equations allows the strength of orbital compression to be controlled, which is obtained by changing the slope of the harmonic potential curve. The physical meaning of this parameter was discussed in the context of the theory of intermolecular interactions in the paper by Zaleśny et al. [18]. Using the simple formula

$$\Delta E_{ex}^{HL} \approx E(\omega) - E(\omega = 0), \tag{12.7}$$

where E_{ex}^{HL} is the exchange-repulsion energy of a studied object enclosed within a chemical cage, whereas $E(\omega)$ and $E(\omega = 0)$ denotes the total energy of the same object in the presence of confining potential and without it, respectively, one can find the value of ω which roughly corresponds to the real chemical situation.

12.2 Electronic Contributions to Longitudinal Electric-Dipole Properties of Atomic and Molecular Systems Embedded in Harmonic Oscillator Potential

The quantum chemical studies of the influence of spatial confinement, represented by spherical or cylindrical harmonic oscillator potential, on the electronic contributions, i.e. resulting from the electron density distortion, to the linear and nonlinear electric properties, have been conducted for many years. Among wide group of systems considered so far one can distinguish atoms: He, Be, ions: H^-, Li^+, F^-, Cl^-, O^{2-}, S^{2-}, molecules: CO, BF, HF, HArF, OCS, ClCCH, LiH, LiF, HCl, CO_2, HCCH, HCN, HCCCN, and molecular complexes: $HCN\cdots(HCN)_n$, $n\in\{1,2,3,4\}$, HCN...HNC, HF...HF [16, 18–33]. Table 12.1 collects data on the behaviour of dipole moment and (hyper)polarizabilities of the listed atomic and molecular systems upon embedding in harmonic oscillator potential (note that we follow here a convention that the dipole moment of all isolated molecules, i.e. without applying confining potential, has positive value). It is worth mentioning that in most studies only the electric properties components along the molecular axis (often assumed to be z-axis), called longitudinal, were considered. Although we are focusing here only on the harmonic oscillator potential it is fair to say that other repulsive potentials were also used in the studies of electric properties of spatially limited chemical objects (see for example [28, 34–36]).

The research conducted thus far leads to the conclusion that the orbital compression substantially changes the optical response of chemical objects. In general, the hyperpolarizabilities are affected the most. Neither dipole moment (μ_z) nor first hyperpolarizability (β_{zzz}) exhibits one particular trend of changes. Their behaviour varies significantly depending on the studied chemical object. OCS and ClCCH molecules are a good illustration of this fact [27]. Based on the calculations carried out for the geometries optimized in the presence of cylindrically symmetric harmonic oscillator potential, it was observed that the spatial restriction more strongly affects the dipole moment of OCS, causing its significant growth. In contrast, μ_z of ClCCH exhibits a drop in value. Furthermore, in the case of ClCCH a change in sign of μ_z was noted. The reason for such a different behaviour of dipole moment was found in the charge redistribution within spatially confined molecules: under the influence of growing orbital compression strength an increase of the charge separation in OCS and the charge shift from the more electronegative chlorine to the less electronegative hydrogen atom of ClCCH occurs. On the other hand, increasing the pressure exerted by the confining harmonic oscillator potential always results in a drop of polarizability (α_{zz}) values. Such a behaviour comes from the fact that polarizability is linearly dependent on the molecular volume of a chemical object, whereas in the presence of confining potentials the electron density undergoes a compression process. Consequently, a reduction of molecular volume

Table 12.1 Changes of electric properties of atomic and molecular systems at equilibrium experimental or optimized (in potential or in vacuum) geometry, embedded in cylindrical or spherical harmonic oscillator potential. Symbols $\searrow$, $\nearrow$ and $\longleftrightarrow$ denote a decrease, an increase and nonmonotonic behaviour of a given property.

System	Geometry	μ_z	α_{zz}	$\lvert\beta_{zzz}\rvert$	γ_{zzzz}	Ref.
	Spherical harmonic oscillator potential					
He	–	–	$\searrow$	–	–	[19]
Be	–	–	$\searrow$	–	$\searrow$	[20]
H^-	–	–	$\searrow$	–	$\searrow$	[19, 21]
Li^+	–	–	$\searrow$	–	–	[19]
F^-	–	–	$\searrow$	–	–	[22]
Cl^-	–	–	$\searrow$	–	–	[22]
O^{2-}	–	–	$\searrow$	–	–	[22]
S^{2-}	–	–	$\searrow$	–	–	[22]
	Cylindrical harmonic oscillator potential					
CO	optimized in potential	$\longleftrightarrow$	–	–	–	[23]
	experimental	$\searrow$	–	–	–	[23]
BF	optimized in potential	$\nearrow$	–	–	–	[23]
	experimental	$\nearrow$	–	–	–	[23]
HF	optimized in potential	$\searrow$	$\searrow$	$\searrow$	–	[24, 25]
	optimized in vacuum	$\searrow$	$\searrow$	$\searrow$	–	[24, 26]
	experimental	$\searrow$	$\searrow$	$\searrow$	–	[16]
HArF	optimized in potential	$\longleftrightarrow$	$\searrow$	$\searrow$	–	[24, 25]
	optimized in vacuum	$\searrow$	$\searrow$	$\searrow$	–	[24]
OCS	optimized in potential	$\nearrow$	$\searrow$	$\longleftrightarrow$	–	[25, 27]
ClCCH	optimized in potential	$\searrow$	$\searrow$	$\searrow$	–	[25, 27]
LiH	optimized in vacuum	$\nearrow$	$\searrow$	$\longleftrightarrow$	–	[26]
	experimental	$\nearrow$	$\searrow$	$\longleftrightarrow$	–	[16, 28]
LiF	optimized in vacuum	$\nearrow$	$\searrow$	$\longleftrightarrow$	–	[26]
	experimental	$\nearrow$	$\searrow$	$\searrow$	–	[16]
HCl	experimental	$\searrow$	$\searrow$	$\searrow$	–	[16]
CO_2	optimized in potential	–	$\searrow$	–	$\searrow$	[30]
HCCH	optimized in potential	–	$\searrow$	–	$\searrow$	[30]
HCN	optimized in potential	$\searrow$	$\searrow$	$\longleftrightarrow$	$\searrow$	[25, 30]
HCCCN	optimized in potential	$\longleftrightarrow$	$\searrow$	$\nearrow$	$\searrow$	[18]
	optimized in vacuum	$\longleftrightarrow$	$\searrow$	$\nearrow$	$\searrow$	[18]
HCN...HCN	optimized in potential	$\searrow$	$\searrow$	$\searrow$	$\searrow$	[32, 33]
HCN...$(HCN)_n$	optimized in vacuum	$\searrow$	$\searrow$	–	–	[31]
HCN...HNC	optimized in potential	–	$\searrow$	$\searrow$	$\searrow$	[32]
HF...HF	optimized in potential	$\searrow$	$\searrow$	$\searrow$	–	[33]

and polarizability occurs. In a similar manner, the second hyperpolarizability (γ_{zzzz}) also decreases under the influence of harmonic oscillator potential.

It should also not be overlooked that the geometric structure of the molecular systems that are subjected to the confining potential naturally affects their optical response. Lo and Klobukowski studied the dipole moment and polarizability for the LiH molecule with different bond lengths in the ground and several excited states, embedded in cylindrical harmonic oscillator potential [29]. They showed that both spatial confinement and geometry modification have substantial impact on the ground and excited state electric properties of LiH. However, the latter effect influences the dipole moment and polarizability to a much larger extent. In the whole range of analyzed LiH's bond lengths the ground state dipole moment increases under the influence of orbital compression. In turn, the behaviour of the excited state dipole moment in the presence of spatial restriction varies depending on the LiH's structure. Unexpectedly, for some values of LiH's bond length (far from the experimental equilibrium geometry) a growth of zz component of the ground and excited state polarizability occurs together with increasing the orbital compression strength. On the other hand, regardless of the assumed bond length, α_{xx} always decreases. Moreover, based on the calculations performed for the HF and HArF molecules, subjected to the cylindrically symmetric harmonic oscillator potential, it was observed that the values of dipole moment as well as polarizability and first hyperpolarizability, obtained for various orbital compression strengths, differ significantly depending on whether the geometry was optimized in vacuum or in potential [24]. As the strength of spatial confinement grows, these differences become larger.

It is well known that the changes in the electron density distribution, resulting from embedding a chemical object in confining potentials, has significant impact not only on electric properties but also on the relative positions of atomic nuclei. The presence of such potentials causes the accumulation of electron density in the interatomic spaces, which consequently leads to the shortening of bonds [5, 6]. Therefore, the geometry optimization in the presence of confining potential is undoubtedly an important aspect of the studies of the spatial confinement phenomenon. Including this process in the investigations allows one to obtain results which more closely reflect those observed in real chemical situation, because in such a situation electric properties are affected by the spatial confinement both directly and indirectly (through the changes in geometry caused by confining medium). Therefore, calculations performed without geometry optimization provide the information only about a direct influence of spatial confinement on the electric properties.

Another interesting element of the studies of molecules within restricted spaces is the symmetry of the spatial confinement and its influence on the optical response. Góra et al. performed calculations of electronic contributions to electric properties of the LiH molecule subjected to spherical and cylindrical harmonic oscillator potential [28]. They showed that increasing the strength of spatial restriction results in a decrease of almost all components of the studied electric properties. The cylindrical potential was found to cause relatively smaller changes of (hyper)polarizability of LiH than its spherical counterpart. Interestingly, the dipole moment of LiH, embedded in cylindrical and spherical confinement, exhibits opposite behaviour in a function of increasing the orbital compression strength, i.e. in the former case μ_z grows, whereas in the latter one a diminishment of μ_z occurs. However, in that work the calculations were performed for only one relative position of LiH and spherical harmonic potential, i.e. placing the bond center of LiH at the origin of a Cartesian coordinate system, which is also a centering point of the potential. It should be noted that the relative position of a studied molecule and applied spherical harmonic potential have strong impact on the properties of this system. This fact was thoroughly analyzed in the work by Chołuj et al. [26]. They performed calculations of electric properties of the LiH, LiF, and HF molecules changing their position (along z-axis) relative to the spherical harmonic oscillator potential. In general, significant differences in the values of electric properties, obtained under spherical and cylindrical confining conditions for each individual orbital compression strength, were observed. The results showed that in the case of LiH, when a more electronegative hydrogen atom is located at the centering point of the spherical harmonic potential or close to it, a growth of μ_z occurs. However, when the H atom is moved further from the origin of coordinate system, the dipole moment values decrease upon increasing the spatial confinement strength. The closer to the centering point of the spherical harmonic potential is the lithium atom of LiH, the greater diminishment of μ_z is found. Similarly diverse behaviour of the electric properties was observed also for the spatially limited LiF and HF molecules. Therefore, this study clearly shows that the analysis of the influence of the topology of confining environment on the molecular electric properties is not that straightforward and that the relative position of a molecule and spherical harmonic oscillator potential should be always carefully considered.

12.3 Vibrational Contributions to Longitudinal Electric-Dipole Properties of Spatially Confined Molecular Systems

The prominent role of molecular vibrations in various nonlinear optical processes is quite well recognized for the unconfined molecules (see for example [37–40]). The root cause is the fact that the nuclear motions contribute to all electronic properties, even if only by means of zero-point vibrational averaging (ZPVA). In the case of second- and higher-order perturbation properties the additional vibrational contributions also come into play, being associated with the shift in equilibrium geometry caused by a perturbing field. Taken together, they are frequently described as the vibrational properties – for instance, the vibrational polarizability by analogy with the electronic polarizability. Though it has been often assumed previously that the vibrational correction could be neglected in comparison to the electronic term, now it is firmly established that the former can be as large as, or even larger than, the latter [39, 41].

Most frequently, the Born-Oppenheimer (BO) approximation is the starting point for calculations of the electric properties of molecules, as the fully nonadiabatic calculation are currently feasible only for diatomic molecules. Within BO approximation the molecular property, by analogy to the total energy, may be divided into terms related to electronic and nuclear degrees of freedom [42]:

$$P = P^e + P^{NR} + P^{curv}. \tag{12.8}$$

In the above equation the P^e stands for the electronic contribution, P^{NR} is due to the geometry relaxation in the presence of electric field, while the remaining contribution (P^{curv}) arises from the change in the shape of potential surface in the presence of electric field. The last term may be also regarded as resulting from the change in ZPVA as a consequence of nuclear relaxation. The P^{NR} contribution usually prevail over P^{curv} one. Two main approaches, allowing the computations of vibrational corrections to electric properties, are the Bishop-Kirtman perturbation theory (BKPT) and the finite-field nuclear relaxation method (FF-NR) [43, 44].

As is apparent from prior discussion, the list of spatially restricted quantum objects whose linear and nonlinear optical response has already been investigated is quite long. Yet it is mainly limited to the purely electronic contributions to static electric properties of these systems. In contrast, the studies on vibrational (hyper)polarizabilities of the spatially limited molecular systems are rare. Undoubtedly, it results from the much larger computational effort required for evaluation of the latter. In fact, the accurate evaluation of the vibrational corrections to the (hyper)polarizabilities of polyatomic molecules still constitutes a challenge for quantum chemistry methods. Studies concerning the spatial confinement influence on the vibrational properties of molecular systems are limited to a few theoretical works performed by Zaleśny et al., in which the cylindrical harmonic oscillator potential was employed to model the orbital compression effect [18, 30, 32]. The most comprehensive analyses are available for a set of linear molecules, encompassing HCN, HCCH, and CO_2 [30], as well as the hydrogen-bonded systems, composed of the HCN molecule or its tautomeric form, i.e. HCN...HCN and HCN...HNC [32]. However, it is fair to mention that some preliminary conclusions were presented also for the HCCCN molecule [18]. Based on the results emerging from the extensive first-principles calculations presented in the cited works several generalizations can be made. First of all, the confinement diminishes the vibrational contributions to the longitudinal (hyper)polarizabilities and its influence increases with increasing the order of vibrational properties. Secondly, the spatial confinement modifies the electronic contribution to (hyper)polarizabilities to a larger extent than the vibrational one. Thus both counterparts should be considered in the interpretation of the electric properties of spatially confined molecular systems. Finally, the breakdown of β^{NR} and γ^{NR} into harmonic and anharmonic contributions, performed employing the Bishop-Kirtman perturbation theory approach, revealed that in the case of single molecules the effect of orbital compression is larger for harmonic than for anharmonic vibrational terms. Hence, the significance of vibrational hyperpolarizability may increase also with the degree of anharmonicity of vibrational motion. Interestingly, the decrease in the static nuclear relaxation first hyperpolarizability of the hydrogen bonded systems is mainly caused by the changes in the harmonic term, while in the case of nuclear relaxation second hyperpolarizability the anharmonic terms contribute more to the drop of this property.

12.4 Two-Photon Absorption in Spatial Confinement

Two-photon absorption (TPA), which represents a third-order nonlinear optical phenomenon, may be described as the electronic excitation of a quantum object induced by the simultaneous absorption of two photons of the same or different energy and, in general, is characterized by several attractive features. Along with the benefits resulting from the application of the TPA phenomenon in the field of spectroscopy (it enables the exploration of spectroscopic states which are one-photon forbidden due to symmetry), there are also a number of technological applications of this resonant nonlinear optical process (three-dimensional optical data storage, high-resolution fluorescence microscopy, fabrication of optoelectronic logical circuits, or nondestructive imaging of biological tissues, just to name a few) [45–47].

The second-order transition moment (S_{ij}^{gf}) constitutes the basic molecular quantity that describes the two-photon absorption process:

$$S_{ij}^{gf} = \sum_k \left[\frac{\langle g|\hat{\mu}_i|k\rangle\langle k|\hat{\mu}_j|f\rangle}{\omega_k - \omega} + \frac{\langle g|\hat{\mu}_j|k\rangle\langle k|\hat{\mu}_i|f\rangle}{\omega_k - \omega} \right]. \tag{12.9}$$

In the above equation $|g\rangle$ and $|f\rangle$ correspond to the initial and final state, respectively, while $|k\rangle$ denotes the intermediate state. Labels i, j stand for the Cartesian coordinates and $\langle i|\hat{\mu}_i|j\rangle$ is the transition moment between states $|i\rangle$ and $|j\rangle$. It is assumed that angular frequencies (ω) satisfy the resonance condition, i.e. for the one source of photons $2\omega = \omega_f$. Usually, the two-photon absorption probability (δ^{gf}) is also discussed in the literature, since this quantity might be related to data extracted from the experimental measurements [48]. According to the procedure introduced by

Monson and McClain the magnitude of orientationally averaged δ^{gf} ($\langle\delta^{gf}\rangle$) in an isotropic medium might be calculated using the following formula [49]:

$$\langle\delta^{gf}\rangle = \frac{1}{30}\sum_{i,j}\left[S_{ii}S_{jj}F + S_{ij}S_{ij}G + S_{ij}S_{ji}H\right], \tag{12.10}$$

where the coefficients F, G, and H depend on the polarization of the incident light beams; for linearly polarized photons $F = G = H = 2$. Note that $\langle\delta^{gf}\rangle$ is often called two-photon absorption strength, by analogy to the oscillator strength (f), which characterizes the one-photon absorption process.

Development of multiphoton based applications relies on the quest for chromophores with large TPA responses, thus considerable efforts are directed toward the design of appropriate molecular species (push-pull dipolar molecular structures, quadrupoles, multichromophoric dendrimeric systems, or nanodots) [45–47]. Moreover, the alteration of bond lengths [50] and environmental effects, especially solvent polarity [51], were shown to significantly influence the TPA strength of molecular systems. According to the results of some experimental studies, the spatial confinement effect also may be considered as another important factor contributing to the changes of TPA strength [52–54]. For example, it was demonstrated that exposing molecular systems to high pressure leads to the reduction of both one- and two-photon absorption responses [52]. On the contrary, an enhancement of TPA activity was reported for different organic molecules confined between the interlayer spaces of clay minerals [53, 54]. In some measure, these findings are in line with those emerging from a theoretical study performed on the HCCCN molecule embedded in a repulsive potential of cylindrical symmetry [18]. Based on the conducted analysis it was found that the absolute value of the second-order transition moment (S_{ij}^{gf}) increases together with the increasing confinement strength. A comprehensive theoretical description of the confinement induced changes in the TPA response has been also presented for the LiH molecule embedded in a two-dimensional harmonic oscillator potential [55]. In particular, the results obtained using the multiconfiguration self-consistent field method together with response function formalism indicate a significant reduction of the two-photon absorption response of lithium hydride at its experimental equilibrium bond length upon confinement. The bond-length dependence of the two-photon absorption strength has also been analyzed in the cited work. According to the obtained results under double perturbation, i.e. when the bond length of the LiH molecule embedded in an external potential is strongly stretched, the TPA strength could increase by several orders of magnitude.

12.5 Conclusions

Summing up the research discussed in this chapter, one can conclude that spatial confinement, modeled by the harmonic oscillator potential, has strong impact on the electronic and vibrational electric properties of chemical objects. The results obtained so far show that increasing the pressure exerted by the confining potential always leads to a decrease of polarizability and second hyperpolarizability. On the other hand, much more diverse behaviour is observed in the case of dipole moment and first hyperpolarizability. Electronic contributions to (hyper)polarizabilities were found to be more strongly reduced by the spatial confinement than its vibrational counterparts, which indicates that the role of vibrational terms in the description of optical response of molecules and molecular complexes is larger when they are enclosed within restricted spaces. Moreover, it was also reported that spatial confinement can significantly affect the two-photon absorption process. Given a huge importance of research on the spatial confinement phenomenon for many fields of science and technology, there is no doubt that the results presented in this chapter are an important source of knowledge and inspiration for further investigations.

References

1 Jaskólski W. Confined many-electron systems. Phys Rep. 1996;27:1–66.
2 Sen KD, editor. Electronic Structure of Quantum Confined Atoms and Molecules. Springer International Publishing; 2014.
3 Piela L. Ideas of quantum chemistry. Elsevier; 2013.
4 Britz DA, Khlobystov AN. Noncovalent interactions of molecules with single walled carbon nanotubes. Chem Soc Rev. 2006;35:637–659.
5 Cammi R, Verdolino V, Mennucci B, Tomasi J. Towards the elaboration of a QM method to describe molecular solutes under the effect of a very high pressure. Chem Phys. 2008;344:135–141.
6 Cammi R, Cappelli C, Mennucci B,Tomasi J. Calculation and analysis of the harmonic vibrational frequencies in molecules at extreme pressure: Methodology and diborane as a test case. J Chem Phys. 2012;137:154112.
7 Michels A, de Boer J, Bijl A. Remarks concerning molecular interaction and their influence in the polarisability. Physica.1937;4:981.
8 Sabin JR,Brändas E, Cruz SA, editors. Advances in Quantum Chemistry: Theory of Confined Quantum Systems. vol. 57–58. Academic Press, Waltham, MA; 2009.
9 Lo JMH, Klobukowski M, Diercksen GHF. Low-Lying Excited States of the Hydrogen Molecule in Cylindrical Harmonic Confinement. Advances in Quantum Chemistry. vol. 48. Academic Press, Waltham, MA; 2005.
10 Ley-Koo E. Recent progress in confined atoms and molecules: Superintegrability and symmetry breakings. Revista Mexicana de Fisica. 2018;64:326–363.
11 Munn RW, Malagoli M, in het Panhuis M. Environmental effects on molecular response in materials for non-linear optics. Synthetic Metals. 2000;109:29–32.
12 Grochala W, Hoffmann R, Feng J, Ashcroft NW. The chemical imagination at work in very tight places. Angewandte Chemie International Edition. 2007;46:3620–3642.
13 Borgoo A, Tozer DJ, Geerlings P,Proft FD. Influence of confinement on atomic and molecular reactivity indicators in DFT. Phys Chem Chem Phys. 2008;10:1406–1410.
14 Korona T, Dodziuk H. Small molecules in C_{60} and C_{70}: which complexes could be stabilized? Journal Chemical Theory and Computation. 2011;7:1476–1483.
15 Kaczmarek A, Bartkowiak W. The influence of the chemical compression on the electric properties of molecular systems within the supermolecular approximation: the LiH molecule as a case study. Phys Chem Chem Phys. 2009;11:2885–2892.
16 Kozłowska J,Zaleśny R, Bartkowiak W. On the nonlinear electrical properties of molecules in confined spaces - from cylindrical harmonic potential to carbon nanotube cages. Chem Phys. 2014;428:19–28.
17 Cammi R. The quantum chemical study of chemical reactions at extreme high pressure by means of the extreme-pressure polarizable continuum model. Annual Reports in Computational Chemistry. 2017;13:117–135.
18 Zaleśny R, Góra RW, Kozłowska J, Luis JM, Ågren H, Bartkowiak W. Resonant and Nonresonant Hyperpolarizabilities of Spatially Confined Molecules: A Case Study of Cyanoacetylene. J Chem Theory Comput. 2013;9:3463–3472.
19 Holka F, Neogrády P, Kellö V, Urban M, Diercksen GHF. Polarizabilities of confined two-electron systems: the 2-electron quantum dot, the hydrogen anion, the helium atom and the lithium cation. Mol Phys. 2005;103:2747.
20 Strasburger K, Naciążek P. Electric dipole hyperpolarizability of the beryllium atom in the gas phase and in spatial confinement. J Phys B: At Mol Opt Phys. 2014;47:025002.

21 Chołuj M,Bartkowiak W, Naciążek P, Strasburger K. On the calculations of the static electronic dipole (hyper)polarizability for the free and spatially confined H^-. J Chem Phys. 2017;146:194301.

22 Holka F, Urban M, Neogrády P,Paldus J. CCSD(T) calculations of confined systems: In-crystal polarizabilities of F^-, Cl^-, O^{2-}, and S^{2-}. J Chem Phys. 2014;141:214303.

23 Chołuj M, Bartkowiak W. Ground-state dipole moment of the spatially confined carbon monoxide and boron fluoride molecules. Chem Phys Lett. 2016;663:84–89.

24 Kozłowska J, Bartkowiak W. The effect of spatial confinement on the noble-gas HArF molecule: structure and electric properties. Chem Phys. 2014;441:83–92.

25 Chołuj M, Kozłowska J, Bartkowiak W. Benchmarking DFT methods on linear and nonlinear electric properties of spatially confined molecules. Int J Quantum Chem. 2018;118:e25666.

26 Chołuj M, Bartkowiak W. Electric properties of molecules confined by the spherical harmonic potential. International Journal of Quantum Chemistry. 2019;119:e25997.

27 Kozłowska J,Roztoczyńska A, Bartkowiak W. About diverse behavior of the molecular electric properties upon spatial confinement. Chem Phys. 2015;456:98–105.

28 Góra RW, Zaleśny R, Kozłowska J, Naciążek P, Roztoczyńska A, Strasburger K, et al. Electric dipole (hyper)polarizabilities of spatially confined LiH molecule. J Chem Phys. 2012;137:094307.

29 Lo JMH, Klobukowski M. Computational studies of one-electron properties of lithium hydride in confinement. Chem Phys. 2006;328:132.

30 Zaleśny R, Góra RW, Luis JM,Bartkowiak W. On the particular importance of vibrational contributions to the static electrical properties of model linear molecules under spatial confinement. Phys Chem Chem Phys. 2015;17:21782–21786.

31 Roztoczyńska A, Kozłowska J, Lipkowski P, Bartkowiak W. Does the spatial confinement influence the electric properties and cooperative effects of the hydrogen bonded systems? HCN chains as a case study. Chem Phys Lett. 2014;608:264–268.

32 Zaleśny R, Chołuj M, Kozłowska J, Bartkowiak W, Luis JM. Vibrational nonlinear optical properties of spatially confined weakly bound complexes. Phys Chem Chem Phys. 2017;19:24276–24283.

33 Kozłowska J, Lipkowski P, Roztoczyńska A, Bartkowiak W. DFT and spatial confinement: a benchmark study on the structural and electrical properties of hydrogen bonded complexes. Phys Chem Chem Phys. 2019;21:17253–17273.

34 Banerjee A, Sen KD, Garza J, Vargas R. Mean excitation energy, static polarizability, and hyperpolarizability of the spherically confined hydrogen atom. J Chem Phys. 2002;116:4054–4057.

35 Sen KD, Garza J, Vargas R, Aquino N. Static dipole polarizability of shell-confined hydrogen atom. Phys Lett A. 2002;295:299–304.

36 Waugh S, Chowdhury A, Banerjee A. On the variation of polarizability and hyperpolarizability of a confined atom with the strength of confinement: a case study of a helium atom. J Phys B: At, Mol Opt Phys. 2010;43:225002.

37 Macak P, Luo Y, Norman P, Ågren H. Electronic and vibronic contributions to two-photon absorption of molecules with multi-branched structures; 2000.

38 Bishop DM, Luis JM, Kirtman B. Vibration and two-photon absorption. Journal of Chemical Physics. 2002;116:9729–9739.

39 Kirtman B, Champagne B, Luis JM. Efficient treatment of the effect of vibrations on electrical, magnetic, and spectroscopic properties. Journal of Computational Chemistry. 2000;21:1572–1588.

40 Drobizhev M, Makarov NS, Tillo SE, Hughes TE, A R. Describing two-photon absorptivity of fluorescent proteins with a new vibronic coupling mechanism. Journal of Physical Chemistry B. 2012;116:1736–1744.

41 Champagne B, Kirtman B. In: Nalwa HS, editor. Handbook of Advanced Electronic and Photonic Materials. vol. 9. Academic San Diego, CA; 2001. p. 63–126.

42 Bishop DM. Molecular vibration and nonlinear optics. Advances in Chemical Physics. vol. 104. John Wiley & Sons, USA; 1998.

43 Kirtman B. Calculation of the Vibrational Contribution to Electronic Properties. AIP Conference Proceedings. 2012;1504:50–62.

44 Reis H, Lauvergnat D, Luis JM, Zaleśny R. In: Frontiers of Quantum Chemistry; 2018. p. 401–429.

45 Pawlicki M, Collins HA, Denning RG, Anderson HL. Two-photon absorption and the design of two-photon dyes. Angew Chem Int Ed. 2009;48:3244.

46 He GS, Tan LS, Zheng Q, Prasad PN. Multiphoton Absorbing Materials: Molecular Designs, Characterizations, and Applications. Chem Rev. 2008;108:1245.

47 Medishetty R, Zaręba JK, Mayer D, Samoć M, Fischer RA. Nonlinear optical properties, upconversion and lasing in metal-organic frameworks. Chem Soc Rev. 2017;46:4976.

48 Norman P, Cronstrand P, Ericsson J. Theoretical study of linear and nonlinear absorption in platinum-organic compounds. Chemical Physics. 2002;285:207–220.

49 Monson PR, McClain WM. Polarization dependence of the two-photon absorption of tumbling molecules with application to liquid 1-chloronaphthalene and benzene. The Journal of Chemical Physics. 1970;53:29–37.

50 Bartkowiak W, Zaleśny R, Leszczyński J. Relation between bond-length alternation and two-photon absorption of a push-pull conjugated molecules: A quantum-chemical study. Chem Phys. 2003;287:103–112.

51 Wielgus M, Zaleśny R, Murugan NA,Kongsted J, Ågren H, Samoć M, et al. Two-photon solvatochromism II: Experimental and theoretical study of solvent effects on the two-photon absorption spectrum of Reichardt's dye. ChemPhysChem. 2013;14:3731–3739.

52 Dreger ZA, Yang G, White JO, Li Y, Drickamer HG. One- and two-photon-pumped fluorescence from Rhodamine B in solid poly(acrylic acid) under high pressure. J Phys Chem B. 1998;102:4380–4385.

53 Kamada K, Tanamura Y, Ueno K, Ohta K, Misawa H. Enhanced two-photon absorption of the chromophores confined in two-dimensional nano-space. Journal of Physical Chemistry C. 2007;111:11193–11198.

54 Suzuki Y, Tenma Y, Nishioka Y, Kawamata J. Efficient nonlinear optical properties of dyes confined in interlayer nanospaces of clay minerals. Chem-Asian J. 2012;7:1170–1179.

55 Kozłowska J, Chołuj M, Zaleśny R, Bartkowiak W. Two-photon absorption of the spatially confined LiH molecule. Phys Chem Chem Phys. 2017;19:7568–7575.

13

A Density Functional Theory Study of Confined Noble Gas Dimers in Fullerene Molecules

Dongbo Zhao[1], Meng Li[2], Xin He[2], Bin Wang[2], Chunying Rong[2], Pratim K. Chattaraj[3], and Shubin Liu[4]

[1] *Institute of Biomedical Research, Yunnan University, Kunming, P.R. China*
[2] *College of Chemistry and Chemical Engineering, Hunan Normal University, Changsha, P. R. China*
[3] *Department of Chemistry and Center for Theoretical Studies, Indian Institute of Technology, Kharagpur, 721302, India*
[4] *Research Computing Centre, University of North Carolina, Chapel Hill, North Carolina 27599-3420, USA*

13.1 Introduction

Confined quantum systems have attracted considerable research interest in the literature largely because of the change of their peculiar physicochemical properties induced by the confinement effect [1–9]. Of many such examples, endohedral fullerenes have been extensively investigated from both epistemological and application viewpoints [1]. Since the first theoretical prediction in 1970 [2], and then experimental identification in 1985 [3], fullerenes have witnessed great success in confining small molecules such as $H_2/2H_2$ [4, 5], NH_3 [6], HF [7], and $H_2O/2H_2O$ [8, 9].

There is another important family of confinements that fullerenes can be employed. They are noble gas (Ng) atoms confined in their cages [10, 11]. Among them, those containing a noble gas dimer (Ng_2) inside the structures (denoted by Ng_2@fullerene) have been particularly challenging. Numerous theoretical studies have been carried out to examine the chemical bonding [12] of Ng_2 inside the cage and exohedral reactivity [13] resulted from the Ng_2 motif. Yet experimental realization of Ng_2@fullerenes has not been reported in the literature.

In this contribution, we summarize our recent studies to computationally revisit this topic and explore the interplay between the interior motif Ng_2 and the exterior fullerene cage within the framework of density functional theory (DFT) [14]. Specifically, we considered the impact of the confinement effect exerted by fullerene molecules onto the Ng_2 dimer from the perspective of structure and reactivity. To that end, we employed a total of nine complexes Ng_2@X (Ng = He, Ne, and Ar; X = C_{50}, C_{60}, and C_{70}) as illustrative examples in this work. In addition, to pinpoint the nature and origin of these structure and reactivity changes, we employed a variety of analytical tools, such as global descriptors from conceptual density functional theory [15, 16], information-theoretic quantities [17, 18], total energy decomposition analysis [14, 19], interaction energy decomposition [20], non-covalent interaction (NCI) analysis [21], and others. These results significantly deepened our understanding about how noble gas reactivity could be impacted and how important role quantum confinement could play in undermining structure and reactivity.

Chemical Reactivity in Confined Systems: Theory, Modelling and Applications,
First Edition. Edited by Pratim Kumar Chattaraj and Debdutta Chakraborty.

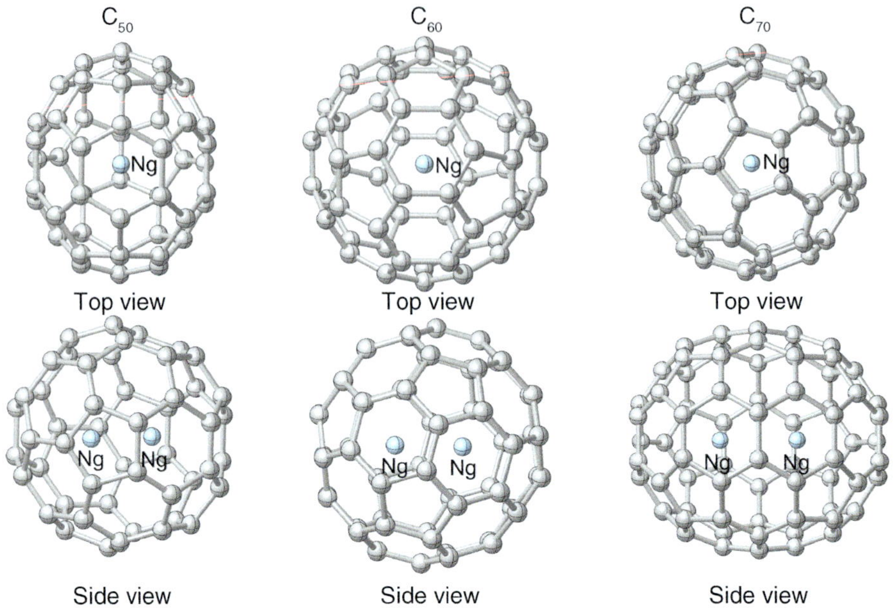

Figure 13.1 Optimized structures of noble gas dimers (Ng_2) confined in fullerene cages.

13.2 Computational Details

A total of nine model endohedral fullerene systems (Figure 13.1) denoted as Ng_2@X, (Ng = He, Ne, and Ar; X = C_{50}, C_{60}, and C_{70}) were investigated to examine the impact of fullerene confinement effect on structure and reactivity properties for Ng_2. For comparison, we also individually evaluated the structural and reactivity properties of these three fullerene molecules and three Ng_2 dimers in the gas phase.

All DFT calculations including structural optimizations, frequency calculations, and natural population analysis were performed with the Gaussian 09[22] package. Unless otherwise stated, we employed ultrafine integration grids (corresponding to 99 radial shells and 590 angular points per shell) and tight self-consistent field convergence (10^{-6} a.u. for energy) to rule out numerical instability in DFT calculations. Additionally, Born-Oppenheimer molecular dynamics (BOMD) simulations were performed for each species at the semi-empirical PM7[23] level to locate the most possible orientation of Ng_2 dimer in a fullerene cage. The temperature was set to be 10^3 K and the trajectory was accumulated for 10^4 steps with a step size of 1 fs, followed by the K-means clustering algorithm with the Cartesian coordinates as the input. We selected 20 BOMD snapshot structures with the lowest electronic energies as the possible structure candidates from the BOMD trajectory. Further structure optimization was carried out at the M06-2X/6-311G(d) [24, 25] level without imposing any symmetry constraints to allow for full variational degree of freedom. Harmonic vibrational frequency calculations were ensued to verify that all the optimized structures were true minima on the potential energy surface (e.g., no imaginary frequencies by visual inspection). The molecular wavefunction information contained in the Gaussian checkpoint files was fed into the Multiwfn 3.6.1[26] program to evaluate the information-theoretic quantities and perform NCI analysis. Suffice to note that with the keyword of *iop*(5/33=1), one can readily obtain all the total energy components from Gaussian single-point calculations. We also employed the ADF

(Amsterdam Density Functional) [27] package to perform the bonding energy decomposition analysis at the same theory level.

13.3 Results and Discussion

We summarize the main findings of our study from the following viewpoints of changes, including structure, bonding, energetics, electronic properties, spectroscopy, and reactivity.

13.3.1 Changes in Structure

Table 13.1 exhibits the interatomic distances of Ng_2 dimers in the gas phase as well as those in fullerenes. One can see that the bond length of Ng_2 dimers is substantially shortened when put in fullerenes. Furthermore, the smaller is the fullerene cage in size, the shorter is the Ng-Ng distance. This phenomenon can possibly be ascribed to the fact that a smaller sized fullerene has smaller interior space, thus forcing the Ng_2 dimer to stay more tightly. This shortened bond length of Ng_2 dimers causes them to be more frustrated inside the fullerene cage, leading to the other changes of these species. From this viewpoint, the structural change due to the confinement is the root cause of the changes in electronic and structure properties as we will show below.

13.3.2 Changes in Interaction Energy

Shown in Table 13.1 are the results of both BSSE (basis set superposition error) [28, 29] corrected and uncorrected interaction energies for all 9 species under study. One can see that the confinement of He_2 in all three fullerenes is energetically favorable, with negative interaction energies. However, for larger Ne_2 and Ar_2 dimers, only the largest fullerene studied, C_{70}, can hold them with negative interaction energies. This observation is understandable because larger Ng_2 dimers tend to occupy more space within a fullerene cage, and smaller fullerenes fail to compensate the energetic cost from electrostatic and other repulsions so their interaction energies are positive.

Table 13.1 Interatomic distances (in Å) and interaction energies[a] (in kcal/mol) at the M06-2X/6-311G(d) level for Ng_2 (Ng = He, Ne, and Ar) confined in C_{50}, C_{60}, and C_{70} fullerenes.

Ng-Ng	vacuum	C_{50}	C_{60}	C_{70}
He-He	2.862	1.835	1.984	2.559
E^{BSSE}_{int}		−1.5	−4.4	−6.7
E_{int}		−2.4	−5.1	−7.4
Ne-Ne	2.685	1.971	2.090	2.557
E^{BSSE}_{int}		15.1	1.3	−8.1
E_{int}		4.7	−7.1	−15.4
Ar-Ar	4.041	2.235	2.352	2.667
E^{BSSE}_{int}		131.4	53.8	−7.1
E_{int}		126.5	49.9	−10.2

a) E^{BSSE}_{int} and E_{int} are the interaction energy between Ng_2 dimer and the fullerene molecule with and without the BSSE (basis set superposition error) correction considered, respectively.

Table 13.2 The chemical bonding analysis results using the ADF package to analyze the bonding energy between the Ng_2 dimer and fullerene fragments.[a)] Units in kcal/mol.

Species	ΔE_{Pauli}	ΔE_{steric}	ΔE_{orb}	ΔE_{disp}	ΔE_{elstat}	ΔE_{bond}
$He_2@C_{50}$	29.6	16.7	−6.1	−6.0	−29.6	4.6
$Ne_2@C_{50}$	16.4	8.9	−4.3	−5.5	−16.4	−1.0
$Ar_2@C_{50}$	8.2	4.2	−3.0	−5.2	−8.2	−4.0
$He_2@C_{60}$	95.0	43.6	−7.6	−11.4	−95.0	24.7
$Ne_2@C_{60}$	50.9	21.4	−4.7	−11.0	−50.9	5.7
$Ar_2@C_{60}$	22.0	7.4	−2.9	−10.6	−22.0	−6.1
$He_2@C_{70}$	431.5	197.8	−31.5	−33.9	−431.5	132.4
$Ne_2@C_{70}$	244.5	103.1	−19.4	−34.8	−244.5	48.9
$Ar_2@C_{70}$	103.2	33.2	−11.5	−34.8	−103.2	−13.2

a) The energy decomposition analysis (EDA) in ADF dissects the interactions that constitute a chemical bond between fragments in a molecule. The total bonding energy ΔE_{bond} consists of contributions from the Pauli repulsion ΔE_{Pauli}, steric interaction ΔE_{steric}, electrostatic attraction ΔE_{elstat}, orbital interactions ΔE_{orb}, and dispersion Pauli repulsion ΔE_{disp}, $\Delta E_{bond} = \Delta E_{Pauli} + \Delta E_{steric} + \Delta E_{orb} + \Delta E_{disp} + \Delta E_{elstat}$.

13.3.3 Changes in Bonding Energy

Table 13.2 shows the results of the total bonding energy ΔE_{bond} analysis [20], with independent contributions from the Pauli repulsion ΔE_{Pauli}, steric interaction ΔE_{steric}, electrostatic attraction ΔE_{elstat}, orbital interactions ΔE_{orb}, and dispersion repulsion ΔE_{disp}, $\Delta E_{bond} = \Delta E_{Pauli} + \Delta E_{steric} + \Delta E_{orb} + \Delta E_{disp} + \Delta E_{elstat}$. From Table 13.2, we have found that the steric effect contributes favorably but is largely compensated and dominated by the opposite electrostatic contribution. This result agrees well with that from total energy partition schemes, whose results will be discussed below.

13.3.4 Changes in Energy Components

Table 13.3 showcases the results from the two schemes of the total energy partition using Ng_2 and fullerene molecules in the gas phase as the references for the 9 Ng_2@fullerene systems studied in this work. One can see that in the first scheme [14], $\Delta E = \Delta T_s + \Delta E_{xc} + \Delta E_e$, both ΔT_s and ΔE_{xc} have positive contributions to ΔE, and it is the electrostatic interaction ΔE_e that contributes negatively to molecular stability. In other words, one has to overcome the large electrostatic repulsion to put an Ng_2 dimer in a fullerene cage. This result is further verified by the second total energy partition scheme [19], $\Delta E = \Delta E_s + \Delta E_e + \Delta E_q$, where we found that in this case both electrostatic and quantum (due to exchange-correlation interactions) effects contribute negatively but the steric hindrance ΔE_s is positive. This result shows that after an Ng_2 dimer is put into the fullerene cage, smaller space occupation is resulted. But to make this happen, one has to overcome large electrostatic and Fermionic quantum repulsions. We have to mention that this total energy difference is quantitatively different from the BSSE corrected interaction energy analysis (see Table 13.1) because their methodologies drastically differ. Yet their general trends are the same.

Figure 13.2 shows the correlation of the total energy difference ΔE with its components. One can observe a strong linear correlation of ΔE with its electrostatic component ΔE_e in Figure 13.2a. This means that it is the electrostatic repulsion that predominates the energetic contribution, in line with the results in Table 13.3. If a two-variable fitting strategy is employed (see Figures 13.2b and 13.2c),

Table 13.3 Numerical results of the two total energy partition schemes using Ng_2 and fullerene in vacuum as the references for the 9 Ng_2@fullerene systems in this work. Units are in kcal/mol.

Species	ΔE	ΔT_s	ΔE_{xc}	ΔE_e	ΔE_s	ΔE_q
$He_2@C_{50}$	0.4	−73.2	−11.9	85.6	−357.7	272.7
$Ne_2@C_{50}$	−3.7	−56.1	−7.9	60.4	−270.6	206.5
$Ar_2@C_{50}$	−7.3	−47.9	−5.5	46.1	−205.3	151.8
$He_2@C_{60}$	14.0	−184.0	−15.3	213.4	−709.5	510.2
$Ne_2@C_{60}$	−2.5	−125.8	−12.1	135.4	−531.2	393.4
$Ar_2@C_{60}$	−15.3	−62.2	−11.4	58.3	−379.2	305.6
$He_2@C_{70}$	202.1	−356.4	−11.8	570.3	−1018.1	649.9
$Ne_2@C_{70}$	97.8	−241.9	−27.8	367.5	−916.1	646.4
$Ar_2@C_{70}$	0.8	−142.9	−43.3	187.0	−838.8	652.5

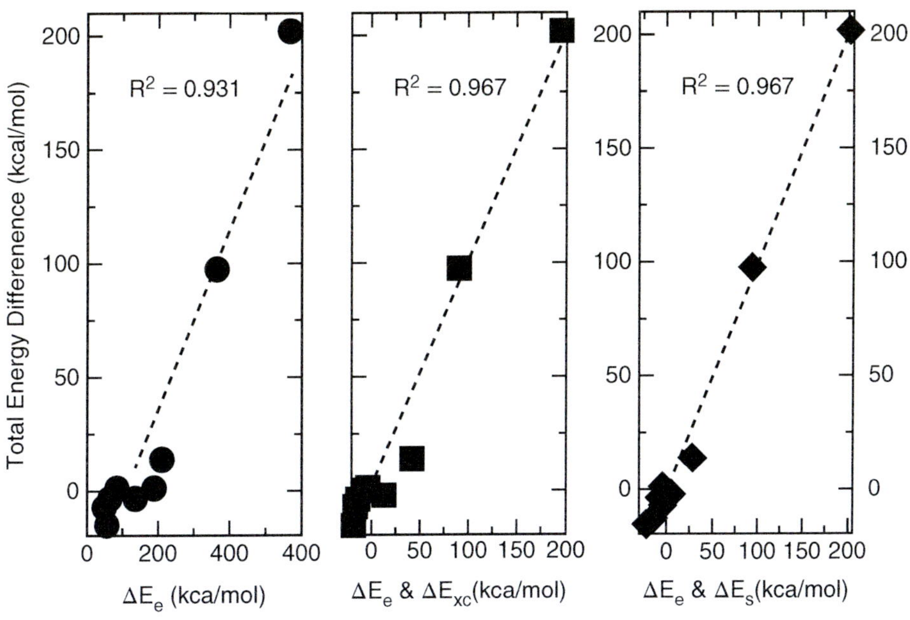

Figure 13.2 Strong linear correlations of the total energy difference ΔE with (a) the electrostatic energy difference ΔE_e, (b) two-variable fitting with ΔE_e and ΔE_{xc}, and (c) two-variable fitting with ΔE_e and ΔE_s.

we found that ΔE_e combining with either ΔE_{xc} or ΔE_s leads to even stronger correlations, indicating that contributions from the exchange-correlation and steric effects are small but cannot be ignored.

13.3.5 Changes in Noncovalent Interactions

To identify weak interactions between Ng_2 and fullerene molecules, the noncovalent interaction (NCI) analysis [21] was conducted, whose results are shown in Figure 13.3. The reduced density

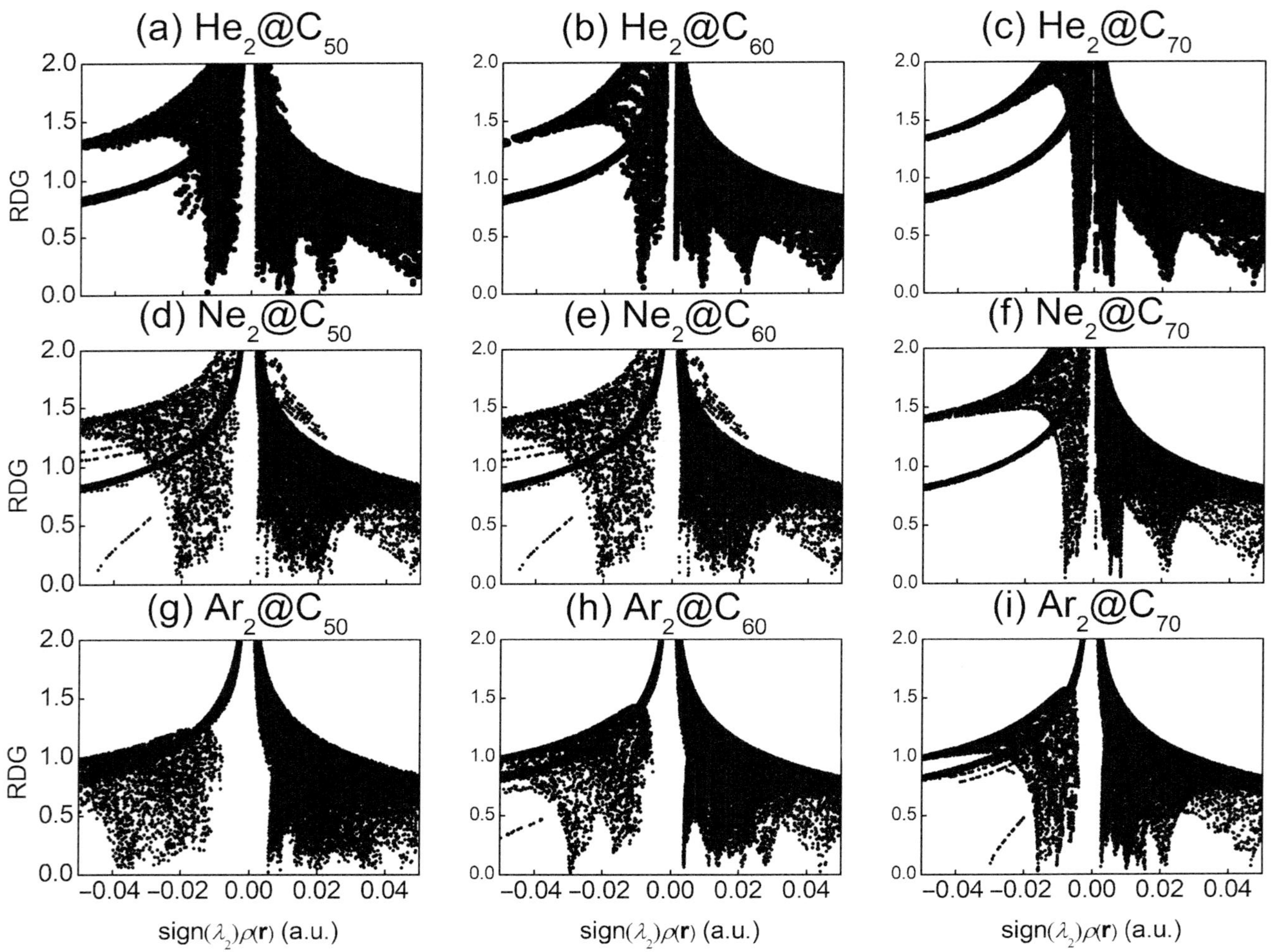

Figure 13.3 Diagrams of the reduced density gradient (RDG) versus sign(λ_2)ρ(r) for 9 endohedral fullerenes.

gradient is large and positive in the regions far from the molecule, but it becomes small in the vicinity between covalent and non-covalent regions. To identify different types of interactions, the sign of density Laplacian is utilized. Plotting low-gradient isosurfaces subject to a further low-density constraint enables real-space visualization of noncovalent interactions. Results in Figure 13.3 illustrate that when He_2 is confined from C_{50} to C_{70}, the "spike" area becomes more discernible, with its value increased from −0.02 to 0 a.u., indicating that there exist more attractive interactions within the complex. The same trend is also observed for Ne_2- and Ar_2-containing species. These results are also in qualitative agreement with the total energy difference analysis results in Table 13.1.

13.3.6 Changes in Information-Theoretic Quantities

Simple density-based quantities from the information-theoretic approach (ITA) can provide insights for different physiochemical processes [18]. In Table 13.4, we examine the results of eight ITA quantities, Shannon entropy (ΔS_S) [30], Fisher information (ΔI_F) [31], Ghosh-Berkowitz-Parr entropy (ΔS_{GBP}) [32], information gain (ΔI_G) [33], 2nd and 3rd orders of absolute Rényi (ΔR_2 and ΔR_3) [34], and relative Rényi entropy ($\Delta^r R_2$ and $\Delta^r R_3$) [35]. Each quantity has its own physiochemical meaning and they represent different yet intrinsic properties of molecular systems. Table 13.5 lists the correlation coefficient of these quantities for either Ng_2 or fullerene type, and Figure 13.4 displays illustrative example of strong linear relationships of the total energy difference ΔE with Shannon entropy, Fisher information and information gain for He_2, Ne_2 and Ar_2 dimers confined in three fullerenes. One can observe strong correlations for Ng_2 dimers across different fullerenes. If fitting a fullerene with three Ng_2 dimers, we cannot often obtain such a strong correlation. Also, when all data points were put together in one plot, none of significant correlations has been observed neither. Strong correlations in Figure 13.4 indicate that ITA quantities could serve as good descriptors of change trends for Ng_2@fullerene systems. Taking He_2 confined in C_{50}, C_{60}, and C_{70} as an example, we see that the binding energy increases with the increased molecular size of a fullerene molecule, so does the Shannon entropy because of the more delocalization nature of the electron density. This is also true for the information gain (a direct measure of the Hirshfeld charge) [36]. Again for He_2 confined from C_{50} to C_{70}, when the information gain difference decreases, so does the Hirshfeld charge difference in Table 13.6.

Table 13.4 Numerical results of eight ITA quantities. Atomic units.

Species	ΔS_S	ΔI_F	ΔS_{GBP}	ΔI_G	ΔR_2	ΔR_3	$\Delta^r R_2$	$\Delta^r R_3$
He_2@C_{50}	−0.674	−4.561	−0.658	−0.035	0.183	0.096	0.596	0.597
Ne_2@C_{50}	−0.470	−3.450	−0.509	−0.027	0.183	0.096	0.597	0.598
Ar_2@C_{50}	−0.293	−2.617	−0.390	−0.024	0.183	0.096	0.598	0.598
He_2@C_{60}	−0.773	−9.045	−1.003	−0.067	2.448	2.227	1.274	1.275
Ne_2@C_{60}	−0.596	−6.772	−0.774	−0.053	2.461	2.246	1.278	1.279
Ar_2@C_{60}	−0.456	−4.835	−0.573	−0.043	2.470	2.261	1.281	1.282
He_2@C_{70}	−1.777	−12.980	−1.943	−0.118	2.979	2.370	1.508	1.510
Ne_2@C_{70}	−1.891	−11.679	−1.864	−0.099	3.025	2.408	1.516	1.517
Ar_2@C_{70}	−1.991	−10.694	−1.848	−0.084	3.062	2.441	1.521	1.523

Table 13.5 Correlation coefficients R^2 values between ΔE and eight ITA quantities for three Ng_2 dimers with different fullerene systems (first three rows) and for three fullerenes with different Ng_2 dimers (last three rows).

R^2	ΔS_S	ΔI_F	ΔS_{GBP}	ΔI_G	ΔR_2	ΔR_3	$\Delta^r R_2$	$\Delta^r R_3$
He_2	1.000	0.998	0.999	0.964	0.999	0.874	0.996	0.997
Ne_2	1.000	0.999	0.999	0.999	1.000	1.000	1.000	1.000
Ar_2	0.998	0.997	0.890	0.999	0.998	0.999	0.995	0.995
C_{50}	1.000	0.769	0.960	0.896	0.479	0.357	0.549	0.549
C_{60}	0.995	0.846	0.970	0.888	0.437	0.316	0.500	0.501
C_{70}	0.676	0.498	0.649	0.443	0.040	0.006	0.065	0.066

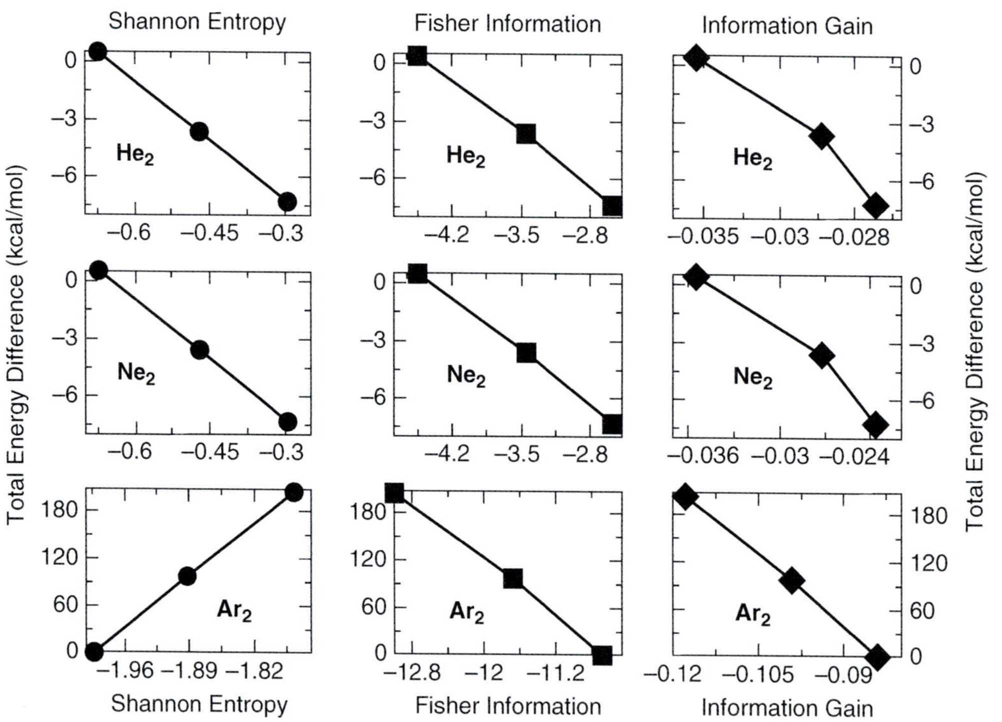

Figure 13.4 Strong linear relationships of the total energy difference ΔE with ITA quantities such as Shannon entropy, Fisher information and information gain for He_2, Ne_2, and Ar_2 confined in C_{50}, C_{60}, and C_{70} fullerenes.

13.3.7 Changes in Spectroscopy

NMR observables, such as chemical shielding and indirect nuclear spin–spin coupling (*J*-coupling) constants, are often employed to explore the interior magnetic environment of fullerene molecules [38, 39]. Shown in Figure 13.5 are the predicted chemical shielding results for isolated Ng_2 dimers in the gas phase and 9 Ng_2@fullerene systems. Also, *J*-couplings for ^3He (with a spin of 1/2) are exhibited in Figure 13.5. It can readily be seen that the trend of chemical shielding is quite similar for He_2, Ne_2, and Ar_2 confined in C_{50}, C_{60} and C_{70}, respectively. In addition, *J*-couplings are

Table 13.6 Hirshfeld charge[a)] on Ng_2 (Ng = He, Ne, and Ar) confined in C_{50}, C_{60}, and C_{70} fullerenes.

Ng-Ng	C_{50}	C_{60}	C_{70}
He-He	0.073	0.056	0.047
Ne-Ne	0.127	0.091	0.067
Ar-Ar	0.180	0.168	0.161

a) Hirshfeld charge is based on the stockholder partition [37] of atoms in molecules.

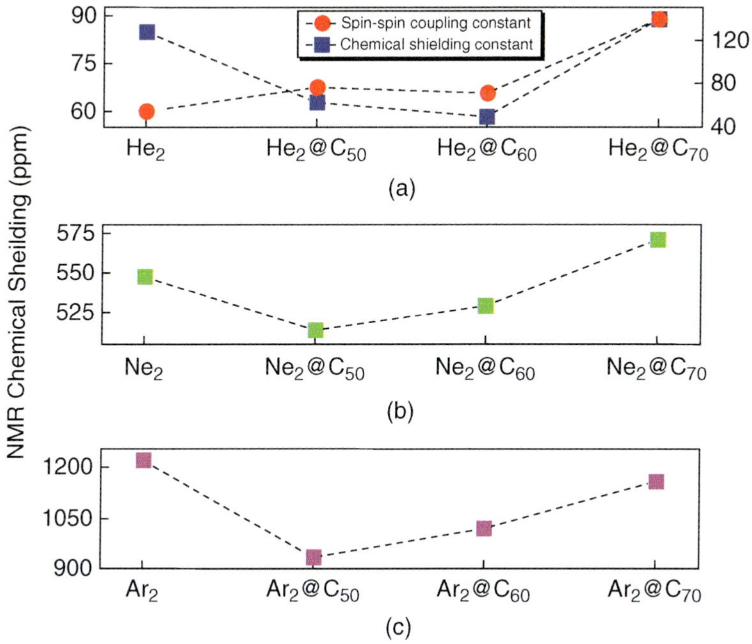

Figure 13.5 NMR parameters predicted at the B3LYP/pcJ-2 level for isolated noble gas dimers and those confined in fullerenes.

closely related to the interatomic distances. For $He_2@C_{50}$ and $He_2@C_{60}$, the predicted interatomic J-couplings for ^3He are very close as can be evidenced by the interatomic distance discrepancy of 0.1 Å. From the NMR results, it is anticipated that stable endohedral fullerenes from Table 13.1 should be identifiable via NMR signals if the compounds can be synthesized experimentally.

13.3.8 Changes in Reactivity

Frontier molecular orbitals (FMOs) are always involved in a chemical reaction. Within this context, we plot the FMOs of $Ar_2@C_{70}$ from both top and side views as an illustration, as shown in Figure 13.6. It is clear that the FMOs are both delocalized over the entire C_{70} motif, without discernible contributions from the Ng_2 dimer. In other words, atomic orbitals from Ng atoms are not implicated in these frontier orbitals. Accordingly, it is anticipated that when chemical reactions occur with the entire system, no noble gas atoms should be involved.

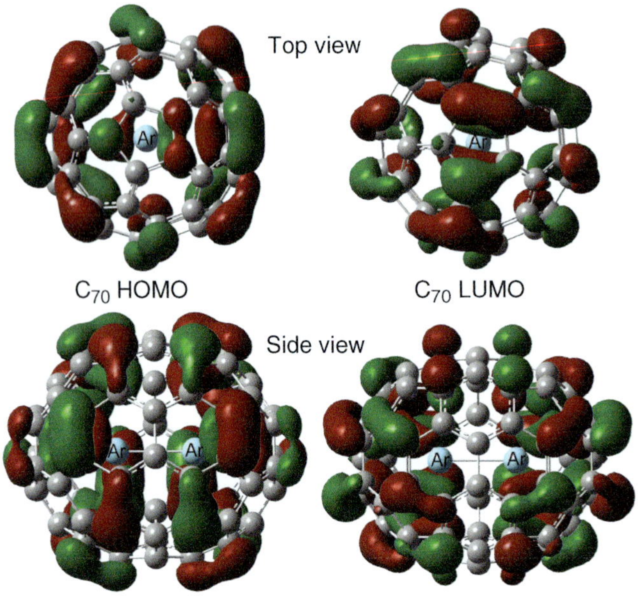

Figure 13.6 HOMO and LUMO contour surfaces of $Ar_2@C_{70}$ molecule. The isovalue is 0.02 a.u.

Table 13.7 Global descriptors from conceptual DFT, including ε_{HOMO} and ε_{LUMO}, chemical potential μ, chemical hardness η, and electrophilicity index ω for the systems studied in this work. Atomic units.

Species	ε_{HOMO}	ε_{LUMO}	μ	η	ω
C_{50}	−0.263	−0.150	−0.206	0.114	0.375
$He_2@C_{50}$	−0.251	−0.135	−0.193	0.115	0.324
$Ne_2@C_{50}$	−0.252	−0.136	−0.194	0.116	0.324
$Ar_2@C_{50}$	−0.256	−0.140	−0.198	0.116	0.337
C_{60}	−0.280	−0.114	−0.197	0.166	0.234
$He_2@C_{60}$	−0.266	−0.100	−0.183	0.166	0.201
$Ne_2@C_{60}$	−0.265	−0.101	−0.183	0.165	0.203
$Ar_2@C_{60}$	−0.263	−0.105	−0.184	0.158	0.215
C_{70}	−0.275	−0.116	−0.195	0.158	0.241
$He_2@C_{70}$	−0.261	−0.102	−0.182	0.159	0.207
$Ne_2@C_{70}$	−0.261	−0.102	−0.182	0.159	0.207
$Ar_2@C_{70}$	−0.262	−0.104	−0.183	0.159	0.211

In Table 13.7, based upon the energies of FMOs (ε_{HOMO} and ε_{LUMO}), we calculated the values of a few well-defined global reactivity descriptors from conceptual DFT [15, 16], including chemical potential μ [$\mu = (\varepsilon_{LUMO} + \varepsilon_{HOMO})/2$] [40], chemical hardness η ($\eta = \varepsilon_{LUMO} - \varepsilon_{HOMO}$) [41], and electrophilicity index ω ($\omega = \mu^2/2\eta$) [42] for the nine systems under study. Chemical hardness is closely related to molecular stability, whereas electrophilicity index corresponds to chemical reactivity of a given system. To be more specific, larger chemical hardness is usually equal to more molecular

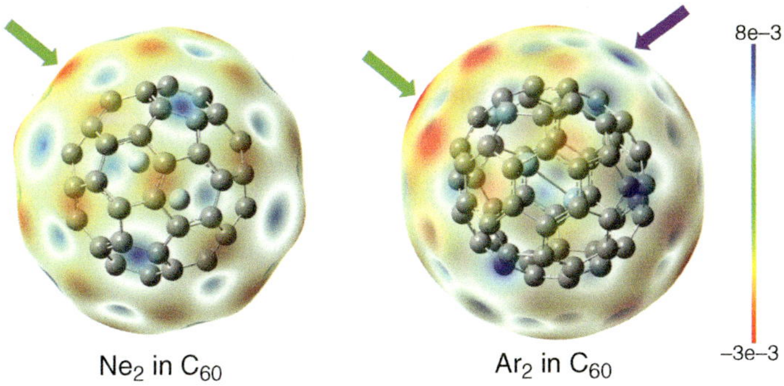

Figure 13.7 The mapped contour surface of the molecular electrostatic potential onto the van der Waals surface for Ne_2 and Ar_2 dimers encapsulated in C_{60} fullerene.

stability. Yet a molecule possessing a smaller value of electrophilicity index becomes more reactive. Full inspection of Table 13.7 leads to the conclusion that Ng_2 in smaller fullerenes are more reactive, whereas in larger fullerenes it often becomes more stable. These results agree well with the conventional chemical intuition.

Chemical reactions normally take place in the vicinity of van der Waals (vdW) surface of a given system. Thus, in Figure 13.7 we have shown the contour surface of the molecular electrostatic potential (MEP) mapped onto the corresponding vdW surface for $Ne_2@C_{60}$ and $Ar_2@C_{60}$. One can easily see that the carbon atoms perpendicular to the Ng-Ng bond are more electronegative. This observation is in line with our Hirshfeld charge results in Table 13.6. The confinement effect renders the Ng atoms more positive and the carbon atoms near the Ng atoms become more negative. In addition, we have revealed before that Hirshfeld charges are good indicators of electrophilicity and nucleophilicity [36], thus Figure 13.7 also showcases that carbon atoms become more nucleophilic, indicating that they are more capable of donating electrons to an electrophile.

Furthermore, in conceptual DFT, Fukui functions [43] can also be harnessed to qualitatively predict nucleophilic and electrophilic attack regions. In Figure 13.8, we have shown these local

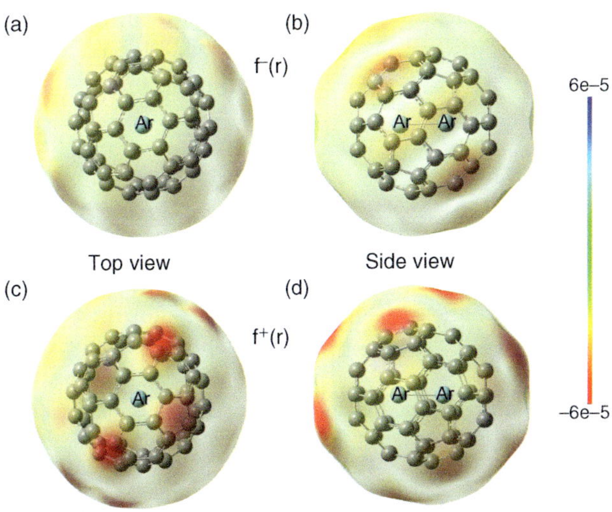

Figure 13.8 Electrophilic (a and b) and nucleophilic (c and d) Fukui functions for Ar_2 dimer encapsulated in C_{60} fullerene.

functions for $Ar_2@C_{60}$, where both electrophilic (Figures 13.8a and 13.8b) and nucleophilic (Figures 13.8c and 13.8d) Fukui functions are exhibited. We found that (i) carbon atoms can be both electrophilic (dark blue areas in Figures 13.8a and 13.8b) and nucleophilic (dark blue regions in Figures 13.8c and 13.8d) in a single fullerene molecule, and (ii) nucleophilic regions in Figures 13.8c and 13.8d are not the same as those of Figure 13.7 (red areas). In other words, using Hirshfeld charge and Fukui charge, we obtain inconsistent results. According to our recent findings [44], this discrepancy is common and not surprising. We have found in a recent study that results from the Hirshfeld charge should be more robust and thus more reliable [44].

13.4 Conclusions

In this contribution, we explored the possibility of making noble gas reactive through quantum confinement in fullerene cages. We have constructed a total of 9 endohedral fullerenes (C_{50}, C_{60}, and C_{70}) confining three Ng_2 dimers (He_2, Ne_2 and Ar_2). Quantum molecular dynamics simulations were employed to explore the potential energy surface of Ng_2 inside the fullerene molecule. We have further examined the optimized structures with a few well-established analytical tools, among which were conceptual density functional theory, information-theoretic approach, total energy decomposition, bonding energy decomposition, non-covalent analysis, and others. Our results show that interatomic distances of Ng_2 inside fullerenes become substantially smaller and noble gas atoms become more electrophilic. Using these analytical tools, we appreciate the nature and origin of these structure and reactivity changes. Our results and conclusions drawn from the present study should provide more understanding and new insights from the viewpoint of the impact of quantum confinement on noble gas structure and reactivity.

Acknowledgments

D.Z. is supported by the startup funding of Yunan University. C.R. acknowledges support from the National Natural Science Foundation of China (No. 21503076) and Hunan Provincial Natural Science Foundation of China (Grant No. 2017JJ3201). P.K.C. would like to thank the DST, New Delhi for the J. C. Bose National Fellowship.

References

1 Popov, A.A.; Yang, S.; Dunsch, L. Endohedral fullerenes. Chem. Rev. 2013, 113, 5989–6113.
2 Osawa, E.. Superaromaticity. Kagaku 1970, 25, 854–863. (in Japanese); *Chem. Abstr.* **1971**, *74*, 75698v.
3 Kroto, H.W.; Heath, J.R.; O'Brien, S.C.; Curl, R.F.; Smalley, R.E. C_{60}: Buckminsterfullerene. Nature 1985, 318, 162–163.
4 Murata, M.; Murata, Y.; Komatsu, K. Synthesis and properties of endohedral C_{60} encapsulating molecular hydrogen. J. Am. Chem. Soc. 2006, 128, 8024–8033.
5 Murata, M.; Maeda, S.; Morinaka, Y.; Murata, Y.; Komatsu, K. Synthesis and Reaction of Fullerene C_{70} Encapsulating Two Molecules of H_2. J. Am. Chem. Soc. 2008, 130, 15800–15801.
6 Whitener, K.E. Jr.; Frunzi, M.; Iwamatsu, S.; Murata, S.; Cross, R.J.; Saunders, M. Putting Ammonia into a Chemically Opened Fullerene. J. Am. Chem. Soc. 2008, 130, 13996–13999.

7 Krachmalnicoff, A.; Bounds, R.; Mamone, S.; Alom, S.; Concistré, M.; Meier, B.; Kouřil, K.; Light, M.E.; Johnson, M.R.; Rols, S.; Horsewill, A.J.; Shugai, A.; Nagel, U.; Rõõm, T.; Carravetta, M.; Levitt, M.H.; Whitby, R.J. The dipolar endofullerene $HF@C_{60}$. Nat. Chem. 2016, 8, 953–957.

8 Kurotobi, K.; Murata, Y. A single molecule of water encapsulated in fullerene C_{60}. Science 2011, 333, 613–616.

9 Zhang, R.; Murata, M.; Aharen, T.; Wakamiya, A.; Shimoaka, T.; Hasegawa, T.; Murata, Y. Synthesis of a distinct water dimer inside fullerene C_{70}. Nat. Chem. 2016, 8, 435–441.

10 Saunders, M.; Jiménez-Vázquez, H.A.; Cross, R.J.; Mroczkowski, S.; Freedberg, D.I.; Anet, F.A.L. Probing the interior of fullerenes by ^3He NMR spectroscopy of endohedral $^3He@C_{60}$ and $^3He@C_{70}$. Nature 1994, 367, 256–258.

11 Saunders, M.; Cross, R.J.; Jiménez-Vázquez, H.A.; Shimshi, R.; Khong, A. Noble gas atoms inside fullerenes. Science 1996, 271, 1693–1697.

12 Krapp, A.; Frenking, G. Is this a chemical bond? A theoretical study of $Ng_2@C_{60}$ (Ng = He, Ne, Ar, Kr, Xe). Chem. Eur. J. 2007, 13, 8256–8270.

13 Fernández, I.; Solà, M.; Bickelhaupt, F.M. Origin of reactivity trends of noble gas endohedral fullerenes $Ng_2@C_{60}$ (Ng = He to Xe). J. Chem. Theory Comput. 2014, 10, 3863–3870.

14 Parr, R.G.; Yang, W. Density Functional Theory of Atoms and Molecules. Oxford: Oxford University Press, 1989.

15 Liu S.B. Conceptual density functional theory and some recent developments. Acta Phys.-Chim. Sin. 2009, 25, 590–600.

16 Geerlings, P.; De Proft, F.; Langenaeker, W. Conceptual density functional theory. Chem. Rev. 2003, 103, 1793–1847.

17 Liu S.B. Information-theoretic approach in density functional reactivity theory. Acta Phys.-Chim. Sin. 2016, 32, 98–118.

18 Rong C.; Wang, B.; Zhao, D.; Liu, S. Information-theoretic approach in density functional theory and its recent applications to chemical problems. WIREs Comput. Mol. Sci. 2019, e1461.

19 Liu S.B. Steric effect: A quantitative description from density functional theory. J. Chem. Phys. 2007, 126, 244103.

20 Bickelhaupt, F.M.; Baerends, E.J. *Kohn-Sham Density Functional Theory: Predicting and Understanding Chemistry*. In: Rev. Comput. Chem.; Lipkowitz, K.B. and Boyd, D.B., Eds.; Wiley-VCH: New York, 2000, Vol. 15, 1–86.

21 Johnson, E.R.; Keinan, S.; Mori-Sánchez, P.; Contreras-García, J.; Cohen, A.J.; Yang, W. Revealing noncovalent interactions. J. Am. Chem. Soc. 2010, 132, 6498–6506.

22 Frisch, M.J.; Trucks, G.W.; Schiegel, H.B.; Scuseria, G.E.; Robb, M.A.; Cheeseman, J.R.; Montgomery, J.A.; Vreven, T., Jr.; Kudin, K.N.; Burant, J.C.; et al. Gaussian 09; Gaussian, Inc.: Pittsburgh, PA, 2009.

23 Stewart, J.J.P. Optimization of parameters for semiempirical methods VI: more modifications to the NDDO approximations and re-optimization of parameters. J. Mol. Model. 2013, 19, 1–32.

24 Zhao, Y.; Truhlar, D.G. The M06 suite of density functionals for main group thermochemistry, thermochemical kinetics, noncovalent interactions, excited states, and transition elements: two new functionals and systematic testing of four M06-class functionals and 12 other functionals. Theor. Chem. Acc. 2008, 120, 215–241.

25 McLean, A.D.; Chandler, G.S. Contracted Gaussian-basis sets for molecular calculations. 1. 2nd row atoms, Z=11–18. J. Chem. Phys. 1980, 72, 5639–5648.

26 Lu, T.; Chen, F. Multiwfn: a multifunctional wavefunction analyzer. J. Comput. Chem. 2012, 33, 580–592.

27 Te Velde, G.T.; Bickelhaupt, F.M.; Baerends, E.J.; Fonseca Guerra, C.; van Gisbergen, S.J.; Snijders, J.G.; Ziegler, T. Chemistry with ADF. J. Comput. Chem. 2001, 22, 931–967.

28 Boys, S.F.; Bernardi, F. Calculation of small molecular interactions by differences of separate total energies – some procedures with reduced errors. Mol. Phys. 1970, 19, 553–566.

29 Simon, S.; Duran, M.; Dannenberg, J.J. How does basis set superposition error change the potential surfaces for hydrogen bonded dimers? J. Chem. Phys. 1996, 105, 11024–11031.

30 Shannon, C.E. A mathematical theory of communication. Bell Syst. Tech. J. 1948, 27, 379–423.

31 Fisher, R.A. Theory of statistical estimation. Proc. Cambridge Philos. Soc. 1925, 22, 700–725.

32 Ghosh, S.K.; Berkowitz, M.; Parr, R.G. Transcription of ground-state density-functional theory into a local thermodynamics. Proc. Natl. Acad. Sci. U.S.A. 1984, 81, 8028–8031.

33 Kullback, S.; Leibler, R.A. On information and sufficiency. Ann. Math. Stat. 1951, 22, 79–86.

34 Rényi, A. Probability theory. Amsterdam: North-Holland, 1970.

35 Nagy, Á.; Romera, E. Relative Rényi entropy and fidelity susceptibility. Europhys. Lett. 2015, 109, 60002.

36 Liu, S.B.; Rong, C.Y.; Lu, T. Information conservation principle determines electrophilicity, nucleophilicity, and regioselectivity. J. Phys. Chem. A 2014, 118, 3698–3704.

37 Hirshfeld, F. Bonded-atom fragments for describing molecular charge densities. Theor. Chim. Acc. 1977, 44, 129–138.

38 Saunders, M.; Jiménez-Vázquez, H.A.; Cross, R.J.; Mroczkowski, S.; Freedberg, D.I.; Anet, F.A.L. Probing the interior of fullerenes by ^{3}He NMR spectroscopy of endohedral $^3He@C_{60}$ and $^3He@C_{70}$. Nature 1994, 367, 256–258.

39 Saunders, M.; Jimenez-Vazquez, H.A.; Cross, R.J.; Billups, W.E.; Gesenberg, C.; Gonzalez, A.; Luo, W.; Haddon, R.C.; Diederich, F.; Herrmann, A. Analysis of isomers of the higher fullerenes by ^{3}He NMR spectroscopy. J. Am. Chem. Soc. 1995, 117, 9305–9308.

40 Parr, R.G.; Donnelly, R.A.; Levy, M.; Palke, W.E. Electronegativity: the density functional viewpoint. J. Chem. Phys. 1978, 68, 3801–3807.

41 Parr, R.G.; Pearson, R.G. Absolute hardness: companion parameter to absolute electronegativity. J. Am. Chem. Soc. 1983, 105, 7512–7516.

42 Parr, R.G.; Szentpaly, L.V.; Liu, S.B. Electrophilicity index. J. Am. Chem. Soc. 1999, 121, 1922–1924.

43 Yang, W.; Parr, R.G.; Pucci, R. Electron density, Kohn–Sham frontier orbitals, and Fukui functions. J. Chem. Phys. 1984, 81, 2862–2863.

44 Wang, B.; Rong, C.; Chattaraj, P. K.; Liu, S. A comparative study to predict regioselectivity, electrophilicity and nucleophilicity with Fukui function and Hirshfeld charge. Theor. Chem. Acc. 2019, 138, 124–132.

14

Confinement Induced Chemical Bonding: Case of Noble Gases

Sudip Pan[1,*], *Gabriel Merino*[2], *and Lili Zhao*[1,*]

[1] *Institute of Advanced Synthesis, School of Chemistry and Molecular Engineering, Jiangsu National Synergetic Innovation Center for Advanced Materials, Nanjing Tech University, Nanjing 211816, China*
[2] *Departamento de Física Aplicada, Centro de Investigación y de Estudios Avanzados, Unidad Mérida.Km 6 Antigua Carretera a Progreso. Apdo., Postal 73, Cordemex, Mérida, Yuc. 97310, México*

14.1 Introduction

Chemical bonding is a fuzzy concept in chemistry defined based on different models since it is neither an experimentally observable quantity nor there is any Hermitian quantum mechanical operator corresponding to this. Each model has its own advantages and disadvantages, but they are very pertinent in explaining the continually cumulative complexity and richness of chemical results in modern era. Among the models, the one proposed by Gilbert Lewis [1] in 1916 involving the heuristic assignment of an electron pair to a chemical bond is the most imperative model to understand the chemical bonding which was formulated through careful scrutiny of structures, stabilities, reactivities and similarities of mainly first octal row of the periodic table. In next few years, it was expanded by Langmuir [2–5] through the introduction of 8, 18, and 32 electron counting rules for *sp* block (main group elements), *spd* block (transition metals) and *spdf* block (lanthanides and actinides), respectively, at a time prior to the advent of quantum chemistry. This model is truly a stroke of genius, which hides unknowingly but still accurately the yet unexplored quantum chemical nature of the chemical bond in its rather simple representation. With time, this model and the associated rules of drawing Lewis structures have been slightly modified; however, the basic characteristic has remained similar till now.

It is really astounding since the formation of covalent bond is a quantum mechanical phenomenon originated from the interference of spatially extended wave functions, being either a (+) or a (−) sign, which was first elucidated by Heitler and London [6] in 1927 through their analysis of the bonding situation in H_2 molecule by employing newly introduced quantum theory by Heisenberg [7] and Schrödinger [8]. The combination of two signs leads to the two possibilities for the new wave function, yielding the bonding and antibonding situations of the interacting electrons, where the positive and negative interferences result in an accumulation and depletion of electronic charge in between interacting atoms, respectively. The square of the wave function gives the amount of electronic charge. The sign of the wave functions is very important to understand the bonding and its essence was correctly realized at that time by Mulliken [9], Hund [10], Lennard-Jones [11], and Hückel [12], and subsequently by Fukui [13] in formulating

*Corresponding authors: ias_sudippan@njtech.edu.cn and ias_llzhao@njtech.edu.cn

Chemical Reactivity in Confined Systems: Theory, Modelling and Applications,
First Edition. Edited by Pratim Kumar Chattaraj and Debdutta Chakraborty.

the frontier molecular orbital (FMO) theory and Woodward and Hoffmann [14] in the orbital symmetry rules for pericyclic reactions. The most important finding emerged from the work of Heitler and London is that "the physical origin of chemical bonding is due to electron-sharing, not due to electron pairing".

In the eminent book *The Nature of the Chemical Bond*, published in 1939, for the first time Pauling [15] successfully showed the connection between the heuristic electron-pair bonding model of Lewis and Heitler-London ansatz for the quantum chemical treatment of H_2, using an electron-pair function for the H-H bond. He put forward the theory of resonating structures to describe molecules. Interestingly, he supported the localized picture of valence bond (VB) theory proposed by Heitler and London but opposed the delocalized picture of molecular orbital (MO) theory. Clearly, he failed to realize the important information that might be extracted from the symmetry of the wave function which makes it superior to the other theory. Later on, with time, various studies [16–19] reported many astonishing and atypical molecules and complexes that possess unusual bonding implying that the proper understanding of bonding situation is quite complicated and the effort to correlate the Lewis representation and the actual physical nature of chemical bond is not always straightforward and valid. This is quite expected given the fact that there are vast variety of materials available in our mother earth constituted of only some limited number of chemical elements, and therefore the chemical bonding is bound to be of infinite varieties imposing different properties.

In addition to the immense complexities already hidden in the bonding, a beautiful exercise to make it even more complicated is to place the target molecule within a cavitand. Depending on the size of the host, the extent of effect of confinement on the bonding can be monitored. The confinement can alter the reactivity as well as different response properties by modifying the wave functions and the associated electronic energy of the orbitals, in comparison to the free molecules [20–23]. The particle-in-a-box problem [24] is a classic example to understand its effect on the changes in electronic energy levels. The energy levels of a free particle are continuous; however, the presence of confinement effect makes the energy levels discrete. Moreover, the energy levels of the particle can be changed by altering the length of the box [25–28]. The numerous reported host-guest complexes represent the real-life examples of confined systems [29].

In this chapter, we will present some examples to show how confinement can even induce chemical bonding in between two noble gas (Ng) atoms in true sense. The Ng chemistry is one of the less developed fields because of their chemical inertness originated from their filled valence electronic shell. Driving by large ionization energy (IE) and small electron affinity (EA) they act as inert towards other elements in the periodic table. Only a strong polarizing center is required to deform rather rigid electron density of Ng and to facilitate a donor-acceptor type of interaction [30–45]. The Ng_2 dimer is essentially unbound in nature, only supported by weak dispersion interaction. The confinement opens an opportunity to force Ng atom to make chemical bond not only with the atoms in the host moiety but also with two Ng atoms. Most of such complexes are not thermochemically stable but kinetically one. They need a large driving force to introduce within the cavitand, but once formed, it is also difficult to make them released. This is amazing that the hurdle to put Ng inside fullerenes was overcome in early nineties when Schwarz and co-workers, Cross and co-workers and by some other groups experimentally prepared He and Ne encapsulated positively charged $Ng@C_{60}^{q+}$ (Ng = He, Ne; q = 1, 2, 3) complexes and their neutral analogues [46–58]. While the charged complexes were formed via high energy bimolecular collision reactions, the $He@C_{60}$ was formed during the preparation of fullerene from graphite in helium environment. The ^3He isotope and Ne inclusion C_{60} complexes were achieved by heating the fullerenes at high temperatures in presence of them. The latter observation led to the proposal of "window mechanism" where the inclusion process was believed to occur via temporary window, created through breakage of one or more C-C bonds of the cage. In fact, a complete set of Ng@C_{60} (Ng = He, Ne, Ar, Kr, Xe) and Ng@C_{70} (Ng = Ne, Ar, Kr, Xe) complexes were also prepared by keeping a sample of fullerenes and a particular Ng at high temperature and pressure condition and were characterized by mass spectroscopy [59]. In due course, attempts were made to make the methodology to insert Ng atoms inside the cage more efficient, and subsequently, the processes involving the strike of the fullerenes with the beam of Ng^+ ion [60], an explosion of a mixture of fullerene and Ng in a confined space [61] and "molecular surgery" [62, 63] were put forward. Even He atom in much smaller cavitand like dodecahedrane, He@$C_{20}H_{20}$ was also experimentally achieved by shooting process [64]. For more information about endohedral fullerenes, the readers are referred to a recent minireview [65].

From computational point of view, to analyze quantum confined systems it is absolutely essential to construct a meaningful and appropriate model which considers the alteration in the electronic wave function because of confinement and appropriate theoretical tools to analyze the bonding situation and response properties. There exist several methods that act as a link between the computed results employing the quantum chemical theory and the heuristic bonding schemes like the Lewis model. To analyze bonding situation, several charge and energy partitioning methods are available. An ideal method should satisfy five conditions viz., "1) mathematically unambiguously defined, 2) derived from accurate quantum chemical calculations, 3) results do not significantly change at different theoretical levels, 4) plausible interpretation of the different terms, and 5) useful for chemical problems," in order to act as a reliable tools to analyze bonding nature [16]. We have mainly used density functional (DFT) theory particularly those functionals which are either parametrized in such a way that they can take care of dispersion interaction or are additionally corrected using - D3 term [66]. For bonding analysis, natural bond orbital (NBO) [67], adaptive natural density partitioning (AdNDP) [68], quantum theory of atoms in molecules (QTAIM) [69], and energy decomposition analyzes (EDA) [70] were performed.

14.2 Computational Details and Theoretical Background

For Ng encapsulated large complexes, we used those density functionals which can take care of dispersion interaction in combination of triple-zeta quality basis set. For Ng inserted molecules, high-level CCSD(T) calculations were performed. Both geometry optimization followed by harmonic frequency calculation were carried out in order to know the nature of the stationary states. A Gaussian program package was employed for these calculations [71].

The NBO method was used for the bonding analysis. This was introduced in the 1980s and in subsequent times it underwent significant alterations and modifications [72]. Hence, sometimes the results obtained from a newer version may be significantly different from an older version of NBO. This method provides us the most reasonable Lewis structure of a molecule, two centered two electrons and three centered two electrons natural bond orbitals, and the corresponding polarities and hybridization of the atoms, Wiberg bond order, and the atomic partial charges. It uses the one-electron density matrix obtained by any reliable quantum chemical methods for the partitioning procedure. Since it employs the density ρ rather than wave function Ψ as the starting point, it works for both DFT and ab initio methods, which is a plus point of this method. However, there are two features of the NBO method which can be debated and sometimes lead to confusing results. The first one is the division of the atomic orbitals into valence and Rydberg space. These two set of orbitals are not treated in equal weightage in the algorithm, which gives better priority to valence orbitals than the Rydberg ones. Since in many cases an atomic center of a molecule can be in excited state, such preselection becomes vague for them and results in a biased outcome. The assignment of atomic orbitals into valence and Rydberg sets is done what the developers of this method seem reasonable. Therefore, NBO results are not "natural" as claimed in the name of the method. Notably, NBO6 and upper version do not consider the p orbitals of transition metal as valence orbitals, which sometimes lead to a completely different bonding scenario. Interestingly, NBO3 considers p orbitals of transition metal as valence orbitals. Therefore, if one compares the NBO data using these two versions, sometimes a significant difference in the results will be obtained. The second critical point is the continuous change in the algorithms of the eight orthogonalization steps in the different versions of NBO, which could sometimes give varying results in moving from one version to another one.

The Adaptive Natural Density Partitioning (AdNDP) method developed by Boldyrev and co-workers [68] is an extension of the NBO method. The NBO method can identify natural orbitals only up to 3c-2e but the AdNDP can describe the bonding in terms of N-center-two-electron bonds where N can be any number less or equal to the total number of atoms in a molecule. This method is very useful for systems having delocalized bonding nature, which are very common in organic and inorganic chemistry.

The Bader's QTAIM method is a topological analysis of the electron density, $\rho(r)$ into atomic basins [69]. This method is mathematically more concrete and evades the ambiguities involved in wave function based partitioning methods. It uses the first ($\nabla\rho(r)$) and second derivatives ($\nabla^2\rho(r)$) of $\rho(r)$ to identify atomic regions and interatomic bond paths, and the atomic basins are separated by the zero-flux surfaces. The latter term can be explained as the gradient vector field whose trajectories of $\nabla\rho(r)$ do not disappear at the nuclei but at the bond critical point (BCP). At BCP, $\nabla^2\rho(r_c)$ has two negative (λ_1, λ_2) and one positive (λ_3) eigenvalues (curvatures). The other critical points which have two positive and one negative curvatures, and all positive curvatures, are called ring and cage critical points, respectively. The unique trajectory of maximum gradient of ρ which starts at a nucleus and terminates to another nucleus going through BCP is called a bond path. Therefore, this method enables the skeletal structure of the molecule to be described in terms of atomic nuclei and chemical bonds, which has turned out to be a very useful method to understand the bonding of molecules [73]. The covalent and non-covalent nature of bonding is characterized by the fulfillment of certain conditions. A high $\rho(r_c)$ and negative $\nabla^2\rho(r_c)$ usually indicate covalent bond where the opposite situation implies non-covalent bonding. However, this is always not true. For heavier than first row elements, it was often found that positive λ_3 dominates over negative $\lambda_1 + \lambda_2$. In these cases, use of energy density $H(r_c)$ is more effective, which is negative for covalent bonds [74].

The bonding situations were analyzed by means of an energy decomposition analysis (EDA) [70] combined with the natural orbitals for chemical valence (NOCV) [75] method by using the ADF 2018.105 program package [76, 77]. In this analysis, if the dispersion-corrected functional is considered, the intrinsic interaction energy (ΔE_{int}) between two fragments can be divided into four energy components as follows:

$$\Delta E_{\text{int}} = \Delta E_{\text{elstat}} + \Delta E_{\text{Pauli}} + \Delta E_{\text{orb}} + \Delta E_{\text{disp}}. \tag{14.1}$$

While the electrostatic ΔE_{elstat} term accounts for the quasiclassical electrostatic interaction between the unperturbed charge distributions of the prepared fragments, the Pauli repulsion ΔE_{Pauli} corresponds to the energy change associated with the transformation from the superposition of the unperturbed electron densities of the isolated fragments to the wavefunction, that follows the Pauli principle through explicit antisymmetrization and renormalization of the production wavefunction. Since we included D3 or D3(BJ), it provides us with the dispersion interaction energy between two interacting fragments. The orbital term $\Delta E_{\rm orb}$ is originated from the mixing of orbitals, charge transfer, and polarization between the isolated fragments, which can be further decomposed into contributions from each irreducible representation of the point group of the interacting system as follows:

$$\Delta E_{orb} = \sum_{r} \Delta E_{r}. \tag{14.2}$$

The EDA-NOCV combination allows the partition of $\Delta E_{\rm orb}$ into pairwise contributions of the orbital interactions, which gives important information about bonding. The charge deformation $\Delta\rho_k(r)$, resulting from the mixing of the orbital pairs $\psi_{\rm k}(\rm r)$ and $\psi_{-\rm k}(\rm r)$ of the interacting fragments, gives the amount and the shape of the charge flow due to the orbital interactions (Eq. 14.3), and the associated energy term $\Delta E_{\rm orb}$ reflects the size of stabilizing orbital energy originated from such interaction (Eq. 14.4).

$$\Delta\rho_{orb}(r) = \sum_{k} \Delta\rho_k(r) = \sum_{k=1}^{N/2} V_k \left[-\psi_{-k}^2(r) + \psi_k^2(r)\right] \tag{14.3}$$

$$\Delta E_{orb} = \sum_{k} \Delta E_k^{orb} = \sum_{k=1}^{N/2} V_k \left[-F_{-k}^{TS} + F_k^{TS}\right] \tag{14.4}$$

This method is found to be excellent for analyzing the bonding situation in many different complexes [78–85].

14.3 The Bonding in He@$C_{10}H_{16}$: A Debate

Helium encapsulated $C_{10}H_{16}$ adamantane, He@adam, was at the center of the debate in a series of papers. In 2004, Haaland and co-workers [86] reported in an *in silico* study an energy minimum inclusion complex, He@adam. The topological analysis of electron density shows that there are four bond paths connecting He and carbon atoms, indicating the presence of He-C chemical bonds (see Figure 14.1). However, He@adam is 154.2 kcal/mol larger in energy than the sum of free adamantane and He. Noting the positive bond rupture energy, the authors discarded the notion of He-C chemical bond. Krapp and Frenking [87] disagreed with their argument by providing some interesting experimentally known species like FN_3, He_2^{2+}, and $HeCCHe^{2+}$ where they exothermically dissociated into FN + N_2, 2He^+, and two HeC^+, respectively [88]. Nevertheless, the 1997 IUPAC definition of chemical bond reads "there is a chemical bond between two atoms or groups of atoms in the case that the forces acting between them are such as to lead to the formation of an aggregation with sufficient stability to make it convenient for the chemist to consider it as an independent molecular species", which does not tell that a chemical bond must be attractive [89]. In a subsequent study, Bader and Fang [90] argued that the interaction between He and carbon atoms connected through the bond path are not repulsive in nature as they are energetically stabilized with respect to the free atoms, but the overall instability arises from the higher energies of the other carbon and from the hydrogen atoms. Moreover, the values of topological descriptors, electron density $\rho(\rm r_c)$, Laplacian of electron density $\nabla^2\rho(\rm r_c)$ and energy density $H(\rm r_c)$ at the bond critical points of He-C bond lie in between of closed-shell and electron-sharing interactions. Therefore, they advocated that He-C connection in He@adam is a true chemical bond, which was again further responded to by Strenalyuk and Haaland [91] with their counterarguments. Hopffgarten and Frenking [92] critically analyzed the views of Haaland and co-workers [86] and Bader and Fang [90], and argued that "the conflicting views arise from different perspectives which are valid

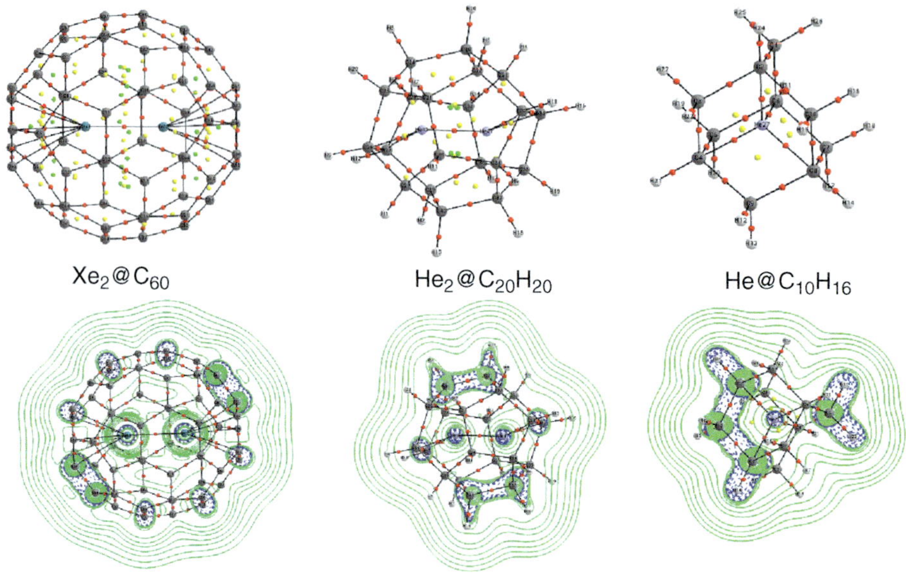

Figure 14.1 Molecular graph of $Xe_2@C_{60}$, $He_2@C_{20}H_{20}$ and $He@C_{10}H_{16}$, and their Laplacian of electron density at a particular plane. Spheres in red, yellow, and green are the bond, ring, and cage critical points. Green solid lines and blue dotted lines show positive and negative region of Laplacian of electron density.

within their own scope of definition". At the end, they thought that "the physical view of chemical bonding advocated by Bader and Fang is more sound and goes deeper when it comes to fundamental laws of physics than the chemical view of Strenalyuk and Haaland which is based on the pragmatic approach that is typical for chemistry".

The argument of Bader and Fang [90] was again criticised by Poater, Solà, and Bickelhaupt [93]. They designed a model system where one of the CH_2 group of He@adam was eliminated and truncated with hydrogen atoms, and, therefore, it opens a free window for helium atom of He@adam. Then, they allowed the modelled system to freely optimize which led to the dissociation of helium, and it made them to conclude that there is not any genuine He-C chemical bonding, which again Krapp and Frenking do not think to be a valid argument as the interatomic interactions in the modelled opened cage are not the same as in the complete complex. Further, Cerpa et al. [94] performed QTAIM analysis on three highly symmetric complexes, $He@C_8H_8$ (O_h), $He@C_{20}H_{20}$ (I_h) and $Ng@C_{60}$ (I_h) to establish that, irrelevant to the corresponding distances, there always exist bond paths between confined Ng and carbon centers although in chemical sense absence of any real chemical bond is expected (see Figure 14.1). They argued that the one-to-one mapping of gradient path/chemical bond would be misleading since the symmetry forced by the external potential will always be followed by the electron density. Schwarz and co-workers [95] finally gave the debate a stop by beautifully presenting the origin of the different viewpoints and by assigning the bond as "confinement bonding".

14.4 Confinement Inducing Chemical Bond Between Two Ngs

In 2007, Krapp and Frenking [87] analyzed the bonding situation in $Ng_2@C_{60}$ (Ng = He-Xe) complex in an elegant study. They found that the confinement elevates the HOMO and LUMO energy level of Ng_2 in confined states in comparison to the free state. Therefore, the confined Ng_2 species would show greater reactivity than free dimer. The computed natural charge shows that the degree of $Ng_2 \rightarrow C_{60}$ charge transfer is rather low for lighter Ng (He-Kr) atoms, but it is remarkably enhanced in the case of Ng = Xe. For the latter case, almost one electron gets shifted from Xe_2 to the C_{60} cage, indicating that it exists as a charge-separated complex, $Xe_2^+@C_{60}^-$. The presence of significant covalent character in Xe-Xe bond of $Xe_2@C_{60}$ is reflected from the considerable negative value (−0.34 Hartree/Å^3) of $H(r_c)$. The Ar-Ar and Kr-Kr bonds also have negative value of $H(r_c)$ (−0.04 and −0.08 Hartree/Å^3, respectively) but somewhat smaller than that in Xe-Xe bond. Therefore, the Xe-Xe bond can undoubtedly be assigned as a genuine chemical covalent bond and the Ar-Ar and Kr-Kr bonds have at least some partial covalent character. Note that for these elements, the essence to have $\nabla^2\rho(r_c)$ negative for covalent bonds cannot be fulfilled since they are heavier than first row elements where this criterion often fails and the $H(r_c)$ value turns out to be a more reliable descriptor. Interestingly, for these cases not only the Ng-Ng bond but also some covalent interaction is found for Ng-C (Ng = Kr, Xe) interaction. This is in the line that the larger size pushes Kr and Xe to go more closely to the host carbon center, eventually making some orbital involvement. Moreover, the inspection of the related MOs reveals that the reactivity of $Ng_2@C_{60}$ (Ng = Ar-Xe) would be considerably different than those of free Ng_2 or C_{60}. This was further confirmed two years later in a comparative reactivity study of $Ng_2@C_{60}$ and C_{60} towards Diels-Alder cycloaddition of 1,3-*cis*-butadiene by Solà and co-workers [96]. They found that $Ng_2@C_{60}$ (Ng = Ar-Xe) shows dramatic enhanced reactivity in comparison to the free C_{60}. Hence, it fulfils the IUPAC definition of chemical bond. The lighter $He_2@C_{60}$ and $Ne_2@C_{60}$ are weakly bonded van der Waals complexes. Wait a moment! The story does not end here. It becomes even more dramatic when Kryachko and co-workers [97] even demanded the formation of a true He-He bond in $He_2@C_{60}$ based on Löwdin's postulate, "a system of electrons and atomic nuclei is said to form a molecule if the Coulombic Hamiltonian **H** – with the center of mass motion removed – has a discrete ground state energy E_0." They evaluated analytically the potential energy curve (PEC) of the He-He dimer encapsulated within C_{60} and then perturbatively solved the corresponding Schrödinger equation and got at least one bound state where two He atoms are bound to each other.

This is in line with the observation obtained in an *ab initio* molecular dynamics study on $Ng_2@C_{60}$ where, within 500 fs time scale, the He_2 was found to undergo translation, vibration, and rotation movements as a single entity, without any random movement (see Figure 14.2) [98]. The heavier Ng_2 also undergoes similar movement but that is restricted by the larger interaction between Ng and carbon cage as such movement needs deformation and reorientation in host case moiety. This observation corroborates with the notion of the presence of some sort of bonding between two He atoms. This situation is comparable with the real-life example when two unwilling partners need to stay in a very small room, and then, realizing that they have to carry on like this, develop some understanding of how to live in a better way by cooperating with each other.

The computational exercise was also made by putting two helium atoms inside a much smaller cage like $C_{20}H_{20}$ that results in the smallest He-He distance (1.265 Å), 0.688 Å shorter than that in $He_2@C_{60}$ and 1.712 Å shorter than that in free He_2 (2.977 Å) [99]. $He_2@C_{20}H_{20}$ was noted to be thermochemically unstable by 169.8 kcal/mol in comparison to the free He atoms and $C_{20}H_{20}$. Although they found a gradient path between two He atoms in QTAIM analysis, they did not find any other supporting indices which can confirm that there is indeed a true chemical bond between two He atoms (see Figure 14.1). They concluded that the short He-He distance is a consequence of the repulsion between He atoms and cage, and that a short internuclear separation does not necessarily conclude the formation of a chemical bond. So, it is only a matter of view,

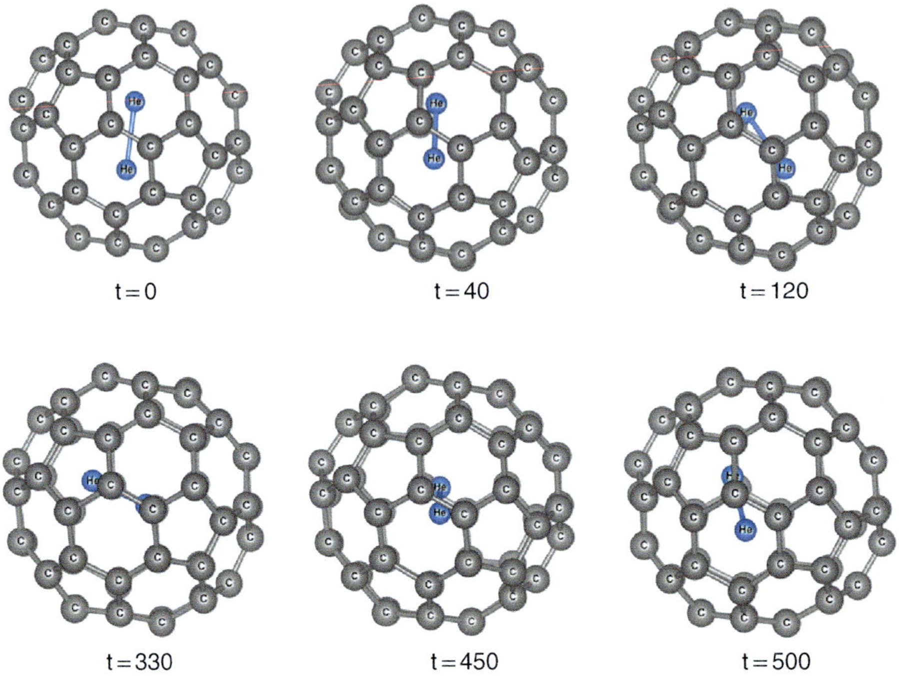

Figure 14.2 Snapshots at different time steps of $He_2@C_{60}$ system. Time in fs. This figure is reproduced from [98], © 2014, Elsevier.

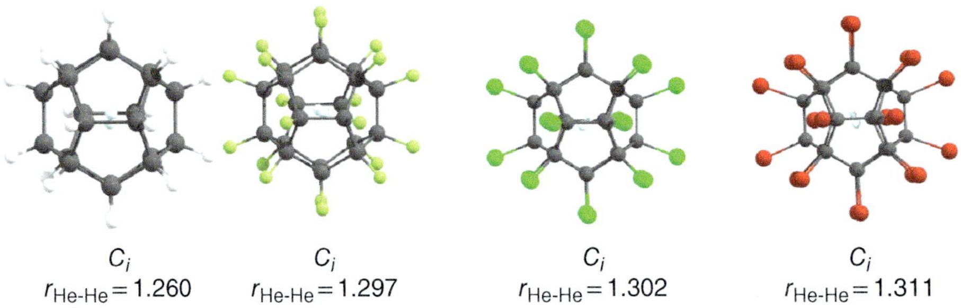

Figure 14.3 Optimized geometries of all the studied $He_2@C_{20}X_{20}$ at ωB97X-D/def2-TZVPP level having C_i point group. This figure is taken from [100], © 2020, John Wiley & Sons.

which changes from scientist to scientist. A very recent study further shows the effect substitution of $C_{20}H_{20}$ cage on He-He bond in $He_2@C_{20}X20$ (X = F, Cl, Br) complexes (see Figure 14.3) [100]. Similar to the $He_2@C_{20}H_{20}$, $He_2@C_{20}X20$ complexes are also thermochemically unstable with respect to spontaneous dissociation into 2He and $C_{20}X20$ but to lesser extent compared to $He_2@C_{20}H_{20}$. Nevertheless, all the complexes are highly kinetically protected. Most importantly, the $H(r_c)$ value for He-He bond, which was positive (0.03 Hartree/Å^3) in $He_2@C_{20}H_{20}$, becomes negative in $He_2@C_{20}X20$ (ranging from −0.05 to −0.07 Hartree/Å^3). Therefore, the substitution with electron-withdrawing group makes an important difference in the bonding situation of He_2 dimer inside the cage. They induce some degree of covalent interaction in between two He atoms. Different reactivity of $He_2@C_{20}X20$ from that of $He_2@C_{20}H_{20}$ is also reflected from that of hardness (HOMO-LUMO energy gap) value.

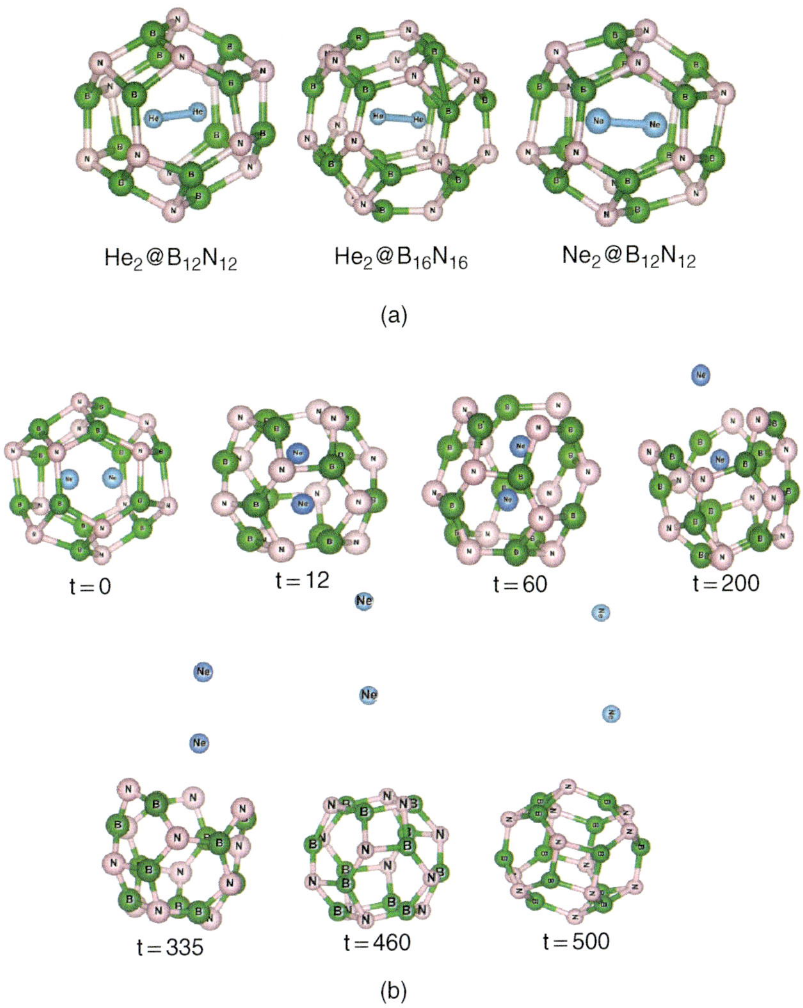

Figure 14.4 (a) The minimum energy geometries of $Ng_2@B_{12}N_{12}$ (Ng = He, Ne) and $He_2@B_{16}N_{16}$ complexes at the M05-2X/6-311+G(d,p) level; (b) The structures of $Ne_2@B_{12}N_{12}$ systems at different time of simulation (t in fs) at 298 K. The He-He distances are 1.306 Å and 1.456 Å in $He_2@B_{12}N_{12}$ and $He_2@B_{16}N_{16}$, respectively, and in $Ne_2@B_{12}N_{12}$ system, the Ne-Ne bond length is 1.608 Å. This figure is taken from [101], © 2014, AIP Publishing.

In another investigation, to check the effect of B-N polarized bonds on Ng atoms in contrast to nonpolarized C-C bonds, Khatua et al. [101] analyzed the structure, stability, and bonding situation of Ng atoms confined within $B_{12}N_{12}$ and $B_{16}N_{16}$ cages through DFT and *ab initio* molecular dynamics study (see Figure 14.4a for the geometries of $Ng_2@B_nN_n$). It is not surprising that Ng encapsulated complexes $Ng@B_nN_n$ (Ng = He-Xe) and $Ng_2@B_nN_n$ (Ng = He, Ne; n = 12, 16) are thermochemically unstable with respect to dissociation and that this instability increases linearly with the increase in the size of Ng. But the molecular dynamics study showed that, except for $Ne_2@B_{12}N_{12}$, the endohedral complexes are kinetically stable against dissociation. Interestingly, molecular dynamics at 298 K on the kinetically unstable $Ne_2@B_{12}N_{12}$ complex supports the "window" mechanism for Ng inclusion-exclusion process where some B-N bonds are broken to open a

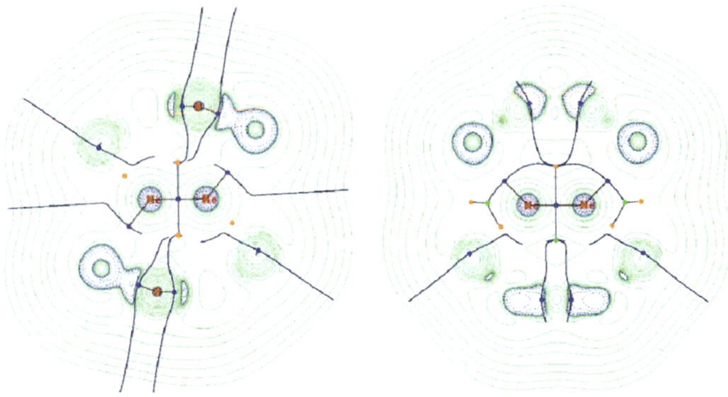

$He_2@B_{12}N_{12}$ $He_2@B_{16}N_{16}$

Figure 14.5 Contour line diagrams of the Laplacian of the electron density of $He_2@B_{12}N_{12}$, and $He_2@B_{16}N_{16}$ systems. This figure is taken from [101], © 2014, AIP Publishing.

temporary window through which both the Ne atoms get released and then the cage gets back its old shape (see Figure 14.4b).

Concentrating only on $He_2@B_nN_n$, the analysis of different properties and bonding indicate that hardness, electrophilicity, polarizability, and HOMO and LUMO energy levels of He_2 in the inclusion complex are significantly different to those in bare cages. Additionally, QTAIM analysis gives maximum electron density path between two He atoms as shown in Figure 14.5 and the corresponding $H(r_c)$ value for He-He bond is −0.06 Hartree/Å^3 in $He_2@B_{12}N_{12}$, whereas in case of $He_2@B_{16}N_{16}$ $H(r_c)$ value is zero. In particular, it can be emphasized that the EDA taking He_2 dimer in the frozen bond distances as those in encapsulated complex gives an orbital interaction ΔE_{orb} value of −5.4 kcal/mol, showing some orbital involvement between two He atoms inside $B_{12}N_{12}$ cage. On the other hand, EDA taking Ng (Ng = He-Xe) or He_2 as one fragment and B_nN_n cage as another fragment implies that there are always significant covalent interactions between Ng or He_2 units and host moiety. Note that in all cases the Pauli repulsion is so strong that it overcompensates the combined attractive effect of attractive Coulombic, orbital, and dispersion interactions. Moreover, during molecular dynamics study, the He_2 was found to move as a single entity. Therefore, all these analyses undoubtedly confirm that at least He_2 inside $B_{12}N_{12}$ cage forms a true chemical bond among themselves.

Pan et al. [102] further investigated the possibility of putting Ng atoms inside B_{40} borospherene [103]. It was noted that B_{40} cavitand is large enough to accommodate He-Rn and He_2-Kr_2. It is remarkable that the interaction between He/Ne and B_{40} is only minutely repulsive in nature (dissociation energy to be −1.8 for He and −7.1 kcal/mol for Ne). As the size of the Ng increases, because of gradually larger repulsion the dissociation energy becomes more negative down the Ng group. Although they are unstable with respect to dissociation, the computed activation free energy barrier is remarkably large, as shown in Figure 14.6. $He_2@B_{40}$ and $Ne_2@B_{40}$ are also kinetically protected enough against dissociation, but the corresponding barrier falls very sharply for $Ar_2@B_{40}$. For $Kr_2@B_{40}$, the corresponding transition state cannot be located.

B_{40} was reported to be a dynamical nanobubble which undergoes continuous conversion between six and seven membered rings [104]. Boron clusters are unique to show such fluxional property because of their ability to form multi-centered delocalized bonds [105–109]. The internal conversion between six and seven membered rings in B_{40} goes via a transition state that involves pushing of one B atom of B_6 ring towards the adjacent B_7 ring. The barrier of bare B_{40} is found

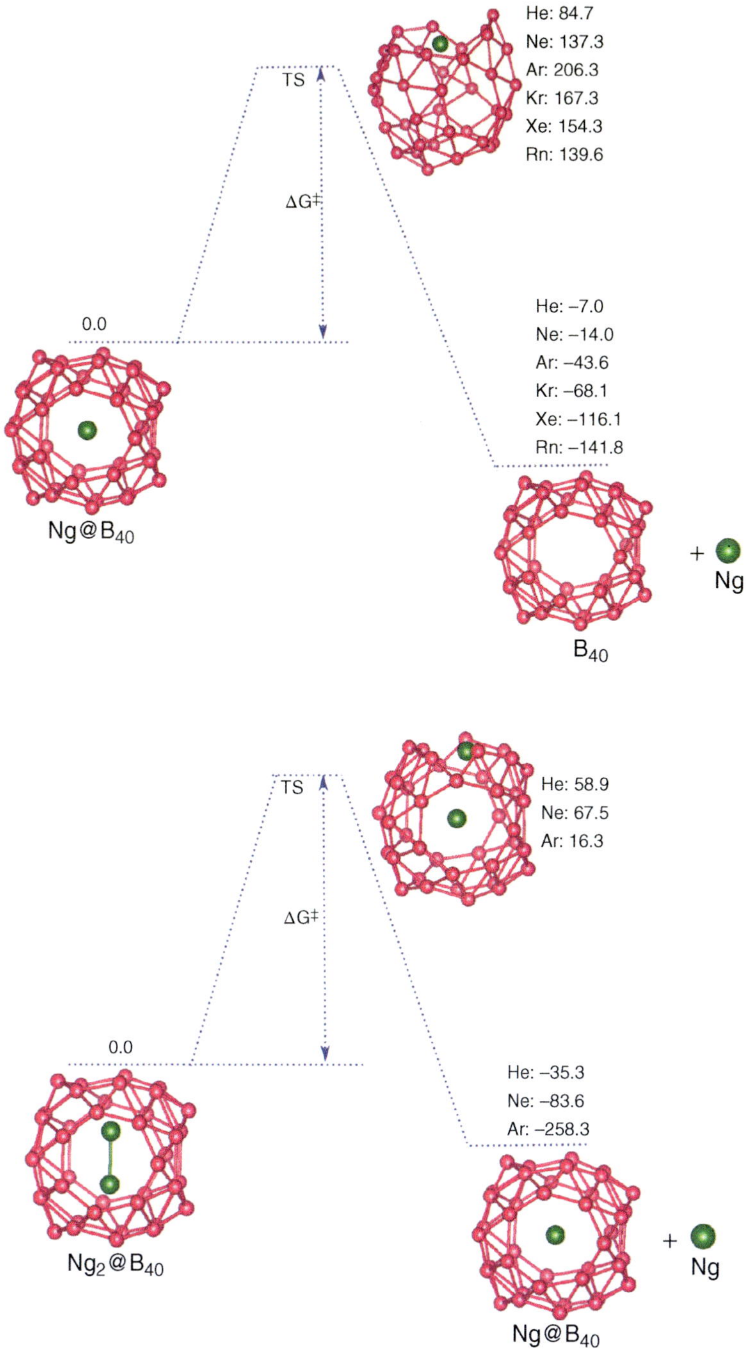

Figure 14.6 Activation free energy barrier ($\Delta G^{\ddagger}$, kcal/mol) for the release of Ng along the B_7 hole computed at the ωB97X-D/def2-TZVP//ωB97X-D/def2-TZVP level. This figure is reproduced from [102], © 2018, Royal Society of Chemistry.

to be 16.4 kcal/mol. The inclusion of Ng inside the cage makes this barrier as 16.0 (Xe) – 18.9 (He) kcal/mol. Therefore, the presence of Ng makes differences in the dynamical phenomenon. Interestingly, the comparative reactivity study taking B_{40} and Xe@B_{40} towards the complexation with $[Fe(\eta^5\text{-}C_5Me_5)]^+$ indicates slightly enhanced reactivity of the latter complex in comparison to the former one. Inspection of the topological descriptors at the Ng-B and Ng-Ng bond critical point implies that Rn-B in Rn@B_{40} and Ar-Ar/B and Kr-Kr/B bonds in Ar_2@B_{40} and Kr_2@B_{40} have some degree of covalent interaction. Ng→B_{40} electron transfer is noted which gradually increases in moving from He to Rn. In fact, for Kr_2 encapsulated case, the resulting complex can be represented as $Kr_2^+B_{40}^-$.

What would the stability and bonding situation of Ng inclusion complexes be if all the sides of the host moiety are not completely closed? Cucurbit[n]urils (CB[n]) are experimentally very well-known to form host-guest complexes [110–115]. The members of this family have a pumpkin-like shape with hydrophobic interior cavity and two sides are opened. The cavity size

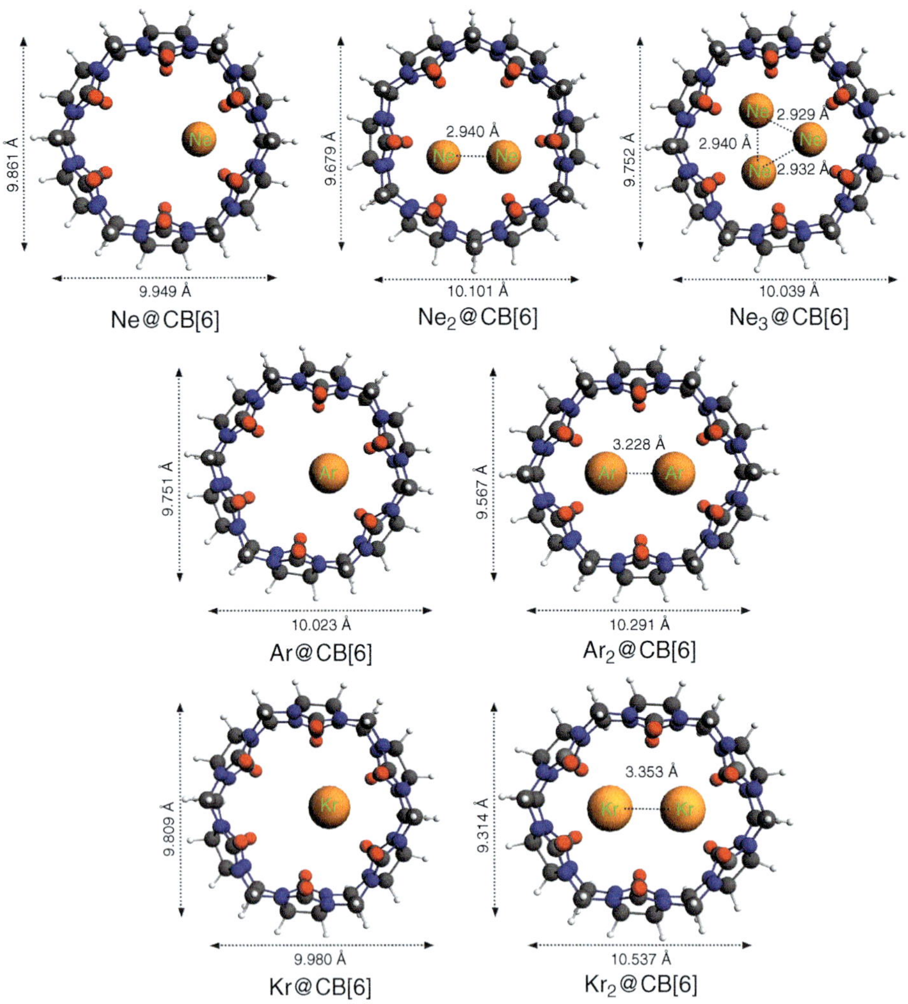

Figure 14.7 The optimized geometries of Ng_n@CB[6] at the ωB97X-D/6-311G(2d,p) level. This figure is reproduced from [116], © 2015, American Chemical Society.

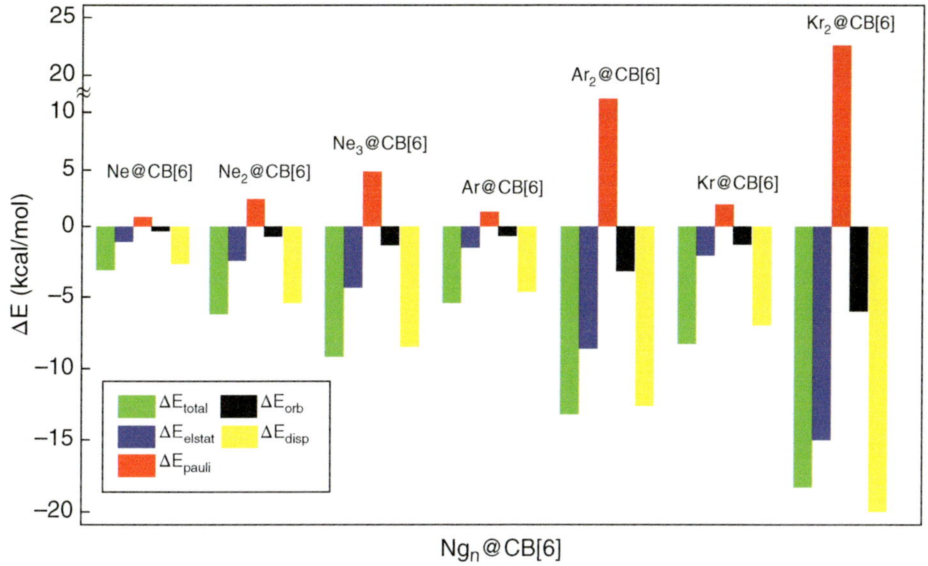

Figure 14.8 EDA results of the Ng_n@CB[6] systems considering Ng_n as one fragment and CB[6] as another fragment at the SSB-D/DZP//ωB97X-D/6-311G(2d,p) level. This figure is reproduced from [116], © 2015, American Chemical Society.

depends on the n value, which represents the number of glycoluril building units. Taking CB[6] as a case, Pan et al. [116] studied the possibility of Ng inclusion complex within it. Calculations showed that CB[6] can accommodate up to three Ne atoms and two Ar and Kr atoms within it (see Figure 14.7). In contrary to the completely caged host, here the Ng atoms are bound in nature as reflected from the positive dissociation energy. Three Ne atoms are bound by a dissociation energy value of 4.3 kcal/mol per Ne atom whereas in the cases of Ar and Kr the first Ng is bound by 4.6 and 7.4 kcal/mol, respectively. However, the second Ng atom is very weakly bound since owing to their larger size, the host moiety needs to undergo significant distortion to accommodate them. He atoms are found to make very weak complexes. The EDA results in Figure 14.8 show that the intrinsic interaction energy is negative in these complexes in which the dispersion interaction has the major contribution followed by electrostatic and orbital interaction. So, in contrast to the previous cases, the presence of two open sides lowers the Pauli repulsion as a result the overall complex becomes stable. Based on QTAIM results, an inspection into the nature of bonding between either two Ng atoms or Ng and a center in host moiety revealed that they are mainly closed-shell type of interaction which corroborates with the EDA data.

14.5 XNgY Insertion Molecule: Confinement in One Direction

Apart from the above cases where cage moiety acts as host and therefore the guest species are confined in all direction, the Ng insertion molecules can also be discussed in the light of the fact that here Ng is forcefully confined between two atoms/groups X and Y. This is a rich family of Ng molecules (mostly Ng = Kr and Xe), which are mostly experimentally realized at low-temperature matrix isolation or predicted to be metastable theoretically [117–119]. HArF is the sole example of an Ar inserted molecule [120], while He and Ne inserted molecules are still waiting for experimental isolation. Similar to the Ng encapsulated complexes, XNgY are only kinetically stable. They are thermochemically unstable with respect to two-body dissociation XNgY → Ng + XY. This is because the insertion of Ng compromises the rather strong X-Y bond strength, which cannot be compensated by the combined strength of X-Ng and Ng-Y bond strength. Therefore, from a computational point of view it is mandatory to check the activation energy barrier for the dissociation XNgY → Ng + XY, which in general involves the X-Ng-Y bending mode. Theoretically Hu and co-workers [121] evaluated the half-life of XNgY molecules depending on the energy barrier height for XNgY → Ng + XY. According to their results, an energy barrier of 6, 13, and 21 kcal/mol would give a half-life in the order of ~10^2 seconds at 100, 200, and 300 K, respectively, for an XNgY molecule. The three-body (3B) dissociation path XNgY → X + Ng + Y can also be sometimes exergonic in nature at room temperature, but this can be stopped by lowering the temperature.

Figure 14.9 depicts some molecules of XNgY type reported theoretically. A thorough *in silico* investigation was devoted by Merino and co-workers [122, 123] to the structure, stability, and bonding situation of HNgY (Y = F, Cl, Br, I, CCH, CN, NC; Ng = Xe, Rn) molecules. Their analysis showed that these molecules might be represented to have the interaction between Ng^+. and $[H\cdots Y]^-$. where they form a polar electron-shared bond. All these molecules are metastable, having a considerable kinetic barrier against the dissociation HNgY → Ng + HY. After that, Pan et al. found that Ng (Xe, Rn) can be inserted within Si-N bond of H_3SiNSi and HSiNSi molecules, forming $H_3SiNgNSi$ and HSiNgNSi (Ng = Xe, Rn), respectively [124]. As expected, the Ng release forming

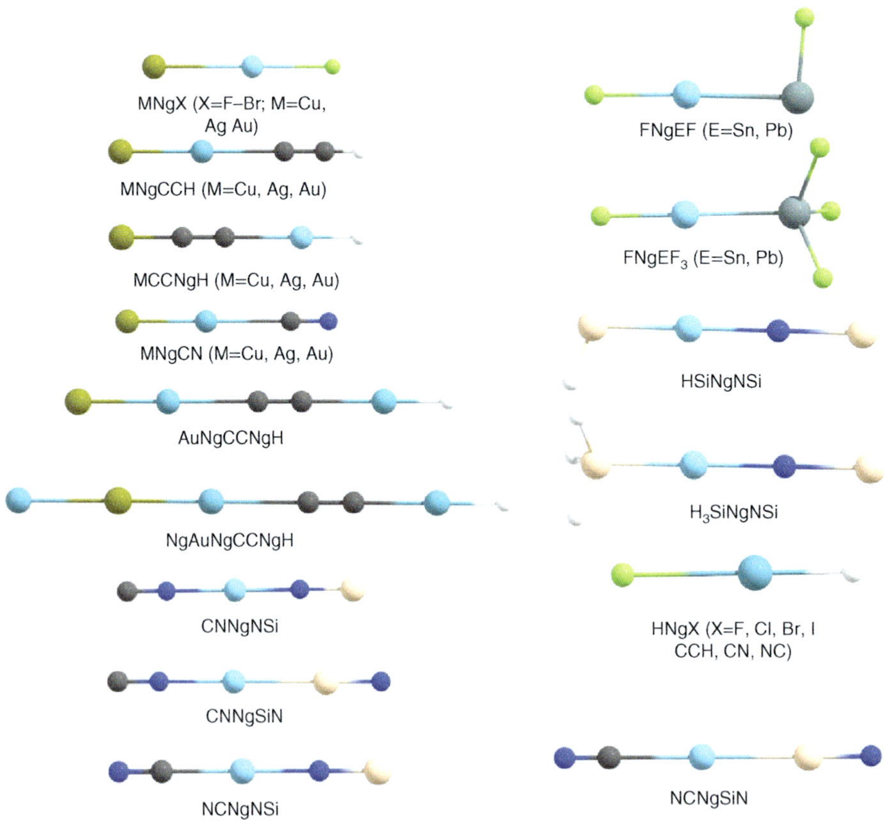

Figure 14.9 The structures of the reported systems of some XNgY type of molecules. This figure is reproduced from [19]. Licensed under CC BY 4.0.

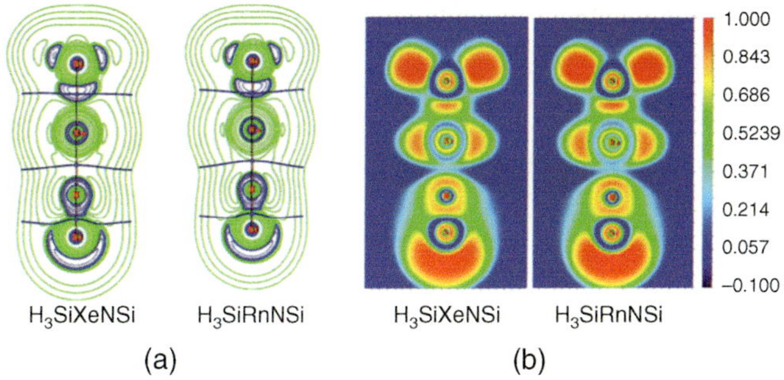

Figure 14.10 Contour plots of (a) Laplacian of electron density and (b) electron localization function of $H_3SiNgNSi$ (Ng = Xe, Rn). This figure is reproduced from [124]. Licensed under CC BY 4.0.

free Ng atom and the parent molecules are highly exergonic in nature at 298 K. Additionally, three-body dissociation producing H_3Si + Ng + NSi or HSi + Ng + NSi are found to be slightly exergonic at room temperature for Ng = Xe, which becomes endergonic with a slight lowering of the temperature. Interestingly, both $\nabla^2\rho(r_c)$ and $H(r_c)$ values are negative for Si-Ng bond along with large electron localization function (ELF), which implies the covalent nature of bonding (see Figure 14.10). The corresponding large Wiberg bond order also corroborates with the covalent notion. The natural charge distribution indicates that these molecules can be better written as $(H_3SiNg)^+(NSi)^-$ and $(HSiNg)^+(NSi)^-$. EDA also supports this assignment where for Si-Ng bond the attractive interaction is dominated by ΔE_{orb}, whereas ΔE_{elstat} contribution is considerable in case of Ng-N bond. The computed kinetic barrier for the two-body dissociation indicates that $H_3SiNgNSi$ is sufficiently protected to be stable in the 250–300 K temperature range, while HSiNgNSi needs a lower temperature range (150–200 K) to be viable.

In another work, Pan et al. reported the first set of molecules, $FNgEF_3$ and FNgEF (E = Sn, Pb; Ng = Kr, Xe, Rn) having E-Ng (E = Sn, Pb) covalent bonds [125]. The $FNgEF_3 \rightarrow Ng + EF_4$ and $FNgEF \rightarrow Ng + EF_2$ are highly spontaneous, but the corresponding energy barrier is noted to be 23.9–49.9 kcal/mol for former cases and 2.2–8.7 kcal/mol in the latter system, with a gradual increase along Kr to Rn. The covalent nature of Ng-E bond and ionic nature of Ng-F can be concluded that from the natural charge distribution, corresponding $H(r_c)$ values, and the EDA computations.

XNgY molecules can generally be represented as $X^+(NgY)^-$. However, an interesting molecule NCNgNSi (Ng = Kr, Xe, Rn) was reported, where both the C-Ng and Ng-N bonds are essentially covalent in nature [126]. Except for NCNgNSi → Ng + CNSiN, all other possible dissociation processes are nonspontaneous in nature. The former process involves a considerable free energy barrier of 25.2–39.3 kcal/mol for Kr to Rn analogues, with a gradual increase along Kr-Rn. The covalent nature for both C-Ng and Ng-N bonds is understood from the sizable WBI value (more than 0.5), negative $H(r_c)$ values, and the EDA computations. Particularly, the latter method is very reliable in this regard where the size of the ΔE_{orb} term can be used as a probe to understand which fragmentation scheme is the best to represent the electronic structure of the whole molecule. The scheme that involves the least charge alteration of the fragments, hence the lowest ΔE_{orb} value is the most suitable one [127, 128]. In the EDA, the consideration of both neutral (electron-shared) and ionic fragments give the lowest ΔE_{orb} value for the former one for both C-Ng and Ng-N bonds. The covalent nature for both the bonds can be further confirmed by the AdNDP analysis that finds a delocalized 3c-2e σ-bond within the C-Ng-N moiety and a totally delocalized 5c-2e σ-bond in these molecules (see Figure 14.11).

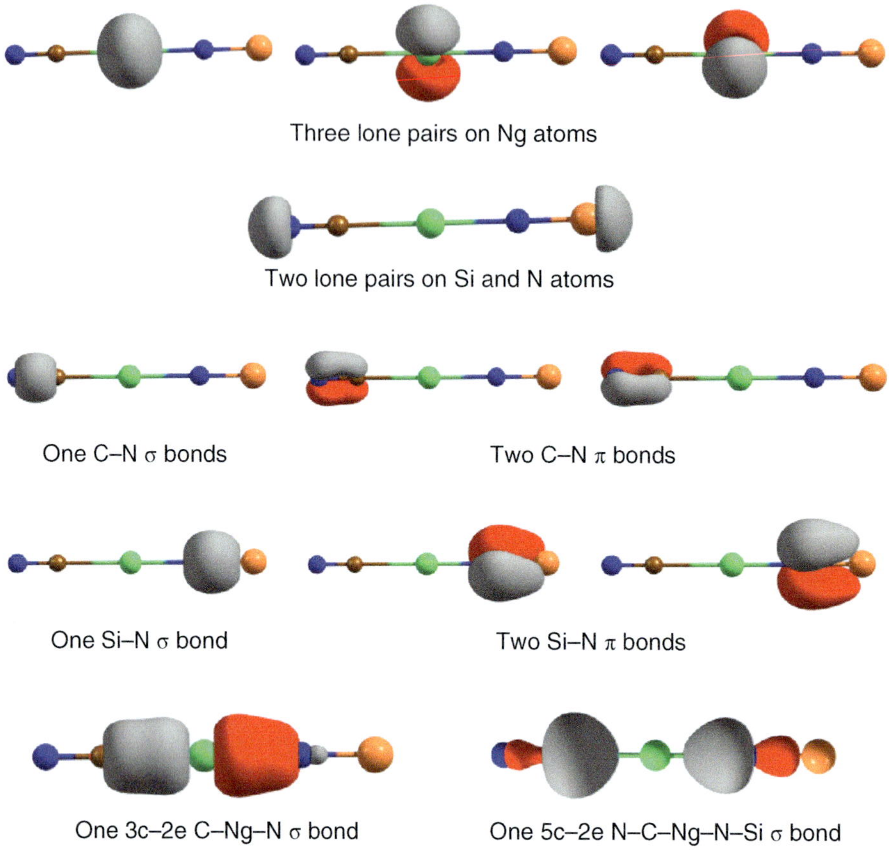

Figure 14.11 The bonding elements recovered by the AdNDP analysis for NCNgNSi (Ng=Kr–Rn) compounds. The occupation numbers in all cases are very closed to 2.00. This figure is reproduced from [126], © 2018, John Wiley & Sons.

Coinage metals (M = Cu, Ag, Au) are found to be effective in forming Ng inserted molecules. It was shown that Ng can be inserted within both M-C and C-H bonds in MCCH; M = Cu, Ag, Au [129, 130]. MNgCCH (Ng = Xe, Rn) represent the first cases to have M-Ng-C bonding unit. While MNgCCH can be expressed as $(MNg)^{+}(CCH)^{-}$ having M-Ng covalent bond, MCCNgH possesses Ng-H covalent bond and the partially covalent C-Ng bond. Furthermore, two Xe inserted analogue, AuXeCCXeH complex and three Xe bound system, XeAuXeCCXeH, might also be viable to be detected in a low-temperature matrix. In a subsequent study, the possibility of Ng insertion in M-C bonds of metal cyanides (MCN) resulting in the formula MNgCN (M = Cu, Ag, Au; Ng = Xe, Rn) is carried out [131]. All these molecules are metastable in nature. These systems show that if an Ng atom is confined between two atoms, it can even form electron-sharing covalent bond(s) with the neighboring atom(s).

14.6 Conclusions

Noble gas (Ng) by virtue of its filled valence electronic shell is inert and reluctant to participate easily with either other elements in the periodic table or between two Ng atoms. A strong polarizing center is required to form a dative bond with Ng where Ng acts as donor. Another way to force Ng atoms to form a chemical bond would be to confine Ng atom(s) within a host cavity. Depending on the size of cavitand and size of the Ng, the degree of covalent bond formation either between Ng and cage centers or between two Ng atoms gets formed. $Xe_2@C_{60}$ is a genuine example where Xe-Xe bond can be confidently termed a covalent bond. Even it does not require a full cage, even forcing an Ng atom to stay within two atoms/groups as in Ng inserted molecules, it is possible to facilitate an electron-sharing covalent bond between Ng and its neighboring centers. Since bonding is a fuzzy concept, it brings a lot of debates. Some criteria support chemical bond, and yet some are not. Sometimes, it also depends on how one particular scientist understands. Therefore, confinement of Ng_2 becomes a playground for the application of different bonding models and each model has their own advantages and limitations which further create debate. This is thus important to enjoy the different flavors and opportunities that confinement can provide. In this tone, the remark of Roald Hoffmann on the concept of the chemical bonding is relevant [132]:

> I think that any rigorous definition of a chemical bond is bound to be impoverishing, leaving one with the comfortable feeling, "yes (no), I have (do not have) a bond," but little else. And yet the concept of a chemical bond, so essential to chemistry and with a venerable history, has life, generating controversy and incredible interest. My advice is this: Push the concept to its limits. Be aware of the different experimental and theoretical measures out there. Accept that at the limits a bond will be a bond by some criteria, may be not others. Respect chemical tradition, relax, and instead of wringing your hands about how terrible it is that this concept cannot be unambiguously defined, have fun with the fuzzy richness of the idea.

Acknowledgements

L.Z. acknowledges the financial support from National Natural Science Foundation of China (Grant No. 21973044), Nanjing Tech University (Grant No. 39837132 and 39837123), and SICAM Fellowship from Jiangsu National Synergetic Innovation Center for Advanced Materials. S.P. thanks Nanjing Tech University for a postdoctoral fellowship.

References

1 Lewis, G.N. (1916). The atom and the molecule. J. Am. Chem. Soc. 38: 762–785.
2 Langmuir, I. (1919). The arrangements of electrons in atoms and molecules. J. Am. Chem. Soc. 41: 868–934.
3 Langmuir, I. (1919). Isomorphism, isosterism and covalence. J. Am. Chem. Soc. 41: 1543–1559.
4 Langmuir, I. (1920). The octet theory of valence and its applications with special reference to organic nitrogen compounds. J. Am. Chem. Soc. 42: 274–292.
5 Langmuir (1921). Types of valence. Science 54: 59–67.
6 Heitler, W. and London, F. (1927). Wechselwirkung Neutraler Atome und Homöopolare Bindung nach der Quantenmechanik. Z. Phys. 44: 455–472.
7 Heisenberg, W. (1925). Über Quantentheoretische Umdeutung Kinematischer und Mechanischer Beziehungen. Z. Phys. 33: 879–893.
8 Schrödinger, E. (1926). Quantisierung als Eigenwertproblem. Ann. Phys. 79: 361–376.

9 Mulliken, R.S. (1932). Electronic structures of polyatomic molecules and valence. II. General considerations. Phys. Rev. 41: 49–71.

10 Hund, F. (1931). Zur Frage der Chemischen Bindung. Z. Phys. 73: 1–30.

11 Lennard-Jones, J.E. (1929). The electronic structure of some diatomic molecules. Trans. Far. Soc. 25: 668–686.

12 Hückel, E. (1931). Quantentheoretische Beiträge zum Benzolproblem. Z. Phys. 70: 204–286.

13 Fukui, K. (1975). Theory of Orientation and Stereoselection. Berlin: Springer Verlag.

14 Woodward, R.B. and Hoffmann, R. (1969). The conservation of orbital symmetry. Angew. Chem., Int. Ed. 8: 781–853.

15 Pauling, L. (1939). The Nature of the Chemical Bond and the Structure of Molecules and Crystals. Ithaca NY: Cornell University Press.

16 Zhao, L., Pan, S., Holzmann, N. et al. (2019). Chemical bonding and bonding models of main-group compounds. Chem. Rev. 119: 8781–8845.

17 Grochla, W. (2007). Atypical compounds of gases, which have been called "noble". Chem. Soc. Rev. 36: 1632–1655.

18 Pan, S., Jana, G., Merino, G., and Chattaraj, P.K. (2019). Noble-noble strong union: gold at its best to make a bond with a noble gas atom. ChemistryOpen 8: 173–187.

19 Saha, R., Jana, G., Pan, S. et al. (2019). How far can one push the noble gases towards bonding? A personal account. Molecules 24: 2933.

20 Sabin, J.R., Brandas, E.J., Eds. Advances in Quantum Chemistry: Theory of Confined Quantum Systems, Academic: Waltham, 2009; Vol. 57.

21 Schettino, V. and Bini, R. (2007). Constraining molecules at the closest approach: chemistry at high pressure. Chem. Soc. Rev. 36: 869–880.

22 Gubbins, K.E., Liu, Y.C., Moore, J.D., and Palmer, J.C. (2011). The role of molecular modelling in confined systems: impact and prospects. Phys. Chem. Chem. Phys. 13: 58–85.

23 Grochala, W., Hoffmann, R., Feng, J., and Ashcroft, N.W. (2007). The chemical imagination at work in very tight places. Angew. Chem. Int. Ed. 46: 3620–3642.

24 Atkins, P.W. and De, P.J. (2006). Atkins' Physical Chemistry. Oxford, U.K.: Oxford University Press.

25 de Groot, S.R. and ten Seldam, C.A. (1946). On the energy levels of a model of the compressed hydrogen atom. Physica 12: 669–682.

26 Sen, K.D., Pupyshev, V.I., and Montgomery, H.E. (2009). Exact relations for confined one-electron systems. Adv. Quantum Chem. 57: 25–77.

27 Buchachenko, A.L. (2001). Compressed atoms. J. Phys. Chem. B 105: 5839–5846.

28 Sarkar, U., Giri, S., and Chattaraj, P.K. (2009). Dirichlet boundary conditions and effect of confinement On chemical reactivity. J. Phys. Chem. A 113: 10759–10766.

29 Popov, A.A., Yang, S., and Dunsch, L. (2013). Endohedral fullerenes. Chem. Rev. 113: 5989–6113.

30 Saha, R., Pan, S., and Chattaraj, P.K. (2017). $NgMCp^+$: Noble gas bound half-sandwich complexes (Ng = He-Rn, M = Be-Ba, Cp = η^5-C_5H_5). J. Phys. Chem. A 121: 3526–3539.

31 Jana, G., Pan, S., and Chattaraj, P.K. (2017). Binding of small gas molecules by metal–bipyridyl monocationic complexes (metal = Cu, Ag, Au) and possible bond activations therein. J. Phys. Chem. A 121: 3803–3817.

32 Pan, S., Saha, R., Mandal, S., and Chattaraj, P.K. (2016). N-heterocyclic carbene, pyridine, isoxazole, furan, noble gas and carbon monoxide supported σ-aromatic cyclic M_3^+ complexes (M = Cu, Ag, Au). Phys. Chem. Chem. Phys. 18: 11661–11676.

33 Pan, S., Ghara, M., Ghosh, S., and Chattaraj, P.K. (2016). Noble gas bound beryllium chromate and beryllium hydrogen phosphate: a comparison with noble gas bound beryllium oxide. RSC Adv. 6: 92786–92794.

34 Saha, R., Pan, S., Mandal, S. et al. (2016). Noble gas supported B_3^+ cluster: formation of strong covalent noble gas–boron bonds. RSC Adv. 6: 78611–78620.

35 Ghara, M., Pan, S., Kumar, A. et al. (2016). A computational study on structure, stability and bonding in noble gas bound metal nitrates, sulfates and carbonates (metal = Cu, Ag, Au). J. Chem. Sci. 128: 1537–1548.

36 Pan, S., Saha, R., Kumar, A. et al. (2016). A noble interaction: an assessment of noble gas binding ability of metal oxides (metal = Cu, Ag, Au). Int. J. Quantum Chem. 116: 1016–1024.

37 Pan, S., Gupta, A., Saha, R. et al. (2015). A coupled-cluster study on the noble gas binding ability of metal cyanides versus metal halides (metal = Cu, Ag, Au). J. Comp. Chem. 36: 2168–2176.

38 Pan, S., Saha, R., and Chattaraj, P.K. (2015). On the stability of noble gas bound 1-tris(pyrazolyl)borate beryllium and magnesium complexes. New J. Chem. 39: 6778–6786.

39 Saha, R., Pan, S., and Chattaraj, P.K. (2015). A comparative study on the noble gas binding ability of BeX clusters (X = SO_4, CO_3, O). J. Phys. Chem. A 119: 6746–6752.

40 Pan, A., Moreno, D., Merino, G., and Chattaraj, P.K. (2014). Stability of the noble gas bound SiH_3^+ clusters. ChemPhysChem 15: 3554–3564.

41 Pan, S., Moreno, D., Cabellos, J.L. et al. (2014). An ab initio study on the stability of $Ng_nBe_2N_2$, $Ng_nBe_3N_2$, and $NgBeSiN_2$ clusters (Ng=He-Rn). ChemPhysChem 15: 2618–2625.

42 Pan, S., Moreno, D., Cabellos, J.L. et al. (2014). In quest of strong Be-Ng bonds among the neutral Ng-Be complexes. J. Phys. Chem. A 118: 487–494.

43 Pan, S., Contreras, M., Romero, J. et al. (2013). $C_5Li_7^+$ and $O_2Li_5^+$ as noble gas trapping agents. Chem. Eur. J. 19: 2322–2329.

44 Pan, S., Jalife, S., Kumar, M. et al. (2013). Structure and stability of $(NG)_nCN_3Be_3^+$ clusters and comparison with $(NG)BeY^{0/+}$ (Ng=Noble Gas and Y=O, S, Se, Te). ChemPhysChem 14: 2511–2517.

45 Pan, S., Jalife, S., Romero, J. et al. (2013). Attractive Xe-Li interaction in Li-decorated clusters. Comput. Theor. Chem. 1021: 62–69.

46 Weiske, T., Böhme, D.K., Hrušák, J. et al. (1991). Endohedral cluster compounds: inclusion of helium within C and C through collision experiments. Angew. Chem. Int. Ed. 30: 884–886.

47 Weiske, T., Hrušák, J., Böhme, D.K., and Schwarz, H. (1991). Formation of endohedral carbon-cluster noble-gas compounds with high-energy bimolecular reactions: $C_{60}He^{n+}$ (n = 1, 2). Chem. Phys. Lett. 186: 459–462.

48 Weiske, T., Wong, T., Krätschmer, W. et al. (1992). The neutralization of HeC in the gas phase: compelling evidence for the existence of an endohedral structure for He@C_{60}. Angew. Chem. Int. Ed. 31: 183–185.

49 Weiske, T., Böhme, D.K., and Schwarz, H. (1991). Injection of helium atoms into doubly and triply charged C_{60} cations. J. Phys. Chem. 95: 8451–8452.

50 Ross, M.M. and Callahan, J.H. (1991). Formation and characterization of $C_{60}He^+$. J. Phys. Chem. 95: 5720–5723.

51 Weiske, T., Hrušák, J., Bohme, D.K., and Schwarz, H. (1992). Endohedral fullerene-noble gas clusters formed with high-energy bimolecular reactions of C_x^{n+} (x = 60, 70; n = 1, 2, 3). Helv. Chim. Acta 75: 79–89.

52 Wong, T., Terlouw, J.K., Weiske, T., and Schwarz, H. (1992). The neutralization—reionization mass spectrum of C_{60}^+. Int. J. Mass. Spectrom. Ion Process. 113: R23–R29.

53 Mowrey, R.C., Ross, M.M., and Callahan, J.H. (1992). Molecular dynamics simulations and experimental studies of the formation of endohedral complexes of buckminsterfullerene. J. Phys. Chem. 96: 4755–4761.

54 Wan, Z., Christian, J.F., and Anderson, S.L. (1992). Ne^{+}+C_{60}: Collision energy and impact parameter dependence for endohedral complex formation, fragmentation, and charge transfer. J. Chem. Phys. 96: 3344–3347.

55 Hrušák, J., Böhme, D.K., Weiske, T., and Schwarz, H. (1992). Ab initio MO calculation on the energy barrier for the penetration of a benzene ring by a helium atom. Model studies for the formation of endohedral He@$C_{60}{}^{+}$ complexes by high-energy bimolecular reactions. Chem. Phys. Lett. 193: 97–100.

56 Christian, J.F., Wan, Z., and Anderson, S.L. (1993). Ne_{+}+C_{60} collisions: The dynamics of charge and energy transfer, fragmentation, and endohedral complex formation. J. Chem. Phys. 99: 3468–3479.

57 Kleiser, R., Sprang, H., Furrer, S. et al. (1993). D: At. Mol. Clusters 28: 89–90.

58 Saunders, M., Jimenez-Vazquez, H.A., Cross, R.J., and Poreda, R.J. (1993). Stable compounds of helium and neon: He@C_{60} and Ne@C_{60}. Science 259: 1428–1430.

59 Saunders, M., Jimenez-Vazquez, H.A., Cross, R.J. et al. (1994). Incorporation of helium, neon, argon, krypton, and xenon into fullerenes using high pressure. J. Am. Chem. Soc. 116: 2193–2194.

60 Shimshi, R., Cross, R.J., and Saunders, M. (1997). Beam implantation: A new method for preparing cage molecules containing atoms at high incorporation levels. J. Am. Chem. Soc. 119: 1163–1164.

61 Peng, R.-F., Chu, S.-J., Huang, Y.-M. et al. (2009). Preparation of He@C_{60} and He_2@C_{60} by an explosive method. J. Mater. Chem. 19: 3602–3605.

62 Murata, M., Murata, Y., and Komatsu, K. (2008). Surgery of fullerenes. Chem. Commun.: 6083–6094.

63 Morinaka, Y., Tanabe, F., Murata, M. et al. (2010). Rational synthesis, enrichment, and ^{13}C NMR spectra of endohedral C_{60} and C_{70} encapsulating a helium atom. Chem. Commun. 46: 4532–4534.

64 Cross, R.J., Saunders, M., and Prinzbach, H. (1999). Putting helium inside dodecahedrane. Org. Lett. 1: 1479–1481.

65 Jalife, S., Arcudia, J., Pan, S., Merino, G. (2020) Endohedral fullerenes confining noble gas atoms. *Chem. Sci.*, submitted.

66 Grimme, S., Antony, J., Ehrlich, S., and Krieg, H.A. (2010). consistent and accurate ab initio parameterization of density functional dispersion correction (DFT-D) for the 94 elements H-Pu. J. Chem. Phys. 132: 154104.

67 Weinhold, F. and Landis, C. (2005). Valency and Bonding, A Natural Bond Orbital Donor – Acceptor Perspective. Cambridge: Cambridge University Press.

68 Zubarev, D.Y. and Boldyrev, A.I. (2008). Developing paradigms of chemical bonding: adaptive natural density partitioning. Phys. Chem. Chem. Phys. 10: 5207–5217.

69 Bader, R.F.W. (1994). Atoms in Molecules: A Quantum Theory. USA: Oxford University Press.

70 Ziegler, T. and Rauk, A. (1977). On the calculation of bonding energies by the Hartree Fock Slater method. Theor. Chim. Acta 46: 1–10.

71 Frisch, M.J., Trucks, G.W., Schlegel, H.B. et al. Gaussian 09, Gaussian, Inc.: Wallingford, CT, 2010.

72 Glendening, E.D., Landis, C.R., and Weinhold, F. (2013). NBO 6.0: Natural bond orbital analysis program. J. Comput. Chem. 34: 1429–1437.

73 Matta, C.F. and Boyd, R.J. (eds.) (2007). The Quantum Theory of Atoms in Molecules. Weinheim: Wiley-VCH.

74 Cremer, D. and Kraka, E. (1984). Chemical bonds without bonding electron density does the difference electron-density analysis suffice for a description of the chemical bond? Angew. Chem. Int. Ed. 23: 627–628.

75 Mitoraj, M. and Michalak, A. (2007). Donor–acceptor properties of ligands from the natural orbitals for chemical valence. Organometallics 26: 6576–6580.

76 ADF2017, SCM, *Theoretical Chemistry*, Vrije Universiteit, Amsterdam, The Netherlands, http://www.scm.com.

77 te Velde, G., Bickelhaupt, F.M., Baerends, E.J. et al. (2001). Chemistry with ADF. J. Comput. Chem. 22: 931–967.

78 Yang, W., Krantz, K.E., Freeman, L.A. et al. (2020). Persistent borafluorene radicals. *Angew. Chem. Int. Ed.*, 59. In: 3850–3854.

79 Saha, R., Pan, S., Merino, G., and Chattaraj, P.K. (2019). Unprecedented bonding situation in viable $E_2(NHB^{Me})_2$ (E = Be, Mg; NHB^{Me} = $(HCN^{Me})_2B$) complexes: neutral E_2 forms a single E-E covalent bond. Angew. Chem. Int. Ed. 58: 8372–8377.

80 Wang, Q., Pan, S., Wu, Y. et al. (2019). Transition-metal chemistry of alkaline-earth elements: the trisbenzene complexes $M(Bz)_3$ (M = Sr, Ba). Angew. Chem. Int. Ed. 58: 17365–17374.

81 Wang, Q., Pan, S., Lei, S. et al. (2019). Octa-coordinated alkaline earth-dinitrogen complexes $M(N_2)_8$ (M = Ca, Sr, Ba). Nat. Commun 10: 3375.

82 Chi, C., Pan, S., Meng, L. et al. (2019). Alkali metal covalent bonding in nickel carbonyl complexes $ENi(CO)_3^-$. Angew. Chem. Int. Ed. 58: 1732–1738.

83 Wu, X., Zhao, L., Jin, J. et al. (2018). Observation of alkaline earth complexes $M(CO)_8$ (M = Ca, Sr, Ba) that mimic transition metals. Science 361: 912–916.

84 Pan, S., Zhao, L., Dias, H.V.R., and Frenking, G. (2018). Bonding in binuclear carbonyl complexes $M_2(CO)_9$ (M = Fe, Ru, Os). Inorg. Chem. 57: 7780–7791.

85 Pecher, L., Pan, S., and Frenking, G. (2019). Chemical bonding in the hexamethylbenzene-SO_2^+ dication. Theor. Chem. Acc. 138: 47.

86 Haaland, A., Shorokhov, D.J., and Tverdova, N.V. (2004). Topological analysis of electron densities: is the presence of an atomic interaction line in an equilibrium geometry a sufficient condition for the existence of a chemical bond? Chem. Eur. J. 10: 4416–4421.

87 Krapp, A. and Frenking, G. (2007). Is this a chemical bond? A theoretical study of $Ng_2@C_{60}$ (Ng=He, Ne, Ar, Kr, Xe). Chem. Eur. J. 13: 8256–8270.

88 Otto, M., Lotz, S.D., and Frenking, G. (1992). Quantum mechanical ab initio studies of the structures and stabilities of halogen azides XN_3 (X= F, Cl, Br, I). Inorg. Chem. 31: 3647.

89 IUPAC Compendium of Chemical Terminology, 2nd edition, 1997. The more recent 2006 edition has the additional sentence: "The principle characteristic of a bond in a molecule is the existence of a region between the nuclei of constant potential contours that allows the potential energy to improve substantially by atomic contraction at the expense of only a small increase in kinetic energy."

90 Bader, R.F.W. and Fang, D.-C.J. (2005). Properties of atoms in molecules: caged atoms and the Ehrenfest force. Chem. Theory Comput. 1: 403–414.

91 Strenalyuk, T. and Haaland, A. (2008). Chemical bonding in the inclusion complex of He in adamantane (He@adam): the origin of the barrier to dissociation. Chem. Eur. J. 14: 10223–10226.

92 Hopffgarten, M.v. and Frenking, G. (2008). Chemical bonding in the inclusion complex of He in adamantane, He@adam: antithesis and complement. Chem. Eur. J. 14: 10227–10231.

93 Poater, J., Solà, M., and Bickelhaupt, F.M. (2006). A model of the chemical bond must be rooted in quantum mechanics, provide insight, and possess predictive power. Chem. Eur. J. 12: 2902–2905.

94 Cerpa, E., Krapp, A., Vela, A., and Merino, G. (2008). The implications of symmetry of the external potential on bond paths. Chem. Eur. J. 14: 10232–10234.

95 Wang, S.-G., Qiu, Y.-X., and Eugen, W.H.S. (2009). Bonding or nonbonding? Description or explanation?"Confinement bonding" of He@ adamantane. Chem. Eur. J. 15: 6032–6040.

96 Osuna, S., Swart, M., and Sola, M. (2009). Reactivity and regioselectivity of noble gas endohedral fullerenes Ng@C_{60} and Ng_2@C_{60} (Ng=He–Xe). Chem. Eur. J. 15: 13111–13123.

97 Nikolaienko, T.Y., Kryachko, E.S., and Dolgonos, G.A. (2018). On the existence of He-He bond in the endohedral fullerene He2@C_{60}. J. Comput. Chem. 39: 1090–1102.

98 Khatua, M., Pan, S., and Chattaraj, P.K. (2014). Movement of Ng_2 molecules confined in a C_{60} cage: an ab initio molecular dynamics study. Chem. Phys. Lett. 610–611: 351–356.

99 Cerpa, E., Krapp, A., Flores-Moreno, R. et al. (2009). Influence of endohedral confinement on the electronic interaction between He atoms: A He_2@$C_{20}H_{20}$ Case Study. Chem. Eur. J. 15: 1985–1990.

100 Jane, G., Chattaraj, P.K. (2020) Effect of substitution on the bonding in He dimer confined within dodecahedrane: a computational study. *J. Comput. Chem.*, submitted.

101 Khatua, M., Pan, S., and Chattaraj, P.K. (2014). Confinement induced binding of noble gas atoms. J. Chem. Phys. 140: 164306.

102 Pan, S., Ghara, M., Kar, S. et al. (2018). Noble gas encapsulated B_{40} cage. Phys. Chem. Chem. Phys. 20: 1953–1963.

103 Zhai, H.-J., Zhao, Y.-F., Li, W.-L. et al. (2014). Observation of an all-boron fullerene. Nat. Chem. 6: 727–731.

104 Martinez-Guajardo, G., Cabellos, J.L., Diaz-Celaya, A. et al. (2015). Dynamical behavior of borospherene: a nanobubble. Sci. Rep. 5: 11287.

105 Moreno, D., Pan, S., Zeonjuk, L.L. et al. (2014). B_{18}^{2-}: a quasi-planar bowl member of the Wankel motor family. Chem. Commun. 50: 8140–8143.

106 Liu, L., Moreno, D., Osorio, E. et al. (2016). Structure and bonding of IrB_{12}^-: converting a rigid boron B12 platelet to a Wankel motor. RSC Adv. 6: 27177–27182.

107 Jalife, S., Liu, L., Pan, S. et al. (2016). Dynamical behavior of boron clusters. Nanoscale 8: 17639–17644.

108 Guo, J.-C., Feng, L.-Y., Wang, Y.-J. et al. (2017). Coaxial triple-layered versus helical $Be_6B_{11}^-$ clusters: dual structural fluxionality and multifold aromaticity. Angew. Chem., Int. Ed. 56: 10174–10177.

109 Saha, R., Kar, S., Pan, S. et al. (2017). A spinning umbrella: carbon monoxide and dinitrogen bound MB_{12}^- clusters (M = Co, Rh, Ir). J. Phys. Chem. A 121: 2971–2979.

110 Lagona, J., Mukhopadhyay, P., Chakrabarti, S., and Isaacs, L. (2005). The cucurbit[n]uril family. A. Angew. Chem. Int. Ed. 44: 4844–4870.

111 Marquez, C., Hudgins, R.R., and Nau, W.M. (2004). Mechanism of host–guest complexation by cucurbituril. J. Am. Chem. Soc. 126: 5806–5816.

112 Pan, S., Mondal, S., and Chattaraj, P.K. (2013). Cucurbiturils as promising hydrogen storage materials: a case study of cucurbit[7]uril. New J. Chem. 37: 2492–2499.

113 Kim, H., Kim, Y., Yoon, M. et al. (2010). Highly selective carbon dioxide sorption in an organic molecular porous material. J. Am. Chem. Soc. 132: 12200–12202.

114 Pan, S., Jana, G., Gupta, A. et al. (2017). Endohedral gas adsorption by cucurbit[7]uril: a theoretical study. Phys. Chem. Chem Phys. 19: 24448–24452.

115 Pan, S., Saha, R., Mandal, S. et al. (2016). Selectivity in gas adsorption by molecular cucurbit[6]uril. J. Phys. Chem. C 120: 13911–13921.

116 Pan, S., Mandal, S., and Chattaraj, P.K. (2015). Cucurbit[6]uril: a possible host for noble gas atoms. J. Phys. Chem. B 119: 10962–10974.

117 Gerber, R. (2004). Formation of novel rare-gas molecules in low-temperature matrices. Annu. Rev. Phys. Chem. 55: 55–78.

118 Pettersson, M., Khriachtchev, L., Lundell, J., Räsänen, M. Noble gas hydride compounds. In: *Inorganic Chemistry in Focus II*, Meyer, G., Naumann, D., Wesemann, L. (Eds.), Wiley-VCH, 2005, pp. 15–34.

119 Khriachtchev, L., Räsänen, M., and Gerber, R.B. (2008). Noble-gas hydrides: new chemistry at low temperatures. Acc. Chem. Res. 42: 183–191.

120 Khriachtchev, L., Pettersson, M., Runeberg, N. et al. (2000). A stable argon compound. Nature 406: 874.

121 Li, T.-H., Liu, Y.-L., Lin, R.-J. et al. (2007). On the stability of noble gas molecules. Chem. Phys. Lett. 434: 38–41.

122 Perez-Peralta, N., Juarez, R., Cerpa, E. et al. (2009). Bonding of xenon hydrides. J. Phys. Chem. A 113: 9700–9706.

123 Juarez, R., Zavala-Oseguera, C., Jimenez-Halla, J.O.C. et al. (2011). Radon hydrides: structure and bonding. Phys. Chem. Chem. Phys. 13: 2222–2227.

124 Pan, S., Saha, R., and Chattaraj, P. (2015). Exploring the nature of silicon-noble gas bonds in $H_3SiNgNSi$ and HSiNgNSi compounds (Ng = Xe, Rn). Int. J. Mol. Sci. 16: 6402–6418.

125 Pan, S., Gupta, A., Mandal, S. et al. (2015). Metastable behavior of noble gas inserted tin and lead fluorides. Phys. Chem. Chem. Phys. 17: 972–982.

126 Pan, S., Jana, G., Ravell, E. et al. (2018). Stable NCNgNSi (Ng=Kr, Xe, Rn) compounds with covalently bound C-Ng-N unit: possible isomerization of NCNSi through the release of the noble gas atom. Chem. Eur. J. 24: 2879–2887.

127 Mondal, S., Cabellos, J.L., Pan, S. et al. (2016) 10-π-Electron arenes à la carte: structure and bonding of the $[E\text{–}(C_nH_n)\text{–}E]^{n-6}$ (E = Ca, Sr, Ba; n = 6–8) complexes. Phys. Chem. Chem. Phys., 18, 11909–11918.

128 Saha, R., Pan, S., Chattaraj, P.K., and Merino, G. (2020). Filling the void: controlled donor-acceptor interaction facilitates the formation of an M-M single bond in the zero oxidation state of M (M = Zn, Cd, Hg). Dalton Trans. 49: 1056–1064.

129 Jana, G., Pan, S., Merino, G., and Chattaraj, P.K. (2018). Noble gas inserted metal acetylides (metal= Cu, Ag, Au). J. Phys. Chem. A 122: 7391–7401.

130 Jana, G., Pan, S., Merino, G., and Chattaraj, P.K. (2017). MNgCCH (M= Cu, Ag, Au; Ng= Xe, Rn): The first set of compounds with M–Ng–C bonding motif. J. Phys. Chem. A 121: 6491–6499.

131 Jana, G., Pan, S., Osorio, E. et al. (2018). Cyanide–isocyanide isomerization: stability and bonding in noble gas inserted metal cyanides (metal= Cu, Ag, Au). Phys. Chem. Chem. Phys. 20: 18491–18502.

132 http://pubs.acs.org/cen/_interactive/explain-bond/index.html.

15

Effect of Both Structural and Electronic Confinements on Interaction, Chemical Reactivity and Properties

Mahesh Kumar Ravva[1], Ravinder Pawar[2], Shyam Vinod Kumar Panneer[3], Venkata Surya Kumar Choutipalli[3], and Venkatesan Subramanian[3]

[1] *Department of Chemistry, SRM University-AP, Andhra Pradesh*
[2] *Department of Chemistry, National Institute of Technology, Warangal*
[3] *Chemical Sciences Division, CSIR-Central Leather Research Institute, Chennai*

15.1 Introduction

Particle in a box is a classic example of quantization through confinement (boundary conditions). The characteristics, ordering, and degeneracy of the energy levels of an unconfined system are significantly affected when subjected to confining potentials. Michels et al. pioneered the work on spherical confinement by enclosing the hydrogen atom inside an impenetrable spherical cavity [1]. Since then, intensive research has been carried out in quantum mechanical confinement, thanks to the advancement of various experimental techniques, which helped in the realization of new areas of research such as quantum computers, nanoelectronics, etc. [2, 3]. In the literature, many theoretical studies have been reported regarding the application of confinement models with hard and soft boundaries on a wide range of systems with different shapes and sizes [4–10].

Endohedral fullerenes are a special class of carbon nanostructures characterized by a fullerene cage encaging atoms, ions, and small molecules in its inner space [11, 12]. The weak interactions are responsible for the stabilization of guest molecules inside the fullerene. As mentioned before, numerous theoretical and experimental investigations have been carried out to understand the geometrical, electrical, and optical properties of endohedral fullerenes [13–15]. Some of the potential applications of these materials include catalysis, energy storage, charge transport, and gas sensors [16]. The exohedral reactivity and regioselectivity of fullerenes are studied by encapsulating various kind of ions [17, 18]. The reactivity between cis-1,3-butadiene and fullerene in the presence and absence of ions are compared [18, 19]. Similar to fullerenes, several investigations have been made to understand the reactivity of sidewalls of carbon nanotubes (CNTs) [20, 21]. Several techniques were developed to functionalize (covalent and non-covalent) sidewall of CNTs. Generally, the outer surface of CNTs is more reactive than the inner surface of CNTs. However, it is shown that the inner space of CNTs provides a new kind of geometrical and electronic confinement to carry out chemical reactions, and thus CNTs are also referred to as nano test tubes [22].

Chemical Reactivity in Confined Systems: Theory, Modelling and Applications,
First Edition. Edited by Pratim Kumar Chattaraj and Debdutta Chakraborty.

15.2 Geometrical Changes in Small Molecules Under Spherical and Cylindrical Confinement

When fullerenes are encapsulated with ions, atoms, or molecules, such fullerenes are called endohedral fullerenes. The space inside the spherical cage is sufficient to host a few atoms/small molecules or their clusters [13, 23–27]. The π-electron delocalization on the surface of fullerene and sphere-like shape of fullerene offers a non-polar spherical confinement. Several experimental and theoretical studies have been carried out to understand the confinement effect by considering fullerene (C_{60}) as a model system. The hollow interior of C_{60} has a unique non-polar confinement environment and atoms/molecules were trapped inside the cage using either physical or chemical methods. This kind of confinement allows the electronic structure of individual molecules to be studied at nano scale. Several strategies have been proposed to introduce guest molecules inside C_{60} [28–32]. A few research groups also proposed "molecular surgery", where fullerene surface is opened with chemical methods. After allowing the guest molecules inside, the surface is closed again with chemical methods. Several experimental and theoretical studies are devoted to characterizing the impact of confinement on geometrical and electronic properties of guest molecules [25, 33, 34]. The bond lengths of diatomic molecules are elongated when they are spatially confined in the fullerene [35]. Several types of diatomic molecules such as H_2, N_2, O_2, F_2, HF, CO, etc., are encapsulated inside fullerene and generally polar molecules are stabilized more compared to non-polar molecules. Recently, Krachmalnicoff et al. demonstrated a synthetic strategy to encapsulate an HF molecule inside C_{60} [35]. Authors have shown that the confined molecules exhibit different vibrational and rotational constants compared to free HF molecule. Ravinder et al. have carried out a detailed analysis on encapsulations of various anions such as F^-, Cl^-, Br^-, OH^-, and CN^- inside fullerenes of different sizes [18]. Results obtained from a combined density functional theory and Born–Oppenheimer molecular dynamics (BOMD) simulations reveal that encapsulation in the larger fullerenes is energetically more favorable than in smaller ones. Even though the small fullerenes do not show much affinity towards these anions, as anions can fluctuate around the center of the cage.

Similar to C_{60}, carbon nanotubes (CNTs) also provide the intriguing opportunity to examine the effect of one-dimensional cylindrical confinement. The extremely high aspect ratio, super hydrophobic graphitic walls, and nanoscale cylindrical interiors of CNTs give rise to many unique applications such as nano filtration, nanoreactors, catalysis, and drug delivery etc. [22, 36–40]. The ultra-efficient transport of gas and water molecules through molecular pipes can be useful in high fluxing separation techniques [41, 42]. The encapsulation of foreign molecules depends on the diameter of CNT, which may change from 0.5–2 nm (in the case of single-walled CNT). The electronic structure and helicity of the tube also show the considerable impact of the host atoms/molecules [43, 44]. Ravinder et al. studied the encapsulation of a fluoride anion in CNT and boron nitride nanotube (BNNT) [20]. It is found that the guest atom could influence the electronic structure of CNT and BNNT upon encapsulation. Furthermore, the results obtained from band structure calculations reveal that band gap of BNNT is significantly altered compared to CNT. This study clearly shows that, upon encapsulation, both host and guest influence each other depending on the nature of interaction between them.

15.3 Hydrogen Bonding Interaction of Small Molecules in the Spherical and Cylindrical Confinement

The behavior of liquids under confinement is different from their corresponding bulk phases [45, 46]. This is mainly due to the local ordering of molecules. Thus, the nature of intermolecular interactions between molecules in the bulk phase is different from that of confined ones. When two to four water molecules were encapsulated inside the fullerene, no hydrogen bonding interactions were observed [47]. The small confined space and highly delocalized π-electron density re-orient the direction of water molecules and they prefer to interact with the surface rather than forming hydrogen bonds. In the case of larger fullerenes, like C_{70}, hydrogen bonding between HF and H_2O molecules are stabilized but they are shortened compared to the hydrogen bonding in free HF-H_2O clusters [48].

Subramanian and co-workers have shown the possibility of formation of one-dimensional water chains inside carbon nanotubes using density functional theory based methods [43]. The optimized geometries of water encapsulated inside nanotubes as obtained at M05-2X/6-31G(d,p) level of theory are shown in Figure 15.1. Interestingly, the shape of the water chains resembles a nanotube which is referred to as ice nanotube (INT) and is made up of one-dimensional stacks of water tetramer. Two hydrogen atoms in each water molecule participated in the hydrogen bonding. Two kinds of hydrogen bonding interactions are observed in ice nanotubes. The first one is the hydrogen bonds that are formed within a tetramer of water. The second type of hydrogen bonds connects these tetramers and leads to one-dimensional stacks of water tubes. The hydrogen bonds which connect the two different rings are weaker compared to the ones formed within the ring. The structural and vibrational features of these clusters resemble that of bulk ice and hence these clusters are called ice nanotubes. Vibrational stretching frequencies of these ice nanotubes are blue-shifted when compared to the vibrational stretching frequencies of bulk ice.

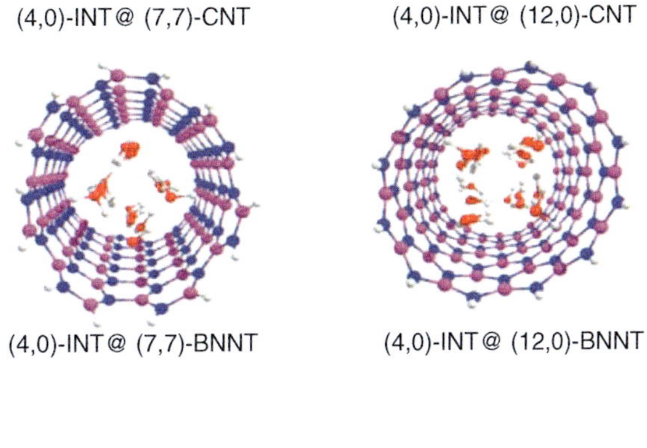

Figure 15.1 Optimized geometries of (4,0)-INT encapsulated inside (7,7) and (12,0) carbon and boron nitride nanotubes.

The role of confinement on the stability of ice nanotubes is gauged by comparing the optimized geometries obtained in the presence and absence of nanotube. Apart from the mechanical confinement, the chemical nature of the tube surface also could impact the hydrogen bonds inside the tube. In order to understand the role of π-electron density on the stability of nanotubes, two different nanotubes with similar diameters made up of carbon and boron nitride nanotubes are considered. It is well known that the π-electron delocalization in CNT is more compared to BNNT. Except in the case of (12,0)-BNNT, in almost all cases the structure is intact (Figure 15.1). It is important to note that the diameter of both armchair (7,7) and zigzag (12,0) nanotubes is nearly the same (9.6 and 9.9 Å, respectively). The interaction energies (IEs) between inner and outer tube are calculated and presented in Table 15.1. The calculated IEs between inner and outer tube reveals that INTs show more binding affinity towards the BNNT compared to the CNT. A significant difference in the IEs is observed between armchair and zigzag tubes.

Table 15.1 Calculated interaction energies (IEs in kcal/mol) between inner and outer tubes using M05-2X and ωB97XD methods employing 6-311++G** and LANL2MB basis sets.

M05-2X		IE of (4,0)-INT
	M05-2X	wB97XD
(7,7)-CNT	−29.84	−54.38
(12,0)-CNT	−39.39	−68.48
(7,7)-BNNT	−46.19	−73.47
(12,0)-BNNT	−44.55	−69.84

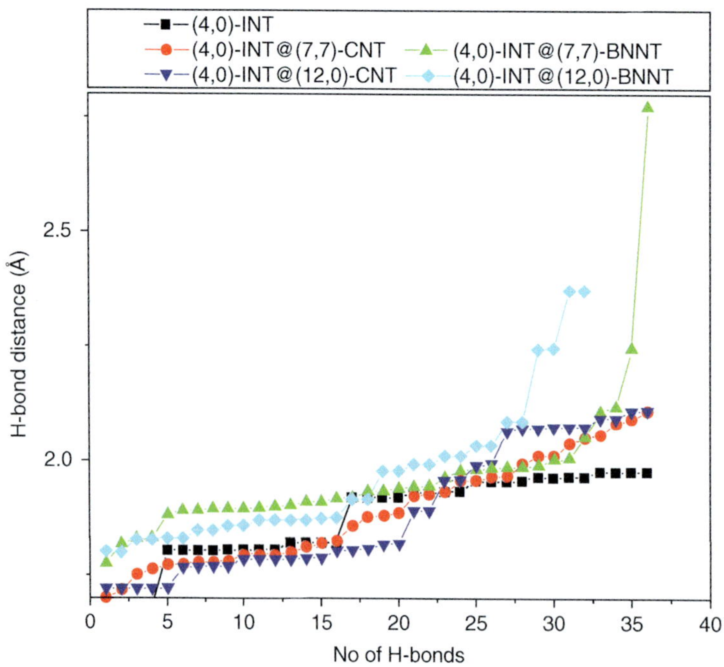

Figure 15.2 The calculated H-bond distances of INT encapsulated inside CNT and BNNT.

Table 15.2 Calculated interaction energies (IEs in kcal/mol) with and without counterpoise correction (CP) of various INTs and HINTs encapsulated CNT and BNNT using M05-2X/6-311++G** level of theory.

	(4,0)-INT	
	CP	No CP
Isolated	−235.48	−260.16
(7,7)-CNT	−226.04	−250.03
(12,0)-CNT	−221.59	−245.68
(7,7)-BNNT	−209.37	−230.97
(12,0)-BNNT	−194.20	−215.00

The geometrical parameters of H-bonds in INTs which are optimized in the presence and absence of nanotube confinement can be compared. It can be seen from Figure 15.2 that the calculated H-bond distances of encapsulated INT reveal that upon confinement all H-bonds are elongated. Particularly, compared to CNT, longer H-bonds are observed in the case of BNNT. The calculated IE of INTs optimized in the presence of nanotube confinement are given in Table 15.2. The IEs of encapsulated tubes are compared with the freely optimized geometries. It is clear that upon encapsulation stability of INT is substantially decreased.

15.4 Spherical and Cylindrical Confinement and Chemical Reactivity

Several synthetic approaches have been developed to access the interiors of CNTs [49–52]. Generally, CNTs are tip-closed when they are synthesized and ex situ approaches such as thermal annealing or acid treatment are necessary to encapsulate the reactants. As a nanoreactor, the reactants enter the tube and products leave the tube once reactions complete inside the tube. In the literature several types of reactions such as syn gas formation, oxidations, halogenations, hydrogenations, and photocatalytic reactions to name a few, are demonstrated inside the CNT [53–58]. All these studies demonstrated that the interior of CNT provide enhanced control over the catalyst and reaction pathway through confinement effects. If one can observe the mechanisms of all these reactions, they can be classified into three major categories. In the first category, reactants enter into CNT and products leave the CNT as soon as the reaction is complete. In the second category, reactants enter into CNT and react with already encapsulated catalyst and leave the tube eventually. In the third category, reactants enter into the tube and undergo phase transition and products may stay inside the tube. Several theoretical and experimental reports have illustrated the above-mentioned mechanisms. The interiors of CNT not only facilitate the reaction but also influence the structural and electronic properties of catalyst particles [59, 60]. The π-electron density of CNT alter the electronic structure of encapsulated catalyst particles. Thus, both the nature of tube and catalyst impact the magnitude and direction of the chemical reactions. Overall, the usefulness of confinement depends on the interaction between reaction intermediates and specific catalyst. We have summarized (in Table 15.3) a broad range of various chemical reactions such as hydrogenation, oxidation, hydroxylation, dehydrogenation, and photo degradation, etc.

Table 15.3 Summary of various chemical reactions carried out within carbon nanotube.

Chemical reaction	Reactant	Catalyst	Result	Ref
Hydrogenation	Benzene	Pd	Higher activity observed	[54]
Hydrogenation	Benzene	Ni	Four to six fold increase in activity observed	[55]
Hydrogenation	Cinnamaldehyde and 1,3-butadiene	Au	90-fold increase	[61]
Oxidation	Methane	Ni	Stable catalyst and high activity	[56]
Oxidation	CO	Au	Lower activity	[62]
Hydroxylation	Benzene	PdV	Higher reaction rates observed	[63]
Oxidative dehydrogenation	Ethylbenzene	CeMn	Enhanced activity	[64]
Photo degradation	Methylene blue	CdS	Activity subtly reduced	[57]

Recently, Khlobystov and co-workers could successfully encapsulate cerium oxide (CeO_2) nanoparticles within hallow carbon nanotubes and also controlled the oxidation of cyclohexene [65]. Raghavachari and co-workers have calculated the energy barriers of a simple SN^2 reaction in the presence and absence of confinement using computational chemistry methods [66]. These calculations reveal that the reaction barrier is enhanced under the nanotube confinement. Authors explained that the high energy barrier is due to large polarizability of CNT. Schlegel and co-workers have also studied the effect of confinement on chemical reactions by considering CNT of different diameter using hybrid density functional theory [67].

In Figure 15.3, the geometries of transition states within various nanotubes such as CNT and BNNTs confinements are shown. Nanotubes with different chiralities are also considered to understand the influence of the diameter on reaction energies. A simple SN^2 reaction of Cl^- exchange is considered as model reaction [68]. The calculated interaction energies between all reactants, TS, and products are energetically stable inside the nanotube. Systematic comparison of reaction energies in the gas phase and inside nanotube reveals that the energy barriers inside the tube are higher than the gas phase. The main reason for the high-energy barrier inside nanotube is due to the presence of nanotube confinement. And the highly delocalized π-electron density introduce polarizability. This reduced the electrostatic interactions between reaction sites which may reduce the kinetics of the Cl^- exchange reaction inside the tube. The trend observed in the reaction barrier for BNNTs varies as gas-phase lower than reaction in pristine tube than reaction in SW defective tube.

15.5 Concluding Remarks

We have summarized our studies on the structural and electronic confinements on interaction, chemical reactivity, and properties of some model systems in this chapter. It is evident that nanotube can function as nano-reactor, which modulates chemical reactions. Study on hydrogen-bonded water molecular clusters inside the nanotube have revealed that confinement

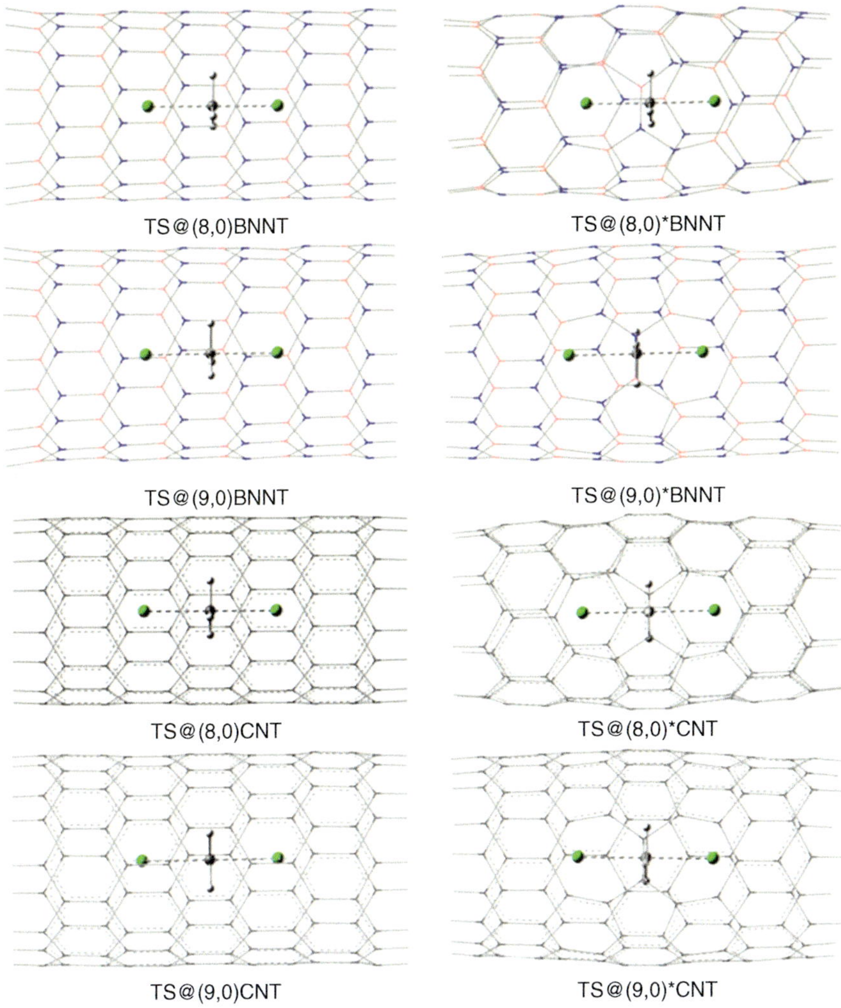

Figure 15.3 Optimized geometries of transition states corresponding to Cl^- exchange reaction in various nanotubes with different chiralities. Reprinted with permission from [57] Copyright 2013 American Chemical Society.

significantly influences the geometry and energetics of hydrogen-bonded clusters. Several experimental and theoretical studies are continuously being pursued to understand the importance of confinement. For example, recent study on the water properties revealed that even though there are only subtle changes in the geometry of H-bonding network due to nano-scale confinement, authors have shown that this re-structuring dramatically changes melting/freezing temperature of water, density, and surface tension [69]. The nano fluid effect is interesting to probe further [70]. It is significant to explore spectroscopic and electrochemical properties under nano-confinement. These studies would pave the way for design and development of energy storage materials [71]. In this interesting perspective article, the author has clearly explained that the two important properties of the electrochemical storage system are electrochemical potential and overpotential, which can be tuned by the effects of nanoscale confinement [71]. Similarly, nanoscale confinement plays a seminal role in biological systems. In a perspective article Hilaire et al. have described the effect of biomolecular crowding due to small molecules molecular constraints, surface packing, and nano-confinement [72].

These studies are highly relevant in the context that cellular environment is highly crowded due to the presence of different macromolecules. It is reported that even small molecular crowding influences the conformational dynamics and stability of proteins. To mimic the cellular environment, specifically engineered nano-confined environment can be designed to unravel interesting biochemical and biophysical questions. In particular, these studies are immensely useful to gain insight into friction, free energy changes, and folding in a crowded cellular environment. In conclusion, the nano-confinement has several applications and prediction and measurement of properties under these nano-confinements provide valuable insights into the development of smart and novel materials.

References

1 A. Michels, J. De Boer, and A. Bijl, Physica 4, 981 (1937).
2 G. Zhang, Science 300, 472 (2003).
3 G. Cuniberti, J. Yi, and M. Porto, Appl. Phys. Lett. 81, 850 (2002).
4 P.O. Fröman, S. Yngve, and N. Fröman, J. Math. Phys. 28, 1813 (1987).
5 S. Yngve, J. Math. Phys. 29, 931 (1988).
6 W. Jaskólski, Phys. Rep. 271, 1 (1996).
7 A.L. Buchachenko, J. Phys. Chem. B 105, 5839 (2001).
8 V.K. Dolmatov, A.S. Baltenkov, J.-P. Connerade, and S.T. Manson, Radiat. Phys. Chem. 70, 417 (2004).
9 J.M.H. Lo, M. Klobukowski, and G.H.F. Diercksen, in *Adv. Quantum Chem.* (Elsevier, 2005), pp. 59–89.
10 J. Gravesen, M. Willatzen, and L.C.L.Y. Voon, Phys. Scr. 72, 105 (2005).
11 A.A. Popov, S. Yang, and L. Dunsch, Chem. Rev. 113, 5989 (2013).
12 M. Yamada, T. Akasaka, and S. Nagase, Acc. Chem. Res. 43, 92 (2010).
13 D.S. Bethune, R.D. Johnson, J.R. Salem, M.S. de Vries, and C.S. Yannoni, Nature 366, 123 (1993).
14 A. Rodríguez-Fortea, A.L. Balch, and J.M. Poblet, Chem. Soc. Rev. 40, 3551 (2011).
15 X. Lu, L. Feng, T. Akasaka, and S. Nagase, Chem. Soc. Rev. 41, 7723 (2012).
16 T. Wang and C. Wang, Acc. Chem. Res. 47, 450 (2014).
17 S. Osuna, M. Swart, and M. Solà, Phys Chem Phys 13, 3585 (2011).
18 P. Ravinder and V. Subramanian, J. Phys. Chem. A 116, 6870 (2012).
19 P. Ravinder and V. Subramanian, Theor. Chem. Acc. 131, 1128 (2012).
20 P. Ravinder, R.M. Kumar, and V. Subramanian, J. Phys. Chem. A 116, 5519 (2012).
21 A.A. Sangolkar and R. Pawar, RSC Adv. 10, 11111 (2020).
22 J. Wang, S. Lu, J. Li, and C. Li, Chem. Commun. 51, 17615 (2015).
23 J. Cioslowski, J. Am. Chem. Soc. 113, 4139 (1991).
24 L. Dunsch and S. Yang, Phys Chem Phys 9, 3067 (2007).
25 K. Kurotobi and Y. Murata, Science 333, 613 (2011).
26 K. Komatsu, Science 307, 238 (2005).
27 A. Krachmalnicoff, M.H. Levitt, and R.J. Whitby, Chem Commun 50, 13037 (2014).
28 M. Murata, Y. Murata, and K. Komatsu, Chem. Commun. 6083 (2008).
29 S.-E. Zhu, F. Li, and G.-W. Wang, Chem. Soc. Rev. 42, 7535 (2013).

30 G.C. Vougioukalakis, M.M. Roubelakis, and M. Orfanopoulos, Chem Soc Rev 39, 817 (2010).
31 L. Gan, D. Yang, Q. Zhang, and H. Huang, Adv. Mater. 22, 1498 (2010).
32 Q. Zhang, T. Pankewitz, S. Liu, W. Klopper, and L. Gan, Angew. Chem. Int. Ed. 49, 9935 (2010).
33 T. Korona, A. Hesselmann, and H. Dodziuk, J. Chem. Theory Comput. 5, 1585 (2009).
34 G.A. Dolgonos and G.H. Peslherbe, Phys Chem Phys 16, 26294 (2014).
35 A. Krachmalnicoff, R. Bounds, S. Mamone, S. Alom, M. Concistrè, B. Meier, K. Kouřil, M.E. Light, M.R. Johnson, S. Rols, A.J. Horsewill, A. Shugai, U. Nagel, T. Rõõm, M. Carravetta, M.H. Levitt, and R.J. Whitby, Nat. Chem. 8, 953 (2016).
36 X. Pan and X. Bao, Acc. Chem. Res. 44, 553 (2011).
37 A. La Torre, M. del C. Giménez-López, M.W. Fay, G.A. Rance, W.A. Solomonsz, T.W. Chamberlain, P.D. Brown, and A.N. Khlobystov, ACS Nano 6, 2000 (2012).
38 W.A. Solomonsz, G.A. Rance, M. Suyetin, A. La Torre, E. Bichoutskaia, and A.N. Khlobystov, Chem. – Eur. J. 18, 13180 (2012).
39 Ihsanullah, Sep. Purif. Technol. 209, 307 (2019).
40 X. Chen, S. Gao, H. Wang, Y. Liu, and Z. Wu, Catal. Commun. 14, 1 (2011).
41 R.H. Tunuguntla, R.Y. Henley, Y.-C. Yao, T.A. Pham, M. Wanunu, and A. Noy, Science 357, 792 (2017).
42 J. Geng, K. Kim, J. Zhang, A. Escalada, R. Tunuguntla, L.R. Comolli, F.I. Allen, A.V. Shnyrova, K.R. Cho, D. Munoz, Y.M. Wang, C.P. Grigoropoulos, C.M. Ajo-Franklin, V.A. Frolov, and A. Noy, Nature 514, 612 (2014).
43 R.M. Kumar, M. Elango, R. Parthasarathi, and V. Subramanian, J. Phys. Chem. A 115, 12841 (2011).
44 J. Shiomi, T. Kimura, and S. Maruyama, J. Phys. Chem. C 111, 12188 (2007).
45 K. Koga, G.T. Gao, H. Tanaka, and X.C. Zeng, Nature 412, 802 (2001).
46 A.I. Kolesnikov, J.-M. Zanotti, C.-K. Loong, P. Thiyagarajan, A.P. Moravsky, R.O. Loutfy, and C.J. Burnham, Phys. Rev. Lett. 93, 035503 (2004).
47 C.N. Ramachandran and N. Sathyamurthy, Chem. Phys. Lett. 410, 348 (2005).
48 R. Zhang, M. Murata, A. Wakamiya, T. Shimoaka, T. Hasegawa, and Y. Murata, Sci. Adv. 3, e1602833 (2017).
49 S.C. Tsang, Y.K. Chen, P.J.F. Harris, and M.L.H. Green, Nature 372, 159 (1994).
50 A.V. Talyzin, I.V. Anoshkin, A.V. Krasheninnikov, R.M. Nieminen, A.G. Nasibulin, H. Jiang, and E.I. Kauppinen, Nano Lett. 11, 4352 (2011).
51 J. Sengupta and C. Jacob, J. Nanoparticle Res. 12, 457 (2010).
52 M.V. Bracamonte, M. Melchionna, A. Stopin, A. Giulani, C. Tavagnacco, Y. Garcia, P. Fornasiero, D. Bonifazi, and M. Prato, Chem. – Eur. J. 21, 12769 (2015).
53 S.A. Miners, G.A. Rance, and A.N. Khlobystov, Chem. Soc. Rev. 45, 4727 (2016).
54 A.M. Zhang, J.L. Dong, Q.H. Xu, H.K. Rhee, and X.L. Li, Catal. Today 93–95, 347 (2004).
55 H. Yang, S. Song, R. Rao, X. Wang, Q. Yu, and A. Zhang, J. Mol. Catal. Chem. 323, 33 (2010).
56 Q. Ma, D. Wang, M. Wu, T. Zhao, Y. Yoneyama, and N. Tsubaki, Fuel 108, 430 (2013).
57 X. Cui, Y. Wang, G. Jiang, Z. Zhao, C. Xu, A. Duan, J. Liu, Y. Wei, and W. Bai, J Mater Chem A 2, 20939 (2014).
58 X. Li, L. Zhang, R.P. Tan, P.-F. Fazzini, T. Hungria, J. Durand, S. Lachaize, W.-H. Sun, M. Respaud, K. Soulantica, and P. Serp, Chem. – Eur. J. 21, 17437 (2015).
59 W. Chen, X. Pan, M.-G. Willinger, D.S. Su, and X. Bao, J. Am. Chem. Soc. 128, 3136 (2006).
60 W. Chen, X. Pan, and X. Bao, J. Am. Chem. Soc. 129, 7421 (2007).
61 X. Zhang, Y.C. Guo, Z. Cheng Zhang, J.S. Gao, and C.M. Xu, J. Catal. 292, 213 (2012).

62 E. Castillejos, R. Chico, R. Bacsa, S. Coco, P. Espinet, M. Pérez-Cadenas, A. Guerrero-Ruiz, I. Rodríguez-Ramos, and P. Serp, Eur. J. Inorg. Chem. 2010, 5096 (2010).

63 H. Zhang, X. Pan, and X. Bao, J. Energy Chem. 22, 251 (2013).

64 C. Li, Y. Sun, and A. Zhang, RSC Adv. 5, 36394 (2015).

65 N. Agasti, M.A. Astle, G.A. Rance, J. Alves Fernandes, J. Dupont, and A.N. Khlobystov, Nano Lett. 20, 1161 (2020).

66 M.D. Halls and K. Raghavachari, Nano Lett. 5, 1861 (2005).

67 J. Phys. Chem. B, 106, 8, 1921–1925 (2002).

68 P. Ravinder and V. Subramanian, J. Phys. Chem. C 117, 5095 (2013).

69 A.W. Knight, N.G. Kalugin, E. Coker, and A.G. Ilgen, Sci. Rep. 9, 8246 (2019).

70 C. Sun, R. Zhou, Z. Zhao, and B. Bai, J. Phys. Chem. Lett. acs. jpclett.0c00591 (2020).

71 M. Fichtner, Phys. Chem. Chem. Phys. 13, 21186 (2011).

72 M.R. Hilaire, R.M. Abaskharon, and F. Gai, J. Phys. Chem. Lett. 6, 2546 (2015).

16

Effect of Confinement on Gas Storage Potential and Catalytic Activity

Debdutta Chakraborty[1], Sukanta Mondal[2], Ranjita Das[3], and Pratim Kumar Chattaraj[4]*

[1] *Department of Chemistry, Katholieke Universiteit, Leuven, Belgium*
[2] *Department of Education, A. M. School of Educational Sciences, Assam University, Silchar, Assam 788011, India*
[3] *Department of Chemistry, University of Houston, Texas 77204, USA*
[4] *Department of Chemistry, Indian Institute of Technology Kharagpur, Kharagpur, India*

16.1 Introduction

Atoms and molecules exhibit fascinating changes in chemical reactivity *vis-à-vis* their corresponding free state counterparts, due to the effect of geometrical confinement. The electronic energy levels, electronic shell filling, orbitals, etc. of a system get affected by the effect of confinement which in turn influences the reactivity and various response properties as compared to the same in the corresponding unconfined system [1–3]. We may consider the case of a particle-in a-box model [4] in order to understand the salient features of the effect of confinement on the changes in electronic energy levels. The energy levels of a free particle appear in the continuum whereas upon introducing the "confining" effects of the box, the energy levels become discrete. Therefore, confinement lies at the heart of quantization. We note that by varying the length of the box, the energy levels of the particle could be changed.

Atoms/molecules encapsulated within organic/inorganic host molecules/materials, represent a "real-life" example of confined systems. It is of interest to analyze the reactivity of atoms and molecules confined within some host molecules as numerous new paradigms *vis-à-vis* the physicochemical properties of the systems under consideration could be unraveled. Therefore, confined quantum systems have been extensively analyzed from both epistemological as well as applied points of view [5–16]. Our focus in this chapter is to highlight the applied aspects of the effect of confinement and how confined regimes could be utilized to solve some challenging real life problems. We highlight two aspects where the effect of confinement could be utilized: (a) sequestration of gas molecules by a particular class of host, clathrate hydrates, and (b) confinement-induced catalysis of some chemical reactions.

We note that encapsulation processes involving organic hosts are significant for various applications as they provide a suitable pathway for sequestering gas molecules. In these host-guest complexes, quantum confinement effects play a crucial role thereby affecting the interaction between the host and the guest. The electron density distribution of the guest gets substantially altered as a result of the geometrical constraint imposed by the host.

*Corresponding author: pkc@chem.iitkgp.ac.in

Chemical Reactivity in Confined Systems: Theory, Modelling and Applications,
First Edition. Edited by Pratim Kumar Chattaraj and Debdutta Chakraborty.

Given the growing need for finding alternative energy materials, hydrogen has emerged as an environment-friendly fuel resource [17–19]. However, finding a suitable hydrogen storage medium at ambient conditions remains a challenge to a large extent. The crucial point herein is whether a potential host can store hydrogen in a reversible manner or not, so that adsorption and desorption can take place in a facile manner [19]. Organic host molecules constitute a suitable candidate for adsorbing hydrogen to this end. Similarly, sequestration of pollutants is crucial for the preservation of our environment. In this context, clathrate hydrates are considered to be very promising candidates for their gas storage potential. Detailed discussions in this regard will be presented in the following sections.

On the other hand, "click chemistry" [20] pioneered by Sharpless constitutes a very useful synthetic methodology. It has been showed that cycloaddition reaction such as Huisgen 1,3 dipolar reaction between alkyne and azide could be catalyzed by introducing copper in the reaction medium [21, 22]. Apart from catalyzing a given cycloaddition reaction, click chemistry can enhance the product conversion as well as the regioselectivity. The presence of copper in the reaction medium, however, limits the applicability of click chemistry in biological medium. Therefore, the quest for finding a copper-free path of catalyzing cycloaddition reactions remains a very interesting research topic. In this context, the impact of geometrical confinement can have significant effect on the kinetic outcome of a given chemical reaction. Mock and co-workers [23, 24] were able to demonstrate (experimentally) that a 1,3 dipolar cycloaddition reaction between 2-azidoethylamine and propargylamine yielding 1-(2-aminoethyl)-4-aminomethyl-1,2,3-triazole, could be accelerated substantially in the presence of an organic host, cucurbit[6]uril (CB[6]). Later on, several experimental studies reported similar findings in the cases of several organic supramolecular hosts [25–27]. Despite these interesting developments, only a few computational studies have been executed pertaining to detailed mechanistic insights until very recently. Crucial question that emerges from these recent developments is whether the presence of any organic or inorganic supramolecular host can promote any given reaction from thermodynamic as well as kinetic points of view. Given the fact that catalyzing a chemical reaction has immense importance as far as various applications are concerned, obtaining an *a priori* guess about the plausible hosts that might facilitate a chemical reaction can be very useful. Therefore, detailed computational investigations using state-of-the-art dispersion-corrected DFT can unravel numerous insights as far as the underlying reaction mechanism is concerned. Detailed discussions in this regard will be presented in the subsequent sections.

16.2 Endohedral Gas Adsorption Inside Clathrate Hydrates

Clathrate hydrate means cage made up of water. However, clathrate hydrates have a long past of research of around 200 years, which has received a substantial attention of the scientific community during the last few decades. It was first reported by Davy to the Royal Society of Chemistry, London, in 1811, that a binary compound containing chlorine and water has a higher melting point in comparison to ice [28]. Later, Faraday found out the composition of the composite reported by Davy to be $Cl_2{*}(H_2O)_{10}$ [29], but till then there was no report on structural integrity of the compound. Thereafter with a one and quarter century silence, in the early 1950s, structure I (*sI*) and structure II (*sII*) clathrate hydrates were identified and characterized [30–32]. Research on cage-like host-guest moiety of water started again. On the basis of size and nature of the guest molecules, there are three different types of crystal structures of gas hydrates: cubic structure I (*sI*), cubic structure II (*sII*), and hexagonal cubic structure (*sH*). These crystals are formed by H-bonded small water cages of pentagonal dodecahedron (5^{12}), tetrakaidecahedron ($5^{12}6^2$), hexakaidecahedron ($5^{12}6^4$), irregular dodecahedron ($4^35^66^3$), and icosahedron ($5^{12}6^8$) geometries [33, 34].

Hydrogen gas storage potential of hydrates is considerable and up to 15.6 mass percentage, which is above the target of DOE [28, 34–36]. Hydrogen hydrates (hydrogen molecule encapsulated hydrates) produce water as the only by-product during the release of hydrogen. Due to this extraordinary feature, clathrate hydrates are not only striking but also cheap hydrogen storage material [28, 35–48]. Existence of hydrogen hydrates was doubtful [46], till the synthesis and characterization of *sII* hydrogen hydrate from a liquid at a pressure of 200 MPa and a temperature of 249 K, by Mao et al. [33]. Though there are a large number of theoretical and experimental studies on the structure of the host-guest system, most reports are still conflicting [44, 47–51]. It was not only hydrogen hydrates got the attention of the researchers, studies on methane hydrates (methane molecule encapsulated hydrates) are also worthwhile and going on worldwide. As methane is a greenhouse gas [52, 53], all those molecular motifs are important which can capture CH_4. Conversely, one can use methane as a fuel for the generation of energy in automotive purpose [54, 55]. Morphologies of hydrogen hydrates and methane hydrates are similar, replacement of hydrogen by methane in the clathrate hydrate host will give methane hydrates. The latter type of compounds, which are also known as fire ice have received more interest in the last 20 years. 184 liters of methane gas can be stored in 1 liter of methane hydrate solid [33, 56]. Seo and Lee et al. studied the kinetics of hydrate formation and dissociation in CH_4/H_2O and $CH_4/C_3H_8/H_2O$ systems and reported that the order of cage formation and the dissociation rate is: $CH_4@5^{12}$ in structure I and II $\gg$ $CH_4@5^{12}6^4$ in structure II > $C_3H_8@5^{12}6^4$ in structure II [57]. A pyridinium-based ionic-long-alkyl-chain-pyridinium-bromide (C12PyBr) is reported recently as a low-dosage-hydrate inhibitor (LDHI) [58]. It is noted that only at 0.1 wt.% inhibitor concentration the inhibition is maximum at a subcooling of 11.2 K, interestingly, which is one order of magnitude lower than that used with Inhibex 101TM [58]. It is reported that in mixed CO-N_2 hydrates the CO sequestration increases when the concentration of N_2 in the gas-hydrate mixture increases; interestingly, this is associated with a structural change of the hydrate crystal from *sI* to *sII* [59]. In a microscale measurement of xenon hydrate nucleation and growth, Sum et al. disclosed the activation energy of hydrate formation (71.11 kJ/mol) and the diffusivity of xenon (ranges from 2.8 $\times 10^{-15}$ to 4.3×10^{-14} m^2/s) for the first time. To achieve more insight and develop comprehensive understanding of inert gas hydrates such data are invaluable [60]. Nowadays research on clathrate hydrates has become significant enough not only from the viewpoint of energy resources but also from the threat of climate change and natural hazards (like tsunami), as well as to enhance knowledge on the gas molecules confined in water clathrate [33, 61, 62]. Recently, in a review article Nguyen et al. discussed gas hydrate formation in confined spaces [63]. Particularly, they discussed that gas hydrate formation can be accelerated by open hydrophobic solids due to tetrahedral ordering of water and increased density of gases at the solid-water interfaces, whereas the reverse phenomenon occurs by open hydrophilic solids due to the development of distorted water assembly and a diminution in gas concentration at the solid-water interface [63]. The effect of confinement on gas hydrate formation is clear from another recent report by Yazaydin et al.; principally, they perceived clear evidence of the effect of confinement when methane hydrate nucleation took place at the center of the silica pore at a very low pressure in comparison to the growth of methane hydrate in the bulk [64]. The development of convex-shaped methane nanobubble is stated [64]. In absence of any confinement, it is reported that growing pressure primes to unvarying mixing between methane and water, furthermore the mixing is convoyed by a subtle improvement of the polarization of involved methane molecule [65]. Recrystallization of Ar hydrate in a confined environment, sapphire anvil cell, is studied by Bini et al. and their findings reveal that the local morphology of the crystal realized in 1 – 1.5 h and the process gets stabilized after 10 h [66]. Magnetic characteristics of oxygen encapsulated clathrate hydrates (O_2@CHs) have been explored by Bu et al. Owing to the unique magnetic properties and characters, O_2@CHs are mentioned as prom**ising** structural units for novel functional icy nanomaterials [67]. A recent study on selected type and size of He@hydrate frameworks put forward the fact of inadequacy of density functional theory (DFT) methods in describing the noncovalent interactions in concert with a relevant role in the formation of clathrates; moreover, it is added that such scantiness in the functionals is ar**ising** as they do not explicitly describe the dispersion effects [68].

In order to understand the small gas molecule trapping ability of water clathrates and associated change in the stability of both the host and guest systems, systematic studies were carried out, starting from the single H_2, CH_4, Ng (He, Ne, Ar) molecule to their clustered counterparts. Moreover, multiscale computational studies were done starting from the classical simulation to dispersion corrected density functional theory studies, so as to understand the structure, stability, and dynamics of these gas molecule encapsulated cage hydrates as well as their possible doped analogues.

16.3 Hydrogen Hydrates

Classical simulations were done using the LAMMPS package [69]. In modelling the hydrogen hydrates with a varied number of H_2 we had used the extended SPC water model [70, 71]. To resemble H_2, an imaginary sphere without any atomic charge was taken for the study [72]. In order to calculate the interaction potential of unlike atoms the Lorentz-Berthelot rule [47, 73] was used. An orthogonal box of size 68 Å × 68 Å × 68 Å was taken, containing 9936 water molecules. Our modelled system was mimicked with the coarse-grained model of Jacobson et al. [75]. Periodic boundary conditions were used for the entire classical molecular dynamics study. Interactions between the component molecules were calculated within a cut-off distance of 10.0 Å. 1.0 fs time step was used for the integrations of the equations of motion with velocity Verlet algorithm. Nose-Hoover thermostat and barostat were employed. To freeze the bond length and bond angle of water molecules and to take care of the long-range Coulombic interactions a fix shake algorithm and a PPPM solver with a precision of 1×10^{-4} were used, respectively. The study revealed the stability gain of the *sI* hydrates in the presences of encapsulated molecular hydrogen. Investigation of obtained radial distribution functions of hydrogen hydrates at different time steps revealed that within the simulated time scale the hydrogen hydrate cage was not ruptured. Interaction energy for the 50% occupancy of the $5^{12}6^2$ channels of *sI* clathrate hydrate by H_2 (see Figure 16.1) becomes negative which thereby indicates metastability of the modelled hydrogen hydrates [76]. In the tetrakaidecahedron cavities a total of 648 hydrogen molecules are confined.

DFT and dispersion corrected DFT studies of the hydrogen hydrates, methane hydrates, as well as noble gas hydrates (noble gas atom encapsulated hydrates) were also done. In doing so, mainly we used B3LYP/6-31G(d), B3LYP/6-311+G(d,p), DFT-D-B3LYP/6-311+G(d,p), and ωB97X-D/6-311+G(d,p) methods [51, 77–80]. In order to understand the kinetic stability of studied gas hydrates and their HF doped analogues, *ab initio* molecular dynamics simulation [81] using Atom Centered Density Matrix Propagation (ADMP) [82–84] technique was done using Gaussian 09 program package [85]. *Ab initio* simulation was performed starting from the coordinates of energy minima forms at different level of theories [77–80]. Different temperatures were considered to check the kinetic stability. In all the cases initial nuclear kinetic energies of the hydrate structures were generated using Boltzmann distribution. Use of a velocity scaling

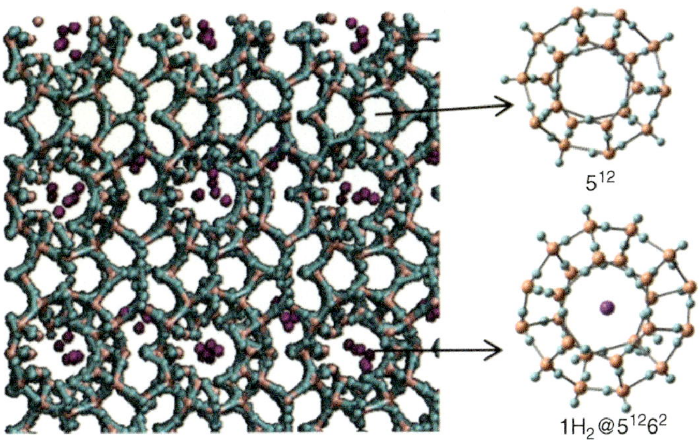

Figure 16.1 A portion of the metastable structure of the *sI* clathrate hydrate which holds a total of 648 H_2 molecules placed only in the $5^{12}6^2$ channels.

thermostat was obvious to maintain the temperature throughout the simulation. A default random number generator seed as implemented in G09 was employed to initiate the initial mass-weighted Cartesian velocity. All simulations were run up to 500 fs. All the DFT computations were done using Gaussian 09 program package [85].

Before doing any rigorous calculation, first of all we checked the stable position of the gas molecules in the vicinity and inside of the water clathrate as well. It was noted that there were two stationary points, one at a point 7 Å away from the center of the cage and another at the center, respectively. But H_2 located at the centroid of the water cage corresponds to a minimum on the potential energy surface while the other structure stands with two imaginary frequencies. Hence, it was obvious to conclude that hydrogen molecule would be present inside the cage instead of being entombed outside the cage [51]. Henceforth, we continued our study with the endohedral encapsulation of different gas molecules. In the assessment of hydrogen hydrates, computed interaction energy, reaction enthalpy, hardness, and electrophilicity values revealed that 5^{12} and $5^{12}6^2$ may store up to two hydrogen molecules and $5^{12}6^8$ may reserve up to six hydrogen molecules [51]. We modeled HF-doped clathrate hydrates by the substitution of one H_2O molecule in the water assembly by one HF molecule [77]. It was found that HF doping increases the stability of clathrate hydrates keeping structural integrity and shape unchanged. From the energetics and conceptual DFT studies we noted that the feasibility of HF doping in 5^{12}, $5^{12}6^2$, $5^{12}6^8$, $5^{12}6^4$, and $4^35^66^3$, as well as for $CH_4@5^{12}$, $CH_4@5^{12}6^8$, $CH_4@5^{12}6^4$, and $CH_4@4^35^66^3$ systems [77, 78]. Moreover, it was also found that HF doping does not alter the size of the clathrate at all significantly [77]. It was pointed out that two hydrogen molecules can be stored in $HF5^{12}$, $HF5^{12}6^2$ systems, whereas the $5^{12}6^8$ clathrate can encapsulate up to six hydrogen molecules (see Figure 16.2) [77]. *Ab initio* molecular dynamics simulation revealed that $HF5^{12}$ and $HF5^{12}6^8$ systems keep on stable up to 500 fs at 200 K, whereas formation of small water cluster was marked from 125 fs in $HF5^{12}6^2$. Significant increment in the hydrogen encapsulation ability of clathrate hydrates due to HF doping was clearly understood. Further, the simulation result revealed that HF-doped hydrogen hydrates, $2H_2@HF5^{12}$, $2H_2@HF5^{12}6^2$, and $6H_2@HF5^{12}6^8$ remain stable up to 500 fs at 200 K. Liberation of one H_2 molecule was pointed out at 298 K from the $6H_2@HF5^{12}6^8$ without any variation in the water molecules and HF assembly. Thus, the kinetic stability of HF doped hydrogen hydrates is more than that of the undoped analogues [77].

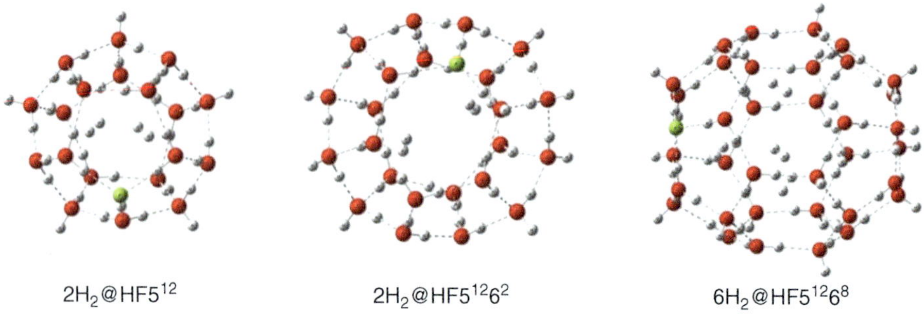

Figure 16.2 Hydrogen encapsulated forms of $HF5^{12}$, $HF5^{12}6^2$, and $HF5^{12}6^8$ at DFT-D-B3LYP/6-311+G(d,p) level of theory. (This figure is reproduced from the work of Chattaraj et al., © 2013, American Chemical Society.)

16.4 Methane Hydrates

A thorough investigation on the optimized geometries of methane hydrates and their HF doped analogues led to a conclusion that the location of methane inside the $5^{12}6^2$ and $HF5^{12}6^2$ results in distorted $CH_4@5^{12}6^2$ and $CH_4@HF5^{12}6^2$, respectively [78], because all thinkable optimizations with different locations of CH_4 molecule inside the water clathrate led to stationary points with imaginary frequencies. Computation via modification of geometry following the displacement vector of imaginary vibration provided their distorted forms. One four-membered and one three-membered H_2O assemblies were noted in the final form of $CH_4@5^{12}6^2$ and $CH_4@HF5^{12}6^2$ (Figure 16.3).

Optimized forms of methane hydrates are provided in Figure 16.4 to help readers understand the two different locations of CH_4 molecule, particularly, dodecahedron and tetrakaidecahedron cavities in sI crystal. Our work corroborates with experimental results showing that in between $5^{12}6^2$ and 5^{12}, methane prefers to get encapsulated by the 5^{12} cage [78, 80, 86]. Particularly, that was noted at the beginning of the nucleation of sI methane hydrates. A likely reason could be the

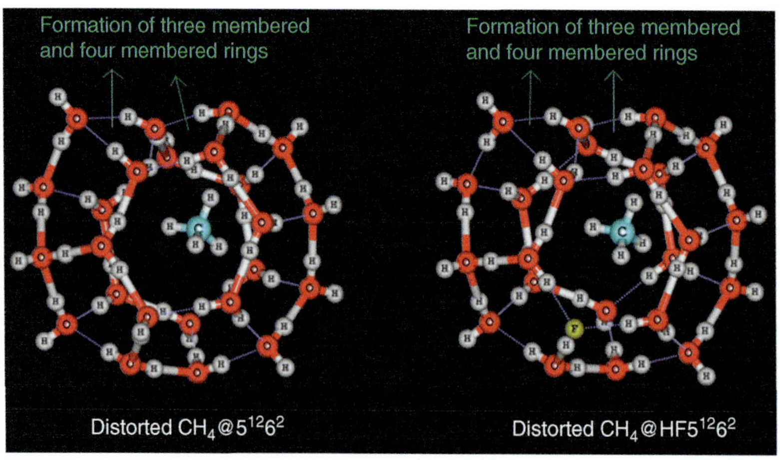

Figure 16.3 Distorted $CH_4@5^{12}6^2$ and $CH_4@HF5^{12}6^2$ methane hydrate forms where three- and four-membered sites are shown. (This figure is reproduced from the work of Chattaraj et al. © 2013, Elsevier.)

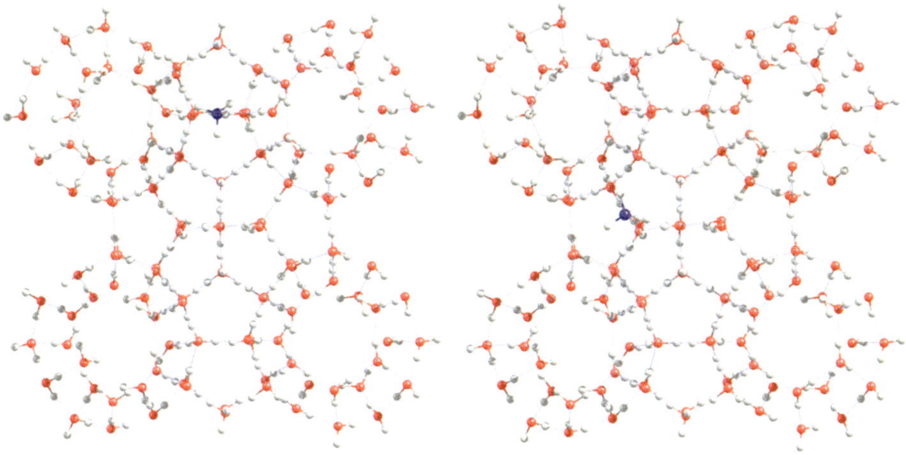

Figure 16.4 ωB97X-D/6-311+G(d,p) structures of sI methane hydrates. CH_4 molecule is at 5^{12} (a) and $5^{12}6^2$ (b) cavities. (This figure is reproduced from the work of Chattaraj et al., © 2018, Elsevier.)

selective host-guest organization in the preliminary stage of *sI*-methane hydrate cluster formation [80, 86]. Geometry and relative stability of small $(H_2O)_5$ and $(H_2O)_6$ clusters present in the clathrate wall, presence of CH_4 guest, integrity, and cavity radii of $(H_2O)_{20}$ and $(H_2O)_{24}$, as well as the weak van der Walls type of forces, particularly dispersion interaction, are major factors responsible for the initial formation of methane encapsulated 5^{12} cavities over $5^{12}6^2$ [80].

16.5 Noble Gas Hydrates

Studies on noble gas hydrates were done considering only He, Ne, and Ar encapsulation in 5^{12}, $5^{12}6^8$, and their HF doped analogues [79]. It was found that 5^{12} and its HF doped analogue can enclathrate up to five helium atoms, three neon atoms, and two argon atoms [79]. However, the 5^{12} and $HF5^{12}$ clathrates can encapsulate only one of each He, Ne, or Ar atom with favorable thermodynamics. In comparison to pentagonal dodecahedron the icosahedron clathrate can host up to nine He atoms whereas its HF doped analogue can enclathrate a maximum of ten He atoms. On the other hand, $5^{12}6^8$ and its HF doped analogue can encage a maximum of six Ne or Ar atoms. It was found that HF doping enhances the noble gas trapping ability, particularly when the size of the atom is large in comparison to other guest noble gas atoms as well as with the decreasing size of the host water cage [79]. Moreover, it was noted that encapsulation of one Ng atom is most favorable due to HF doping. Electron density analysis revealed that the interactions between Ng atoms and the constituent of the HF doped hydrate cage wall are purely noncovalent in nature. In particular, noncovalent nature found in the Ng–O, Ng–F, and Ng–Ng interactions [79]. Dynamics study revealed that at 298 K the pentagonal dodecahedron clathrate and its HF doped analogue can encage one He, one Ne, or one Ar atom up to 500 fs [79]. Whereas the $5^{12}6^8$ and its HF doped analogue can host up to eight and ten He atoms at 225 and 298 K, respectively, up to 500 fs. Although, in the case of HF doped analogue slight distortion in the cage wall was noted. Both the $5^{12}6^8$ and $HF5^{12}6^8$ can host up to six Ne atoms at 225 and 150 K temperature, respectively, up to 500 fs [79]. Although, it was noted that $5^{12}6^8$ and $HF5^{12}6^8$ can host six Ar atoms up to 500 fs, but at 298 K clathrate wall integrity burst in to the formation of four-membered and three- membered water assemblies [79].

16.6 Confinement Induced Catalysis of Some Chemical Reactions

Reaction cavities are considered to play an important role in accelerating the rate of enzyme catalyzed reactions. Enzyme catalyzed reactions are highly regioselective. However, the mechanistic details of enzyme catalyzed reactions are not very well understood. It is generally believed that favorable conformation between the reacting moieties upon encapsulation inside the cavities, plays an important role in the enzyme catalyzed reactions [87]. It is suggested that stabilization of the transition state inside the cavity also plays an important role in accelerating enzyme-catalyzed reactions.

Mock and co-workers showed that by introducing CB[6] into the reaction medium, the 1,3 dipolar cycloaddition reaction between alkyne (substrate1) and alkyl azide (substrate2) could be accelerated [23, 24]. The product (1,4 disubstituted triazole) was formed in a regioselective manner. The experiment conducted by Mock and co-workers represents an example of artificial enzyme-mimetic reaction. Mock and co-workers argued that CB[6] at first randomly binds both substrates to produce 1:1 host-guest system, substrate1/substrate2@CB[6]. Both of these 1:1 complexes can bind a second substrate, yielding a ternary complex. The ternary complex can subsequently produce the product triazole. The rate-limiting step was argued by Mock and co-workers to be the release of the product from the host. The reaction mechanism, however, becomes complex due to the substrate inhibition of the host moiety. Therein, the 1:1 complex such as substrate1@CB[6] can bind another substrate1, thereby leading to formation of a homoternary complex instead of a heteroternary complex. Therefore, the possibility of the intended cycloaddition reaction gets hindered. Mock's group subsequently analyzed the kinetic outcome of a similar reaction in presence of bulkier substituent group within the substrates. In order to provide a rationale behind the observed features, Mock and co-workers compared the binding strengths of the binary with that of the ternary complexes. It was believed that CB[6] helps in increasing the local concentration of the reactants. The cavity of CB[6] can also facilitate the formation of the ternary complex in an optimum manner, thereby accelerating the rate of the reaction as compared to the free state reaction. Stoddart's group [26] showed that Mock and co-workers' rationale behind the catalytic role of CB[6] was indeed correct. They undertook experiments where the same substrates utilized by Mock et al. were employed in the presence of larger homologues of CB family, namely CB[7] and CB[8]. Even though CB[7]/CB[8] can form ternary complexes with the reactants, no triazole formation was noted. Therefore, the crucial role of the confining effect of CB[6] cavity was demonstrated. Despite the limitations of this path in view of the diminished catalytic turnover *vis-à-vis* product inhibition, these developments were further explored by several researchers. We highlight a few representative examples in this regard [88–94]. Sanders et al. used a porphyrin trimer in order to catalyze a Diels–Alder reaction. Bols et al. employed a bridged cyclodextrin moiety in order to facilitate the oxidation of benzyl alcohol. Rebek et al. utilized many molecular cavities which were used to catalyze reactions. Raymond et al. showed that Nazarov cyclization reaction could be catalyzed by a $Ga_4L_6^{12-}$ self-assembled nanocage. New reaction schemes were proposed by Raymond et al. in order to address the catalytic turnover issue. In addition to the thermal reactions, several photochemical reactions were also shown to get accelerated by employing several hosts.

Computational chemistry can provide additional insights into the rationale behind the observed experimental features. In most of the cases under consideration, the host and the guests interact in a non-covalent manner. Therefore, dispersion-corrected DFT could be employed in order to model the processes mimicking confinement induced reactions. Due to the large system size of the supramolecular hosts, utilization of *ab-initio* wave function **based** methodologies become prohibitive, thereby rendering DFT as a method of choice. These host-guest complexes can attain many conformations. However, given the large size of the systems under consideration, a comprehensive exploration of the concerned potential energy surfaces (PES) becomes computationally demanding even at the DFT level of theory. The relative orientation of the reactants inside the cavitands plays an important role in determining the kinetic outcome of the reaction. Therefore, it is necessary to explore the PES of the host-guest complexes in a careful and rigorous way.

In order to provide insights into the results obtained by Mock et al.'s experiments, Maseras and co-workers performed DFT-based calculations [95, 96]. Their analysis revealed important insights into the 1,3 dipolar cycloaddition reaction occurring inside CB[6]. They demonstrated that the rate limiting step in the catalytic cycle is the release of the product from the host. They had also accounted for the observed regioselectivity in the concerned reaction. Li and co-workers [97] employed Monte Carlo simulations and DFT computations in order to elucidate the reaction mechanism of the reaction between p-quinone and cyclohexadiene inside a self-assembled capsule. They demonstrated that the encapsulation of the reactants into the capsule is driven by a non-covalent interaction between the host and the guests. They were able to show that the relative position and conformation of the reactants inside the host leads to various reactivity patterns due to the variation in host-guest interaction. A crucial point that emerged from this study is the ability of the host in bringing down the energy barrier associated with the cycloaddition reaction. Calvaresi and co-workers [98] employed a QM/MM investigation in order to understand the kinetics of the bromination reaction of N-phenylacetamide inside a carbon nanotube (CNT). They have shown that the confining effect of the CNT enables an almost regioselective yield of the para substituted product in addition to the acceleration in the rate of the reaction. Our group has tried to understand how different confining regimes affect the thermodynamic and kinetic outcomes of some model Diels-Alder reactions [99, 100]. To this end, a classic [4+2] cycloaddition reaction between 1,3-butadiene and ethylene to yield cyclohexene was studied inside two organic hosts, $ExBox^{+4}$ and CB[7] (in the gas phase) (Figures 16.5 and 16.6).

In the free gaseous state, the cycloaddition reaction between 1,3-butadiene and ethylene takes place slowly at ambient temperature and pressure conditions. Upon encapsulation of the reactants inside $ExBox^{+4}$, the reaction becomes slower and thermodynamically less favorable as compared to the unconfined reaction (Table 16.1).

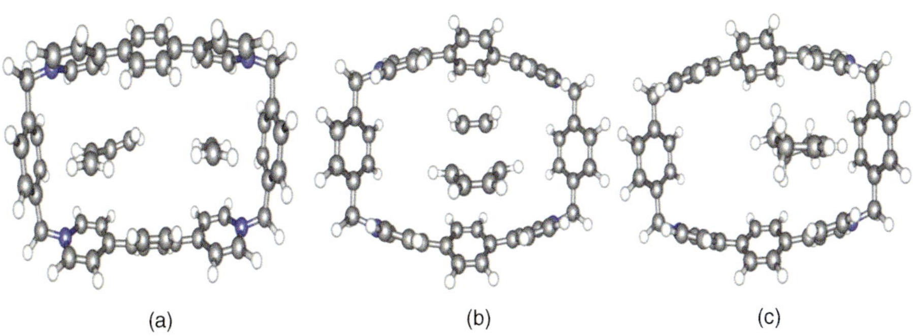

Figure 16.5 Geometrical structures of (a) reactants, (b) TS, (c) product inside $ExBox^{+4}$ respectively in the case of reaction between 1,3 butadiene and ethylene. (This figure is reproduced from [99] © 2017, John Wiley & Sons.)

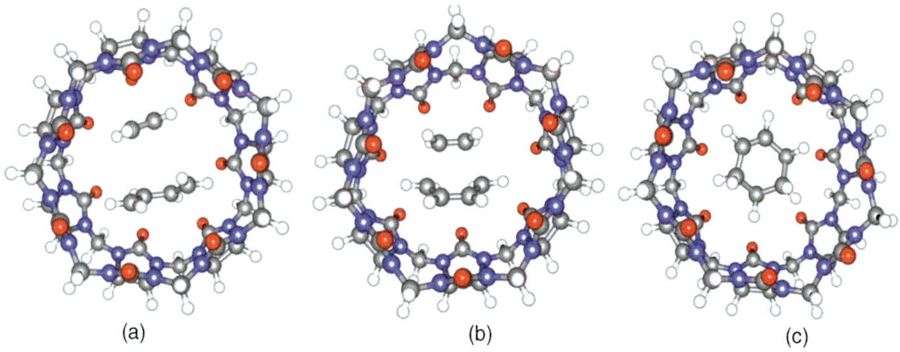

Figure 16.6 Geometrical structures of (a) reactants, (b) TS, (c) product inside CB[7] respectively in the case of reaction between 1,3 butadiene and ethylene. (This figure is reproduced from [99], © 2018, John Wiley & Sons.)

Table 16.1 Free energy change (ΔG, kcal/mol) and reaction enthalpy change (Δ*H*, kcal/mol) at 298.15 K and one atmospheric pressure for the overall reaction between 1,3 butadiene and ethylene at the confined and unconfined states, Free energy and enthalpy of activation ($\Delta G\ddagger/\Delta H\ddagger$, kcal/mol) at 298.15 K and one atmospheric pressure, rate constant (k, sec^{-1}) associated with the processes. Computed at the wb97xd/6-311G(d,p) level of theory. (This figure is reproduced from [99], © 2017, John Wiley & Sons.)

Systems	ΔG	ΔH	ΔG‡	ΔH‡	k
$Reaction_{Free}$	−39.37	−46.74	27.32	20.98	$5.80*10^{-08}$
$Reaction_{ExBox}^{+4}$	−36.51	−41.44	32.81	30.21	$5.51*10^{-12}$
$Reaction_{CB}[7]$	−40.02	−42.75	24.13	21.65	$1.27*10^{-05}$

The reason behind this could be attributed to the increase in the entropic cost associated the aforementioned reaction inside $ExBox^{+4}$ moiety. The guests are forced to interact with the pyridinium rings of the host in a parallel displaced fashion. Therefore, the host prohibits the reactants to attain the favorable conformation so that the reaction gets facilitated. The reactants therefore need to reorganize themselves inside the host so as to take part in the cycloaddition reaction. CB[7] host, on the other hand, helps the reactants to attain a suitable conformation. As a result of which the entropic cost associated with the pre-organization of the reactants inside CB[7] gets reduced as compared to that in the free state reaction. Therefore, the concerned reaction becomes thermodynamically and kinetically more favorable inside CB[7]. In addition to this geometrical effect, due to the effect of confinement, the orbital interaction between the HOMO of the diene and LUMO of the dienophile becomes more feasible as compared to that in the free state (Table 16.2). It should be noted that the considered reaction is of normal electron demand type.

More importantly, confinement imposed by CB[7] stabilizes the transition state involved in the cycloaddition reaction. This factor also plays an important role in deciding the thermodynamic and kinetic outcomes. To this end, topological descriptors within the purview of Atoms-in-Molecules (AIM)[148] theory were employed in order to decipher the nature of bonding at the transition states. For the benefit of the reader, we highlight some salient features of the AIM theory. Within the premise of AIM theory [148], bonding interaction is classified **based** on the magnitude of electron density ($\rho(r_c)$) around a concerned bond critical point (BCP), sign of the Laplacian of electron density ($\nabla^2\rho(r_c)$) as well as the local electron energy density ($H(r_c)$). If $H(r_c) < 0$ and $\nabla^2\rho(r_c) > 0$, then

Table 16.2 HOMO-LUMO gaps (in eV) for the diene and dienophile for the various systems under consideration. Computed at the wb97xd/6-311G(d,p) level of theory.Computed at the wb97xd/6-311G(d,p) level of theory. (This figure is reproduced from [99], © 2017, John Wiley & Sons.)

Systems	Energy Gap (Dienophile$_{(LUMO)}$ -Diene$_{(HOMO)}$)	Energy Gap (Diene$_{(LUMO)}$ -Dienophile$_{(HOMO)}$)
Reactant (Free state)	10.88	10.93
Reactant@ExBox^{+4}	7.26	7.69
Reactant@CB[7]	9.44	10.54

Table 16.3 Electron density descriptors (in a.u.) at the bond critical points (BCP) present in between 1,3 butadiene and ethylene moieties at the transition state (TS) in the free state as well as at the Guest@ExBox^{+4}/CB[7] moieties. (This figure is reproduced from [99], © 2017, John Wiley & Sons.)

Systems	BCP	$\rho(r_c)$	$\nabla^2\rho(r_c)$	$H(r_c)$	$-G(r_c)/V(r_c)$
$TS_{(Free)}$	C-C$_{(forming)}$	0.0514	0.0459	−0.0095	0.6885
	C-C$_{(forming)}$	0.0514	0.0459	−0.0095	0.6885
$TS_{(ExBox^{+4})}$	C-C$_{(forming)}$	0.0512	0.0466	−0.0094	0.6916
	C-C$_{(forming)}$	0.0516	0.0464	−0.0096	0.6887
$TS_{(CB[7])}$	C-C$_{(forming)}$	0.0499	0.0452	−0.0088	0.6955
	C-C$_{(forming)}$	0.0541	0.0445	−0.0107	0.6708

the bonding could be classified as having partly covalent character. For a covalent type bonding, the ratio of local kinetic energy density ($G(r_c)$) and local potential energy density ($V(r_c)$) should be less than 1. Results showed that the developing C-C bond at the transition state involved in the reaction inside CB[7] acquires more covalent character as compared to that in the other two cases under consideration (Table 16.3).

In order to investigate whether CB[7] can promote other Diels-Alder reactions or not, DFT-based computation was undertaken [100]. To this end, dienes, viz., furan, thiophene, cyclopentadiene, benzene, and a classic dienophile, ethylene, were employed (Figure 16.7).

Results indicated that all the aforementioned Diels-Alder reactions become thermodynamically and kinetically more favorable inside CB[7] as compared to the corresponding free state reactions, at ambient temperature and pressure conditions (Table 16.4).

The main reason behind these observations could be rationalized **based** on the formation of a suitable encounter complex of the reactants inside CB[7] thereby reducing the entropic cost in bringing the reactants together as well as the stabilization of the transition states involved in the reaction due to the effect of confinement. A similar rationale was provided by Himo and co-workers [101] in their recent computational study while explaining the experimentally observed rate acceleration and selectivity of 1,3 Huisgen cycloaddition reaction observed inside the cavitand synthesized by Rebek. It is to be noted, however, that a straightforward extrapolation

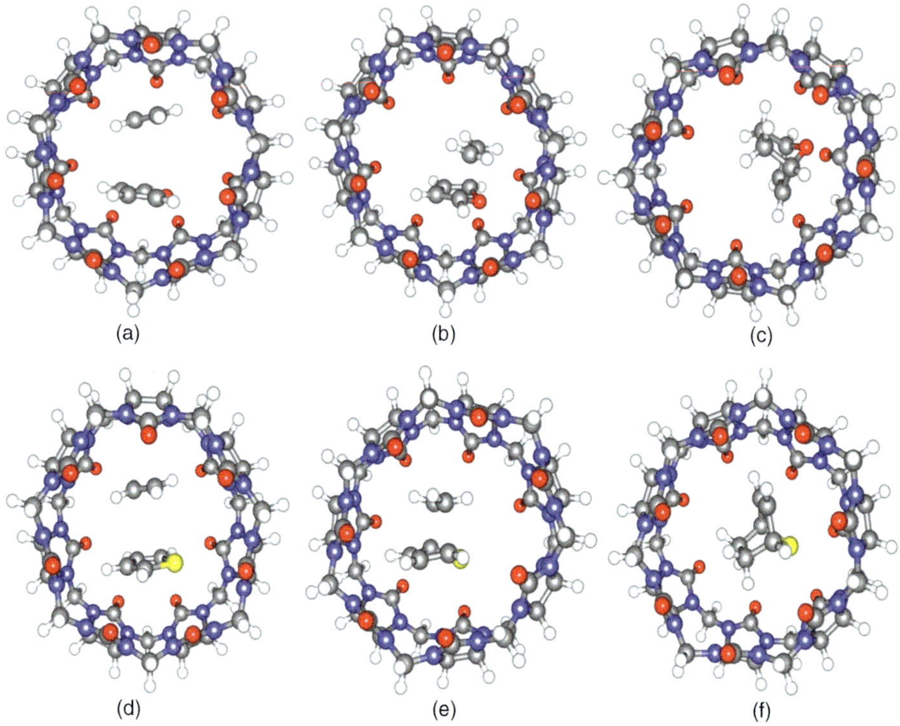

Figure 16.7 Geometrical structures of (a) reactants, (b) TS, (c) product for the case of reaction in between furan and ethylene inside CB[7] and (d) reactants, (e) TS, (f) product for the case of reaction in between thiophene and ethylene inside CB[7] respectively. (This figure is reproduced from [100], © 2018, John Wiley & Sons.)

Table 16.4 Free energy change (ΔG, kcal/mol) and reaction enthalpy change (Δ*H*, kcal/mol) at 298.15 K and one atmospheric pressure for the overall reaction at the confined and unconfined states, Free energy and enthalpy of activation (Δ*G*‡/Δ*H*‡, kcal/mol) and the rate constant (k, sec^{-1}) associated with the processes at 298.15 K and one atmospheric pressure. Computed at the wb97xd/6-311G(d,p) level of theory. Reaction between benzene, furan, cyclopentadiene and thiophene with ethylene has been represented as reaction1, reaction2, reaction3 and reaction4 respectively. (This figure is reproduced from [100], © 2018, John Wiley & Sons.)

Systems	Δ*G*	Δ*H*	Δ*G*‡	Δ*H*‡	k
$Reaction1_{Free}$	17.95	3.63	49.95	36.93	$1.50*10^{-24}$
$Reaction2_{Free}$	2.54	−11.28	37.07	24.19	$4.14*10^{-15}$
$Reaction3_{Free}$	−13.52	−27.70	32.52	19.46	$9.05*10^{-12}$
$Reaction4_{Free}$	3.56	−10.35	45.66	32.71	$2.10*10^{-21}$
$Reaction1_{CB}[7]$	13.21	9.51	43.93	41.15	$3.90*10^{-20}$
$Reaction2_{CB}[7]$	−5.39	−9.26	30.34	27.70	$3.59*10^{-10}$
$Reaction3_{CB}[7]$	−17.67	−21.28	25.92	23.03	$6.15*10^{-7}$
$Reaction4_{CB}[7]$	−1.90	−5.07	39.60	36.69	$5.83*10^{-17}$

of the computationally obtained Gibbs free energy changes associated with a chemical reaction to the experimentally observed kinetics data is difficult to obtain. Maseras and co-workers [102] employed microkinetic modelling methodology in order to correlate computationally obtained data with that of experiments. The concentrations of all species involved in a given reaction over time can be obtained by assigning a reaction rate to each one of them and providing initial concentrations. The kinetic model can provide insights into the variation of concentration of different moieties over time and therefore provide an account of the chemical kinetics. Maseras' group developed a kinetic model that can take into account the different **behavior** of the host-guest complexes in solution. Their developed model can follow the reaction mechanism from start to end, thereby providing a comprehensive account of the process.

Based on the aforementioned developments, it can be stated that confinement-induced facilitation of chemical reactions deserves further attention from both experimental and computational viewpoints. Particular emphasis is to be placed on developing ways of limiting product inhibition, artificially limiting non-productive alignment of the reactants inside the hosts, etc. It needs to be examined whether one could utilize any external perturbation to modulate the arrangement of the reactants in a desired way. Even though gas phase DFT calculations can provide some qualitative understanding of these complex processes, the role of the solvents is not taken into account in gas phase DFT analysis. Since most of the host-guest chemistry discussed above takes place in some solvent medium, it is not very clear how and to what extent solvents influence the outcome. Since, the cavity of many organic hosts are hydrophobic, it is unlikely that solvents can participate in the "confined" reactions in an explicit way. However, solvents can affect the outcome of the reaction in an implicit way by means of long-range electrostatic interactions with the encapsulated species. To this end, multiscale computational analysis should be considered. This aspect deserves further careful scrutiny.

16.7 Outlook

Based on the results discussed herein, it becomes quite clear that confinement can bring about diverse changes in chemical reactivity [103]. Confinement could be utilized to design efficient reaction vessels. Confined environments could be helpful in order to store environmentally as well as biologically important molecules. In this regard, computational chemistry can assist the experimental endeavors by providing suitable design principles for the realization of the interesting new chemistry at the confined state. Gas-phase DFT calculations can provide qualitative insights into many situations, however, utilization of multi-scale modelling techniques is required in order to address complex scenarios present in real-life experiments. Associated computational and methodological developments should be perused so that hitherto unforeseen chemistry occurring at the confined space could be explored.

Acknowledgements

PKC would like to thank DST, New Delhi, for the J.C. Bose National Fellowship. SM thanks the University Grants Commission, New Delhi for UGCBSR Research Start-Up-Grant (No. F.30-458/2019(BSR)).

References

1 Sabin, J.R., Brandas, E.J., and Cruz, S.A. (2009). Advances in Quantum Chemistry: Theory of Confined Quantum Systems, vol. 57–58. Waltham: Academic.

2 Schettino, V. and Bini, R. (2007). Constraining molecules at the closest approach: chemistry at high pressure. Chemical Society Reviews 36 (6): 869–880.

3 Gubbins, K.E., Liu, Y.C., Moore, J.D. et al. (2011). The role of molecular modeling in confined systems: impact and prospects. Physical Chemistry Chemical Physics 13 (1): 58–85.

4 Atkins, P.W. and De, P.J. (2006). Atkins' Physical Chemistry. Oxford: Oxford University Press.

5 Filippov, A.E., Klafter, J., and Urbakh, M. (2001). Confined molecules under shear: from a microscopic description to phenomenology. Physical Review Letters 87 (27): 275506.

6 Pupyshev, V.I. and Bobrikov, V.V. (2004). The confined diatomic molecule problem. International Journal of Quantum Chemistry 100 (4): 528–538.

7 Gonzalez, J., Molina, A., Soto, C.M. et al. (2012). Detection of interaction between redox centers of surface confined molecules by means of cyclic voltammetry and differential staircase voltcoulommetry. Journal of Electroanalytical Chemistry 664 (1): 53–62.

8 Walsh, C.A., Yuan, J., and Brown, L.M. (2000). A procedure for measuring the helium density and pressure in nanometre-sized bubbles in irradiated materials using electron-energy-loss spectroscopy. Philosophical Magazine A 80 (7): 1507–1543.

9 Ivanov, M.Y. and Corkum, P.B. (1993). Generation of high-order harmonics from inertially confined molecular ions. Physical Review A 48 (1): 580–590.

10 Sarsa, A. and Le Sech, C. (2011). Variational Monte Carlo method with Dirichlet boundary conditions: application to the study of confined systems by impenetrable surfaces with different symmetries. Journal of Chemical Theory and Computation 7 (9): 2786–2794.

11 Sebastianelli, F., Xu, M., Elmatad, Y.S. et al. (2007). Hydrogen molecule in the small dodecahedral cage of a clathrate hydrate: quantum five-dimensional calculations of the coupled translation–rotation eigenstates. The Journal of Physical Chemistry C 111 (6): 2497–2504.

12 Sarkar, U., Giri, S., and Chattaraj, P.K. (2009). Dirichlet boundary conditions and effect of confinement on chemical reactivity. The Journal of Physical Chemistry A 113 (40): 10759–10766.

13 Chattaraj, P.K. and Sarkar, U. (2003). Effect of spherical confinement on chemical reactivity. The Journal of Physical Chemistry A 107 (24): 4877–4882.

14 Chattaraj, P.K. and Sarkar, U. (2003). Chemical reactivity of the spherically confined atoms. Chemical Physics Letters 372 (5–6): 805–809.

15 Sarkar, U., Khatua, M., and Chattaraj, P.K. (2012). A tug-of-war between electronic excitation and confinement in a dynamical context. Physical Chemistry Chemical Physics 14 (5): 1716–1727.

16 Khatua, M. and Chattaraj, P.K. (2013). Molecular reactivity dynamics in a confined environment. Physical Chemistry Chemical Physics 15 (15): 5588–5614.

17 Rosi, N.L., Eckert, J., Eddaoudi, M. et al. (2003). Hydrogen storage in microporous metal-organic frameworks. Science 300 (5622): 1127–1129.

18 Han, S.S., Furukawa, H., Yaghi, O. et al. (2008). Covalent organic frameworks as exceptional hydrogen storage materials. Journal of the American Chemical Society 130 (35): 11580–11581.

19 Jena, P. (2011). Materials for hydrogen storage: past, present, and future. The Journal of Physical Chemistry Letters 2 (3): 206–211.

20 Kolb, H.C., Finn, M.G., and Sharpless, K.B. (2011). Click chemistry: diverse chemical function from a few good reactions. Angewandte Chemie International Edition 40 (11): 2004–2021.

21 Rostovtsev, V.V., Green, L.G., Fokin, V.V. et al. (2002). A stepwise Huisgen cycloaddition process: copper(i) catalyzed regioselective "ligation" of azides and terminal alkynes. Angewandte Chemie International Edition 41 (14): 2596–2599.

22 Tornoe, C.W., Christensen, C., and Meldal, M.P. (2002). Eptidotriazoles on solid phase: [1,2,3]-triazoles by regiospecific copper(I)-catalyzed 1,3-dipolar cycloadditions of terminal alkynes to azides. The Journal of Organic Chemistry 67 (9): 3057–3064.

23 Mock, W.L., Irra, T.A., Wepsiec, J.P. et al. (1983). Cycloaddition induced by cucurbituril. A case of Pauling principle catalysis. The Journal of Organic Chemistry 48 (20): 3619–3620.

24 Mock, W.L., Irra, T.A., Wepsiec, J.P. et al. (1989). Catalysis by cucurbituril. The significance of bound-substrate destabilization for induced triazole formation. The Journal of Organic Chemistry 54 (22): 5302–5308.

25 Ballester, P., Fujita, M., and Rebek, J. (2015). Molecular containers. Chemical Society Reviews 44 (2): 392–393.

26 Hou, X., Ke, C., and Stoddart, J.F. (2016). Cooperative capture synthesis: yet another playground for copper-free click chemistry. Chemical Society Reviews 45 (14): 3766–3780.

27 Vallavoju, N. and Sivaguru, J. (2014). Supramolecular photocatalysis: combining confinement and non-covalent interactions to control light initiated reactions. Chemical Society Reviews 43 (12): 4084–4101.

28 Davy, H. (1811). The Bakerian Lecture. On some of the combinations of oxymuriatic gas and oxygene, and on the chemical relations of these principles, to inflammable bodies. Philosophical Transactions of the Royal Society of London 101: 1–35.

29 Faraday, M. and Davy, H. (1823). XIV. On fluid chlorine. Philosophical Transactions of the Royal Society of London 113: 160–165.

30 Clausen, W.F. (1951). Suggested structures of water in inert gas hydrates. The Journal of Chemical Physics 19 (2): 259–260.

31 Clausen, W.F. (1951). A second water structure for inert gas hydrates. The Journal of Chemical Physics 19 (11): 1425–1426.

32 Muller, H.R. and Stackelberg, V.M. (1952). Zur Struktur der Gashydrate. Naturwissenschaften 39: 20–21.

33 Sloan, E.D. (1998) Clathrate Hydrates of Natural Gases, 2nd edn (eds M. Dekker) New York.

34 Mao, W.L., Mao, H., Alexender, F. et al. (2002). Hydrogen clusters in clathrate hydrate. Science 297 (5590): 2247–2249.

35 Mao, W.L. and Mao, H.K. (2004). Hydrogen storage in molecular compounds. Proceedings of the National Academy of Sciences of the United States of America 101 (3): 708–710.

36 Vos, W.L., Finger, L.W., Hemley, R.J. et al. (1994). Novel H_2-H_2O clathrates at high pressures. Physical Review Letters 71 (19): 3150.

37 Powell, H.M. (1948). The structure of molecular compounds. Part IV. Clathrate compounds. Journal of the Chemical Society: 61–73.

38 Davidson, D.W. and Frank, F.E. (1973). Water: A Comprehensive Treatise. New York: Plenum Press New York.

39 Hammerschmidt, E.G. (1934). Formation of gas hydrates in natural gas transmission lines. Industrial & Engineering Chemistry 26 (8): 851–855.

40 Deaton, W.M., Frost, E.M. (1949) Gas hydrates and their relation to the operation of natural-gas pipe lines. U.S. Bur. Mines Monograph, 8.

41 van der Waals, J.H., Platteeuw, J.C. (1958) Clathrate solutions. Advances in Chemical Physics, 2, 1–57.

42 Somayazulu, M.S., Finger, L.W., Hemley, R.J. et al. (1996). High-pressure compounds in methane-hydrogen mixtures. Science 271 (5254): 1400–1402.

43 Loubeyre, P., Letoullec, R., and Pinceaux, J.P. (1994). Compression of $Ar(H_2)_2$ up to 175 GPa: A new path for the dissociation of molecular hydrogen? Physical Review Letters 72 (60): 1360–1363.

44 Lokshin, K.A., Zhao, Y., He, D. et al. (2004) Structure and dynamics of hydrogen molecules in the novel clathrate hydrate by high pressure neutron diffraction. Physical Review Letters, 93 (12), 125503(1–4).

45 Mao, W. and Mao, H.K. (2002). USA Patent: 6735960.

46 von Stackelberg, M. (1949). Feste gashydrate. Naturwissenschaften 36 (11): 327–333.

47 Patchkovskii, S. and Tse, J. (2003). Thermodynamic stability of hydrogen clathrates. Proceedings of the National Academy of Sciences of the United States of America 100 (25): 14645–11450.

48 Sluiter, M.F., Adachi, H., Belosludov, R.V. et al. (2004). *Ab initio* study of hydrogen storage in hydrogen hydrate clathrates. Materials Transactions 45 (5): 1452–1454.

49 Alavi, S., Ripmeester, J.A., and Klug, D.D. (2005). Molecular-dynamics study of structure II hydrogen clathrates. The Journal of Chemical Physics 123 (2): 024507.

50 Inerbaev, T.M., Belosludov, V.R., Belosludov, R.V. et al. (2006). Dynamics and equation of state of hydrogen clathrate hydrate as a function of cage occupation. Computational Materials Science 36 (1–2): 229–233.

51 Chattaraj, P.K., Bandaru, S., and Mondal, S. (2011). Hydrogen storage in clathrate hydrates. The Journal of Physical Chemistry A 115 (2): 187–193.

52 Khalil, M.A.K. (1999). Non-CO_2 greenhouse gases in the atmosphere. Annual Review of Environment and Resources 24 (1): 645–661.

53 Howarth, R.W., Santoro, R., and Ingraffea, A. (2011). Methane and the greenhouse-gas footprint of natural gas from shale formations. Climatic Change 106: 679–690.

54 Murray, E.P., Tsai, T., and Barnett, S.A. (1999). A direct-methane fuel cell with a ceria-based anode. Nature 400: 649–651.

55 Tina, D., Lev, S., Yaghi, O.M. et al. (2004). Design of new materials for methane storage. Langmuir 20 (7): 2683–2689.

56 Max, D.M. (2003). Natural Gas Hydrate in Oceanic and Permafrost Environments. Kluwer Academic Publishers 0-7923-6606-9.

57 Truong-Lam, H.S., Seo, S.D., Kim, S. et al. (2020). In situ Raman study of the formation and dissociation kinetics of methane and methane/propane hydrates. Energy Fuels 34 (5): 6288–6297.

58 Romero-Martínez, A., Hernández-Guerrero, E., Ramirez-Jaramillo, E. et al. (2020). Synthesis and evaluation of a pyridinium-based ionic liquid-type surfactant as a new low-dosage methane hydrate inhibitor. Energy Fuels 34 (8): 9243–9251.

59 Petuya, C., Patt, A., Simon, J.M. et al. (2020). Molecular selectivity of CO-N_2 mixed hydrates: Raman spectroscopy and GCMC studies. The Journal of Physical Chemistry C 124 (22): 11886–11891.

60 Zhao, J., Liang, H., Yang, L. et al. (2020). Growth kinetics and gas diffusion in formation of gas hydrates from ice. The Journal of Physical Chemistry C 124 (24): 12999–13007.

61 Sloan, E.D. (2003). Fundamental principles and applications of natural gas hydrates. Nature 426: 353–359.

62 Koh, C.A., Sloan, E.D., Sum, A.K. et al. (2012). Fundamentals of Materials for Energy and Environmental Sustainability. In: Cambridge University Press and Materials Research Society. UK: Cambridge.

63 Nguyen, N.N., Galib, M., and Nguyen, A.V. (2020). Critical review on gas hydrate formation at solid surfaces and in confined spaces—why and how does interfacial regime matter? Energy Fuels 34 (6): 6751–6760.

64 Yu, K.B. and Yazaydin, A.O. (2020). Does confinement enable methane hydrate growth at low pressures? Insights from molecular dynamics simulations. The Journal of Physical Chemistry C 124 (20): 11015–11022.

65 Pruteanu, C.G., Robinson, V.N., Ansari, N. et al. (2020). Squeezing oil into water under pressure: inverting the hydrophobic effect. The Journal of Physical Chemistry Letters 11 (12): 4826–4833.

66 Fanetti, S., Scelta, D., and Bini, R. (2020). Growth dynamics of crystalline ar hydrate. The Journal of Physical Chemistry C 124 (18): 10159–10166.

67 Zhang, H., Huang, H., Gao, L. et al. (2020). Magnetic dioxygen clathrate hydrates: a type of promising building blocks for icy crystalline materials. The Journal of Physical Chemistry C 124 (19): 10669–10678.

68 Yanes-Rodríguez, R., Arismendi-Arrieta, D.J., and Prosmiti, R. (2020). He inclusion in ice-like and clathrate-like frameworks: a benchmark quantum chemistry study of guest–host interactions. Journal of Chemical Information and Modeling 60 (6): 3043–3056.

69 Plimpton, S.J. (1995). Fast parallel algorithms for short-range molecular dynamics. Journal of Computational Physics 117 (1): 1–19.

70 Teleman, O., Jonsson, B., and Engstrom, S. (1987). A molecular dynamics simulation of a water model with intramolecular degrees of freedom. Journal Molecular Physics 60 (1): 193–203.

71 Mizan, T.I., Savage, P.E., and Ziff, R.M. (1994). Molecular dynamics of supercritical water using a flexible SPC model. The Journal of Physical Chemistry 98 (49): 13067–13076.

72 Klauda, J.B. and Sandler, S.I. (2003). Phase behavior of clathrate hydrates: a model for single and multiple gas component hydrates. Chemical Engineering Science 58 (1): 27–41.

73 Hirschfelder, J.O., Curtiss, C.F., and Bird, R.B. (1954). The Molecular Theory of Gases and Liquids. New York: Wiley.

74 Allen, M.P. and Tildesley, D.J. (1996). Computer Simulation of Liquids. Oxford: Oxford University Press.

75 Jacobson, L.C., Waldemar, H., and Valeria, M. (2009). Thermodynamic stability and growth of guest-free clathrate hydrates: a low-density crystal phase of water. The Journal of Physical Chemistry B 113 (30): 10298–10307.

76 Mondal, S., Ghosh, S., and Chattaraj, P.K. (2013). A molecular dynamics study on sI hydrogen hydrate. Journal of Molecular Modeling 19: 2785–2790.

77 Mondal, S., Giri, S., and Chattaraj, P.K. (2013). Possibility of having HF-doped hydrogen hydrates. The Journal of Physical Chemistry C 117 (22): 11625–11634.

78 Mondal, S., Giri, S., and Chattaraj, P.K. (2013). Methane hydrates and their HF doped analogues. Chemical Physics Letters 578: 110–114.

79 Mondal, S. and Chattaraj, P.K. (2014). Noble gas encapsulation: clathrate hydrates and their HF doped analogues. Physical Chemistry Chemical Physics 16: 17943–17954.

80 Mondal, S., Goswami, T., Jana, G. et al. (2018). A possible reason behind the initial formation of pentagonal dodecahedron cavities in sI-methane hydrate nucleation: A DFT study. Chemical Physics Letters 691: 415–420.

81 Marx, D. and Hutter, J. (2000). Ab initio molecular dynamics: theory and implementation. In: Proceedings of Modern Methods and Algorithms of Quantum Chemistry (ed. J. Grotendorst), 301–449. Jülich, NIC Series: John von Neumann Institute for Computing.

82 Schlegel, H.B., Millam, J.M., Iyengar, S.S. et al. (2001). Ab initio molecular dynamics: Propagating the density matrix with Gaussian orbitals. The Journal of Chemical Physics 114 (22): 9758–9763.

83 Iyengar, S.S., Schleegl, H.B., Millam, J.M. et al. (2001). Ab initio molecular dynamics: Propagating the density matrix with Gaussian orbitals. II. Generalizations based on mass-weighting, idempotency, energy conservation and choice of initial conditions. The Journal of Chemical Physics 115 (22): 10291–10302.

84 Schlegel, H.B. (2002). Ab initio molecular dynamics: Propagating the density matrix with Gaussian orbitals. III. Comparison with Born–Oppenheimer dynamics. The Journal of Chemical Physics 117 (19): 8694.

85 (a) Frisch, M.J. et al. (2009). Gaussian 09, revision A.1, Gaussian. Pittsburgh, PA: Inc. (b) Frisch, M.J. et al. (2010). Gaussian 09, revision C.01, Gaussian. Wallingford, CT: Inc.

86 Sun, Q. and Zheng, H.F. (2006). In situ Raman spectroscopic study of the dissociation of methane hydrate at high pressure. Applied Spectroscopy 60 (9): 985–988.

87 Ringe, D. and Petsko, G.A. (2008). How enzymes work. Science 320 (5882): 1428–1429.

88 Marty, M., Clyde-Watson, Z., Twyman, L.J. et al. (1998). Acceleration of a hetero-Diels–Alder reaction by cyclic metalloporphyrin trimers. Chemical Communications (20): 2265–2266.

89 Marinescu, L.G. and Bols, M. (2006). Very high rate enhancement of benzyl alcohol oxidation by an artificial enzyme. Angewandte Chemie International Edition 45 (28): 4590–4593.

90 Wyler, R., de Mendoza, J., and Rebek, J. (1993). A synthetic cavity assembles through self-complementary hydrogen bonds. Angewandte Chemie International Edition 32 (12): 1699–1701.

91 Meissner, R.S., Rebek, J., and de Mendoza, J. (1995). Autoencapsulation through intermolecular forces: a synthetic self-assembling spherical complex. Science 270 (5241): 1485–1488.

92 Heinz, T., Rudkevich, D.M., and Rebek, J. (1998). Pairwise selection of guests in a cylindrical molecular capsule of nanometre dimensions. Nature 394: 764–766.

93 Johnson, D.W. and Raymond, K.N. (2001). The self-assembly of a $[Ga_4L_6]^{12-}$ tetrahedral cluster thermodynamically driven by host–guest interactions. Inorganic Chemistry 40 (20): 5157–5161.

94 Hastings, C.J., Pluth, M.D., Bergman, R.G. et al. (2010). Enzymelike catalysis of the Nazarov cyclization by supramolecular encapsulation. Journal of the American Chemical Society 132 (20): 6938–6940.

95 Carlqvist, P. and Maseras, F. (2007). A theoretical analysis of a classic example of supramolecular catalysis. Chemical Communications 2007: 748–750.

96 Goehry, C., Besora, M., and Maseras, F. (2015). Computational study on the mechanism of the acceleration of 1,3-dipolar cycloaddition inside cucurbit[6]uril. ACS Catalysis 5 (4): 2445–2451.

97 Xu, L., Hua, W., Hua, S. et al. (2013). Mechanistic insight on the Diels–Alder reaction catalyzed by a self-assembled molecular capsule. The Journal of Organic Chemistry 78 (8): 3577–3582.

98 Marforio, T.D., Bottoni, A., Giacinto, P. et al. (2017). Aromatic bromination of N-phenylacetamide inside CNTs. Are CNTs real nanoreactors controlling regioselectivity and kinetics? A QM/MM investigation. The Journal of Physical Chemistry C 121 (49): 27674–27682.

99 Chakraborty, D., Das, R., and Chattaraj, P.K. (2017). Does confinement always lead to thermodynamically and/or kinetically favorable reactions? A case study using Diels-Alder reactions within $ExBox^{+4}$ and CB[7]. ChemPhysChem 18 (16): 2162–2170.

100 Chakraborty, D. and Chattaraj, P.K. (2018). Confinement induced thermodynamic and kinetic facilitation of some Diels-Alder reactions inside a CB[7] cavitand. Journal of Computational Chemistry 39 (3): 151–160.

101 Daver, H., Harvey, J.N., Rebek, J.J. et al. (2017). Quantum chemical modeling of cycloaddition reaction in a self-assembled capsule. Journal of the American Chemical Society 139: 15494–15503.

102 Goehry, C., Besora, M., and Maseras, F. (2018). Computational description of a Huisgen cycloaddition inside a self-assembled nanocapsule. The European Journal of Organic Chemistry 18: 2103–2109.

103 Chakraborty, D. and Chattaraj, P.K. (2019). Bonding, reactivity, and dynamics in confined systems. The Journal of Physical Chemistry A 123: 4513–4531.

17

Engineering the Confined Space of MOFs for Heterogeneous Catalysis of Organic Transformations

Tapan K. Pal[1†], Dinesh De[2†], and Parimal K. Bharadwaj[3]*

[1] *Department of Science, School of Technology, Pandit Deendayal Petroleum University, Gandhinagar-382421, Gujarat*
[2] *Department of Basic Science, Vishwavidyalaya Engineering College, Lakhanpur, CSVTU, Chhatisgarh-497116*
[3] *Department of Chemistry, Indian Institute of Technology Kanpur, Kanpur 208016, Uttar Pradesh (India) & Department of Chemistry, Indian Institute of Technology Bombay,*

17.1 Introduction

Metal-organic frameworks (MOFs) are an emerging class of porous materials with fascinating structural features [1] such as shape and size of pores, architectural flexibility [2], ability to exchange metal ions/ligands in some cases [3] without catastrophic breakdown of the structure, and so on. While numerous MOFs have been synthesized in the last two decades for potential applications concerning energy and environmental sustainability [4], the high BET surface area, large inter-connected channels, activation of guests by the linkers and metal nodes alike are some of the attributes MOFs have to behave as excellent heterogeneous catalysts [5]. MOFs can participate in the catalytic reaction by using reactive metal nodes (or unsaturated metal center, UMCs) as Lewis acid (Scheme 17.1), the Lewis base present in the organic linker or even the cooperative effect exerted by both of them. Besides, metal complexes or nanoparticles anchored within the cavity of MOFs can exhibit catalytic activities [6]. This present article provides an account of catalytic activities of MOFs although it is not exhaustive and we might have missed some of important literature on the subject.

One of the first examples of heterogeneous catalysis by a MOF was reported in 1994 by Fujita and co-workers [7] who showed facile cyanosilylation reactions between aldehydes and silyl cyanides. Since then, a large number of groups have attempted various organic transformation reactions catalyzed by MOFs. These porous materials help in the organic transformations in different ways as schematically illustrated below.

17.2 Catalysis at the Open Metal Sites

The existence of open metal sites (OMSs) in porous MOFs is very attractive since it can strongly interact and subsequently activate the substrate for catalysis. A wide variety of both transition and main group metal ions can function in this role. This OMS can be present in the as-synthesized MOF or can be generated *in situ* by removing loosely metal-bound solvent molecules.

*Corresponding author: pkb@iitk.ac.in
†Authors contributed equally

Chemical Reactivity in Confined Systems: Theory, Modelling and Applications,
First Edition. Edited by Pratim Kumar Chattaraj and Debdutta Chakraborty.

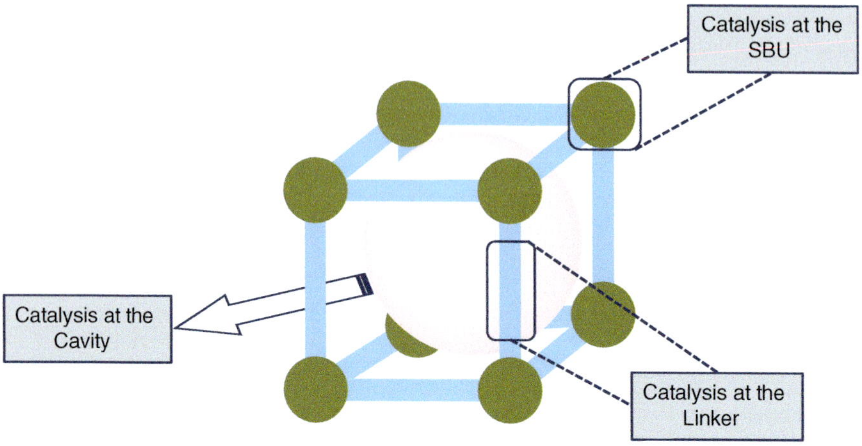

Scheme 17.1 Schematic presentation for the illustration of different modes of catalysis in MOFs.

17.2.1 MOFs Endowed with Open Metal Site(s)

One of the first reports in this category came from the laboratory of Fujita et al. [7], who reported a 2D framework constructed from 4,4′-bpy and $Cd(NO_3)_2$ (Figure 17.1). The empty coordination sites on the metal center could catalyze the cyanosilylation reactions of aromatic aldehydes with trimethylsilylnitrile.

Lanthanide ions are capable of exhibiting high coordination numbers and can have empty coordination sites. The MOF, **$\{[Gd_2(L)_3(dmf)_4]\cdot 4DMF\cdot 3H_2O\}_n$** [8], was synthesized solvothermally by reacting a dicarboxylate ligand (L = 2,2′,6,6′-tetranitro-4,4′-biphenyldicarboxylate) with $Gd(NO_3)_3$ $6H_2O$. A single crystal of this porous MOF could be used for carrying out both cyanosilylation and Knoevenagel condensation reactions inside the pores without losing crystallinity (Figure 17.2). The Gd(III) ion activated the substrate aldehyde via direct interactions, leading to facile reactions.

It was found that two MOFs with imidazolate [9] and pyrimidinolate [10] as linkers and Cu(II) as the node could increase the coordination number of the metal on substrate binding without

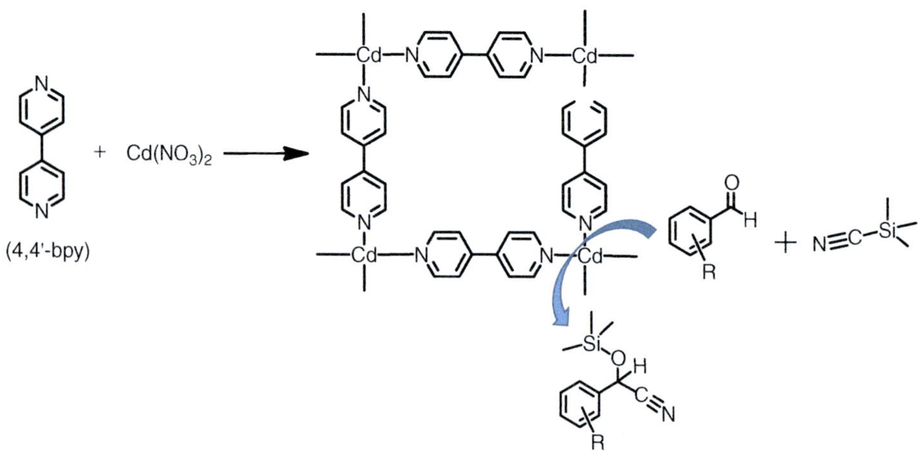

Figure 17.1 Structure of 2D network constructed from 4,4′-bpy and Cd^{2+} to show cyanosilylation reaction. Counter ions are omitted for clarity.

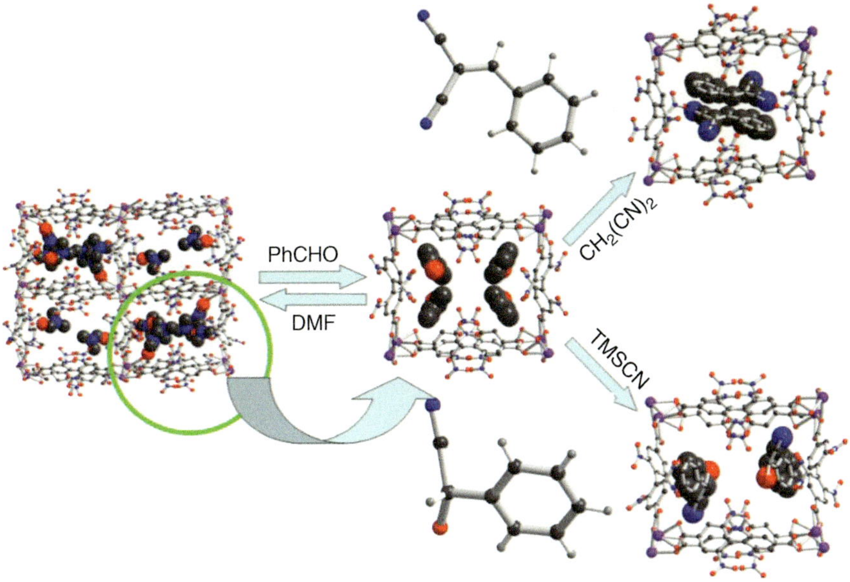

Figure 17.2 Cyanosilylation and Knoevenagel condensation reactions observed in a single crystal.

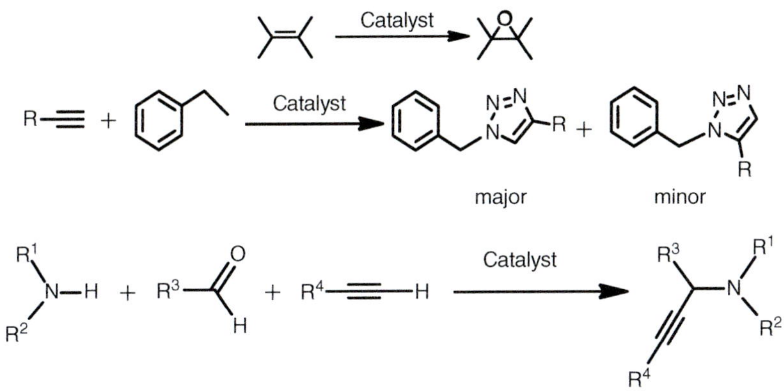

Scheme 17.2 Oxidation of alkanes (top), 1,3-dipolar cycloaddition reaction (middle), and three-component coupling reactions of aldehydes, terminal alkynes, and amines (bottom). Source: Based on Luz, I., Llabré, F. X., Xamena, S. I., and Corma, A. (2012) Bridging homogeneous and heterogeneous catalysis with MOFs: Cu-MOFs as solid catalysts for three-component coupling and cyclization reactions for the synthesis of propargylamines, indoles and imidazopyridines. J. Catal., 285, 285–291.

disturbing the structural integrity. This property made the two MOFs as prominent heterogeneous catalysts for oxidation of activated alkanes in the presence of hydroperoxide [11], 1,3-dipolar cycloaddition of azides to terminal alkynes (*click* reactions) [12], or three-component coupling of aldehydes, alkynes, and amines (A^3 reaction) (Scheme 17.2) [13].

Another Cu(II)-MOF, [Cu**L**$_2$] (**L** = *N*-(4-pyridyl)-D,L-valine), had been synthesized using a valine-derived linker (Figure 17.3a) [14]. In this structure, each Cu(II) ion displayed distorted square-planar geometry bonded with two carboxyl oxygen atoms and two pyridyl nitrogens in a *trans* orientation with the axial sites remaining uncoordinated. Due to the available axial sites, the MOF exhibited facile catalytic activity in the cross-coupling reaction of arylboronic acids

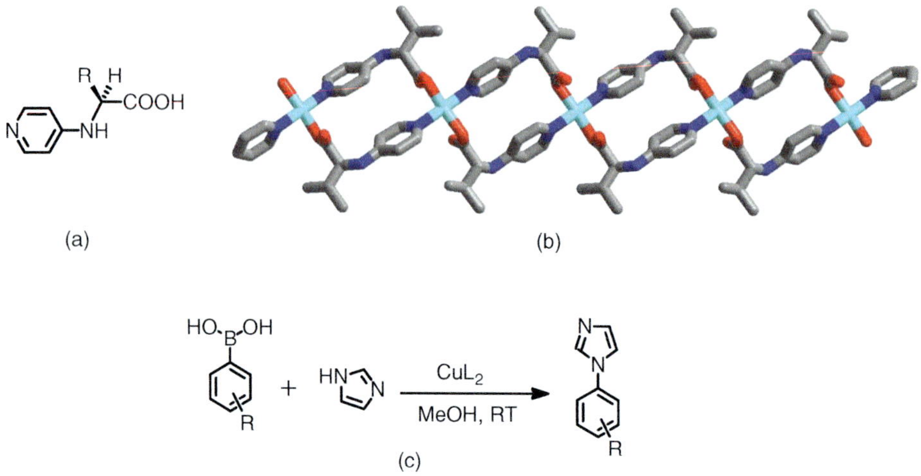

Figure 17.3 (a) The linker L = *N*-(4-pyridyl)-D,L-valine, (b) the 1D polymeric chain in CuL_2, and (c) catalytic application of CuL_2 in cross-coupling of arylboronic acids with imidazole. Source: Wu, C.-D., Li, L., and Shi, L.-X. (2009) Heterogeneous catalyzed aryl–nitrogen bond formations using a valine derivative bridged metal–organic coordination polymer. Dalton Trans., 6790–6794. © 2009, Royal Society of Chemistry.

with imidazole. A moderate to excellent yields (55 to 97%) were obtained for different substituted arylboronic acids.

A microporous sodalite-type MOF, $Mn_3[(Mn_4Cl)_3(BTT)_8(CH_3OH)_{10}]_2$ (H_3BTT = 1,3,5 – benzenetristetrazol -5- yl) had been reported [15]. The structure formed channels where two different types of Mn(II) centers were detected. These Mn(II) centers had vacant binding sites exposed toward the center of the channel that were readily accessible to substrates (Figure 17.4). This MOF

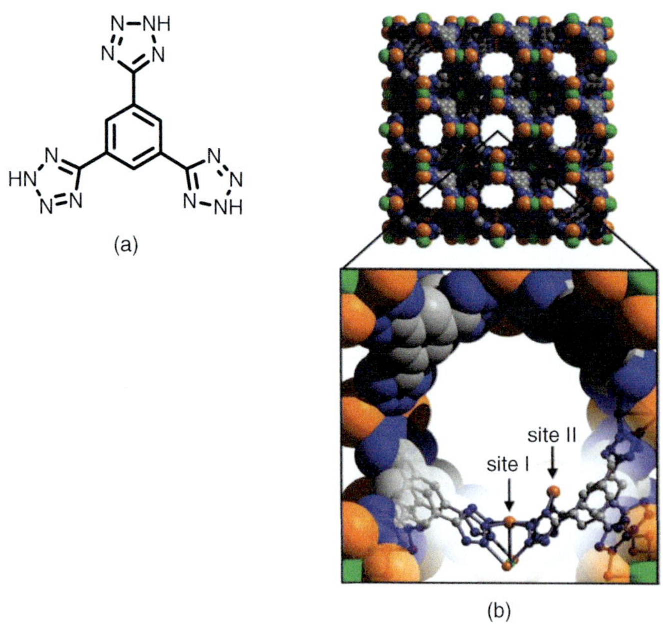

Figure 17.4 Structure of the (a) ligand and (b) MOF showing two different types of Mn^{II} sites situated at the channel surface.

could catalyze size selective cyanosilylation of aromatic aldehydes and the Mukaiyama-aldol reactions. The cyanosilylation reaction was almost quantitative while 63% yield was observed at room temperature in the case of the Mukaiyama-aldol reaction.

17.2.2 Removal of Volatile Molecules From Metal Nodes to Perform Catalysis

Perhaps the easiest way to have OMSs in MOFs is by removal of labile metal bound solvent molecules without collapsing the overall structure. These OMSs are inclined to accept electron density from substrate molecules during the reaction, i.e., the metal will behave as a Lewis acid center.

In 2005, Férey and his group reported **MIL-101(Cr)** [16] with the formula, $Cr_3X(H_2O)_2O(bdc)_3$ (X = F, OH; bdc = 1,4-benzenedicarboxylate). The framework consisted of trimeric Cr(III)-SBU (SBU = secondary building unit) chelated by 1,4-benzenedicarboxylate where two Cr ions were connected with terminal water molecules (Figure 17.5).

Subsequently, these metal-bound water molecules in **MIL-101(Cr)** could be removed by heating under vacuum and the resulting MOF could be utilized as a Lewis acid catalyst in various organic reactions. It was found to catalyze cyanosilylation of aldehydes [17]. Oxidation of tetralin to 1-tetralone with high yields and selectivity were achieved using tert-butylhydroperoxide (TBHP) or acyl peroxy radicals (generated by *in situ* decomposition of trimethyl acetaldehyde) and O_2 as oxidants [18]. Similarly, in the presence of TBHP, cyclohexene (CyH), α-pinene, and limonene could be readily oxidized to corresponding α,β-unsaturated ketones [19]. Cycloaddition reaction between carbon dioxide and an epoxide leading to the formation of cyclic carbonates under solvent-free condition could be achieved with tetrabutylammonium bromide (TBAB) as the co-catalyst [20].

Again **MIL-100(Fe)** [21] ($Fe_3F(H_2O)_2$-O(btc)2 (1,3,5-btc = benzene tricarboxylic)) had been successfully used for Friedel–Crafts reaction of benzene with benzyl chloride to afford diphenyl methane quantitatively within minutes. And **MIL-100(Sc)** ($Sc_3(OH)(H_2O)_2O(btc)_2$) had been utilized [22] for several C–C bond forming reactions [23], viz. (i) Friedel–Crafts type Michael addition, (ii) intermolecular carbonyl *ene* reaction, and (iii) ketimine and aldimine formation.

A cobalt containing MOF, **Co-MOF-74** (also known as **CPO-27-Co**) [24] had been synthesized from Co(II) and 2,5-dioxido-1,4-benzenedicarboxylate ($dobdc^{4-}$) that after activation could be used for the cycloaddition of CO_2 to styrene epoxide (Figure 17.6) [25], forming cyclic carbonate.

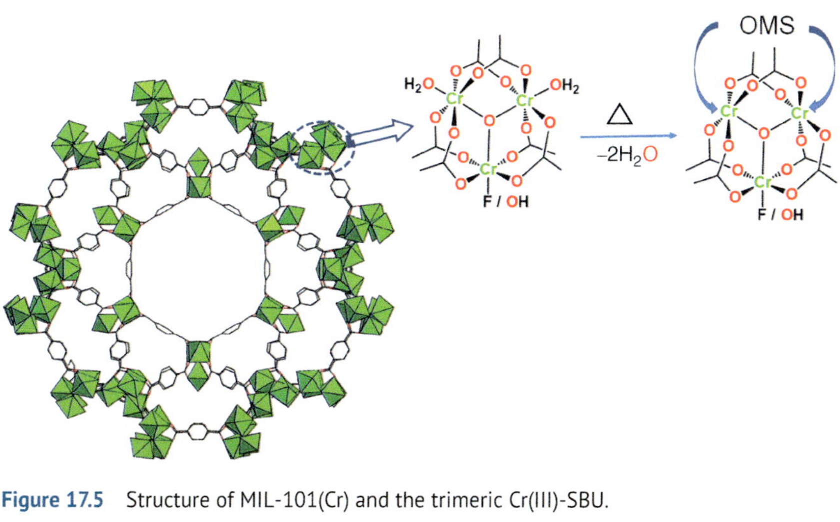

Figure 17.5 Structure of MIL-101(Cr) and the trimeric Cr(III)-SBU.

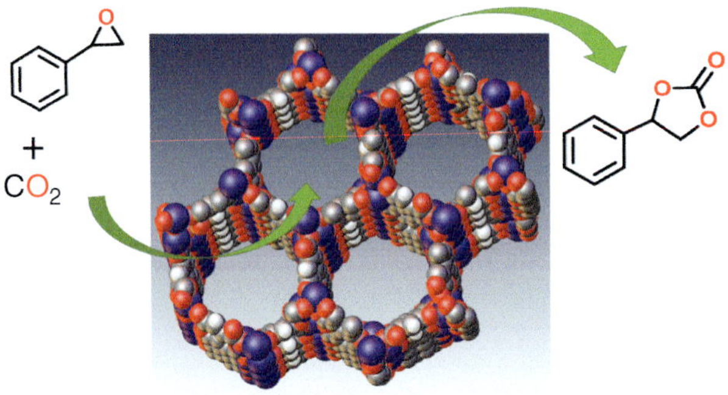

Figure 17.6 Structure of Co-MOF-74 and illustration of its catalytic application in cycloaddition of CO_2 to epoxides.

Subsequently, **Fe-MOF-74** or **CPO-27-Fe** and its magnesium-diluted analogue, **$Fe_{0.1}Mg_{1.9}$(dobdc)** were used for C–H bond activation of ethane and convert it into ethanol and acetaldehyde using nitrous oxide as the terminal oxidant [26]. De-solvated **Fe_2(dobdc)** had coordinatively unsaturated square pyramidal iron(II) centers directed towards the hexagonal channels. In the presence of N_2O (O atom transfer agent), it produced approximately 60% η^1-O and 40% η^1-N coordination (Figure 17.7), subsequently on heating at 60 °C generated highly reactive iron(IV)–oxo species (Scheme 17.3). When N_2O:ethane:Ar mixture (10:25:65) on flowing over the framework

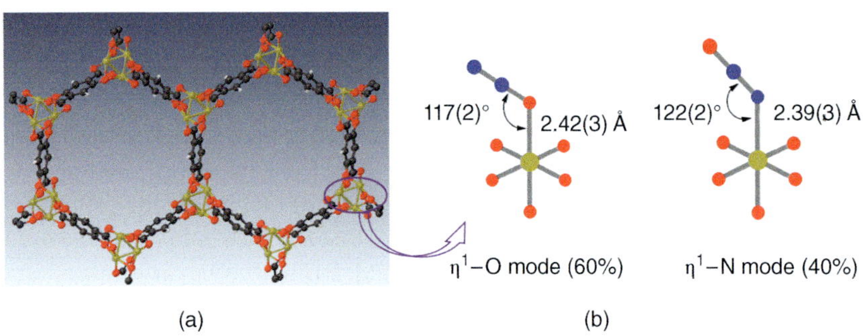

Figure 17.7 (a) Structure of Fe_2(dobdc), showing hexagonal channels decorated with exposed iron(II) sites, (b) N_2O binding modes in Fe_2(dobdc), obtained from powder neutron diffraction study. Fe, C, N, and O are represented by orange, grey, dark blue, and red spheres respectively; H atoms are omitted for clarity.

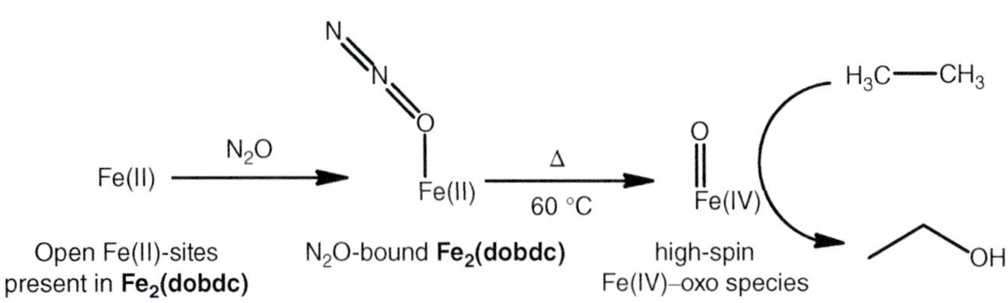

Scheme 17.3 Schematic presentation for the generation of Fe(IV)–oxo species in Fe_2(dobdc) which reacts with the strong C–H bonds of ethane and converts it to ethanol.

at 75 °C formed a mixture of compounds including ethanol, acetaldehyde, diethyl ether and other ether oligomers. But the magnesium-diluted analogue, **$Fe_{0.1}Mg_{1.9}$(dobdc)** afforded ethanol and acetaldehyde in a 10:1 ratio under the same conditions with better yield (60% with respect to iron).

There are several other examples of such MOFs known in the literature. These MOFs after activation through removal of metal bound solvent molecules had been used in Knoevenagel condensation [27], Mukaiyama-aldol condensation [28], Friedel–Crafts benzylation [29], cycloaddition of CO_2 to epoxides [30], cross-dehydrogenative coupling reaction [31], isomerization of α-pinene oxide and conversion of citronellal into isopulegol [32], Biginelli reaction [33], Henry reaction [34], Click reaction and three-component coupling of amines, aldehydes and alkynes [35], Pechmann reactions [36], Strecker reaction [37], and so on. This list is ever-increasing and detailed descriptions will be out of the scope of this chapter.

17.2.3 Catalysis at the Metal node Post Transmetalation

Post-synthetic transmetalation is a fascinating technique to afford catalytically active MOFs that are difficult to synthesize *ab initio*. Thus, **Zn-MFU-4*l*** could be easily transmetalated with Ni^{2+} to afford **Ni-MFU-4*l*** (Figure 17.8a) [38]. **Ni-MFU-4*l*** was highly active for ethylene dimerization

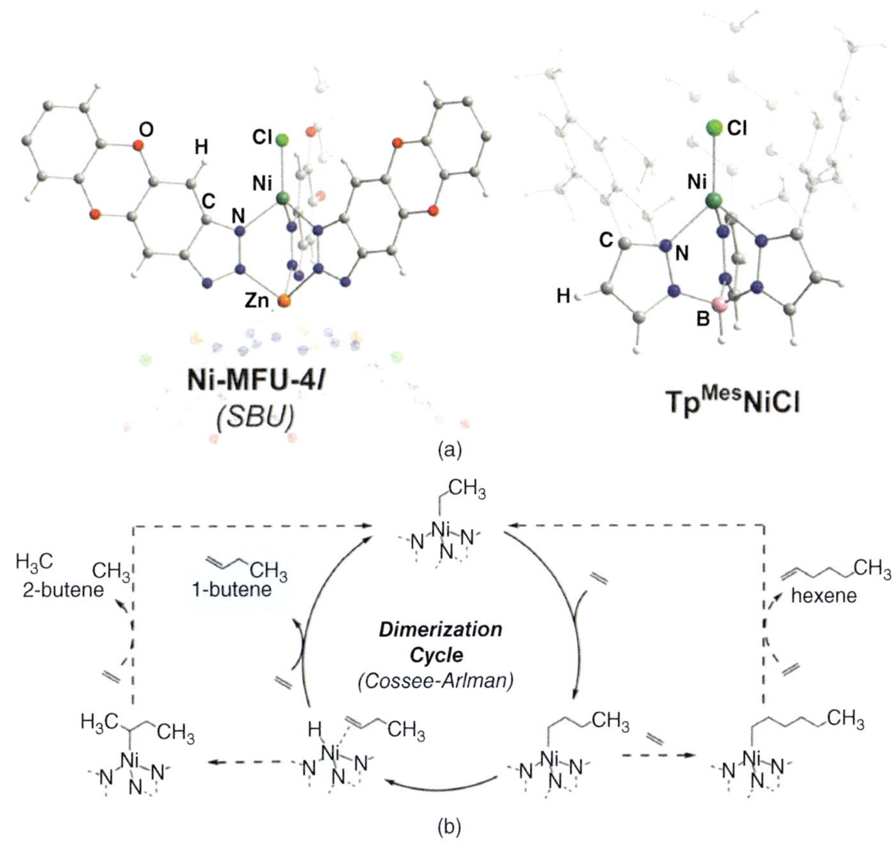

Figure 17.8 (a) Structure of active sites present in Ni-MFU-4*l*(left) and Tp^{Mes}NiCl (right). (b) Proposed ethylene dimerization mechanism. Source: Metzger, E. D., Brozek, C. K., Comito, R. J., and Dincă, M. (2016) Selective dimerization of ethylene to 1-butene with a porous catalyst. ACS Cent. Sci.2, 148–153. © 2016, American Chemical Society.

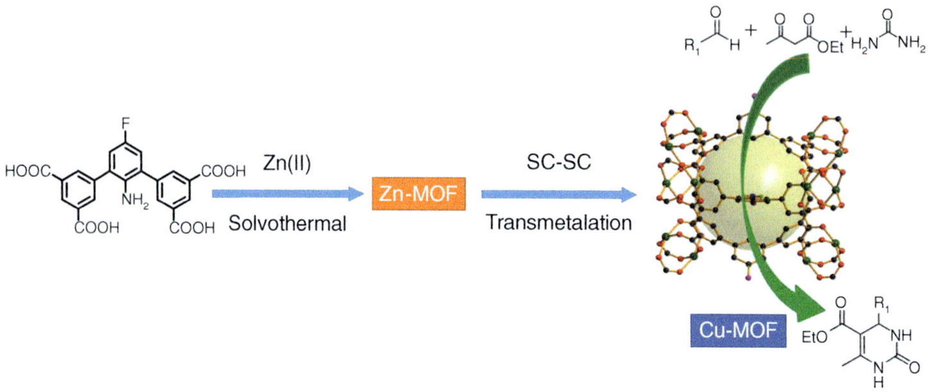

Figure 17.9 Synthesis of Cu-MOF from Zn-MOF by a SC-SC transmetalation process to show the Biginelli reaction. Source: Pal, T. K., De, D., Senthilkumar, S., Neogi, S. and Bharadwaj, P. K. (2016) A partially fluorinated, water stable Cu(II)–MOF derived via transmetalation: significant gas adsorption with high CO_2 selectivity and catalysis of Biginellireactions. Inorg. Chem., 55, 7835–7842. © 2016, American Chemical Society

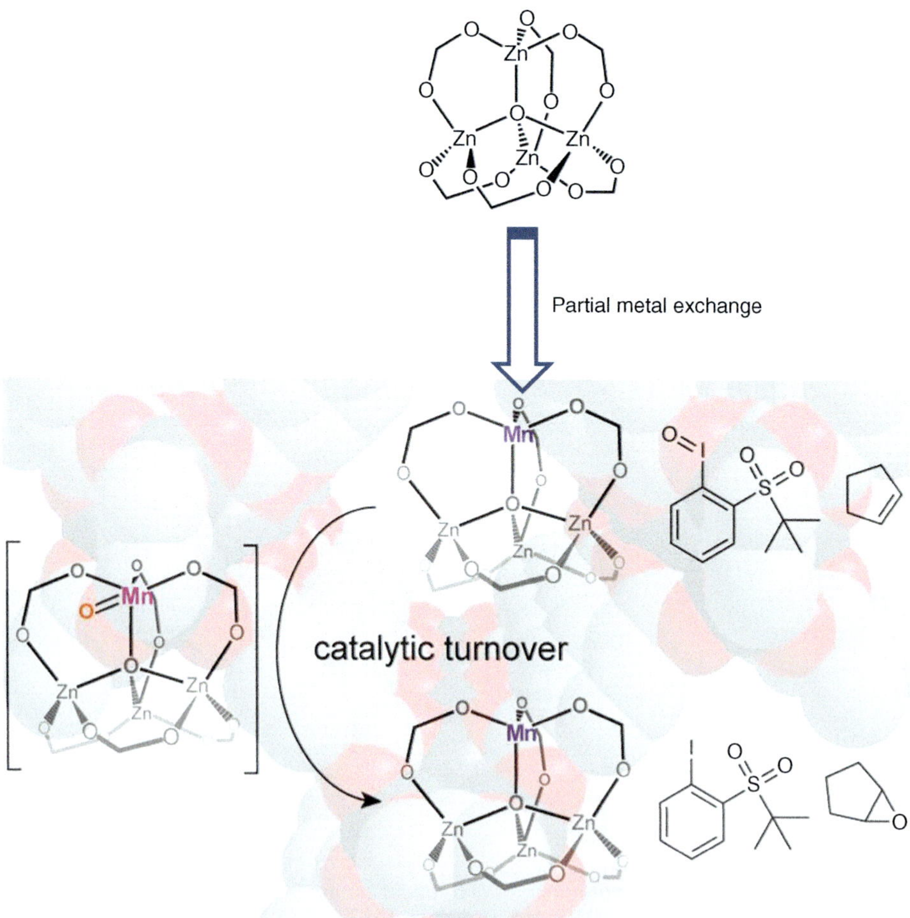

Figure 17.10 Partial exchange of Zn_4O SBU in MOF-5 to obtain $MnZn_3O$ SBU and the illustration of catalytic activity in epoxidation of cyclic alkenes.

to produce 1-butene (TOF: 41,500 h^{-1}, selectivity: 96%) under optimal conditions compared to the homogeneous catalyst **$[Tp^{Mes}Ni]^+$** (Tp^{Mes} = HB(3-mesitylpyrazolyl)$_3$) [TOF 34,600 h^{-1}]. The pristine MOF, **Zn-MFU-4*l*** did not exhibit any catalytic activity in alkene dimerization. The Cossee–Arlman mechanism was proposed for high yields of butene (Figure 17.8b).

A Cu(II)-MOF was obtained from the isostructural Zn(II)-MOF via SC-SC transmetalation (Figure 17.9) [39]. The activated Cu(II)-MOF exhibited excellent catalytic activity in the Biginelli coupling reactions affording high yields of dihydroprimidinones.

One of the Zn(II) ions of Zn_4O SBU in MOF-5 had been substituted with Mn(II) ion. In the presence of ^{t}BuSO$_2$PhIO, the Mn(II) site was transformed into Mn(IV)-oxo intermediate [40]. The high-spin Mn(IV)-oxo intermediate could catalyze the epoxidation of cyclic alkenes to form epoxides exclusively (Figure 17.10).

17.3 Functionalization in the MOF to Furnish Catalytic Site

In the forgoing discussion, we have shown that it is possible to have or generate open coordination sites on the metal ions of certain MOFs by the thermal removal of labile solvent molecules for successful catalysis. Alternative strategies for hosting a catalytic site in the MOF involve the following possibilities.

17.3.1 Attaching the Catalytically Active Moieties to the Metal Nodes (SBU)

Recent years have witnessed efforts of using the open coordination positions in the metal nodes as anchoring sites for grafting suitable organocatalysts. In activated MIL-101(Cr), proline derivatives [(S)-N-(pyridin-3-yl)-pyrrolidine-2-carboxamide (3-ppc) and (S)-N-(pyridin-4-yl)-pyrrolidine-2-carboxamide (4-ppc)] had been covalently connected through the pyridine group (Figure 17.11) to afford C-MIL-101(Cr) [41]. The resulting MOFs showed activity in asymmetric aldol reactions between different aldehydes and ketones in good to excellent yields (60–90%) with fair to good enantio-selectivity for the R-isomers (*ee* 55–80%).

Similarly, NU-1000-bpy-NiCl$_2$ synthesized from NU-1000 (Figure 17.12) by incorporation of (bpy)Ni^{II} moiety [42] exhibited excellent catalytic activity in the gas phase ethylene-dimerization in the presence of Et$_2$AlCl at ambient temperature (conversion 95%). The dimerization activity was higher than the corresponding homogeneous analogue (bpy)NiCl$_2$. The catalyst showed reusability up to at least three times.

17.3.2 Preconceived Catalytic Site into the Linker

MOFs with linkers incorporating organocatalysts will be like immobilizing a homogeneously active catalyst in a designed environment. Telfer's group incorporated the *boc*-protected proline moiety to have (*S*)-pro-boc to construct IRMOF-pro-boc [43] that could be deprotected to form IRMOF-pro (Figure 17.13). This MOF afforded the chiral aldol product with 29% *ee* upon reacting acetone with 4-nitrobenzaldehyde.

Another porous MOF, L$_{Cu}$PRO (Figure 17.14) [44] in the presence of the co-catalyst imidazole synergistically catalyzed the Baylis–Hillman reaction between α,β-unsaturated carbonyl compounds and aromatic aldehydes. A maximum of 75% yield was obtained within 24 h of reaction between methyl vinyl ketone and 4-nitrobenzaldehyde.

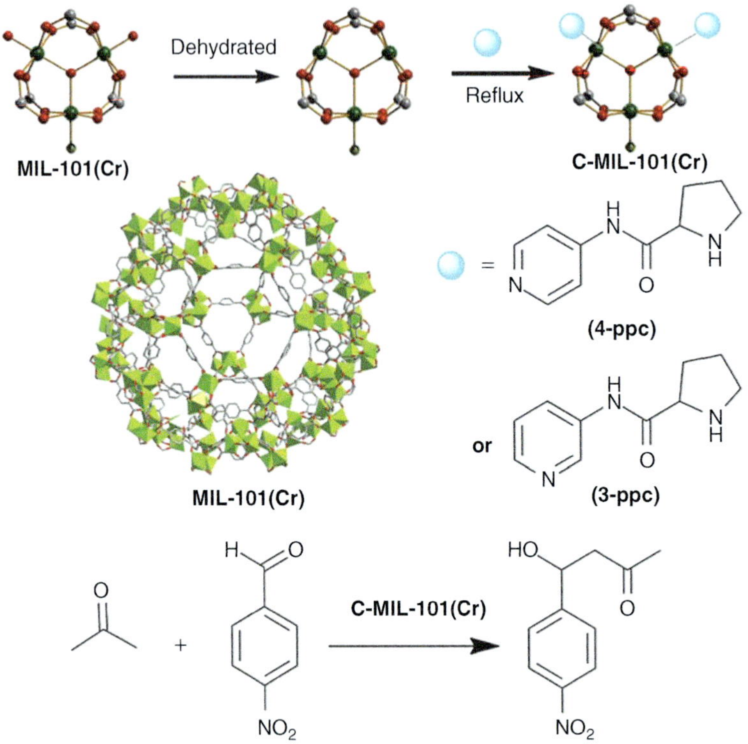

Figure 17.11 Schematic presentation for the installation of L-proline-derived organocatalysts to MIL-101(Cr) to generate C-MIL-101(Cr) and its catalytic application in aldol reaction.

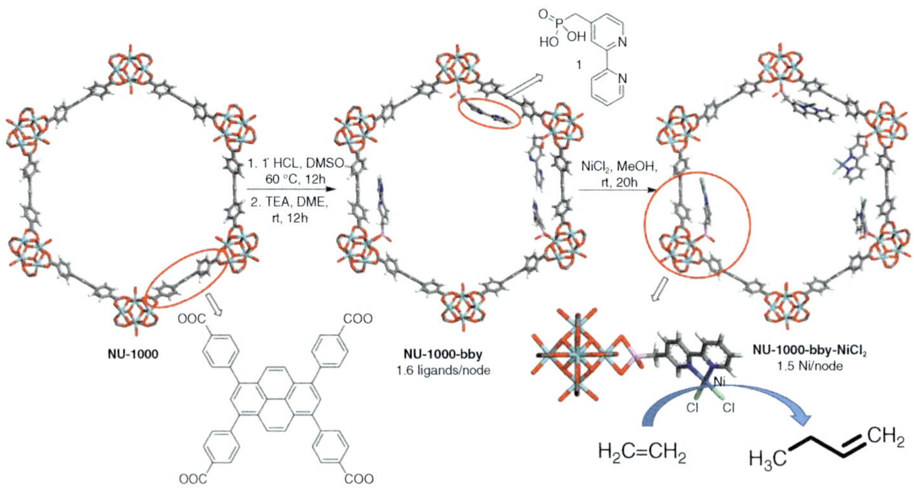

Figure 17.12 Schematic presentation for the preparation of NU-1000-bpy-$NiCl_2$ to illustrate the ethylene dimerization reaction.

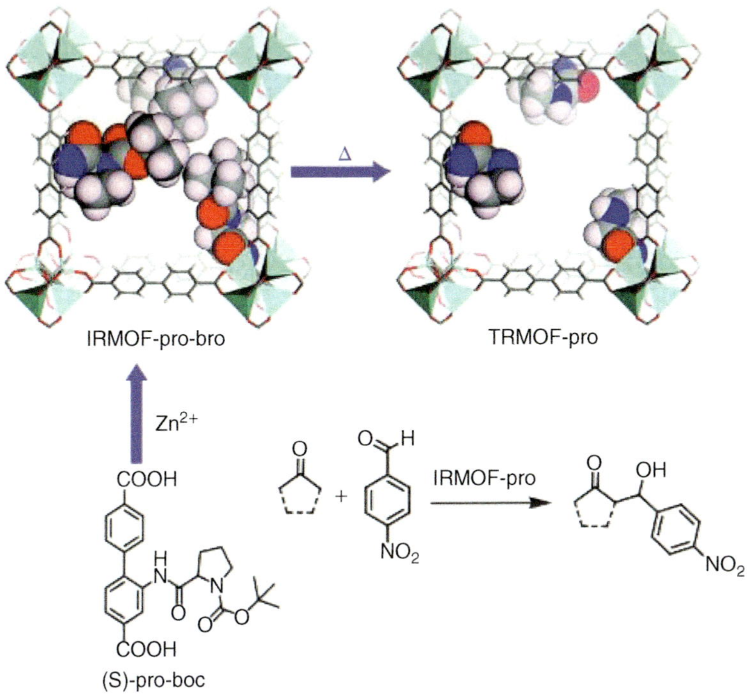

Figure 17.13 Synthesis of IRMOF-pro from IRMOF-pro-boc for the catalytic application in aldol reaction.

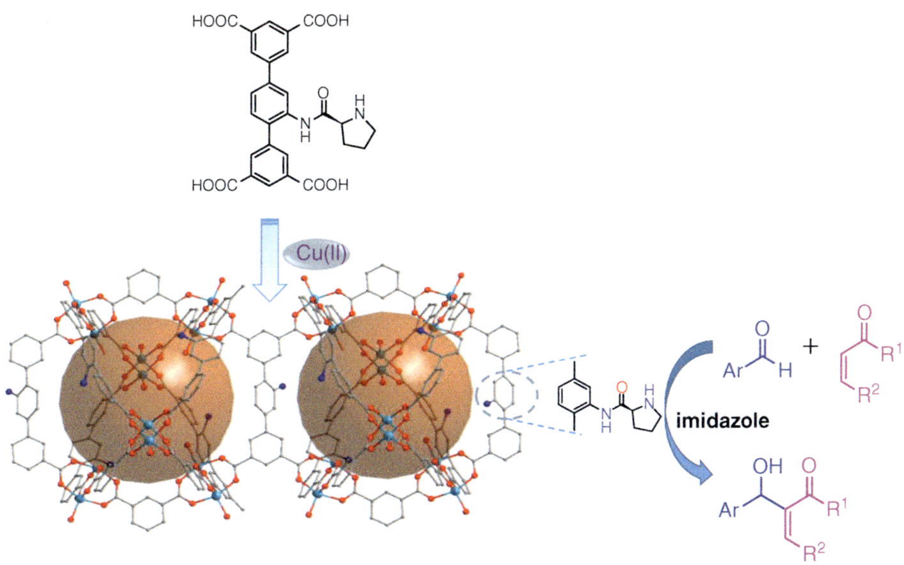

Figure 17.14 Synthesis of L_{Cu}PRO and its application in Baylis–Hillman reaction. Source: De, D., Pal, T. K., and Bharadwaj, P. K. (2016) A porous Cu(II)-MOF with proline embellished cavity: cooperative catalysis for the Baylis-Hillman reaction. Inorg. Chem., 55, 6842–6844. © 2016, American Chemical Society

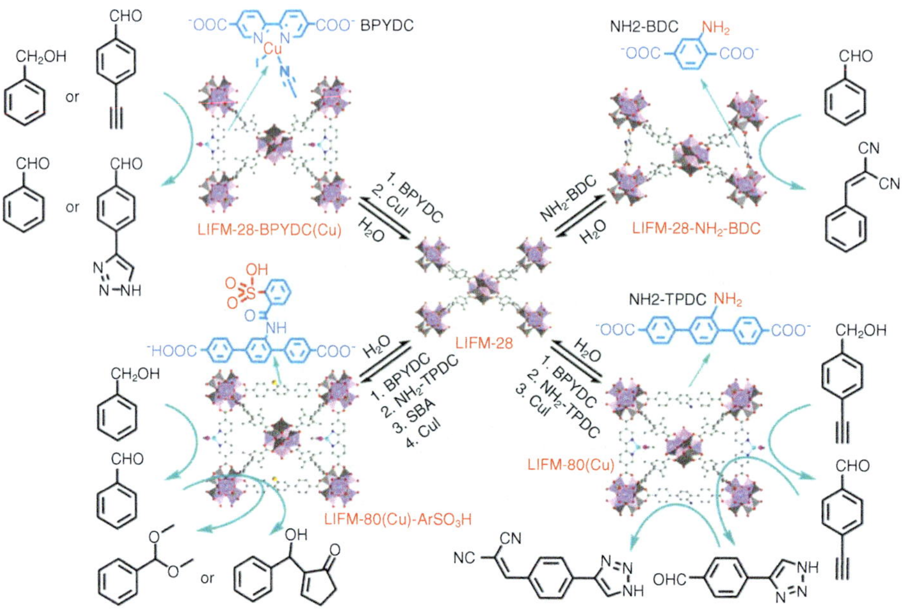

Figure 17.15 Reversible construction of MOF catalysts from the parent MOF LIFM-28for different catalytic applications. – $ArSO_3H$ group are presented by orange ball and Zr_6-SBUs are presented by purple polyhedrons. H and F are omitted for clarity.

It was possible to reversibly install and remove several catalytically active functional ligands in the same primitive MOF, LIFM-28 [45] leading to different heterogeneous catalysts (Figure 17.15) such as alcohol oxidation, Knoevenagel condensation, click, acetal, and Baylis–Hillman reactions.

A MOF built with polytopic urea and 4,4′ -bipyridine linkers with Zn(II) (NU-601) (Figure 17.16) was reported [46]. This framework was found to be an active hydrogen-bond-donor catalyst for Friedel–Crafts reactions between pyrroles and nitroalkenes.

An *N*-heterocyclic carbene (NHC) containing ligand had been incorporated into the parent MOF **UiO-68-NH$_2$** through post-synthetic ligand exchange strategy to afford UiO-68-NHC (Figure 17.17) [47]. Using **UiO-68-NHC**, the hydrosilylation of CO_2 with diphenylsilane was preformed to yield silyl methoxide. Followed by a mild hydrolysis, a nearly quantitative yield of methanol was achieved under mild conditions (1 atm CO_2, RT).

17.3.3 Post Synthetic Modification of the Linker

Yaghi and coworkers reported **IRMOF-3** ($Zn_4O(bdc\text{-}NH_2)_3$) decorated with pendant-NH_2 groups [48]. The amine groups were accessible for further functionalization (Figure 17.18). The Schiff base condensation reaction had been performed with salicylidene (sal) moiety to afford **IRMOF-3-sal** followed by metal chelation with $V(O)(acac)_2{\cdot}H_2O$ (acac = acetylacetonate) to form **IRMOF-3-sal-V** (Figure 17.18) [49]. **IRMOF-3-sal-V** catalyzed the oxidation of cyclohexene with TBHP (40% conversion of cyclohexene in THF at 60 °C for 72 h). Another covalent post-synthesis to prepare **IRMOF-3[Au]** was reported [50]. **IRMOF-3[Au]** was then employed to catalyze the multicomponent domino coupling reaction of *N*-tosyl protected ethynylaniline, paraformaldehyde and piperidine (Figure 17.18).

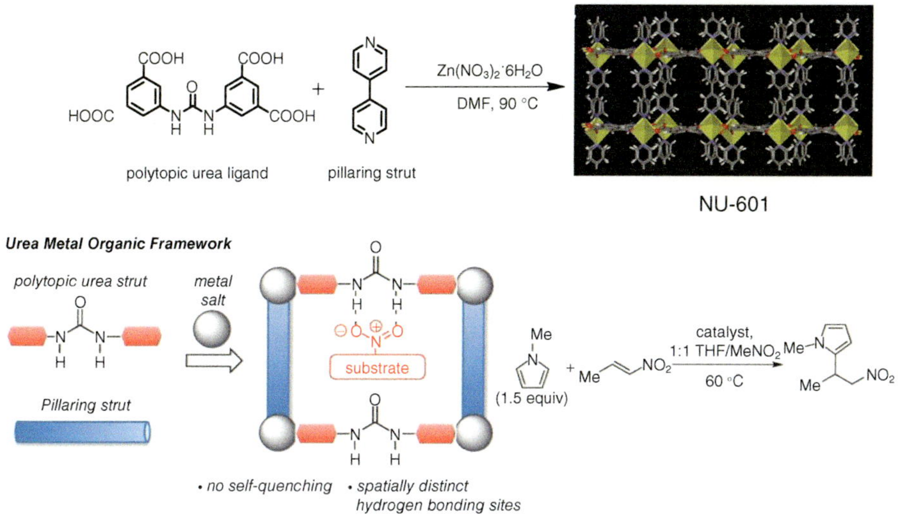

Figure 17.16 Synthesis of NU-601 (top) and urea MOF strategy for substrate activation (bottom). Source: Roberts, J. M., Fini, B. M. Sarjeant, A. A., Farha, O. K., Hupp, J. T., and Scheidt, K. A. (2012) Urea metal–organic frameworks as effective and size-selective hydrogen-bond catalysts, J. Am. Chem. Soc. 134, 3334–3337. © 2012, American Chemical Society.

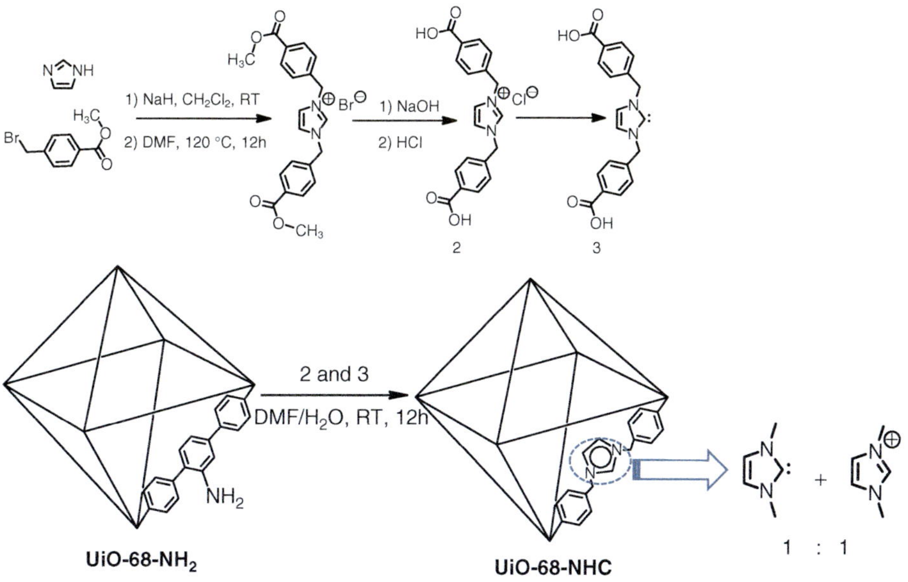

Figure 17.17 UiO-68-NHCwas synthesized by post synthetic ligand exchange from the parent MOF UiO-68-$NH_2$with *in-situ* generated *N*-heterocyclic carbine 3 from an imidazolium precursor 2. Source: Zhang, X., Sun, J., Wei, G., Liu, Z., Yang, H., Wang, K., and Fei, H. (2019) In situ generation of an *N*-heterocyclic carbene functionalized metal–organic framework by postsynthetic ligand exchange: efficient and selective hydrosilylation of CO_2. Angew. Chem. Int. Ed. 58, 2844–2849. © 2019, John Wiley & Sons.

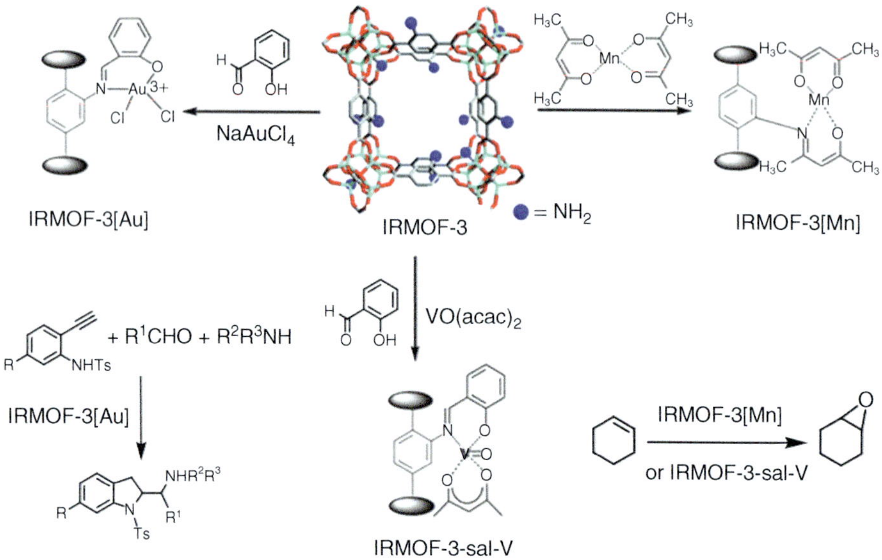

Figure 17.18 Post-functionalization of IRMOF-3 with three different metal complexes of Mn (right), Au (left), V (middle), and illustration of their catalytic application.

Again, **IRMOF-3[Mn]** was prepared by attaching manganese(II) acetylacetonate complex to **IRMOF-3** [51]. It catalyzed the epoxidation of alkenes (conversion 68%) using molecular oxygen and trimethylacetaldehyde as the oxidant (Figure 17.18).

In a major overhaul of the cavity in **NH_2-MIL-101(Cr)**, both phenylene-mono(oxamate) (PMA) and ethylenediamine (EDA) were complexed with Cu(II), forming **NH_2-MIL-101(Cr)-PMA-CuEDA** (Figure 17.19) [52]. It exhibited high catalytic activity in the *click* reaction involving benzyl azide with phenylacetylene (100% conversion) and in one-pot three-component coupling reaction to form propargylamine (60% conversion).

The MOF **UiO-66(L3)** had been synthesized by post-synthetic "click" reaction between **UiO-66-N_3** and 4-ethynyl-2,2′-bipyridine (**L3**) [53]. **UiO-66(L3)** in combination with $Ni(COD)_2$, and PPh_3 functioned as Ni-immobilized heterogeneous catalyst for Suzuki–Miyaura coupling reaction between bromobenzene and phenylboronic acid, in the presence of K_2CO_3 in CH_3CN at 65 °C for 12 h (Figure 17.20). A maximum of 96% yield of the product was achieved with a good recyclability (recycled at least up to seven times).

When the photocatalyst EOSIN-Y was connected to the free NH_2 group of **UiO-66-NH_2** (Scheme 17.4) [54] the resulting **EY@UiO-66-NH_2** was found to be quite an effective photocatalyst for the C-H activation of tertiary amines in oxidative cyanation reactions with excellent yields (maximum 96%). The catalytic activity of **EY@UiO-66-NH_2** was even better than the homogeneous EOSIN-Y and could be recycled up to 10 cycles.

17.3.4 MOFs with Linkers Having Coordinated Metal Ions (Metalloligands)

When a metal ion bound to a linker and the resulting metalloligand can form the MOF where the metal ion of the metalloligand moiety could be responsible for the catalytic activity and the other metal ions maintain structural integrity. Various metallo-porphyrin and -salen moieties have been used (Scheme 17.5) for building MOF catalysts. Several Zn-/Mn-/Co-/Pd-/Fe-porphyrin-based

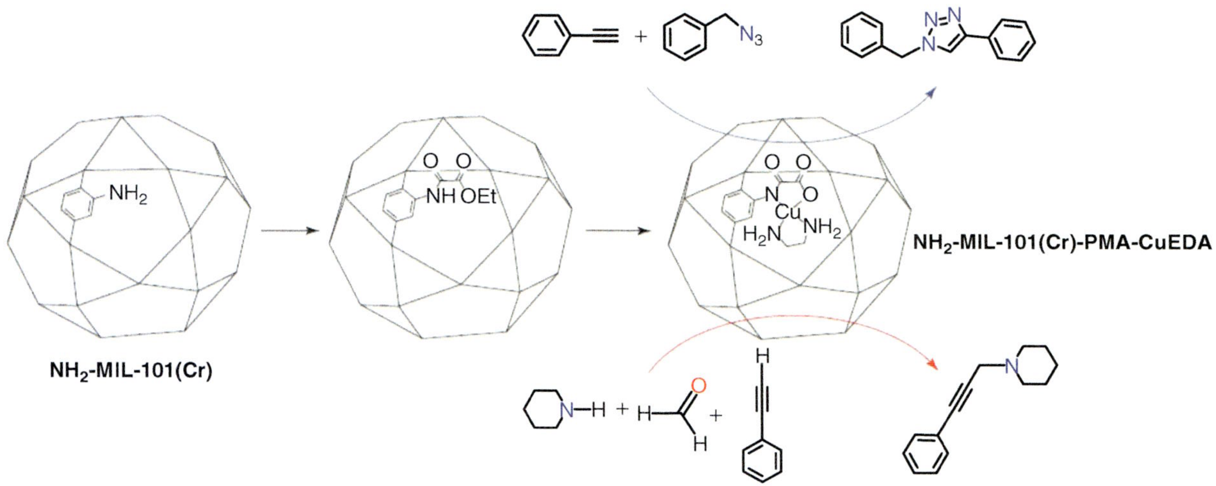

Figure 17.19 Post-synthetic modification of -NH_2 groups in NH_2-MIL-101(Cr) to form heterogeneous Cu-based catalyst. Source: Based on Juan-Alcaniz, J., Ferrando-Soria, J., Luz, I., Serra-Crespo, P., Skupien, E., Santos,V.P., Pardo, E., LlabresiXamena, F.X., Kapteijn, F., and Gascon, J. (2013) The oxamate route, a versatile post-functionalization for metal incorporation in MIL-101(Cr): Catalytic applications of Cu, Pd, and Au. J. Catal., 307, 295–304.

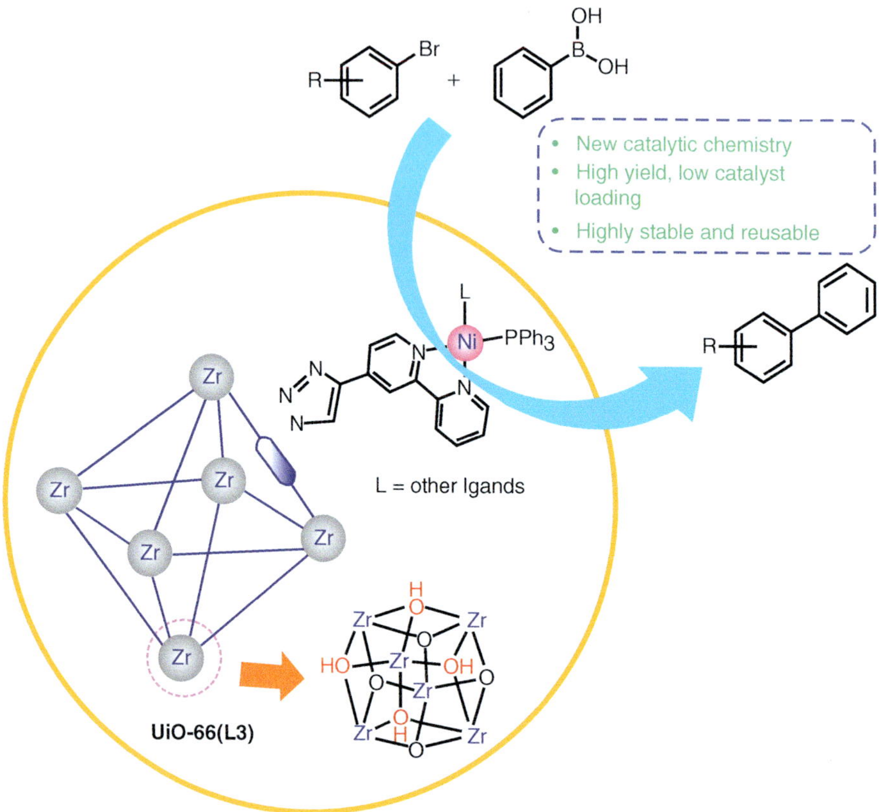

Figure 17.20 Suzuki-Miyaura cross coupling reaction catalyzed by Ni-immobilized UiO-66(L3).

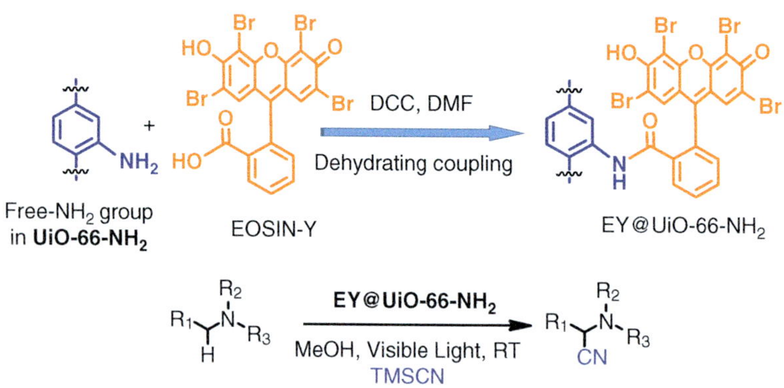

Scheme 17.4 Post-synthetic modification of free amine in pristine framework of UiO-66-$NH_2$with EOSIN-Y dye to form EY@UiO-66-NH_2 (above) for cyanation reaction (below). Source: Based on Kumar, G., Solanki, P., Nazisha, M., Neogi, S., Kureshya, R. I., and Khan, N. H. (2019) Covalently hooked EOSIN-Y in a Zr(IV) framework as visible-light mediated, heterogeneous photocatalyst for efficient C-H functionalization of tertiary amines. J. Catal., 371, 298–304.

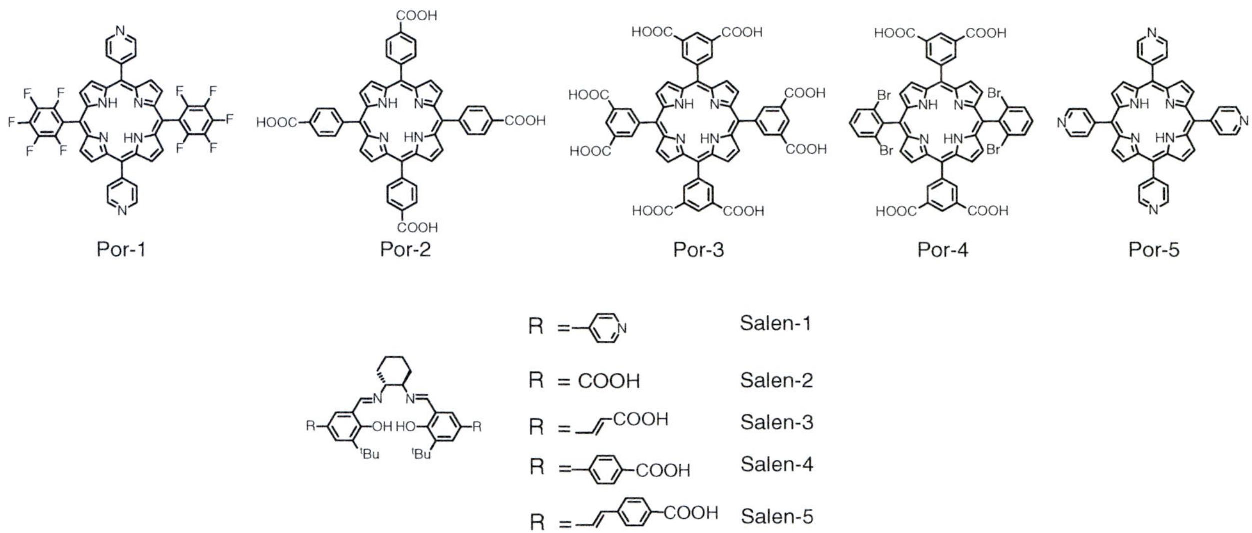

Scheme 17.5 Schematic presentation of various porphyrin and salen containing metalloligand for the synthesis of MOFs.

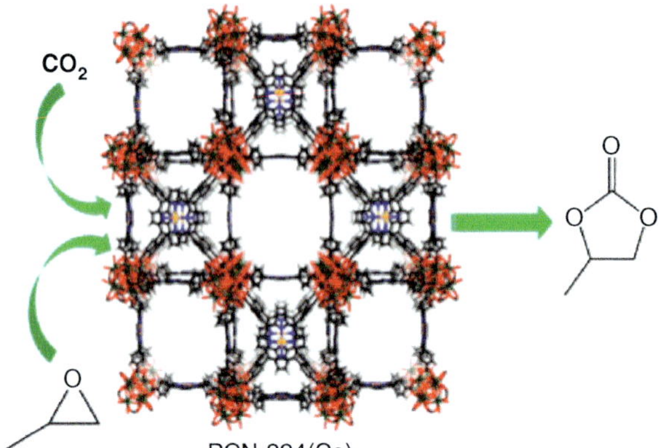

Figure 17.21 Structure of PCN-224(Co) and the schematic presentation of fixation of CO_2.

MOFs (Por-MOFs) have been reported by Nguyen [55], Hupp [56], Ma [57], Wu [58], and Zhou [59] groups for catalytic applications in acyl-transfer reaction, epoxidation of olefin, cycloaddition of CO_2 to epoxides etc. One example is illustrated below.

A series of highly stable MOFs with the formula of Zr_4(M-por-2) (denoted as **PCN-224**; M = no metal, Ni, Co, Fe) had been reported [60]. The framework has 6-connected Zr_6 clusters. **PCN-224(Co)** was used in the chemical fixation of CO_2 with epoxides to form cyclic carbonate. The TON and TOF values were 518 and 129 h^{-1}, respectively after the third run, signifying its reusability (Figure 17.21).

Like porphyrin, salen ligand had been used to have $Zn_2(bpdc)_2$(Mn-salen-1) [61] and Zn_2(tcpb)(Mn-salen-1) [62] (bpdc: biphenyl dicarboxylate; tcpb: tetrakis(4-carboxyphenyl) benzene). Both showed excellent catalytic activity for the asymmetric epoxidation of 2,2-dimethyl-2*H*-chromene in the presence of 2-(*tert*-butylsulfonyl)iodosylbenzene as oxidant. A series of isoreticular Mn-salen-derived chiral MOFs had also been constructed [63] using the Mn-salen ligands shown in Scheme 17.5 with Zn_4SBUs that showed asymmetric epoxidation of unfunctionalized olefins.

In another report, three Zr_6-SBU containing MOFs, **BPV-MOF-Ir**, **mBPV-MOF-Ir**, and **mPT-MOF-Ir** had been reported (Figure 17.22) [64]. These were highly active catalysts for tandem hydrosilylation/*ortho*-silylation of aryl ketones and aldehydes, tandem dehydrocoupling/*ortho*-silylation reactions of *N*-methylbenzyl amines, and borylations of aromatic C–H bonds at the Ir center.

The metalloligand Cp*Rh(5,5′-dcbpy)Cl(Cp* = pentamethylcyclopentadiene) was grafted via post-synthetic ligand exchange into the MOF **UiO-67** to afford **Cp*Rh@UiO-67** (Figure 17.23) [65]. This exhibited photocatalytic CO_2 reduction to formate with a TON of 47.

17.4 MOFs as Bifunctional Catalyst

Apart from the catalytic activity shown by open metal sites or Lewis basic sites, the catalytic activity due to the presence of bifunctionalities in MOF offered synergistic effect towards catalytic reaction [66]. Thus, a polystyrene (PS) nanosphere was first synthesized, having –COOH terminals, which was then reacted with zinc nitrate in the presence of 2-methyl imidazole,

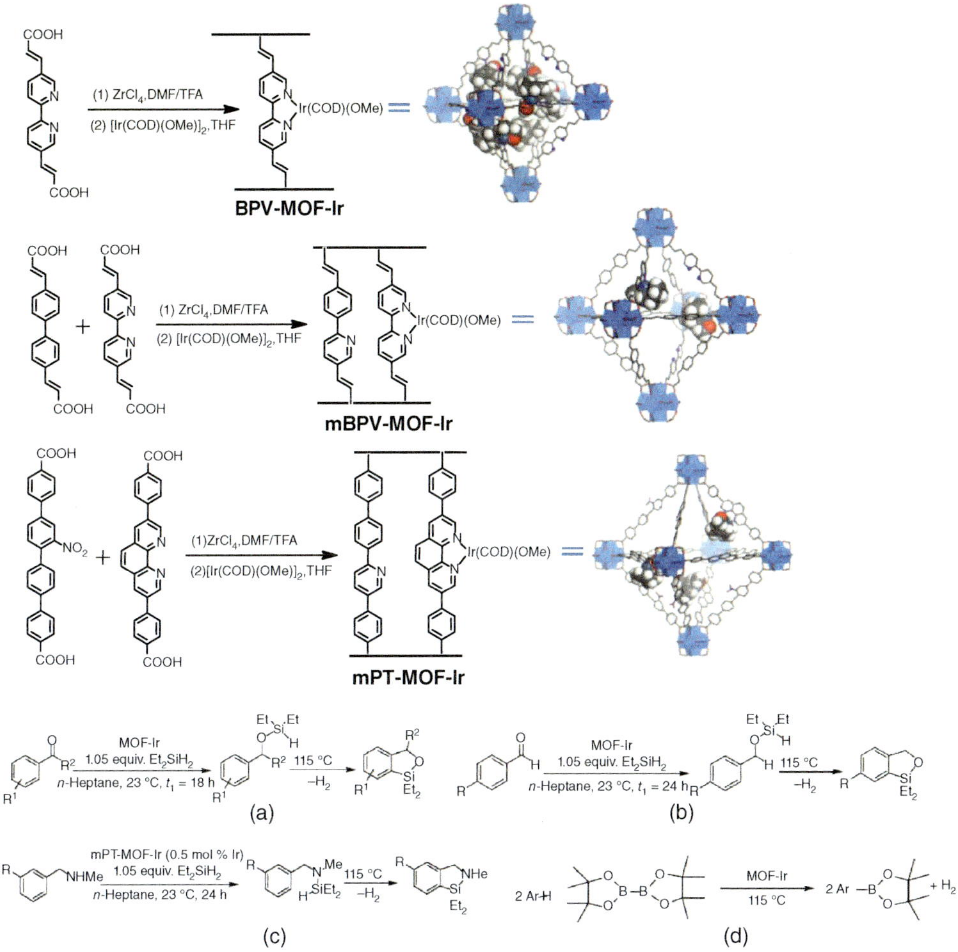

Figure 17.22 Synthesis of various Zr-MOFs chelating with Ir-based catalyst (top). Tandem hydrosilylation of (a) ketones, and (b) aldehyde and intramolecular *ortho*-silylation of benzylic silylethers to prepare benzoxasiloles, (c) tandem dehydrocoupling of N-methylbenzyl amines and intramolecular *ortho*-silylation of (hydrido)silyl amines to azasilolanes, (d) C-H borylation of arenes. Source: Manna, K., Zhang, T., Greene, F.X., and Lin, W. (2015) Bipyridine- and phenanthroline-based metal-organic frameworks for highly efficient and tandem catalytic organic transformations via directed C-H activation. J. Am. Chem. Soc., 137, 2665–2673. © 2015, American Chemical Society.

which led to the formation of **PS-ZIF-8** composite, and after dipping in toluene the PS was removed to give **ZIF-8-H** (Scheme 17.6). This could catalyze the [3+3] cycloaddition reaction [67] between 1,3-cyclohexandione and α,β-unsaturated aldehyde (Scheme 17.7). The reaction between 3-methyl-2-butenal with 1,3-cyclohexanedione showed 89 % conversion and 99.9 % selectivity at 20 °C in dichloromethane within 24 h.

The catalyst was versatile and successfully converted several α,β-unsaturated aldehyde compounds to the corresponding desired product.

The MOF **MIL-101-Cr** had been sequentially derivatized to Brønsted acidic MIL-101-Cr-SO_3H and then Lewis acidic@Brønsted acidic MOF, **MIL-101-Cr-SO_3H·Al(III)** (Scheme 17.8).

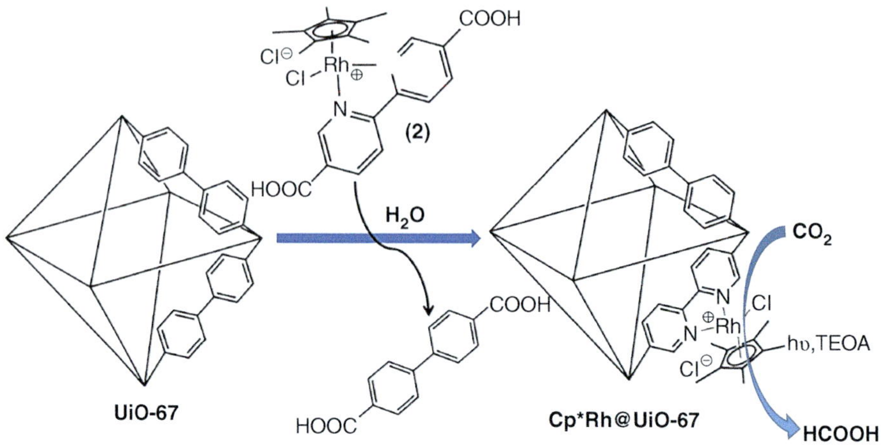

Figure 17.23 Synthetic scheme of Cp*Rh@UiO-67 and its catalytic application in CO_2 reduction to formate. Source: Chambers, M. B., Wang, X., Elgrishi, N., Hendon, C. H., Waksh, A., Bonnefoy, J., Canivet, J., Quadrelli, E. A., Farrusseng, D., Mellot-Draznieks, C., and Fontecave, M. (2015) Photocatalytic carbon dioxide reduction with rhodium-based catalysts in solution and heterogenized within metal–organic frameworks. ChemSusChem, 8, 603–608. © 2015, John Wiley & Sons.

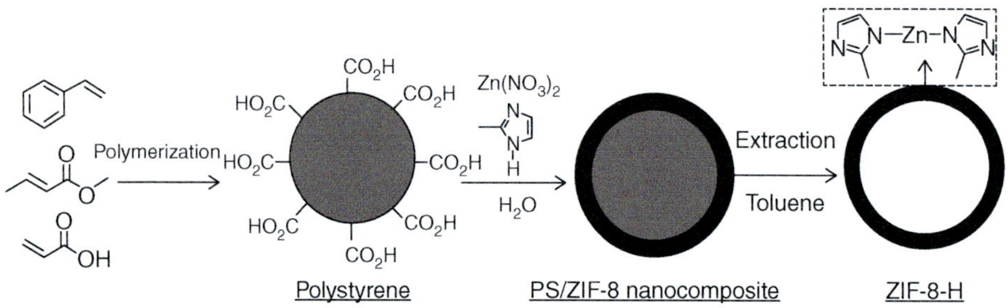

Scheme 17.6 The synthesis of ZIF-8-H.

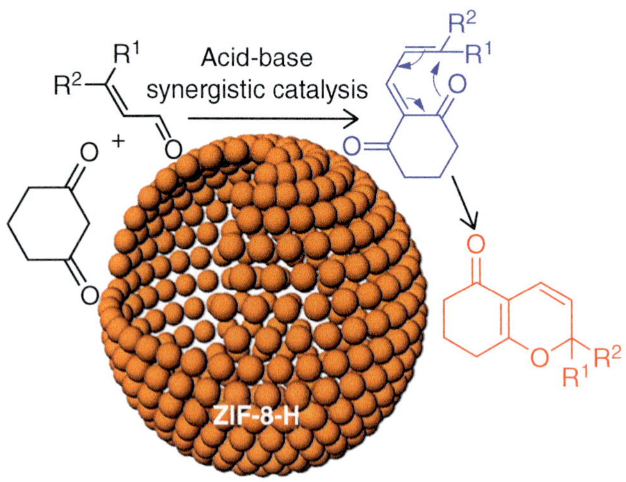

Scheme 17.7 Cyloaddition catalytic reaction by ZIF-8-H.

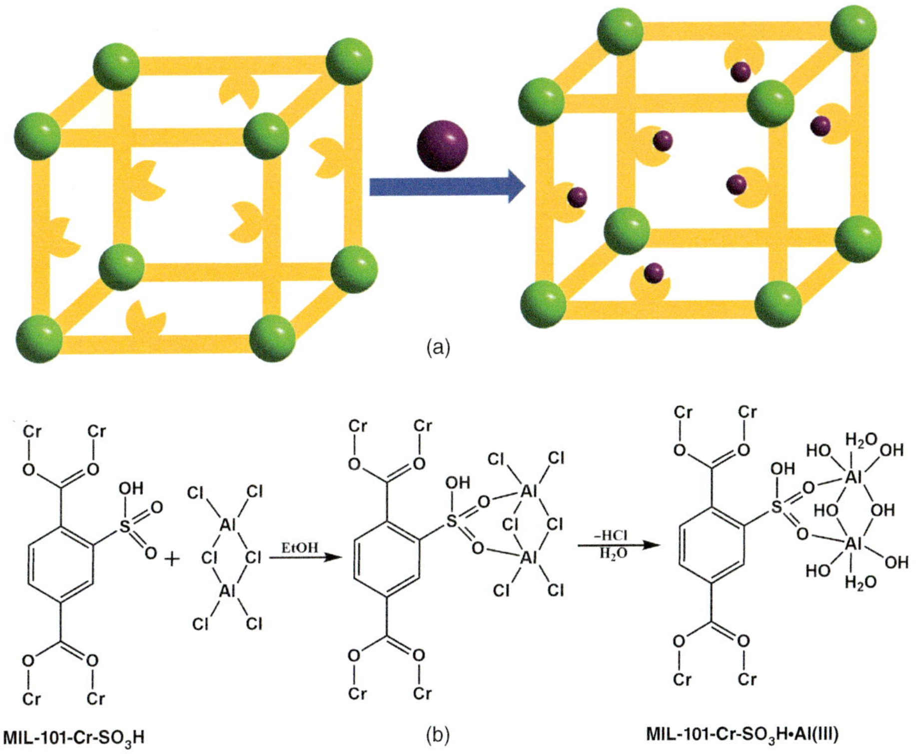

Scheme 17.8 (a) The inclusion of Lewis acidic center (violet ball) and (b) proposed reaction mechanism during inclusion of acidic center within MIL-101-Cr-SO_3H·Al(III).

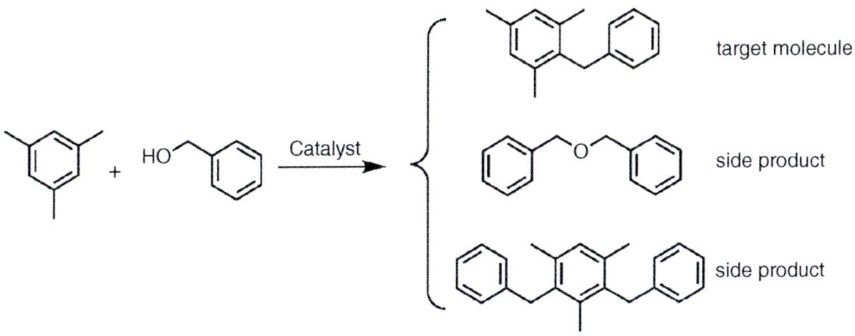

Scheme 17.9 Benzylation reaction between 1,3,5-trimethylbenzene and benzyl alcohol.

The catalyst **MIL-101-Cr-SO_3H·Al(III)** could perform fixed-bed reaction (benzylation) between mesitylene and benzyl alcohol (Scheme 17.9) and showed selectivity >99% of the desired compound, 2-(benzyl)-1,3,5-trimethyl benzene after 2 h with 100% benzyl alcohol conversion. The catalytic efficiency was superior to the benchmark H-Beta and HMOR [68] besides some other MOFs (**MIL-100-Fe**, **HKUST-1**, and **PW@MIL-100-Cr**) (benzyl alcohol conversion <20%) under the same experimental conditions. When the same reaction was performed individually with either **MIL-101-Cr-SO_3H** or **Al_2O_3** very little conversion happened, indicating the presence of open metal site solely not sufficient for this catalytic reaction. Furthermore, the physical mixture between them displayed very low yield.

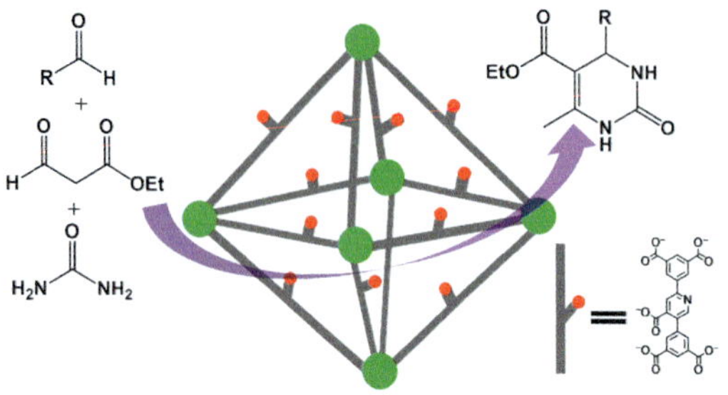

Scheme 17.10 Biginelli catalytic reaction by M-DDIA.

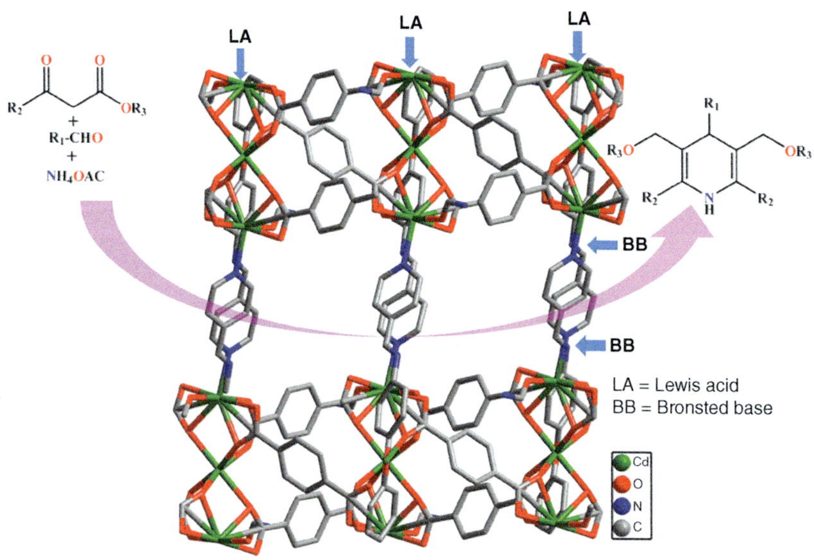

Scheme 17.11 Hantzsch catalytic reaction by TMU-33.

This concept of Lewis acid and Brønsted functionalities in MOF had been used for the one pot three-component Biginelli (Scheme 17.10) [69] and Hantzsch reactions (Scheme 17.11) [70].

Another interesting cooperative catalyst with bifunctional Cu(I) in a MOF having Zr_{12} SBU was synthesized, which catalyzed the conversion of CO_2 to ethanol (C2 product) through hydrogenation reaction (Scheme 17.12) [71].

As discussed earlier, tandem catalytic reactions are always welcome to cut down the synthetic step. A bifunctional core-shell MOF-COF composite, MOF@COF (**PCN-222-Co@TpPa-1**) (Scheme 17.13) stabilized via $\pi\cdots\pi$ interactions had acid-base bifunctional sites that could catalyze tandem deacetalization followed by Knoevenagel condensation reaction [72].

The composite material had Lewis acidic sites (Co(II) and Zr(IV) clusters) and TpPa-1 Brønsted basic site allowing one pot cascade deacetalization–Knoevenagel reaction (Scheme 17.14).

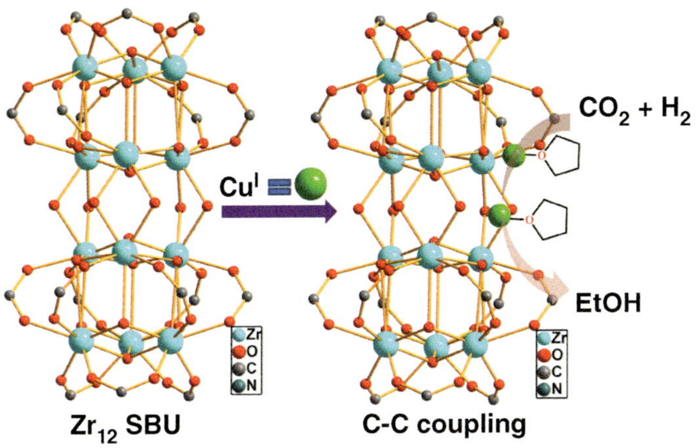

Scheme 17.12 Representation for the conversion of CO_2 to ethanol.

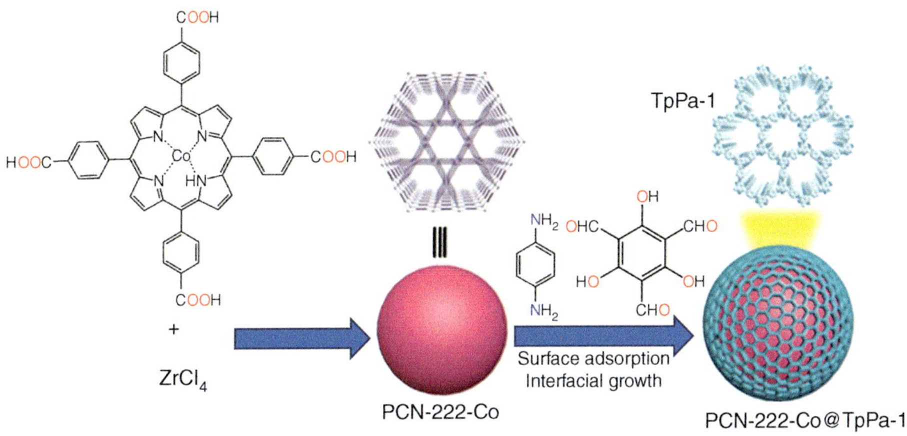

Scheme 17.13 The synthetic path of PCN-222-Co@TpPa-1. Source: Gao, M.-L., Qi, M. –H., Liu, L., and Han, Z.-B. (2019) An exceptionally stable core–shell MOF/COF bifunctional catalyst for a highly efficient cascade deacetalization–Knoevenagel condensation reaction. Chem. Commun., 6377–6380. © 2019, Royal Society of Chemistry

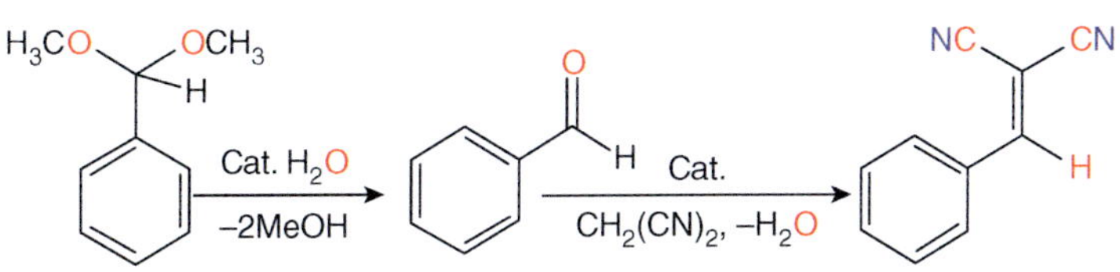

Scheme 17.14 One-pot tandem deacetalization–Knoevenagel reaction by PCN-222-Co@TpPa-1.

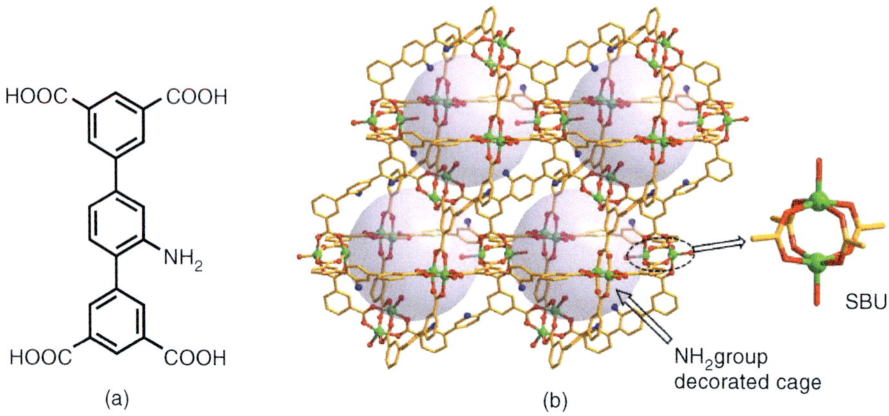

Figure 17.24 (a) Tetracarboxylate linker, (b) the pore is highly decorated with amine group. Source: De, D. T. K. Pal, S. Neogi, S. Senthilkumar, Debasree Das, S. Sen Gupta, P. K. Bharadwaj, (2016) A versatile Cu(II) metal–organic framework exhibiting high gas storage capacity with selectivity for CO_2: conversion of CO_2 to cyclic carbonate and other catalytic abilities, Chem. Eur. J. 22, 3387–3396. © 2016, John Wiley & Sons.

Use of only **PCN-222-Co** or **TpPa-1** could not execute the cascade reaction. However, the same reaction could be done by using **PCN-124 MOF** [73] or **Zr-MOF-NH$_2$** under UV light [74]. Another Lewis acid–Brønsted base combined MOF, **NH$_2$-MIL-101(Al)** performed tandem heterogeneous Meinwald rearrangement–Knoevenagel condensation [75]. The cycloaddition reaction of CO_2 to epoxide generally catalyzed by either Lewis acid or Lewis base (*vide supra*). However, the bifunctional MOFs could do the same reaction very smoothly [76]. A Cu(II)-MOF built with a linear tetracarboxylic acid ligand having pendant amine group (Figure 17.24) [77] formed a porous structure. High thermal stability, generation of Lewis acidic copper site, and the Lewis basic amino group in the cavity showed excellent catalytic activity in the cycloaddition of CO_2 to epoxide in the presence of TBAB as co-catalyst.

When the bent tetracarboxylate ligand was used (Figure 17.25) [78], the activated Cu(II)-MOF showed excellent catalytic ability of cycloaddition of CO_2 to epoxides at room temperature in the

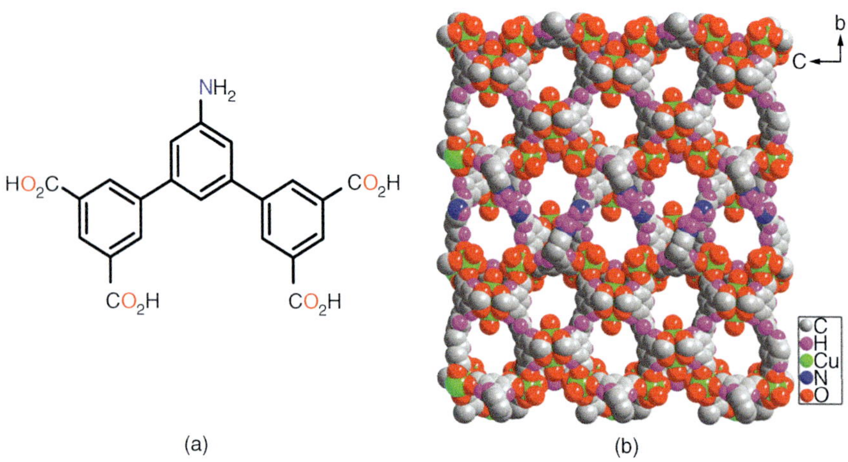

Figure 17.25 (a) Schematic diagram of the ligand, (b) the pore along *a*-axis. Source: Sharma, V., De, D., Saha, R., Das, R., Chattaraj, P. K. and Bharadwaj, P. K. (2017) A Cu(II)-MOF capable of fixing CO_2 from air and showing high capacity H_2 and CO_2 adsorption, Chem. Commun., 53, 13371–13374. © 2017, Royal Society of Chemistry.

presence of TBAB. It is noteworthy that the CO_2 present in the atmosphere could be successfully converted to the cyclic carbonate upon passing the atmospheric air into the MOF solution in the presence of TBAB.

17.5 Impregnation/Encapsulation of Nanoparticles in the MOF Cavity for Catalysis

Through supramolecular engineering, MOFs can encapsulate biomolecules, various organic compounds, metal nanoparticles, etc. [79]. The high surface area and chance of functionalization in the MOFs can also help to grow or graft the metal nanoparticle very softly without hampering the MOF structure where metal nanoparticle dispersion, size and shape of the cavity are important parameters for having stable and efficient materials [80]. A variety of metal nanoparticles like Pd, Ni, Ag, Au, Pt, TiO_2 had been impregnated into MOFs [81]. The C-C bond formation is an important reaction to bring on biaryl units formation in industry as well as in several organic synthesis. In this regard, Sonogashira, Suzuki, Heck etc. are the important reactions [82]. In an early report, the MOF, **MIL-101Cr-NH$_2$** had been used for the preparation of **Pd@ MIL-101Cr-NH$_2$** with different loading of 4, 8, 12, and 16 wt % to carry out the Suzuki–Miyura coupling reactions (Scheme 17.15) in water with Cs_2CO_3 as a base for 6 h [83].

Among all the Pd loaded catalyst, the catalyst with 8 wt % **Pd@ MIL-101Cr-NH$_2$** showed highest activity with quantitative yield. TEM images of the Pd loaded catalyst showed that nanoparticle sizes were broader in 12 and 18 wt % loading (Figure 17.26) and the diffusion of the reacting substrate in to the cavity was difficult due to pore jamming. However, recyclability of the catalyst posed problems due to leaching of Pd and more investigations would be necessary for effective catalysis with such systems.

In another report, gold nanoparticle encapsulated MOF was synthesized for catalyzing one pot three component coupling reaction between paraformaldehyde, piperidine, and phenylacetylene

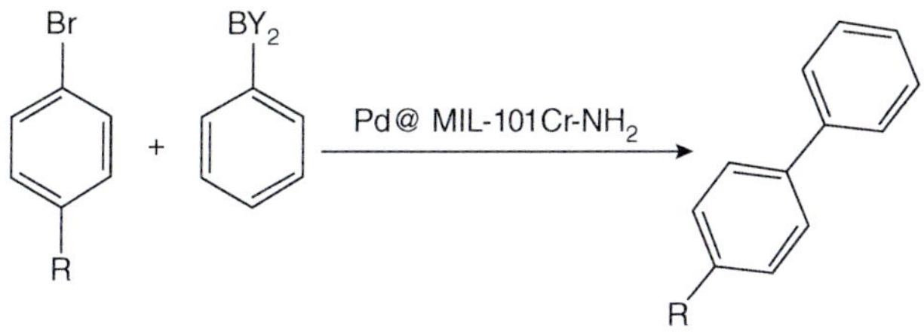

Scheme 17.15 Suzuki–Miyaura reactions by Pd@MIL-101Cr-NH$_2$.

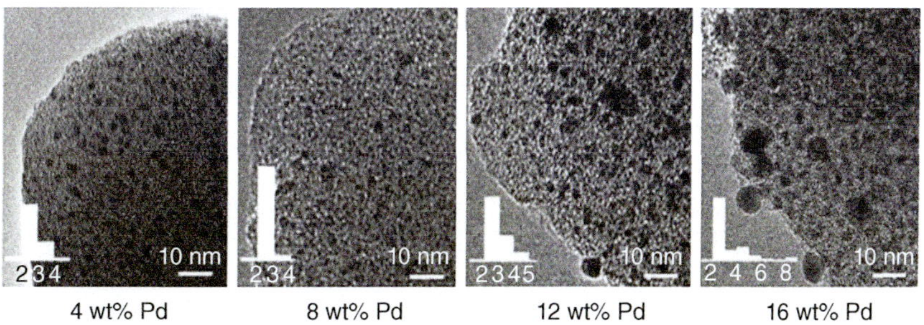

Figure 17.26 Different amount of palladium nanoparticles loading on MIL-101Cr-NH$_2$.

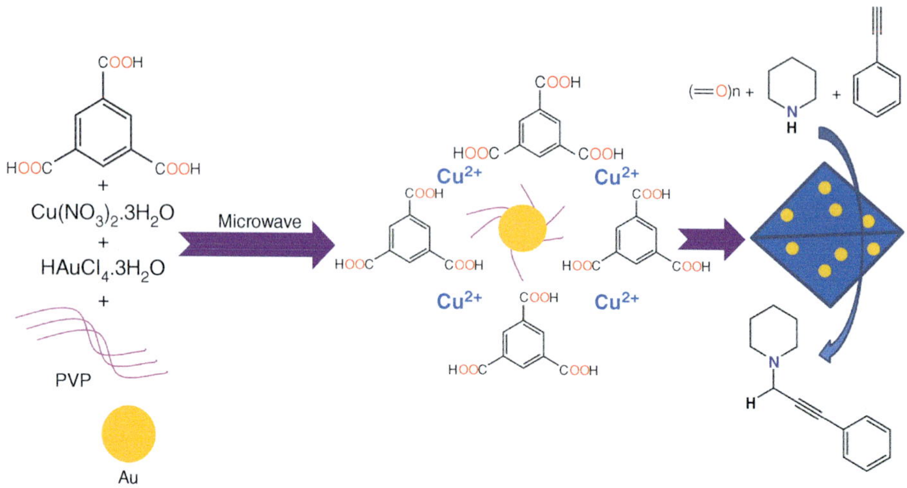

Scheme 17.16 Schematic representation for the preparation of ultrafine Au/MOF-199. Source: Jiang, Y., Zhang, X., Dai, X., Zhang, W., Sheng, Q., Zhuo, H., Xiao, Y., and Wang, H. (2017) Microwave-assisted synthesis of ultrafine Au nanoparticles immobilized on MOF-199 in high loading as efficient catalysts for a three-component coupling reaction. Nano Research, 10, 876–889. © 2017, Springer Nature

(Scheme 17.16) [84]. The synthesis of ultrafine gold nanoparticles in **Au@MOF-199** was achieved by microwave irradiation with different loading of Au (0.22%, 1.82%, 3.25%, and 5.13%).

Kinetics of the catalytic reaction showed that the Au loaded catalyst performed much better than the **MOF-199**. Figure 17.27a showed that Au loaded MOF, 0.22 and 1.82 wt % displayed (0.25 mmol formaldehyde and 50 mg catalyst) formaldehyde conversion of 15% and 45% after 5 min and increased to 86% and 93% after 30 min. Under similar reaction conditions, **MOF-199** gave 7% conversion after 20 min that slightly increased after 30 min.

In order to remove the transportation problem, the amount of catalyst reduced and reactant amount increased, which exhibited that the aldehyde conversion 10, 6, and 5% by 5.13%,

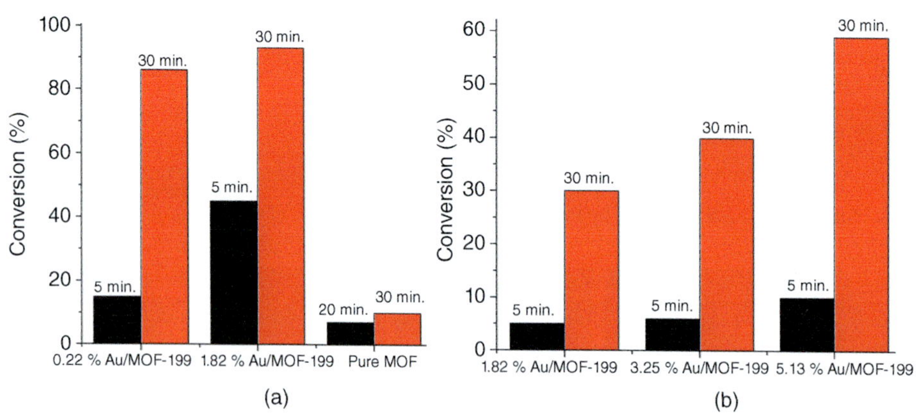

Figure 17.27 Formaldehyde conversion in A^3 reaction by (a) 0.25 mmol HCHO and 50 mg catalyst and (b) 0.25 mmol HCHO and 5 mg catalyst.

3.25%,1.82% wt% Au/MOF-199 at 5 min and aldehyde conversion augmented to 59, 40, and 30% after 30 min (Figure 17.27b) at 120 °C in 1,4-dioxane solvent. The TOFs and TONs of 5.13%, 3.25%, 1.82% Au/**MOF-199** at 5 min are 738, 713, 717 h^{-1} and 363, 365, 358 respectively. By using silver nanoparticles the similar three component coupling done by Verpoort and co-workers with >99 % conversion and yield >96 % under nitrogen atmosphere in dichloromethane solvent [85]. Tandem catalytic reaction by nanoparticle@MOFs are very rare [86]. He, Li and co-workers first showed the tandem catalytic reaction by Pd@MIL-101 for the production of methyl isobutyl ketone (MIBK) from acetone in one pot route [87]. The production of MIBK is executed industrially in three steps: condensation, dehydration, and hydrogenation. Also another catalyst, zirconium phosphate, and palladium-doped resins can do the same reaction with 35% acetone conversion. However, in both cases high-pressure hydrogen (50 bar) was required and is the main disadvantage. But by using Pd@MIL-101 and at only 7.5 bar of hydrogen pressure 60% acetone conversion and 90.2% selectivity of MIBK was obtained. The presence of highly acidic part in the catalyst accountable for the condensation and dehydration step whereas the Pd particles execute the last (hydrogenation) step. The cascade catalytic reaction by bimetallic nanoparticle@MOFs is also rarely available in the literature. In this regard an interesting cascade catalytic reaction was reported by Xu, Jiang and coworkers [88]. They had synthesized multifunctional PdAg alloys, PdAg@MIL-101 through double solvent procedure and performed one pot cascade catalytic reaction for the synthesis of desired secondary aryl amines by the nitrobenzene hydrogenation and reductive amination of benzaldehydes (Figure 17.28).

The efficacy of the bimetallic nanoparticle was compared with the commercially available catalyst Pd/C and physical mixture between **Pd@MIL-101** and **Ag@MIL-101**. The bimetallic nanoparticle showed higher selectivity (90%) for the secondary arylamine, the physical mixture and Pd/C exhibited 60% and 31% respectively. This result indicates the Lewis acidic nature of the MOF along with the hydrogenation takes place by palladium nanoparticles where the silver nanoparticles directed towards the selectivity (Figure 17.28). In a recent report, a well-dispersed ultra-small nickel nanoparticle could be installed within the MOF, UiO-66-NH_2 (Scheme 17.17) and performed CO_2 fixation reaction with epoxide to cyclic carbonate [89].

At 70 °C for 6 h and 1 MPa CO_2 pressure the MOF, **Ni@ZrOF** could convert CO_2 to cyclic carbonate with styrene epoxide in 99% selectivity and 98% yield. The presence of Zr^{4+} and Ni° site along the basic amino group displayed for such excellent catalytic performance.

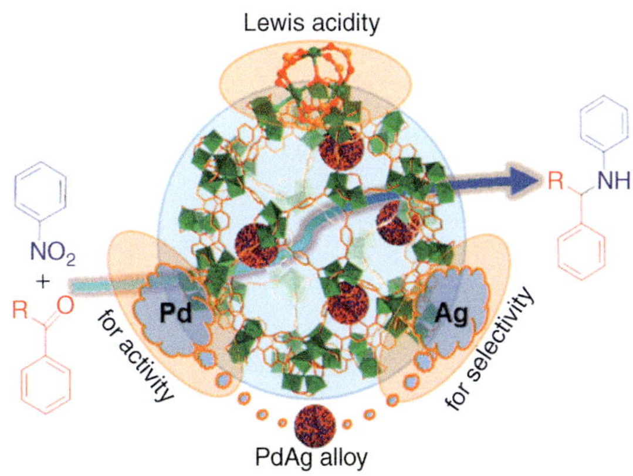

Figure 17.28 Cascade reaction of aldehydes to secondary amine by nitrobenzene hydrogenation and reductive amination of benzaldehydes.

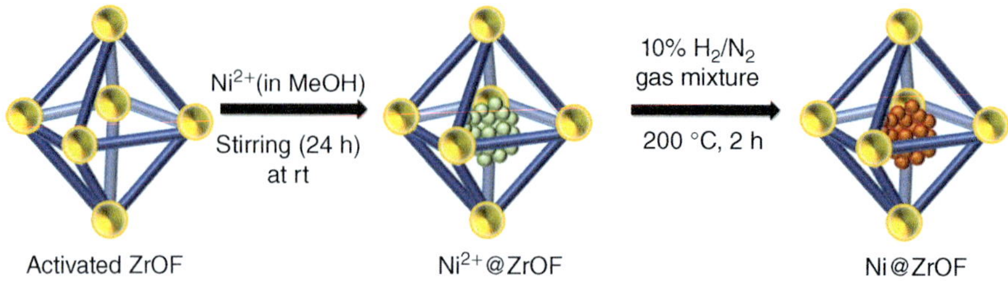

Scheme 17.17 Solution impregnation method for the preparation of Ni@ZrOF. Source: Singh, M., Solanki, P., Patel, P., Mondal, A., and Neogi, S. (2019) Highly active ultra small Ni nanoparticle embedded inside a robust metal–organic framework: remarkably improved adsorption, selectivity, and solvent-free efficient fixation of CO_2. Inorg. Chem., 58, 8100–8110. © 2019, American Chemical Society

17.6 Engineering Homochiral MOFs for Enantioselective Catalysis

A homo chiral MOF, $\mathbf{[Zn_4O(L_1)_{3/2}]\cdot 16H_2O\cdot 4THF}$, was synthesized solvothermally by reacting a carboxylate ligand, **L1** with Zn(II) ion in a mixture of solvent (DMSO, THF and water) at 100 °C (Scheme 17.18) [90]. The structure consisted of Zn_4O SBUs forming a D3 symmetric cage (Figure 17.29). After removal of solvent molecules by heating it was treated with n-BuLi in toluene whereupon the hydrogen atom of the hydroxyl group present in the framework was replaced by the lithium ion. Through choice of proper molar ratio of 1:2 and 1:4 between the MOF (per formula unit) and n-BuLi, two daughter products, **MOF-Li1** and **MOF-Li2**, were isolated. These two were able to catalyze asymmetric cyanohydrin reaction between benzaldehyde with Me3SiCN. However, the catalytic activity of **MOF-Li1** was much greater compared to **MOF-Li2**. The pure S enantiomeric **MOF-Li1** converted an aromatic aldehyde to the R enantiomer with 96% *ee*. However, the presence of electron withdrawing substituent (4-nitrobenzaldehyde, 4-chlorobenzaldehyde) in the aldehyde the conversion was quantitative, with *ee* up to >99%. The catalytic reactions occurred within the pores of the catalyst and not on the surface.

Two highly stable and porous Zr (IV) based MOFs $\mathbf{[Zr_6O_4(OH)_8(H_2O)_4(L_2)_2]}$ (spiro-1) and $\mathbf{[Zr_6O_4(OH)_8(H_2O)_4(L_3)_2]}$ (spiro-2) (Figure 17.29) with the same ***sjt*** topology (Figure 17.30),

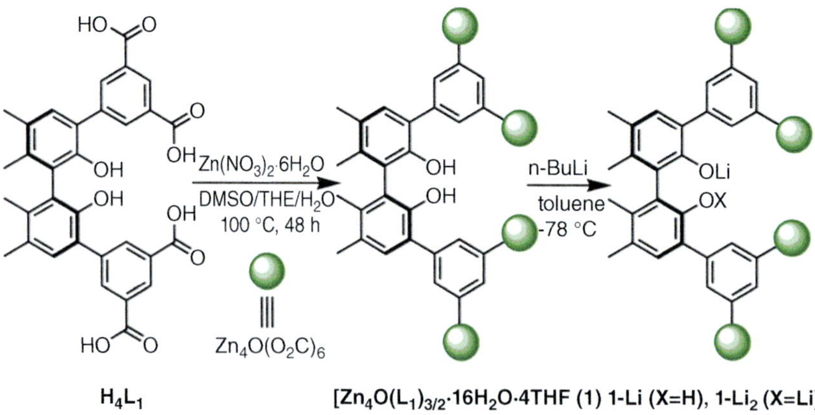

Scheme 17.18 Schematic representation for MOF formation followed by lithiation. Source: Mo, K., Yang, Y., and Cui, Y. (2014) A homochiral metal-organic framework as an effective asymmetric catalyst for cyanohydrin synthesis, J. Am. Chem. Soc., 136, 1746–1749. © 2014, American Chemical Society

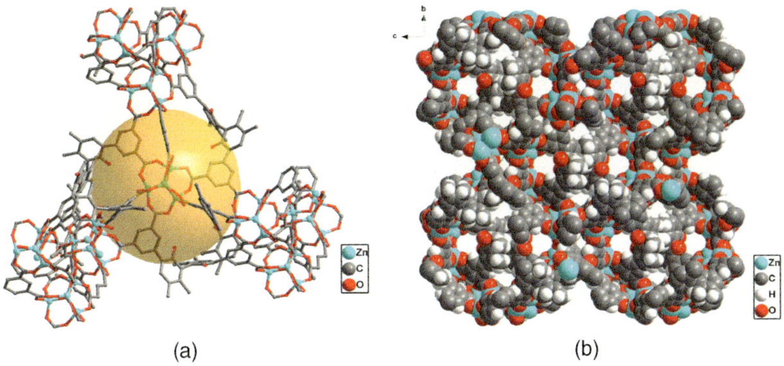

Figure 17.29 (a) Ten Zn_4O clusters and (b) 3D structure along an axis.

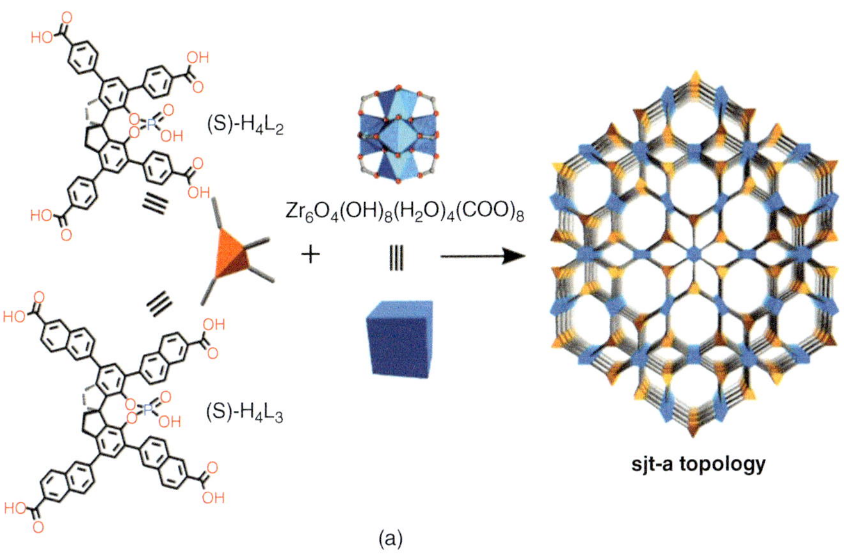

Figure 17.30 Same *sjt* topology having Zr_6 clusters formed from two different ligands. Source: Gong, W., Chen, X., Jiang, H., Chu, D., Cui, Y., and Yan Liu (2019) Highly stable Zr(IV)-based metal-organic frameworks with chiral phosphoric acids for catalytic asymmetric tandem reactions. J. Am. Chem. Soc., 141, 7498–7508. © 2019, American Chemical Society

and containing dangling phosphoric acid group for asymmetric tandem catalytic reactions were synthesized [91]. The void space provided by the spiro-2 (91.7 %) was larger than the spiro-1 (73.2 %) as calculated by PLATON. Notably, the void space of spiro-2 was the highest among Zr-MOFs.

The presence of free high acidic functional group and the Lewis acidic Zr(IV) sites helped in performing multiple component tandem enantioselective asymmetric Friedel–Crafts, and diacetilization-acetilization reactions with high yields. Both the frameworks were able to perform tandem reaction for the conversion of cyclic acetals to the corresponding 2,3-dihydroquinazolinones with satisfactory yields and considerable enantiomeric excess (Scheme 17.19). For instance, when R=4-H and catalyst (S)-Spiro-1 then the yield of the desired compound was found to be 95%

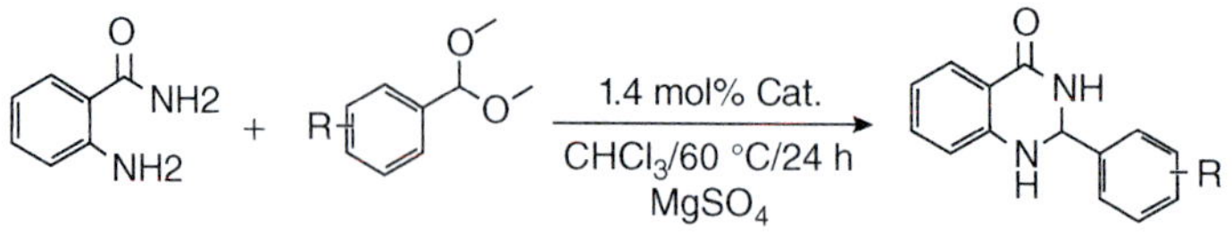

Scheme 17.19 The tandem reaction between 2-aminobenzamide with arylaldehydes dimethyl acetal to 2,3-dihydroquinazolinones. Source: Sawano, T., Thacker, N. C., Lin, Z., McIsaac, A. R., and Lin, W. (2015) Robust, chiral, and porous BINAP-based metal–organic frameworks for highly enantioselective cyclization reactions. J. Am. Chem. Soc., 137, 12241–12248. © 2015, American Chemical Society.

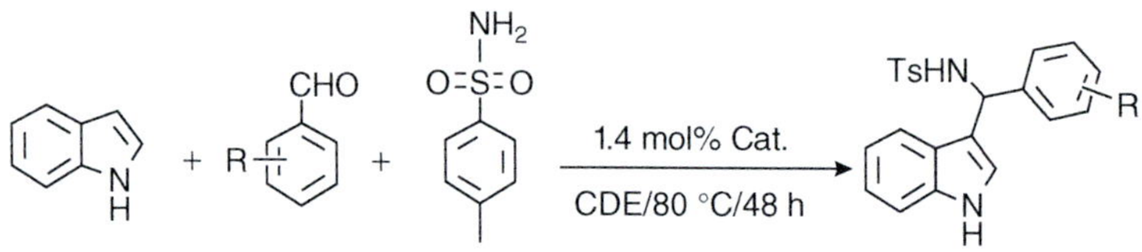

Scheme 17.20 Three component cascade enantioselective Friedel–Crafts reaction.

with 92% *ee*. Meanwhile the catalyst (S)-Spiro-2 showed 98% yield and 84% *ee* both after 15 h under similar experimental conditions. The catalysts were also capable of converting different substituted acetals (4-Me, 4-F, 4-Br, 3-NO2, 3-Cl etc.) to the corresponding dihydroquinazolinones compounds.

The versatility of these MOFs was further proved by the one pot three component tandem Friedel–Crafts reaction between aldehyde, indoles, and p-toluenesulpfonamide (Scheme 17.20). The aromatic aldehydes having electron donating and withdrawing substituents (4-H, 4-Me, 4-F, 4-Br, 4-CF_3 etc.) proceeded smoothly to the desired Friedel–Crafts product with yield 83–95 % and 95–99% *ee* by (S)-Spiro-1. The efficacy of the heterogeneous system was much more compared to the homogeneous catalytic system, ((S)-H_4L_2 and $ZrOCl_2$). When a sterically hindered reactant was used in the above two reactions the conversion was very low, indicating that the catalytic reaction exclusively took place inside the cavity and not on the surface of the MOFs.

Using BINAP with an aromatic carboxylate group connected via acetylenic linkage at either end, two MOFs were synthesized **(BINAP-MOF (I)** or **BINAP-dMOF (II)**). The **BINAP-MOF (I)** was synthesized from the ligand H_2L_4 with $ZrCl_4$ whereas the **BINAP-dMOF (II)** was synthesized from the mixture of ligands H_2L_4 and H_2L_5 (Figure 17.31) [92].

Metalation of the BINAP-MOFs with rhodium salt ($Rh(nbd)_2BF_4$ and $[Rh(nbd)Cl]_2/AgSbF_6$, nbd= norbornadiene) led to the newly formed MOFs (**I·Rh** and **II·Rh**). These were able to catalyze enantioselective alder-ene cyclo isomerization, reductive cyclization of 1,6-enynes, and Pauson–Khand cyclization of 1,6-enynes with carbon monoxide with high yields and high *ee* (Figure 17.32). It had been observed that the alder-ene cyclization activity was enhanced when less or non-coordinating anion (SF_6^- i.e. I·Rh(SbF_6)) compared to the coordinating anion (Cl^- i.e. I·RhCl) was used, signifying that the reaction proceeded through the formation of Rhoda cycle intermediate (Figure 17.32). When **I·Rh** was used for the asymmetric Pauson–Khand cyclization of sterically encumbered 1,6-enynes with cinnamaldehyde (as a source of carbon monoxide) it led to <5% yield due to steric hindrance between MOF and the rhodacycle intermediate. However, the same reaction went smoothly with **II·Rh** (67 % yield and 80% *ee*) due to large channel present (Figure 17.33).

Several chiral MOFs having BINOL backbone had also been reported which performed asymmetric enantioselective catalytic reactions with excellent yield and enantiomeric excess [93].

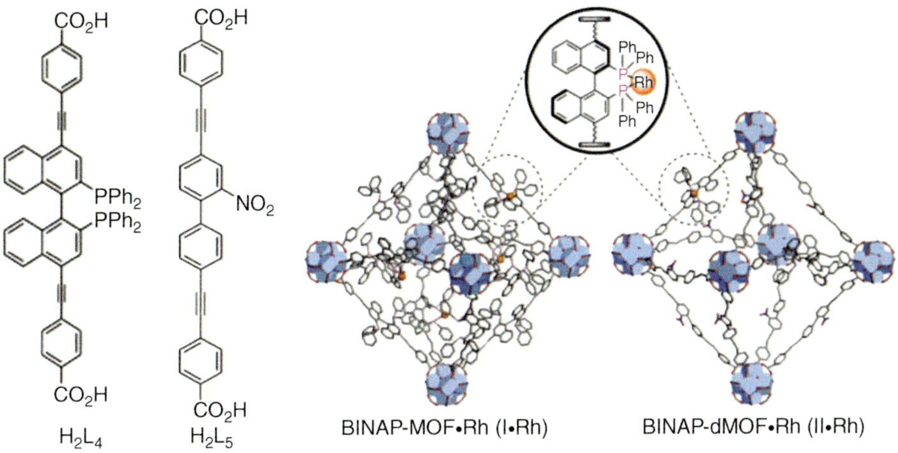

Figure 17.31 The ligands H_2L_4 and H_2L_5 and the rhodium metalation within the MOF.

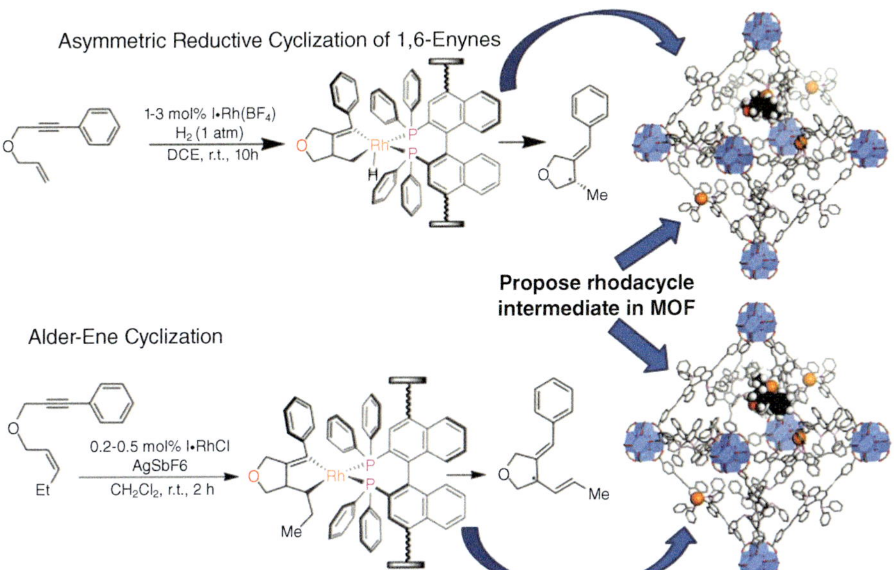

Figure 17.32 The enantioselective Alder-ene and reductive cyclization catalytic reaction by rhodium metalated MOF.

A polyoxometalate-based chiral MOF [94] had been isolated that catalyzed the asymmetric conversion of olefins to their dihydrodroxylation products. The enantiopure Ni(II) based MOF **Ni-PYI1** was constructed solvothermally by reacting $Ni_2H[BW_{12}O_{40}]$, L-tert-butoxycarbonyl-2-(imidazole)-1-pyrrolidine (L-BCIP, (PYI is pyrrolidin-2-ylimidazole)), and 4,4′-bipyridine (bpy) (Scheme 17.21). The framework Ni-PYI1 could catalyze the conversion of styrene to its dihydroxy compounds, (R)-phenyl-1,2-ethanediol (Scheme 17.21) in the presence of hydrogen peroxide (15%) in dichloromethane at 40 °C with 75% conversion and >95% *ee*. The versatility of this framework was also observed in the conversion of 3-chlorovinylbenzene, 2-chlorovinylbenzene, and 4-chlorovinylbenzene to the corresponding dihydroxy compounds with conversion range 75–79% and good enantioselectivity. However, large olefins, 3,5-di-tert-butyl-4′-vinylbiphenyl, afforded

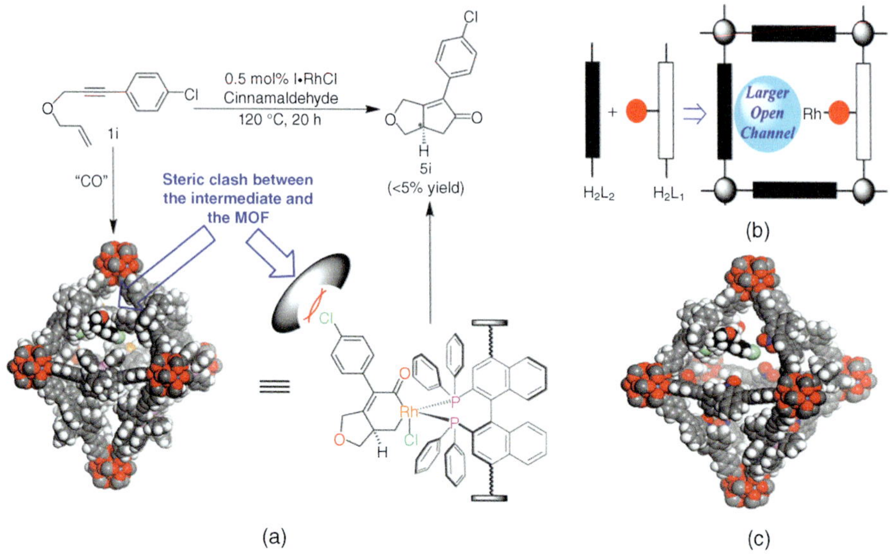

Figure 17.33 (a) The steric clash between rhodacycle intermediate and the MOF in I·Rh, (b) large cavity space in II·Rh due to linker H_2L_5 and (c) bicyclic rhodacycle intermediate in the larger cavity of II·Rh (space fill model).

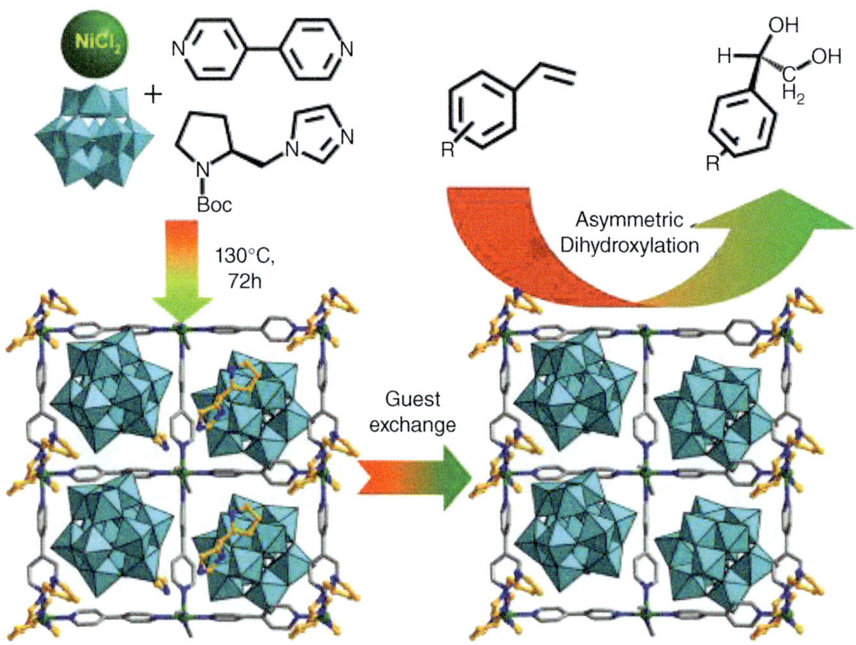

Scheme 17.21 The solvothermal synthesis of Ni-PYI1 and asymmetric dihydroxylation of olefin by the same MOF.

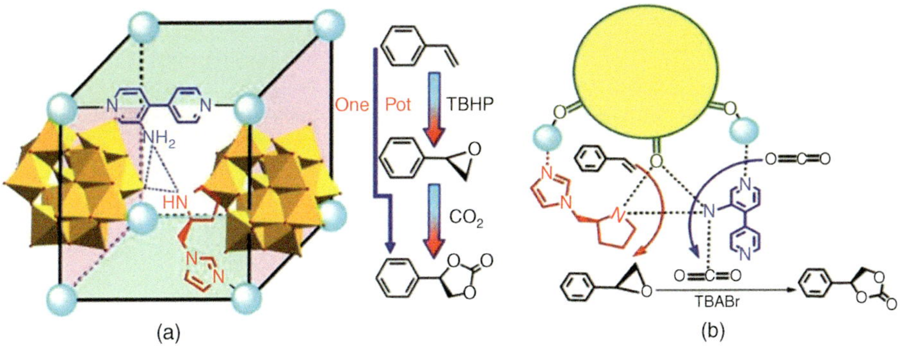

Figure 17.34 (a) Representation of tandem catalytic reaction by MOF component and (b) possible mechanism for cascade reaction from styrene to styrene carbonate. Source: Han, Q., Qi, B., Ren, W., He, C., Niu, J., and Duan, C. (2015) Polyoxometalate-based homochiralmetal-organic frameworks for tandem asymmetric transformation of cyclic carbonates from olefins. Nature Comm. 6, 10007–10008. Licensed under CC BY 4.0.

poor yields. The homogeneous catalytic system, L-PYI, and $Ni_2H[BW_{12}O_{40}]$ showed significantly lower catalytic activity.

The synthesis of chiral cyclic carbonate from cycloaddition of CO_2 with epoxide by chiral MOFs had been attempted by several workers in the field. As an example, polyoxometalated homochiral MOF (**ZnW-PYI1**, PYI is pyrrolidin-2-ylimidazole) was synthesized solvothermally by reacting $Zn(NO_3)_2{\cdot}6H_2O$, L-BCIP, $TBA_4W_{10}O_{32}$ and NH_2-BPY (Figure 17.34) [95]. The framework was able to catalyze the conversion of chiral epoxide (S or R) to the asymmetric cyclic carbonate (>90% *ee*) with yield >99% in the presence of the co-catalyst TBAB, 0.5MPa of CO_2 pressure, for 48 h at 323 K. Interestingly, the framework can efficiently convert styrene to phenyl(ethylene carbonate) in tandem process with high yield and excellent *ee* under 5 bar CO_2 pressure, t-butylhydroperoxide for 96 h at 323 K. For example, the desired carbonate (R)-phenyl (ethylencarbonate) was accomplished in 92 % yield with 80% enantimeric excess. The presence of -NH_2 moiety in the framework increased the CO_2 concentration around the reactive center and the presence of $ZnW_{12}O_{40}{}^{6-}$ activate the reacting component significantly, leading to the excellent selectivity and yield in the asymmetric one pot tandem reaction.

17.7 Conclusion

The chemical industry represents a vital part of the economy in many industrialized and developing countries. However, the manufacture of chemical products also leads to enormous quantities of environmentally harmful waste. Both national and international organizations have recognized the importance of cleaner syntheses for environmental protection. In this context, generation of large structures with void spaces resembling zeolites for catalysis is definitely a progress toward the right direction. Solid heterogeneous catalysts offer many process engineering advantages compared to homogeneous processes including their non-corrosiveness, the wide range of temperatures and pressures that can be applied, and the easier separation of substrates and products from the catalyst. The chapter is aimed at highlighting the recently developed methods by which the coordination space in MOFs can be engineered for catalyzing different organic transformations. Future goals in this area would be designing of the coordination space in terms of shapes and sizes for selectively enclathrate substrates for regio- as well as stereo-selective organic transformations that are different from the usual thermodynamic route. It follows, therefore, that enormous potential exists in the area of MOFs as heterogeneous catalysts.

Acknowledgements

Partial financial supports received from the DST and the MNRE, New Delhi, India (to PKB) are gratefully acknowledged. TKP acknowledges to PDPU/ORSP/R&D/SRP/2019-20/1361/1 and DD acknowledges to the "TEQIP Collaborative Research Scheme (CRS)".

References

1 Moulton, B. and Zaworotko, M.J. (2001). From molecules to crystal engineering: supramolecular isomerism and polymorphism in network solids. Chem. Rev. 101 (6): 1629–1658.
2 Morris, R.E. and Wheatley, P.S. (2008). Gas storage in nanoporous materials. Angew. Chem. Int. Ed. 47: 4966–4981.
3 Pal, T.K. (2019). Syntheses, structures and topology variations of metal organic frameworks built from a semi-rigid tetracarboxylate ligand. Chem. Select 4: 536–542.
4 (a) Batten, S.R. and Robson, R. (1998). Interpenetrating nets: ordered. periodic entanglement., Angew. Chem. Int. Ed. Engl. 37: 1461–1494. (b) Eddaoudi, M., Moler, D.B., Li, H. et al. (2001). Modular chemistry: secondary building units as a basis for the design of highly porous and robust metal–organic carboxylate frameworks. Acc. Chem. Res. 34: 319–330. (c) Evans, O.R. and Lin, W. (2002). Crystal engineering of NLO materials based on metal–organic coordination networks. Acc. Chem. Res. 35: 511–522. (d) Lee, S., Mallik, A.B., Xu, Z. et al. (2005). Crystallized frameworks with giant pores: are there limits to the possible? Acc. Chem. Res. 38: 217–225.
5 Farrusseng, D., Aguado, S., and Pinel, C. (2009). Metal–organic frameworks: opportunities for catalysis. Angew. Chem. Int. Ed. 48: 7502–7513.
6 (a) Canivet, J., Vandichel, M., and Farrusseng, D. (2016). Origin of highly active metal- organic framework catalysts: defects? Defects! Dalton Trans. 45: 4090–4099. (b) Fang, Z., Bueken, B., De Vos, D.E., and Fischer, R.A. (2015) Defect-engineered metal–organic frameworks. Angew. Chem. Int. Ed., 54, 7234–7254; (c) Chizallet, C., Lazare, S., Bazer-Bachi, D. et al. (2010). Catalysis of transesterification by a nonfunctionalized metal-organic framework: acido-basicity at the external surface of ZIF-8 probed by FTIR and ab initio calculations. J. Am. Chem. Soc. 132: 12365–12377. (d) Kozachuk, O., Luz, I., Llabrési Xamena, F.X. et al. (2014). Multifunctional, defect-engineered metal–organic frameworks with ruthenium centers: sorption and catalytic properties. Angew. Chem. Int. Ed. 53: 7058–7062. (e) Sun, Z.G., Li, G., Liu, L.P., and Liu, H.O. (2012). Au nanoparticles supported on Cr-based metal-organic framework as bimetallic catalyst for selective oxidation of cyclohexane to cyclohexanone and cyclohexanol. Catal. Commun. 27: 200–205. (f) Esken, D., Turner, S., Lebedev, O.I. et al. (2010). Au@ZIFs: Stabilization and encapsulation of cavity-size matching gold clusters inside functionalized zeolite imidazolate frameworks ZIFs. Chem. Mater. 22: 6393–6401. (g) Liu, H.L., Liu, Y.L., Li, Y.W. et al. (2010). Metal–organic framework supported gold nanoparticles as a highly active heterogeneous catalyst for aerobic oxidation of alcohols. J. Phys. Chem. C 114: 13362–13369. (h) Muller, M., Turner, S., Lebedev, O.I. et al. (2011). Au@MOF-5 and Au/MOx@MOF-5 (M = Zn, Ti; x = 1, 2): preparation and microstructural characterization. Eur. J. Inorg. Chem.: 1876–1887. (i) Juan-Alcaniz, J., Ferrando-Soria, J., Luz, I. et al. (2013). The oxamate route, a versatile post-functionalization for metal incorporation in MIL-101(Cr): catalytic applications of Cu, Pd, and Au. J. Catal. 307: 295–304. (j) Leus, K., Dendooven, J., Tahir, N. et al. (2016). Atomic layer deposition of Pt nanoparticles within the cages of MIL-101: a mild and recyclable hydrogenation catalyst. Nanomaterials 6: 45.

7 Fujita, M., Kwon, Y.J., Washizu, S., and Ogura, K. (1994). Preparation, clathration ability, and catalysis of a two-dimensional square network material composed of cadmium(II) and 4,4′-bipyridine. J. Am. Chem. Soc. 116: 1151–1152.

8 Das, R.K., Aijaz, A., Sharma, M.K. et al. (2012). Direct crystallographic observation of catalytic reactions inside the pores of a flexible coordination polymer. Chem. Eur. J. 18: 6866–6872.

9 Masciocchi, N., Bruni, S., Cariati, E. et al. (2001). Extended polymorphism in copper(II) imidazolate polymers: a spectroscopic and XRPD structural study. Inorg. Chem. 40: 5897–5905.

10 Tabares, L.C., Navarro, J.A.R., and Salas, J.M. (2001). Cooperative guest inclusion by a zeolite analogue coordination polymer. sorption behavior with gases and amine and group 1 metal salts. J. Am. Chem. Soc. 123: 383–387.

11 Luz, I., Leó, A., Boronat, M. et al. (2013). Selective aerobic oxidation of activated alkanes with MOFs and their use for epoxidation of olefins with oxygen in a tandem reaction. Catal. Sci. Technol. 3: 371–379.

12 Luz, I., Llabré, F.X., Xamena, S.I., and Corma, A. (2010). Bridging homogeneous and heterogeneous catalysis with MOFs: "Click" reactions with Cu-MOF catalysts. J. Catal. 276: 134–140.

13 Luz, I., Llabré, F.X., Xamena, S.I., and Corma, A. (2012). Bridging homogeneous and heterogeneous catalysis with MOFs: Cu-MOFs as solid catalysts for three-component coupling and cyclization reactions for the synthesis of propargylamines, indoles and imidazopyridines. J. Catal. 285: 285–291.

14 Wu, C.-D., Li, L., and Shi, L.-X. (2009). Heterogeneous catalyzed aryl–nitrogen bond formations using a valine derivative bridged metal–organic coordination polymer. Dalton Trans., 6790: 6794.

15 Horike, S., Dinc ã, M., Tamaki, K., and Long, J.R. (2008) Size-selective Lewis acid catalysis in a microporous metal-organic framework with exposed Mn^{2+} coordination sites. J. Am. Chem. Soc., 130, 5854- 5855.

16 Férey, G., Mellot-Draznieks, C., Serre, C. et al. (2005). A chromium terephthalate-based solid with unusually large pore volumes and surface area. Science 309: 2040–2042.

17 Henschel, A., Gedrich, K., Kraehnert, R., and Kaskel, S. (2008). Catalytic properties of MIL-101. Chem. Commun.: 4192–4194.

18 Kim, J., Bhattacharjee, S., Jeong, K.E. et al. (2009). Selective oxidation of tetralin over a chromium terephthalate metal organic framework, MIL-101. Chem. Commun.: 3904–3906.

19 Maksimchuk, N.V., Kovalenko, K.A., Fedin, V.P., and Kholdeeva, O.A. (2010). Heterogeneous Selective Oxidation of Alkenes to α,β-Unsaturated Ketones over Coordination Polymer MIL-101. Adv. Synth. Catal. 352: 2943–2948.

20 Zalomaeva, O.V., Chibiryaev, A.M., Kovalenko, K.A. et al. (2013). Cyclic carbonates synthesis from epoxides and CO_2 over metal–organic framework Cr-MIL-101. J. Catal. 298: 179–185.

21 Horcajada, P., Surble, S., Serre, C. et al. (2007). Synthesis and catalytic properties of MIL-100 (Fe), an iron (III) carboxylate with large pores. Chem. Commun.: 2820–2822.

22 Mowat, J.P.S., Miller, S.R., Slawin, A.M.Z. et al. (2011). Synthesis, characterisation and adsorption properties of microporous scandium carboxylates with rigid and flexible frameworks. Microporous Mesoporous Mater. 142: 322–333.

23 Mitchell, I., Gonzalez-Santiago, B., Mowat, J.P.S. et al. (2013). Remarkable Lewis acid catalytic performance of the scandium trimesate metal organic framework MIL-100(Sc) for C–C and C=N bond-forming reactions. Catal. Sci. Technol. 3: 606–617.

24 Dietzel, P.D.C., Morita, Y., Blom, R., and Fjellvag, H. (2005). An in situ high-temperature single-crystal investigation of a dehydrated metal–organic framework compound and field-induced magnetization of one-dimensional metal–oxygen chains. Angew. Chem., Int. Ed. 44: 6354–6358.

25 Cho, H.-Y., Yang, D.-A., Kim, J. et al. (2012). A dual-walled cage MOF as an efficient heterogeneous catalyst for the conversion of CO_2 under mild and co-catalyst free conditions. Catal. Today 185: 35–40.

26 Xiao, D.J., Bloch, E.D., Mason, J.A. et al. Crocella', V., Yano, J., Bordiga, S., Truhlar, D.G., Gagliardi, L., Brown, C.M., and Long, J.R. (2014) Oxidation of ethane to ethanol by N_2O in a metal–organic framework with coordinatively unsaturated iron(II) sites. Nature Chemistry 6: 590–595.

27 Pal, T.K., De, D., Neogi, S. et al. (2015). Significant gas adsorption and catalytic performance by a robust Cu(II)–MOF derived through SC-SC transmetalation of a thermally less stable Zn(II)–MOF. Chem. Eur. J. 21: 19064–19070.

28 Horike, S., Dinca, M., Tamaki, K., and Long, J.R. (2008). Size-selective Lewis acid catalysis in a microporous metal-organic framework with exposed Mn^{2+} coordination sites. J. Am. Chem. Soc. 130: 5854–5855.

29 Horcajada, P., Surble, S., Serre, C. et al. (2007). Synthesis and catalytic properties of MIL-100(Fe), an iron(III) carboxylate with large pores. Chem. Commun.: 2820–2822.

30 Li, P.–.Z., Wang, X.–.J., Liu, J. et al. (2017). Highly effective carbon fixation via catalytic conversion of CO_2 by an acylamide-containing metal–organic framework. Chem. Mater. 29: 9256–9261.

31 Sharma, V., De, D., and Bharadwaj, P.K. (2018). A multifunctional metal–organic framework for oxidative C–O coupling involving direct C–H activation and synthesis of quinolines. Inorg. Chem. 57: 8195–8199.

32 Vermoortele, F., Vandichel, M., Van de Voorde, B. et al. (2012). Electronic effects of linker substitution on Lewis acid catalysis with metal–organic frameworks. Angew. Chem. Int. Ed. Engl. 51: 4887–4890.

33 (a) Wang, M., Xie, H., Wu, C.D., and Wang, Y.G. (2009). From one to three: a serine derivate manipulated homochiral metal-organic framework. Chem. Commun.: 2396–2398. (b) Pal, T.K., De, D., Senthilkumar, S. et al. (2016). A partially fluorinated, water stable Cu(II)–MOF derived via transmetalation: significant gas adsorption with high CO_2 selectivity and catalysis of Biginelli reactions. Inorg. Chem. 55: 7835–7842. (c) Verma, A., De, D., Tomar, K., and Bharadwaj, P.K. (2017). An amine functionalized metal–organic framework as an effective catalyst for conversion of CO_2 and Biginelli reactions. Inorg. Chem. 56: 9765–9771. (d) Gupta, A.K., De, D., Tomar, K., and Bharadwaj, P.K. (2018). A Cu(II) metal–organic framework with significant H_2 and CO_2 storage capacity and heterogeneous catalysis for the aerobic oxidative amination of $C(sp^3)$–H bonds and Biginelli reactions. Dalton Trans. 47: 1624–1634.

34 (a) Gupta, A.K., De, D., and Bharadwaj, P.K. (2017). A NbO type Cu(II)-MOF showing efficient catalytic activity in Friedländer and Henry reactions. Dalton Trans. 46: 7782–7790. (b) Gupta, M., De, D., Pal, S. et al. (2017). A porous two-dimensional Zn(II)-coordination polymer exhibiting SC-SC transmetalaton with Cu(II): efficient heterogeneous catalysis for the Henry reactions and detection of nitro explosives. Dalton Trans. 46: 7619–7627.

35 (a) Tuci, G., Rossin, A., Xu, X., Ranocchiari, M., Bokhoven, J.A. v., Luconi, L., Manet I., Melucci, M., Giambastiani, G. (2013) "Click" on MOFs: a versatile tool for the multimodal derivatization of N_3-decorated metal organic frameworks. Chem. Mater., 25, 11, 2297–2308; (b) Yang, T., Cui, H., Zhang, C. et al. (2013). Porous metal–organic framework catalyzing the three-component coupling of sulfonyl azide, alkyne, and amine. Inorg. Chem. 52: 9053–9059.

36 Gupta, A.K., De, D., Katoch, R. et al. (2017). Synthesis of a NbO type homochiral Cu(II) metal–organic framework: ferroelectric behavior and heterogeneous catalysis of three-component coupling and Pechmann reactions. Inorg. Chem. 56: 4698–4706.

37 (a) Gupta, M., De, D., Tomar, K., and Bharadwaj, P.K. (2017). From Zn(II)-carboxylate to double walled Zn(II)-carboxylato phosphate MOF: change in the framework topology, capture and conversion of CO_2 and Strecker reaction. Inorg. Chem. 56: 14605–14611.

38 Metzger, E.D., Brozek, C.K., Comito, R.J., and Dincă, M. (2016). Selective dimerization of ethylene to 1-butene with a porous catalyst. ACS Cent. Sci. 2: 148–153.

39 Pal, T.K., De, D., Senthilkumar, S. et al. (2016). A partially fluorinated, water stable Cu(II)–MOF derived via transmetalation: significant gas adsorption with high CO_2 selectivity and catalysis of Biginelli reactions. Inorg. Chem. 55: 7835–7842.

40 Stubbs, A.W., Braglia, L., Borfecchia, E. et al. (2018). Selective catalytic olefin epoxidation with Mn^{II}-exchanged MOF-5. ACS Catal. 8: 596–601.

41 Banerjee, M., Das, S., Yoon, M. et al. (2009). Postsynthetic modification switches an achiral framework to catalytically active homochiral metal–organic porous materials. J. Am. Chem. Soc. 131: 7524–7525.

42 Madrahimov, S.T., Gallagher, J.R., Zhang, G. et al. (2015). Gas-phase dimerization of ethylene under mild conditions catalyzed by MOF materials containing (bpy)Ni(II) complexes. ACS Catal. 5: 6713–6718.

43 Lun, D.J., Waterhouse, G.I.N., and Telfer, S.G. (2011). A general thermolabile protecting group strategy for organocatalytic metal-organic frameworks. J. Am. Chem. Soc. 133: 5806–5809.

44 De, D., Pal, T.K., and Bharadwaj, P.K. (2016). A porous Cu(II)-MOF with proline embellished cavity: cooperative catalysis for the Baylis-Hillman reaction. Inorg. Chem. 55: 6842–6844.

45 Cao, C.-C., Chen, C.–.X., Wei, Z.–.W. et al. (2019). Catalysis through dynamic spacer installation of multivariate functionalities in metal–organic frameworks. J. Am. Chem. Soc. 141: 2589–2593.

46 Roberts, J.M., Fini, B.M., Sarjeant, A.A. et al. (2012). Urea metal–organic frameworks as effective and size-selective hydrogen-bond catalysts. J. Am. Chem. Soc. 134: 3334–3337.

47 Zhang, X., Sun, J., Wei, G. et al. (2019). In situ generation of an N-heterocyclic carbene functionalized metal–organic framework by postsynthetic ligand exchange: efficient and selective hydrosilylation of CO_2. Angew. Chem. Int. Ed. 58: 2844–2849.

48 Roswell, J.L.C. and Yaghi, O.M. (2006). Effects of functionalization, catenation, and variation of the metal oxide and organic linking units on the low-pressure hydrogen adsorption properties of metal–organic frameworks. J. Am. Chem. Soc. 128: 1304–1315.

49 Ingleson, M.J., Barrio, J.P., Guilbaud, J.B. et al. (2008). Framework functionalisation triggers metal complex binding. Chem. Commun.: 2680–2682.

50 Zhang, X., Llabrési Xamena, F.X., and Corma, A. (2009). Gold(III)–metal organic framework bridges the gap between homogeneous and heterogeneous gold catalysts. J. Catal. 265: 155–160.

51 Bhattacharjee, S., Yang, D.A., and Ahn, W.S. (2011). A new heterogeneous catalyst for epoxidation of alkenes via one-step post-functionalization of IRMOF-3 with a manganese(II) acetylacetonate complex. Chem. Commun.: 3637–3639.

52 Juan-Alcaniz, J., Ferrando-Soria, J., Luz, I. et al. (2013). The oxamate route, a versatile post-functionalization for metal incorporation in MIL-101(Cr): Catalytic applications of Cu, Pd, and Au. J. Catal. 307: 295–304.

53 Elumalai, P., Mamlouk, H., Yiming, W. et al. (2018). Recyclable and reusable heteroleptic nickel catalyst immobilized on metal-organic framework for Suzuki-Miyaura coupling. ACS Appl. Mater. Interfaces 10: 41431–41438.

54 Kumar, G., Solanki, P., Nazisha, M. et al. (2019). Covalently hooked EOSIN-Y in a Zr(IV) framework as visible-light mediated, heterogeneous photocatalyst for efficient C-H functionalization of tertiary amines. J. Catal. 371: 298–304.

55 Shultz, A.M., Farha, O.K., Hupp, J.T., and Nguyen, S.T. (2009). A catalytically Active, permanently microporous MOF with metalloporphyrin struts. J. Am. Chem. Soc. 131: 4204–4205.

56 Farha, O.K., Shultz, A.M., Sarjeant, A.A. et al. (2011). Active-site-accessible porphyrinic metal–organic framework materials. J. Am. Chem. Soc. 133: 5652–5655.

57 Meng, L., Cheng, Q., Kim, C. et al. (2012). Crystal engineering of a microporous, catalytically active fcu topology MOF using a custom-designed metalloporphyrin linker. Angew. Chem. Int. Ed. 51: 10082–10085.

58 Zou, C., Zhang, Z., Xu, X. et al. (2012). A multifunctional organic–inorganic hybrid structure based on Mn^{III}–porphyrin and polyoxometalate as a highly effective dye scavenger and heterogeneous catalyst. J. Am. Chem. Soc. 134: 87–90.

59 Feng, D., Gu, Z.-Y., Li, J.-R. et al. (2012). Zirconium-metalloporphyrin PCN-222: mesoporous metal–organic frameworks with ultrahigh stability as biomimetic catalysts. Angew. Chem. Int. Ed. 51: 10307–10310.

60 Feng, D., Chung, W.-C., Wei, Z. et al. (2013). Construction of ultrastable porphyrin Zr metal–organic frameworks through linker elimination. J. Am. Chem. Soc. 135: 17105–17110.

61 Cho, S.-H., Ma, B., Nguyen, S.T. et al. (2006). A metal–organic framework material that functions as an enantioselective catalyst for olefin epoxidation. Chem. Commun.: 2563–2565.

62 Shultz, A.M., Farha, O.K., Adhikari, D. et al. (2011). Selective surface and near-surface modification of a noncatenated, catalytically active metal-organic framework material based on Mn(salen) struts. Inorg. Chem. 50: 3174–3176.

63 (a) Song, F., Wang, C., Falkowski, J.M. et al. (2010). Isoreticular chiral metal–organic frameworks for asymmetric alkene epoxidation: tuning catalytic activity by controlling framework catenation and varying open channel sizes. J. Am. Chem. Soc. 132: 15390–15398. (b) Song, F., Wang, C., and Lin, W. (2011). A chiral metal–organic framework for sequential asymmetric catalysis. Chem. Commun.: 8256–8258. (c) Falkowski, J.M., Wang, C., Liu, S., and Lin, W. (2011). Actuation of asymmetric cyclopropanation catalysts: reversible single-crystal to single-crystal reduction of metal–organic frameworks. Angew. Chem., Int. Ed. 50: 8674–8678. (d) Falkowski, J.M., Liu, S., Wang, C., and Lin, W. (2012). Chiral metal–organic frameworks with tunable open channels as single-site asymmetric cyclopropanation catalysts. Chem. Commun.: 6508–6510.

64 Manna, K., Zhang, T., Greene, F.X., and Lin, W. (2015). Bipyridine- and phenanthroline-based metal-organic frameworks for highly efficient and tandem catalytic organic transformations via directed C-H activation. J. Am. Chem. Soc. 137: 2665–2673.

65 Chambers, M.B., Wang, X., Elgrishi, N. et al. (2015). Photocatalytic carbon dioxide reduction with rhodium-based catalysts in solution and heterogenized within metal–organic frameworks. ChemSusChem 8: 603–608.

66 Huang, Y.-B., Liang, J., Wang, X.-S., and Cao, R. (2017). Multifunctional metal–organic framework catalysts: synergistic catalysis and tandem reactions. Chem. Soc. Rev.: 126–157.

67 Zhang, F., Wei, Y., Wu, X. et al. (2014). Hollow zeolitic imidazolate framework nanospheres as highly efficient cooperative catalysts for [3+3] cycloaddition Reactions. J. Am. Chem. Soc. 136: 13963–13966.

68 Li, B., Yiming, K.L., Zhang, J. et al. (2015). Metal–organic framework based upon the synergy of a Brønsted acid framework and Lewis acid centers as a highly efficient heterogeneous catalyst for fixed-bed reactions. J. Am. Chem. Soc. 137: 4243–4248.

69 Zhao, S.–.Y., Chen, Z.-Y., Wei, N. et al. (2019). Highly efficient cooperative catalysis of single-site Lewis acid and Brønsted acid in a metal-organic framework for the Biginelli reaction. Inorg. Chem. 58: 7657–7661.

70 Rouhania, F. and Morsali, A. (2017). Highly effective Brønsted base/Lewis acid cooperative catalysis: a new Cd metal-organic framework for synthesis of Hantzsch 1, 4-DHPs at ambient temperature. New J. Chem. 41: 15475–15484.

71 An, B., Li, Z., Song, Y. et al. (2019). Cooperative copper centres in a metal–organic framework for selective conversion of CO_2 to ethanol. Nature Catalysis 2: 709–717.

72 Gao, M.-L., Qi, M.–.H., Liu, L., and Han, Z.-B. (2019). An exceptionally stable core–shell MOF/COF bifunctional catalyst for a highly efficient cascade deacetalization–Knoevenagel condensation reaction. Chem. Commun.: 6377–6380.

73 Park, J., Li, J.-R., Chen, Y.-P. et al. (2012). A versatile metal–organic framework for carbon dioxide capture and cooperative catalysis. Chem. Commun.: 9995–9997.

74 Toyao, T., Saito, M., Horiuchi, Y., and Matsuoka, M. (2014). Development of a novel one-pot reaction system utilizing a bifunctional Zr-based metal–organic framework. Catal. Sci. Technol. 4: 625–627.

75 Srirambalaji, R., Hong, S., Natarajan, R. et al. (2012). Tandem catalysis with a bifunctional site-isolated Lewis acid–Brønsted base metal–organic framework, NH_2-MIL-101(Al). Chem. Commun.: 11650–11652.

76 (a) Kathalikkattil, A.C., Roshan, R., Tharun, J. et al. (2016). A sustainable protocol for the facile synthesis of zinc-glutamate MOF: an efficient catalyst for room temperature CO_2 fixation reactions under wet conditions. Chem. Commun.: 280–283. (b) Zhen, X.G., Chen, Z.C., Cheng, J.B., and Duan, H.C. (2016). New rht-type metal–organic frameworks decorated with acylamide groups for efficient carbon dioxide capture and chemical fixation from raw power plant flue gas. ACS Appl. Mater. Interfaces 846: 31746–31756. (c) Liang, L., Liu, C., Jiang, F. et al. (2017). Yuan. D., and Hong, M. Carbon dioxide capture and conversion by an acid-base resistant metal-organic framework, Nature Comm. 8: 1233–1234. (d) Zhao, D., Liu, X.-H., Guo, J.-H. et al. (2018). Porous metal–organic frameworks with chelating multiamine sites for selective adsorption and chemical conversion of carbon dioxide. Inorg. Chem. 57: 2695–2704. (e) He, H., Sun, Q., Gao, W. et al. (2018). A stable metal–organic framework featuring a local buffer environment for carbon dioxide fixation. Angew. Chem. Int. Ed. 57: 4657–4662. (f) Wei, L.Q. and Ye, B.H. (2019). Efficient conversion of CO_2 via grafting urea group into a $[Cu_2(COO)_4]$-based metal-organic framework with hierarchical porosity. Inorg Chem. 58: 4385–4393. (g) Li, Y., Zhang, X., Xu, P. et al. (2019). The design of a novel and resistant Zn(PZDC)(ATZ) MOF catalyst for the chemical fixation of CO_2 under solvent-free conditions. Inorg. Chem. Front. 6: 317–321. (h) Parmar, B., Patel, P., Pillai, R.S. et al. (2019). Efficient catalytic conversion of terminal/internal epoxides to cyclic carbonates by porous Co(ii) MOF under ambient conditions: structure–property correlation and computational studies. J. Mater. Chem. A 7: 2884–2894.

77 De, D.T.K., Pal, S., Neogi, S.S. et al. (2016). A versatile Cu(II) metal–organic framework exhibiting high gas storage capacity with selectivity for CO_2: conversion of CO_2 to cyclic carbonate and other catalytic abilities. Chem. Eur. J. 22: 3387–3396.

78 Sharma, V., De, D., Saha, R. et al. (2017). A Cu(II)-MOF capable of fixing CO_2 from air and showing high capacity H_2 and CO_2 adsorption. Chem. Commun. 53: 13371–13374.

79 (a) Lin, Z.-J., Zheng, H.–.Q., Chen, J. et al. (2018). Encapsulation of phosphotungstic acid into metal–organic frameworks with unable window sizes: screening of PTA@MOF catalysts for efficient oxidative desulfurization. Inorg. Chem. 57: 13009–13019. (b) Majewski, M.B., Howarth, A.J., Li, P. et al. (2017). Enzyme encapsulation in metal–organic frameworks for applications in catalysis. CrystEngComm. 19: 4082–4091. (c) Li, G., Zhao, S., Zhang, Y., and Tang, Z. (2018). Metal–organic frameworks encapsulating active nanoparticles as emerging composites for catalysis: recent progress and perspectives. Adv. Mater. 30 (51): 1800702.

80 Dhakshinamoorthy, A. and Garcia, H. (2012). Catalysis by metal nanoparticles embedded on metal-organic frameworks. Chem. Soc. Rev. 41: 5262–5284.

81 Moon, H.R., Lim, D.W., and Suh, M.P. (2013). Fabrication of metal nanoparticles in metal-organic frameworks. Chem. Soc. Rev. 42: 1807–1824.

82 (a) Miyaura, N. and Suzuki, A. (1995). Palladium-catalyzed cross-coupling reactions of organoboron compounds. Chem. Rev. 95: 2457–2483. (b) Beletskaya, I.P. and Cheprakov, A.V. (2000). The Heck reaction as a sharpening stone of palladium catalysis. Chem. Rev. 100: 3009–3066. (c) Chinchilla, R. and Nájera, C. (2011). Recent advances in Sonogashira reactions. Chem. Soc. Rev. 40: 5084–5121.

83 Pascanu, V., Yao, Q., Gmez, A.B. et al. (2013). Sustainable catalysis: rational Pd loading on MIL-101Cr-NH_2 for more efficient and recyclable Suzuki–Miyaura reactions. Chem. Eur. J. 19: 17483–17493.

84 Jiang, Y., Zhang, X., Dai, X. et al. (2017). Microwave-assisted synthesis of ultrafine Au nanoparticles immobilized on MOF-199 in high loading as efficient catalysts for a three-component coupling reaction. Nano Research 10: 876–889.

85 Kassymova, M., Mahieu, A. de, Chaemchuen, S., Demeyere, P., Mousavi, B., Zhuiykov, S., Yusubovc, M.S., and Verpoort, F. (2018) Post-synthetic modified MOF for A^3-coupling reaction of aldehyde, amine, and alkyne. Catal. Sci. Technol., 8, 4129–4140.

86 (a) Cirujano, F.G., Xamena, F.X.L.I., and Corma, A. (2012). MOFs as multifunctional catalysts: one-pot synthesis of menthol from citronellal over a bifunctional MIL-101 catalyst. Dalton Trans. 41: 4249–4254. (b) Zhao, M., Deng, K., He, L.C. et al. (2014). Core–shell palladium nanoparticle@ metal–organic frameworks as multifunctional catalysts for cascade reactions. J. Am. Chem. Soc. 136: 1738–1741.

87 Pan, Y.Y., Yuan, B.Z., Li, Y.W., and He, D.H. (2010). Multifunctional catalysis by Pd@MIL-101: one-step synthesis of methyl isobutyl ketone over palladium nanoparticles deposited on a metal–organic framework. Chem. Commun. 46: 2280–2282.

88 Chen, Y.-Z., Zhou, Y.-X., Wang, H. et al. (2015). Multifunctional PdAg@MIL-101 for one-pot cascade reactions: combination of host–guest cooperation and bimetallic synergy in catalysis. ACS Catal. 5: 2062–2069.

89 Singh, M., Solanki, P., Patel, P. et al. (2019). Highly active ultra small Ni nanoparticle embedded inside a robust metal–organic framework: remarkably improved adsorption, selectivity, and solvent-free efficient fixation of CO_2. Inorg. Chem. 58: 8100–8110.

90 Mo, K., Yang, Y., and Cui, Y. (2014). A homochiral metal-organic framework as an effective asymmetric catalyst for cyanohydrin synthesis. J. Am. Chem. Soc. 136: 1746–1749.

91 Gong, W., Chen, X., Jiang, H. et al. (2019). Highly stable Zr(IV)-based metal-organic frameworks with chiral phosphoric acids for catalytic asymmetric tandem reactions. J. Am. Chem. Soc. 141: 7498–7508.

92 Sawano, T., Thacker, N.C., Lin, Z. et al. (2015). Robust, chiral, and porous BINAP-based metal–organic frameworks for highly enantioselective cyclization reactions. J. Am. Chem. Soc. 137: 12241–12248.

93 (a) Wu, C.-D., Hu, A., Zhang, L., and Lin, W. (2005). A homochiral porous metal–organic framework for highly enantioselective heterogeneous asymmetric catalysis. J. Am. Chem. Soc. 127: 8940–8941. (b) Wu, C.-D. and Lin, W. (2007). Heterogeneous asymmetric catalysis with homochiral metal-organic frameworks: network-structure-dependent catalytic activity. Angew. Chem. Int. Ed. 46: 1075–1078.

94 Han, Q., He, C., Zhao, M. et al. (2013). Engineering chiral polyoxometalate hybrid metal-organic frameworks for asymmetric dihydroxylation of olefins. J. Am. Chem. Soc. 135: 10186–10189.

95 Han, Q., Qi, B., Ren, W. et al. (2015). Polyoxometalate-based homochiralmetal-organic frameworks for tandem asymmetric transformation of cyclic carbonates from olefins. Nature Comm. 6: 10007–10008.

18

Controlling Excited State Chemistry of Organic Molecules in Water Through Incarceration

V. Ramamurthy

Department of Chemistry, University of Miami, Coral Gables, FL 33124, USA

18.1 Introduction

Chemistry, as we know it, is concerned with reactions in various media such as gas phase, solvents, solids, and biological systems [1]. The principles that we learn based on the behavior of molecules in gas phase are applied to that in solutions with subtle variations taking into account the effects of solvents. The same reactions when they occur in a biological system such as in a protein medium require further modifications due to restricted environments and weak interactions imposed by the environment [2]. Thus the behavior of molecules become more selective as the medium become more restrictive – gas phase to solution to protein to solids. In the same order the reactive molecules also lose their freedom, the ΔS become more negative. Clearly the restriction imposed by the medium has an influence on the chemical behavior of molecules. Recognizing this, we have been interested in exploring the excited state behavior of molecules in various constrained environments such as crystals, micelles, and host-guest supramolecular assemblies [3–14]. This review concerns our recent contributions on one such host-guest supramolecular capsular assembly.

The literature is replete with studies in cyclodextrins (CD) [10, 15], cucurbiturils (CB), calixarenes (CA), and related cavitand hosts [16–19]. All these hosts, although they solubilize an organic molecule in aqueous medium, partially expose them to water. While they facilitate conducting reactions of organic molecules in water, they do not offer total incarceration. This review summarizes our work with a cavitand known as octa acid (OA) [11, 20], originally synthesized by Gibb and Gibb [21, 22], that is soluble in water under basic conditions (pH > 8.5). Its structure and internal dimensions are provided in Figures 18.1 and 18.2. Similar to CD, CB, and CA, the commonly familiar hosts, the interior of the cavitand OA is hollow and can accommodate hydrophobic molecules. However, unlike CD, CB, and CA, the two ends are of different dimensions, with one of them being too narrow even for an oxygen molecule to pass through. More importantly, cavitand OA spontaneously forms a capsule, an assembly of two molecules (Figure 18.1 bottom) in the presence of a guest. This unique feature attracted our attention. Several water-soluble three-dimensional hosts have been reported during the last decade [23–30]. Each one has its own advantages and disadvantages in terms of light absorption, photostability, availability of space for accommodation of reactant, and products of reactions and availability. We were attracted to OA capsule due to its ability to control and provide selectivity similar to crystals [6, 14] and zeolites [9, 12] while being more flexible and softer like micelles [13].

Email: murthy1@miami.edu

Chemical Reactivity in Confined Systems: Theory, Modelling and Applications,
First Edition. Edited by Pratim Kumar Chattaraj and Debdutta Chakraborty.

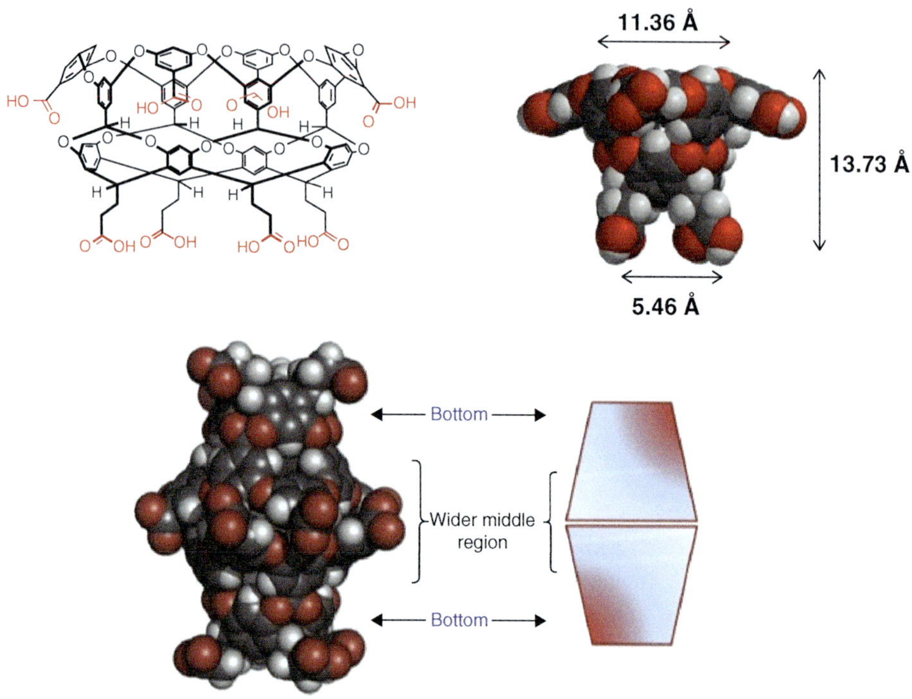

Figure 18.1 Top: Structure and dimensions of cavitand octa acid. Bottom: CPK model of the capsular assembly of two molecules of octa acid.

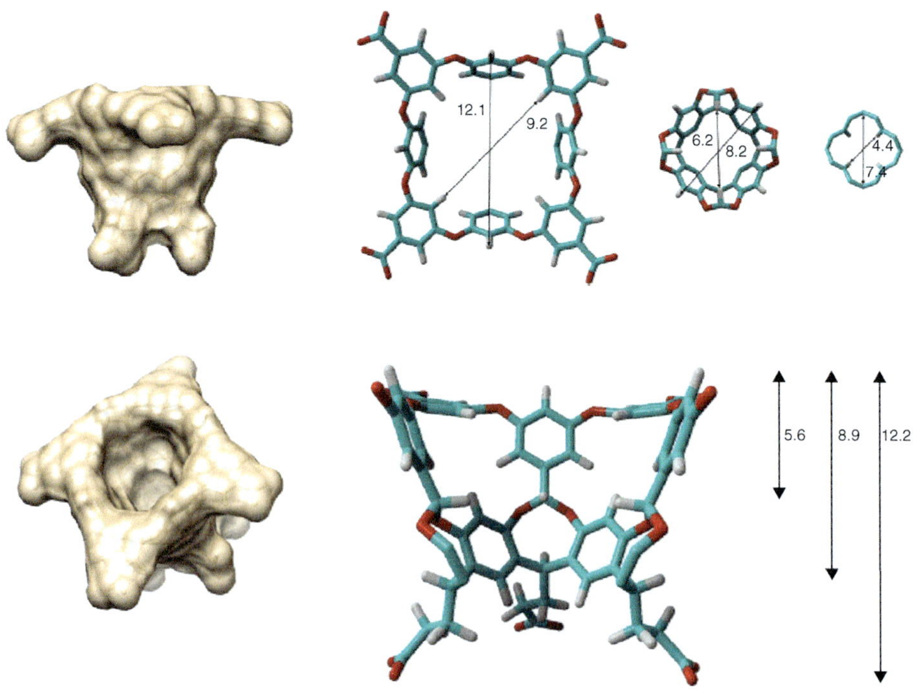

Figure 18.2 A cartoon presentation of the cavity of the cavitand OA. Entrance diameter and depth of the cavity presented.

We have been addressing the following questions concerning the behavior of organic molecules encapsulated within the above OA capsule: (a) How extensively the capsule would alter the excited state chemistry and physics of included guest molecules [31–54]? (b) Can a molecule present outside the capsule interact in terms of spin, electron, and energy with the encapsulated guest molecule [55–66]? (c) Would the ground and excited state dynamics of molecules be altered by OA encapsulation [67–78]? (d) Can the capsule be used to deliver molecules of interest through phototriggering methodology [79–84]? (e) Will the capsule be stable on active (TiO_2 and gold nanoparticle) and inert (silica, clay, Zr phosphate) surfaces [60, 85–94]? Would the functionalized surfaces be effective in energy and electron transfer prompted processes? (f) Will reactive intermediates and strained molecules be stable within the capsule [95–97]? The current review provides a glimpse of our efforts in exploring OA capsule as a medium for altering the ground and excited state behavior of organic molecules. Our original publications provide details and readers are directed to them for an in-depth understanding of the answers to the above questions.

18.2 Complexation Properties of OA

OA, depending on the guest, forms different types of host–guest complexes in borate buffer solution (pH 8.7) (Figure 18.3) [76]. It forms a 1:1 (host to guest) complex with a molecule that has hydrophobic body and hydrophilic head group (e.g. adamantyl substituted with an ammonium group at 1-position). When the molecule is fully hydrophobic with no hydrophilic groups it forms either a 2:1 or a 2:2 complex (adamantane, adamantyl ester etc.). Which of the two, 2:1 or 2:2 complex, would be formed depends on the size of the guest molecule and concentrations of the host

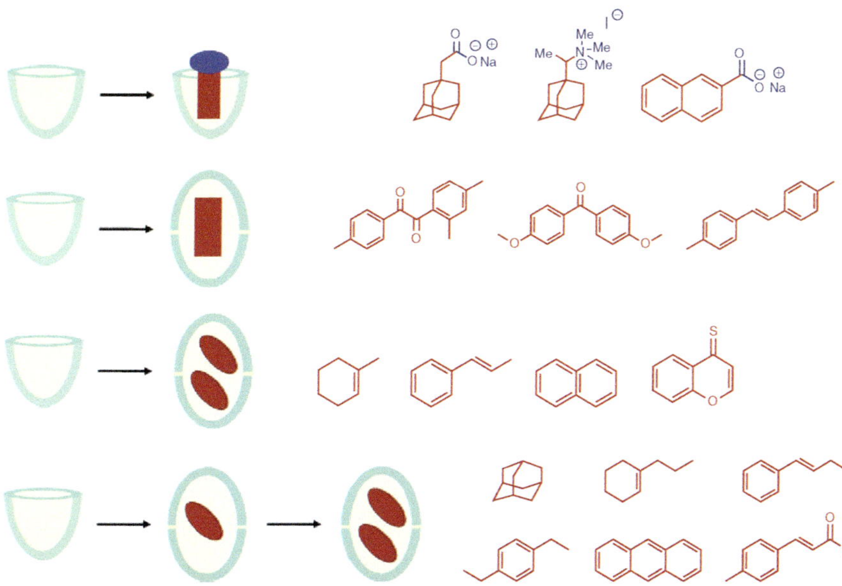

Figure 18.3 Host-guest complex formation with various. Nature and stoichiometry of the complex depend on the guest molecule. Source: Based on Jayaraj, N., Zhao, Y., Parthasarathy, A., Porel, M., Liu, R. S. H., Ramamurthy, V. (2009) Nature of Supramolecular Complexes Controlled by the Structure of the Guest Molecules: Formation of Octa Acid Based Capsuleplex and Cavitandplex. Langmuir, 25, 10575–10586.

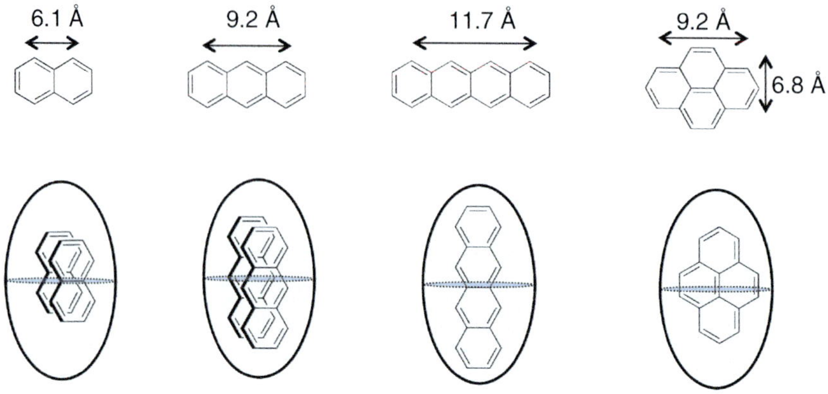

Figure 18.4 Nature of OA complex with aromatic molecules. Note molecules longer than tetracene and wider than pyrene could be included within OA capsule.

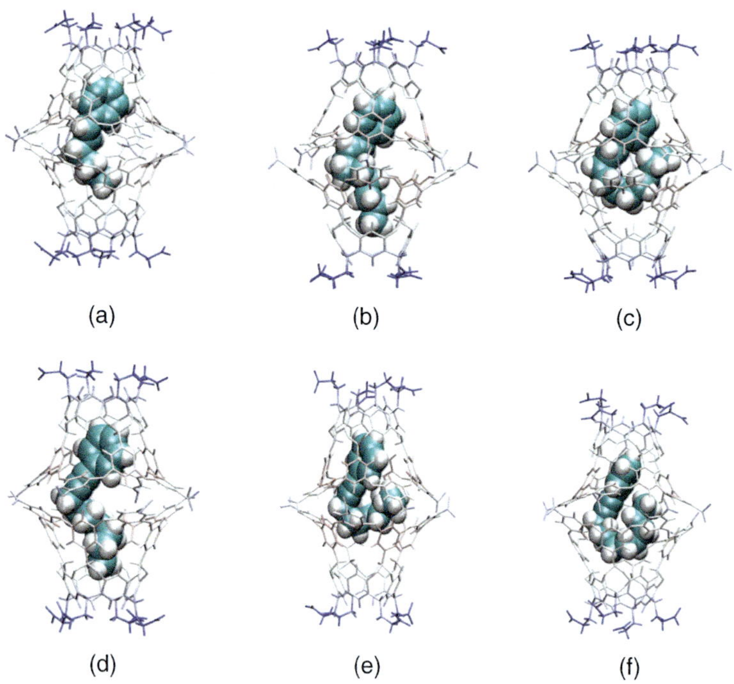

Figure 18.5 The most representative structures obtained from MD simulation after performing a cluster analysis (GROMACS, OPLS-AA force field) showing the orientation of different guest molecules inside OA capsule. (a)–(c) 1-phenyl-2- alkenes and (d)-(f) 1-phenyl-2- alkynes.

and the guest in solution. ^{1}H NMR titration experiments help define the complex and follow the dynamics of complex formation. Perusal of rigid aromatic molecules listed in Figure 18.4 provides information concerning the limitations of the guest molecules that could be included within OA capsule. Molecules longer than tetracene (> 12 Å) and wider than pyrene (> 7 Å) could not be included within the OA capsule. Also, it is clear that smaller molecules such as naphthalene and anthracene form a 2:2 complex while larger and bulkier molecules such as tetracene and pyrene form a 2:1 complex. At the same time, one should note that molecules with a long flexible alkyl chain could be included [75]. MD simulated structures of such molecules shown in Figure 18.5 suggest that these molecules fold their chain and stay within the capsule. Flexible molecules as long as 17 Å could be readily included within OA capsule. The inclusion of guests within OA is achieved by mildly stirring or shaking the required amounts of OA and guest in borate buffer for 10–15 mts. ^{1}H NMR spectroscopy is an ideal tool to follow the host-guest complex formation and 2D-NMR spectra are valuable to gain an insight into the structure of the host-guest complexes.

To gain an understanding of the forces that control complexation, ΔH and ΔS values were measured by isothermal calorimetry (ITC) for a few selected guests (adamantly series and naphthyl series) that form 1:1 complex [78]. The conclusion from this study is that all guest inclusion is driven by negative ΔH. Depending on the guest the ΔS is either positive or negative. The negative ΔH is attributed to weak van der Walls (adamantly series) and π–π interaction (naphthyl series) with the walls of the OA capsule. In all cases, independent of whether ΔS is positive or negative, the overall ΔG is negative suggesting the weak interactions between the host cavity and the guest drive the complexation.

18.3 Properties of OA capsule

The capsule although soluble in aqueous medium, internal polarity of the capsule is detected to be similar to that of benzene [77]. Internal polarity was monitored by several photophysical probes such as pyrene, coumarins, pyrenealdehyde etc. (Figure 18.6a). The change in I_1/I_3 (ratio of intensities of the emission bands 1 and 3), a measure of the polarity difference between water and within

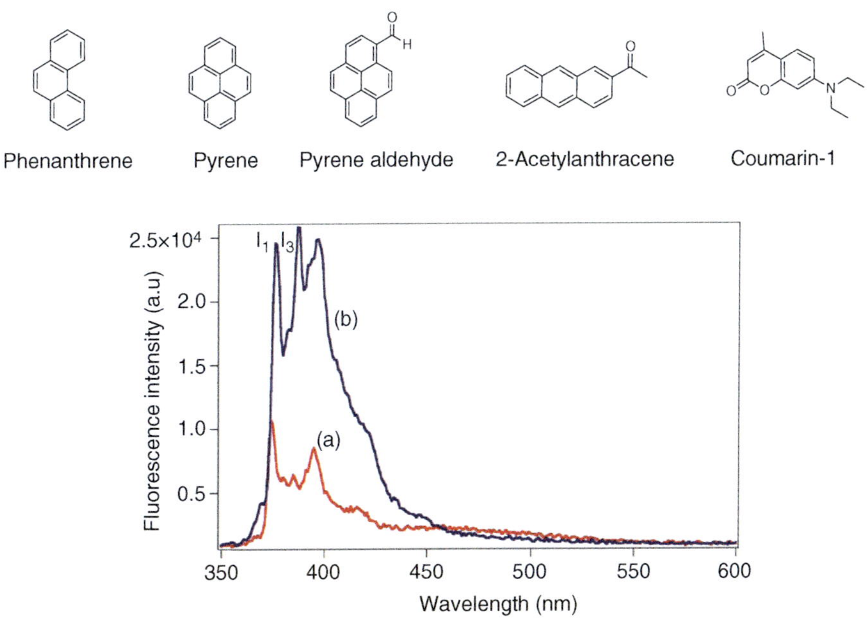

Figure 18.6 Top: Structures of organic molecules used as polarity probes. Bottom: Fluorescence emission spectra in borate buffer (pH~9) of (a) 1×10^{-5} M pyrene and (b) pyrene@$(OA)_2$. Note the relative intensities of I_1 to I_3 in the two cases. Excitation wavelength: 320 nm.

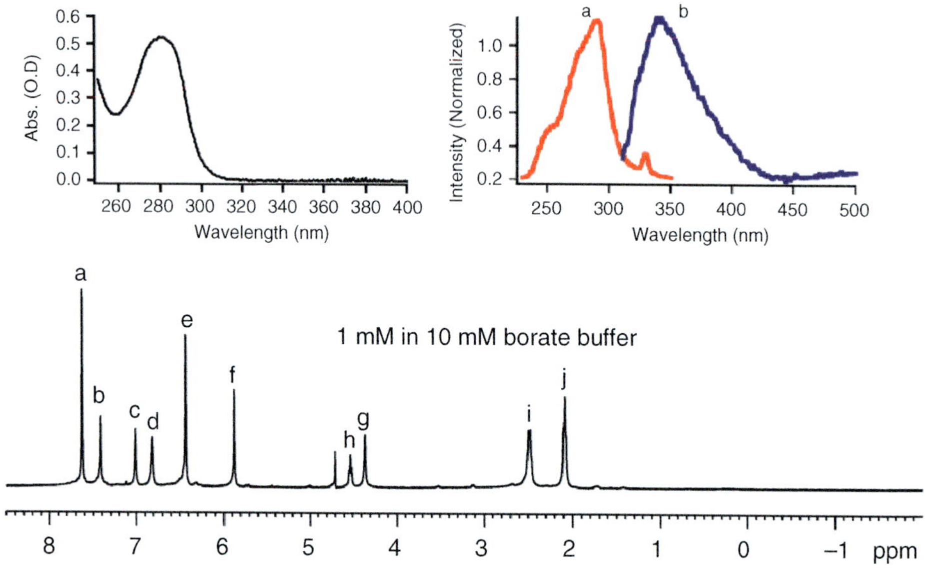

Figure 18.7 top: Absorption, excitation and emission spectra of OA. Bottom: ^{1}H NMR spectrum of OA in borate buffer is also included.

OA, is shown in Figure 18.6b. Absence of water within OA cavity was also confirmed by EPR coupling constants of nitroxides that are used as spin probes. Both EPR active and fluorescent probes indicate interior of the OA capsule does not have any water molecule. Thus it is important to be aware that the guest molecules that are solubilized in water with the help of OA remain in a hydrophobic aromatic-like environment.

Use of OA as the host for photochemical reactions requires one to be aware of its excited state properties [64]. OA absorbs in the region 230 to 320 nm and emits in between 320 to 430 nm (Figure 18.7) It can also act as a triplet energy donor. This is illustrated with examples where it sensitizes triplet reactions of included guests (Figure 18.8). With the help of probe reactant molecules the triplet energy of OA is estimated to be 73 kcal mol^{-1}. Furthermore, OA is a good electron donor with an he oxidation potential of ~1.5 eV. For examples when acceptors are excited in presence of OA their fluorescence are quenched suggesting that OA is a good electron donor (Figure 18.9). Thus if one is interested in examining excited state reactions of encapsulated guests one should be aware that OA itself can act as an electron transfer agent and it is not an inert reaction cavity.

18.4 Dynamics of Encapsulated Guests

Unlike molecules in crystals, guest molecules within a capsule are not frozen. The guest molecules within a capsule undergo various types of motions while at the same time they are restricted with respect to that in isotropic solvent medium. The freedom of guest molecules depends on its structure, how well they fit, what type of weak forces hold them inside the capsule and how much free space is around them within the capsule. Various types of possible motions a guest can undergo within a capsule are illustrated in a cartoon fashion in Figure 18.10 [74]. These are rotations along long and short axes, sliding motion if there are two molecules within a capsule, and the capsule opening and closing.

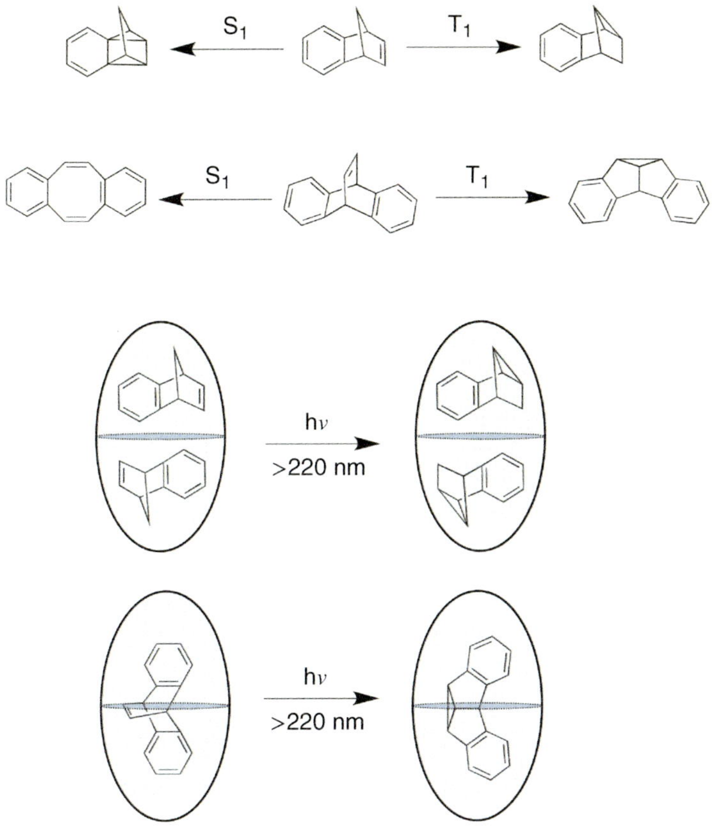

Figure 18.8 Examples of triplet sensitization by OA. As illustrated above encapsulated molecules give different products from S_1 and T_1. Within OA they give only products from T_1.

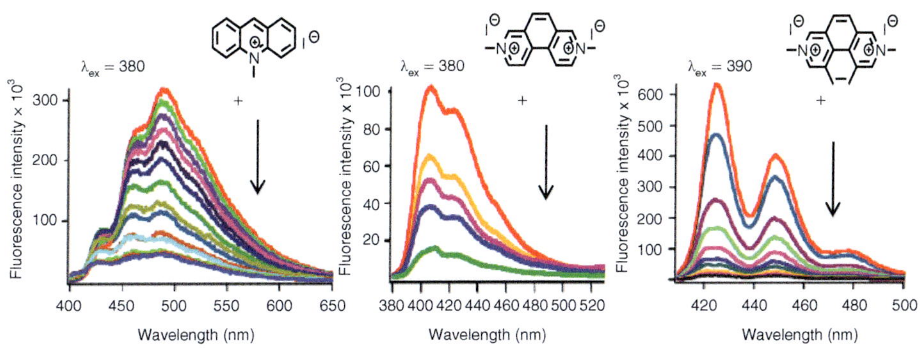

Figure 18.9 Examples of electron transfer by OA to excited acceptors. Fluorescence quenching of electron acceptors by OA.

How easily the guest molecule tumbles within a capsule has been probed with EPR active nitroxide molecules as guests [74]. While the EPR coupling constants is a reflection of the micropolarity of the immediate environment, the rotational correlation time of the probe nitroxide provides information regarding the rotation of the probe within the OA capsule. In Figure 18.11 the rotational correlation time of the probe NET-6 in water and within OA capsule are plotted with respect to the

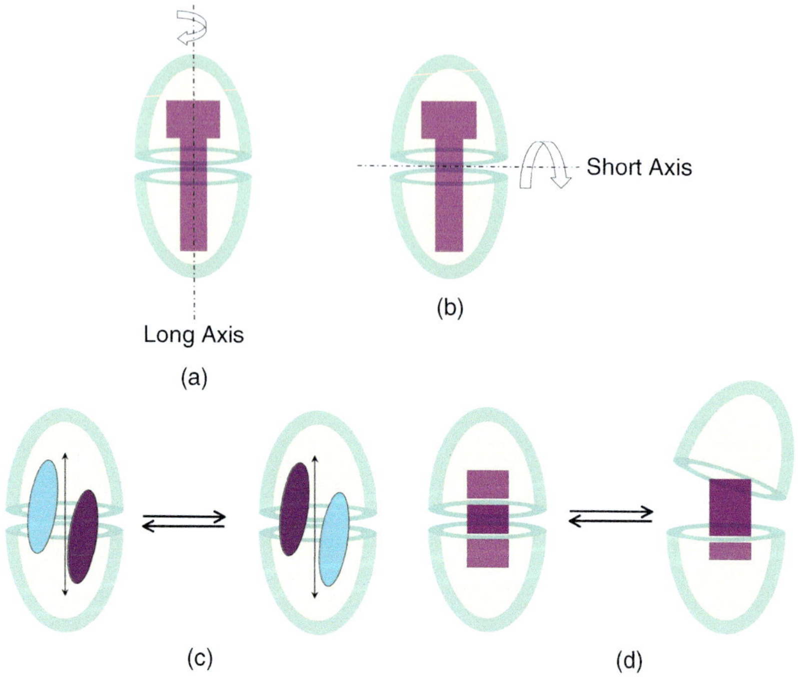

Figure 18.10 Various motions a guest and capsule could undergo in solution. The motions are rotations along long and short axes (a) and (b), sliding motion if there are two molecules within a capsule (c) and the capsule opening and closing (d). Source: Kulasekharan, R., Jayaraj, N., Porel, M., Choudhury, R., Sundaresan, A. K., Parthasarathy, A., Ottaviani, M. F., Jockusch, S., Turro, N. J., Ramamurthy, V. (2010) Guest rotations within a capsuleplex probed by NMR and EPR techniques. Langmuir, 26, 6943–6953. © 2010, American Chemical Society

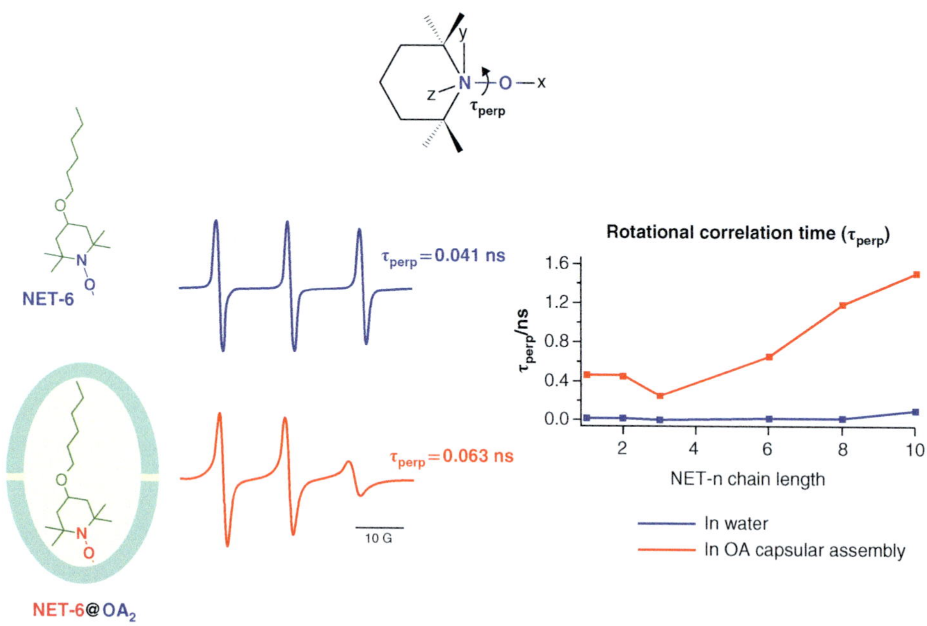

Figure 18.11 EPR signals of NET-6 in water and within OA capsule. A plot of rotational correlation time for several O-alkyl nitroxides estimated by simulation of the spectra is also shown.

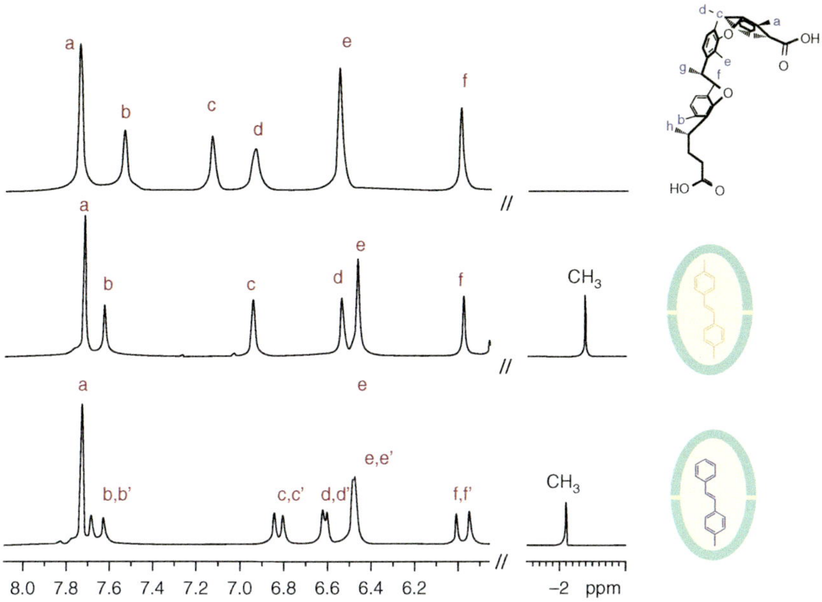

Figure 18.12 Top: Partial ^{1}H NMR spectrum for free OA. Middle: Partial ^{1}H NMR spectrum for 4,4′-dimethyl stilbene@OA_2. Bottom: Partial ^{1}H NMR spectrum for 4-dimethyl stilbene@OA_2.

alkyl chain length (the alkyl group of the probe NET varied from methyl to decyl group). Clearly the rotational correlation time within OA capsule is longer than in water indicating that the tumbling motion is slowed within the capsule. As shown in the plot the rotational correlation time depends on the length of the alkyl substituent. While in water the rotational correlation time is independent of the chain length, within the capsule it depends on the length of the alkyl substituent. Clearly with longer molecules there is restriction for rotation in EPR time scale.

^{1}H NMR spectra also provide information concerning the rotational motions of guests, host as well as total capsule [74]. As illustrated in Figure 18.12, close analysis of the signals corresponding to the host OA ($H_a - H_f$ with $\delta > 5$ ppm in the spectra) provide information about the rotation of the guest along long and short axis (Figure 18.10). Cavitand OA is made up of the repetition of the panel shown in Figure 18.12 top first line. If the molecule is frozen along the long axis four independent NMR signals for four panels would be expected. If the guest is asymmetric and do not tumble within the capsule one would expect eight independent NMR signals for eight panels belonging to top (4) bottom (4) cavitands. In the spectra shown in Figure 18.12 (middle) there is only one signal for magnetically equivalent hydrogens of OA, suggesting the guest molecule (4,4′-dimethyl stilbene) freely rotates around the long axis. On the other hand, in the case of 4-dimethyl stilbene two signals appear, one for the top and the other for the bottom cavitand, suggesting that although this molecule freely rotates along the long axis it does not freely tumble along the short axis.

With a non-symmetric guest the two halves of the capsule (made up of two OA) are magnetically different because the molecular frame of the guest that occupies the two halves are not the same. If the guest does not tumble along the short axis, making the two halves identical, the chemically equivalent hydrogens of the capsule on the two halves will have different chemical shifts. If they tumble fast in the NMR time scale along the short axis (Figure 18.10), there would be a single signal for chemically equivalent hydrogens on the two halves. In Figure 18.13 the NMR spectra for three host-guest complexes are presented. In all cases a single signal for chemically equivalent hydrogens

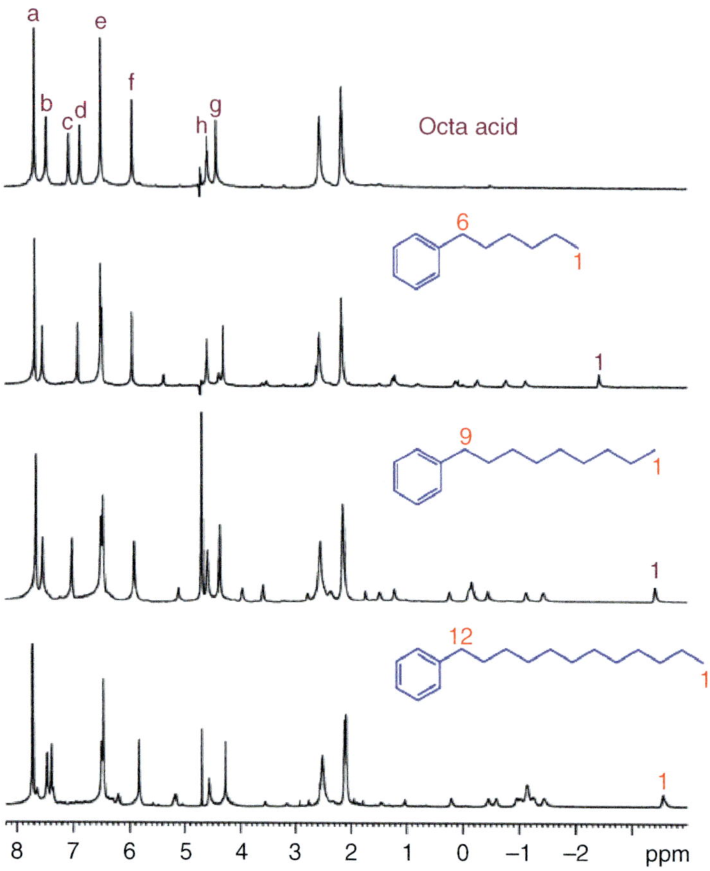

Figure 18.13 Top: Partial ^{1}H NMR spectrum for free OA. Next three: Partial ^{1}H NMR spectrum of 1-phenyl alkanes of various lengths.

is noted. This is an indication of the guest molecule tumbling freely in the NMR time scale along the short axis. This is to be contrasted with 4-dimethyl stilbene@OA_2 discussed above (Figure 18.12).

Clearly whether a molecule is able to tumble depends on the structure [74]. Apparently a co-ordinated motion is required for the tumbling. As illustrated in Figure 18.13 the hydrogens of the top and bottom parts of the capsule have a single signal suggesting that these long flexible molecules are able to freely rotate along the short axis within the capsule. Such information is useful in interpreting the dynamics of molecules in a confined space.

As one would expect, the tumbling motion depends on the temperature [74]. In general higher temperature facilitates tumbling. In Figure 18.14, the ^{1}H NMR spectra of the OA-guest complex at various temperatures are provided. This molecule is anchored at one end of the capsule through C–H---π interaction between the CH_3 group of the guest and the aryl groups present at the narrow corner of the host. At room temperature (25 °C), the spectrum provides an indication of difficulty the molecule is having to tumble; signals are not cleanly split (spectrum "b" in the figure). However, at lower temperature (5 °C) rotation is slowed and the host signals are split (spectrum "a" in the figure). On the other hand at higher temperature (55 °C) the tumbling seems fast, evident from the appearance of a single signal for the two halves of the OA capsule (spectrum "c" in the figure). From analysis of the temperature dependent spectra the ΔH and ΔS for the tumbling are estimated to be

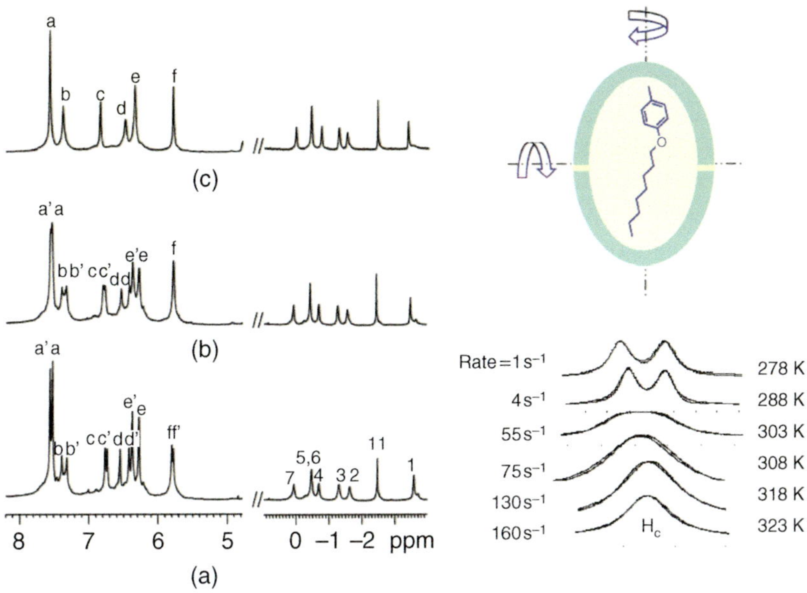

Figure 18.14 Partial ^{1}H NMR spectra of guest@OA_2 (500 MHz, [OA] = 1mM) in 10 mM borate buffer (D_2O) at (a) 5 °C, (b) 25 °C, and (c) 55 °C (OA peaks are marked as a–f, guest peaks are marked as 1–11). Also, NMR simulations (WINDNMR-Pro) for the H_c resonance in guest @OA_2 and Erying plot.

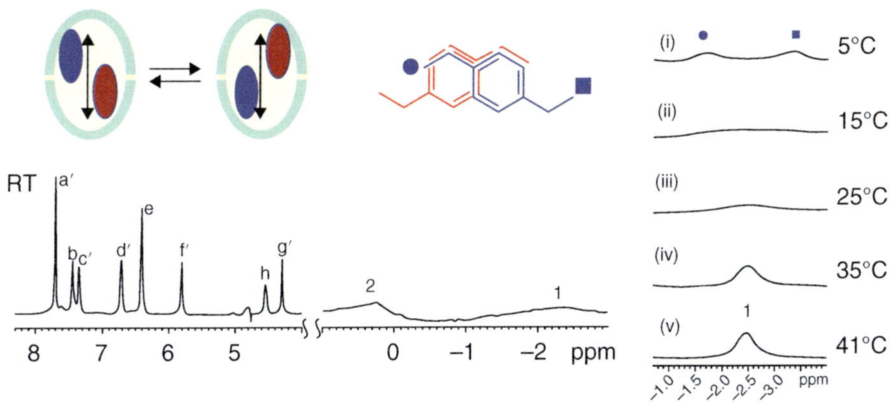

Figure 18.15 Left: Partial ^{1}H NMR spectrum of 1,4-diethylbenze included within capsule at room temperature. Right: ^{1}H NMR spectra of the methyl groups of the ethyl substituent at various temperatures.

9.7 kcal mol^{-1} and −18.5 eu. The large negative ΔS suggests that the molecular motion is restricted within the capsule. Similar analysis of the ^{1}H NMR spectra of a large number of molecules with different structural features suggest that tumbling motion that makes the two halves of the capsule identical depends on the guest molecule.

As shown in Figure 18.3 aromatic molecules such as naphthalene and anthracene form 2:2 complexes [76]. In these structures the two aromatic molecules that are not held together by any interactions may slide one over the other. Whether it can happen will depend on the restriction provided by the capsule. In Figure 18.10c slithering (sliding) motion of the encapsulated molecules is illustrated. Such a motion while in solution is normal and does not call for any special attention, but when the space is restricted it may become important. The molecule shown in Figure 18.15 is symmetrical but when they stack and remain static the two ethyl groups, although chemically equivalent, are not magnetically equivalent. If they slide (slither) fast on the NMR time scale the two groups would become magnetically equivalent; if they don't, each would be unique with different chemical shifts. We exploited this feature to probe whether the slithering motion is permitted within OA capsule. In Figure 18.15 the two ethyl groups present at the two ends are indicated with a filled circle and square. Interestingly, while at 5 °C there are two signals for the methyl group, at 35 °C there is a single signal. This suggests that slithering motion that is facile and non-unique in solution become restricted within the OA capsule. Thus, once again the motions of molecules within the confined space of OA capsule become restricted. At the same time it is important to note that unlike crystals the molecules are not frozen inside the capsule. How much freedom the guest molecule experiences depends on its size and shape and how well they fit within the capsule.

18.5 Dynamics of Host-Guest Complex

One of the most important questions concerning the stability of OA capsule is whether it remains stable in the time scale of the spectral measurements. In general one could think of the capsule remaining completely closed (Figure 18.16a), partially open-and-close (Figure 18.16b) and assemble-disassemble (Figure 18.16c) in the time scale of measurements. These possibilities would be expected to occur in different time scales. Results on the dynamics of capsular complexes monitored through NMR and photophysical measurements are briefly discussed below [67].

The first question is whether the capsule would open and close (breathing motion) within the excited state lifetime of encapsulated guest. This is addressed by quenching the excited state of the encapsulated guest by ground state oxygen, which has a low singlet energy (22 kcal mol^{-1}). It is well known that the excited singlet and triplet states of organic molecules are quenched by ground state oxygen provided it can access them. Since oxygen is outside the capsule in water it can quench only if the capsule opens enough to access the excited molecule. The rate constants for quenching of excited guests estimated from quenching of their emission by oxygen are tabulated in Figure 18.17. Quenching was confirmed by generation of singlet oxygen identified by its emission (see Figure 18.17). The rate constants were estimated by monitoring the decay of excited triplets either by their absorption or emission (phosphorescence) in presence of oxygen. It is important to note in Figure 18.17 that molecules whose triplet lifetime is longer than 17 μs are quenched by oxygen, and the ones shorter than 10 μs are not quenched. This suggests that the capsule opens, at least enough for oxygen to enter (Figure 18.16b, in the time scale of 10 μs).

As expected, the capsule opening depends on the nature of the guest. This information was revealed by coumarins as guests [68]. Emission and absorption maxima of coumarins depend on the polarity of the environment where it is present (Figure 18.18) [98]. As mentioned above

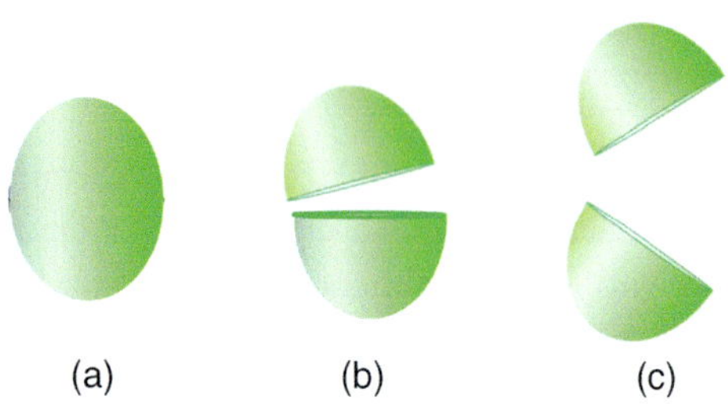

Figure 18.16 The carton representations of the capsule in various forms, closed, partially open, and fully open.

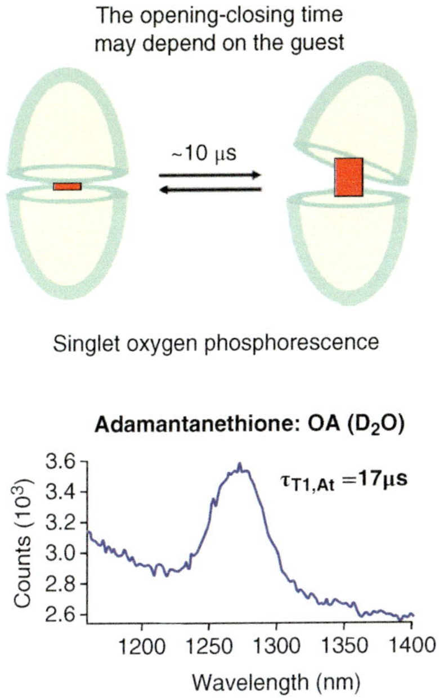

Triplet/Singlet senzitizers	Lifetime within OA (D_2O) / μs	$k_q(O_2)/M^{-1}s^{-1}$
Fluorenone[a]	726	3×10^5
4,4'-Dimethylbenzil[a]	596	8.7×10^7
Benzil[a]	164	1.3×10^7
Thiofenchone[a]	187	1.2×10^6
Camphorthione[a]	65	2.0×10^7
Adamantanethione[a]	17	2.3×10^7
Xanthione[a]	4.7	-[c]
4,4'-Dimethoxythio benzophenone[a]	0.62	-[c]
Pyrene[b]	0.34	-[c]
Phenanthrene[b]	0.05	-[c]

[a]Triplet life time, [b]Singlet life time and [c]$k_q<10^5M^{-1}s^{-1}$

Figure 18.17 Right: The rate constants for quenching of excited state (singlet or triplet) of encapsulated guests by oxygen. Left bottom: emission from singlet oxygen upon quenching. Left top: Cartoon representation of capsule partial opening and its time constant.

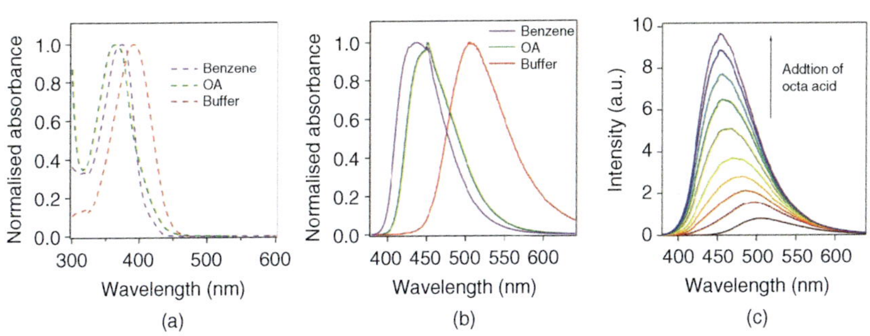

Figure 18.18 (a) Normalized absorption spectra and (b) normalized emission spectra of C-152 in benzene (purple), C-152@$(OA)_2$ (green), C-152 in buffer (red); (c) Fluorescence titration spectra (λ_{ex} = 370 nm) of C-152 with addition of OA. Source: Based on Jones, G. I., Jackson, W. R., Choi, C. Y., Bergmark, W. R. (1985) Solvent Effects on Emission Yield and Lifetimes for Coumarin Laser Dyes. Requirements for a Rotatory Decay Mechanism. J Phys Chem, 89, 294–300.

micro-polarity of the capsular interior is non-polar while outside is polar (water). If the capsule opens during the excited state lifetime of the guest, water molecules would enter the interior and increase the micropolarity of the capsule. Monitoring the solvation dynamics of coumarins (Figure 18.19) in ultrafast time scale has divulged that the capsule is fully closed at least up to 3 ns when the guests fit well within the capsule (Figure 18.19). As illustrated in Figure 18.19, the emission maxima remains constant up to 3 ns, the time up to which the spectra were recorded.

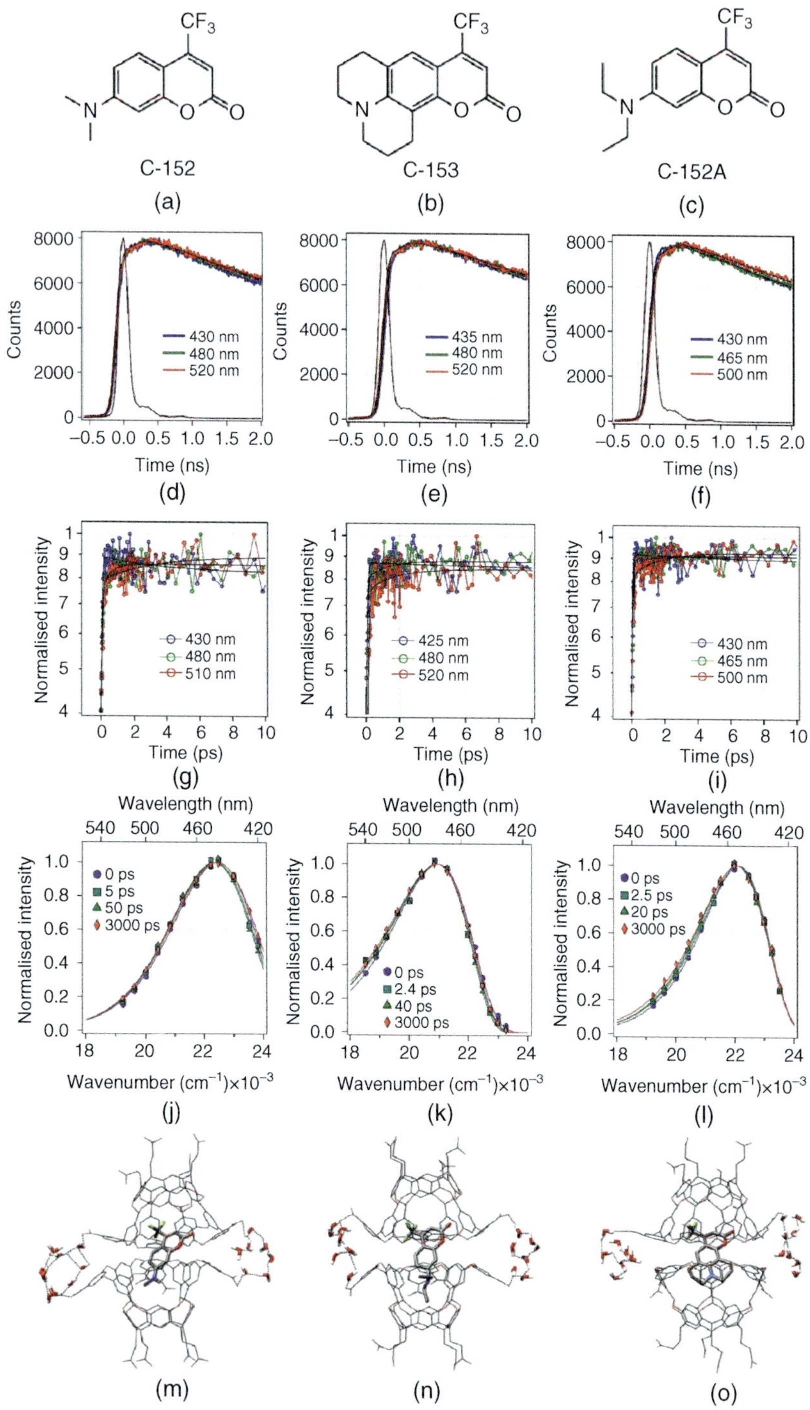

Figure 18.19 (a)–(c): Structures of three small coumarins. (d)–(f) Representative TCSPC decay profile at some selected wavelengths. (g)–(i) Representative ultrafast fluorescence transients at some selected wavelengths (j)–(l): Time-resolved emission spectra (TRES) constructed from the combination of femtosecond fluorescence up-conversion and TCSPC method. No change in the spectra with time. (m)–(o) Molecular dynamics simulated structures host-guest complexes. Water molecules are away from the capsule and there is no interaction between the encapsulated coumarin and water molecules.

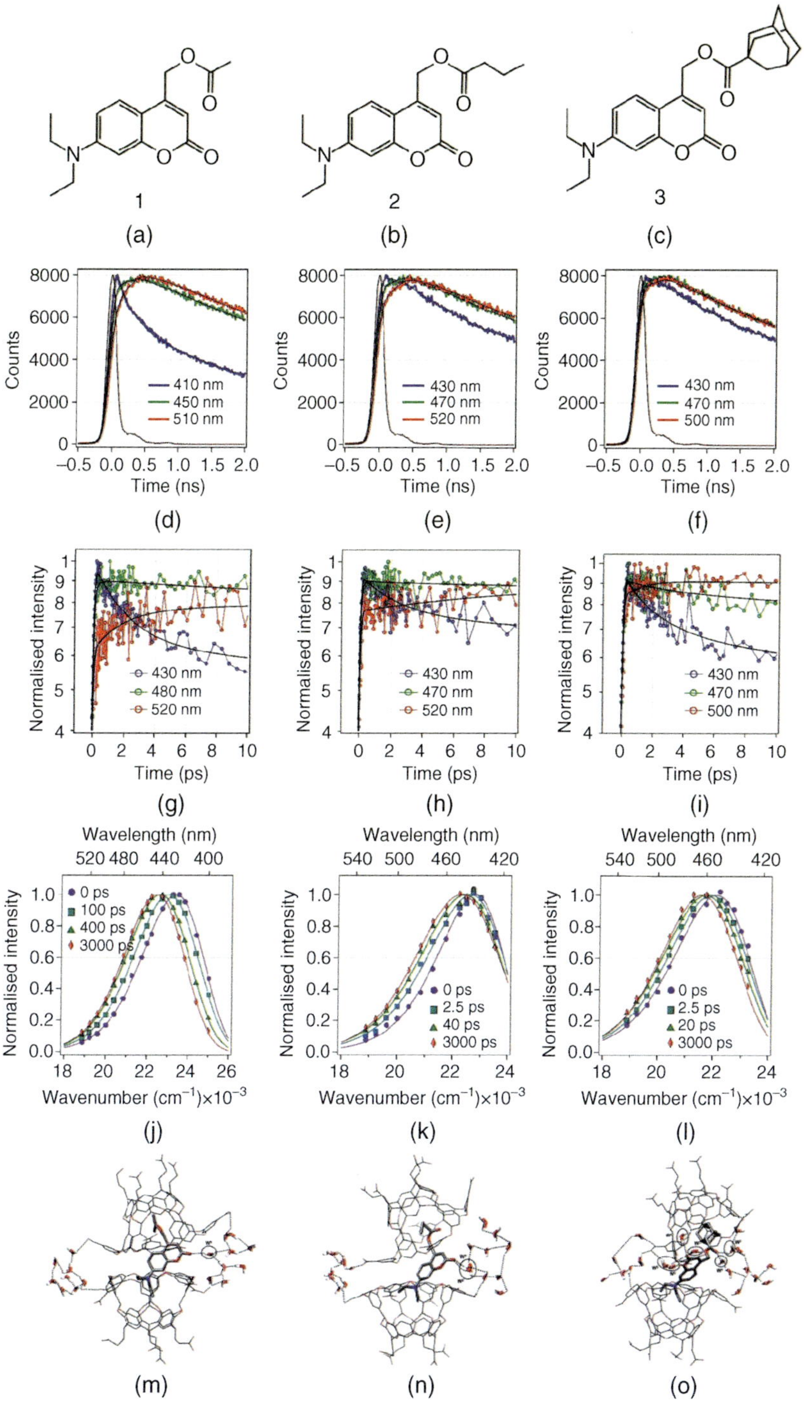

Figure 18.20 (a)–(c): Structures of three larger coumarins. (d)–(f) Representative TCSPC decay profile at some selected wavelengths. (g)–(i) Time-resolved emission spectra (TRES) constructed from the combination of femtosecond fluorescence up-conversion and TCSPC method. The spectra shifts with time. (m)–(o) Molecular dynamics simulated structures of host-guest complexes. Water molecules seep into the capsule even in ground state. There are interactions between the encapsulated coumarin and water molecules.

On the other hand, when the guest coumarins are larger and does not fit well (Figure 18.20) the capsule opens within 3 ns time scale. The MD simulated structures show that these molecules are in contact with exterior water molecules even in equilibrated structures. Importantly, upon excitation the emission maxima shifts with time indicating water molecules enter the capsule and surround the emitting coumarins slowly. This can happen only if the capsule opens even in ps time scale. Most likely the opening is triggered by the water molecules that are already in touch with the guest coumarins. Comparison of Figures 18.19 and 18.20 clearly confirms the capsule opening depends on the size and hydrophobicity of the guest molecules. The capsule that is slightly ajar in the ground state (Figure 18.20, m, n and o) starts opening upon excitation in the ps time scale. When the guest is hydrophobic and fits well it will spend all its excited state lifetime within the capsule with no contact with water molecules.

The final question in the context of capsule dynamics is whether the capsule as a whole is stable, if not how long it takes for the capsule to fully disassemble and assemble. Information regarding this was sought by 2D-NMR spectroscopy using amphiphilic benzylidene-3-methylimidazolidinone molecules as guests [69]. In these capsuleplexes, although the guest does not tumble, the two OA molecules that form the capsule become magnetically equivalent because the two OA molecules that form the capsule exchange their positions in the NMR time scale. This is equivalent to the content of the capsule remaining stationary while the capsule swirls around it. This is different from the guest tumbling within the capsule and making the two halves magnetically equivalent. In this case the guest remains stationary while the two parts of the host exchange making the two halves magnetically equivalent.

Possible modes of cavitand exchange in the absence and presence of excess OA in the NMR timescale are shown in Figure 18.21. In the figure, the two cavitands that form the capsule are color coded. When the capsule is stable the two chemically equivalent hydrogens for example g and g′ would have different chemical shifts. On the other hand when the two cavitands that form the capsule exchange their position hydrogens g and g′ would have the same chemical shift. In the 2D ROSEY spectra g and g′ would show correlation provided in the time scale of mixing the two hydrogens because of exchange they become equivalent. When excess OA is present along with capsule there can be an exchange with the excess OA molecules as well when the capsule disassembles and reassembles. In Figure 18.21 cavitands 1 and 2 of the capsule and free cavitand are color coded to illustrate the possible modes of three-way exchange occurring during disassembly and assembly. The three cartoons represent the occurrence under three different conditions: H:G 2:1, H:G 3:1, and H:G 7:1.

Two-dimensional NMR, ROESY, and NOESY correlation data suggest that in 300 ms time scale, the two halves of the capsule exchange between themselves and with free OA. In Figure 18.22 the guest-host structure is shown above the ROESY spectra. In this if there is no exchange the two methyl groups of the guest should show correlation only with one of the two cavitands. If the cavitands exchange their position, the methyl groups would show correlation with the protons of both cavitands. As shown in Figure 18.22 the two methyl groups of the guest correlate with both g protons of the capsule. This is a clear evidence for the exchange of cavitands that form a capsule. Such an exchange should facilitate exchange between the hydrogens of the cavitands on either side also. As expected, "g" hydrogens of the two cavitands that form a capsule show correlation (Figure 18.21). These suggest that the two cavitands exchange their position during the NMR time scale. In other words, the capsule disassembles and reassembles in the 300 ms.

How easily the cavitands that form the capsule exchange depends on the hydrophobicity of the guest molecule. In Figure 18.23 the 2D ROSEY correlations between the two halves of the capsule

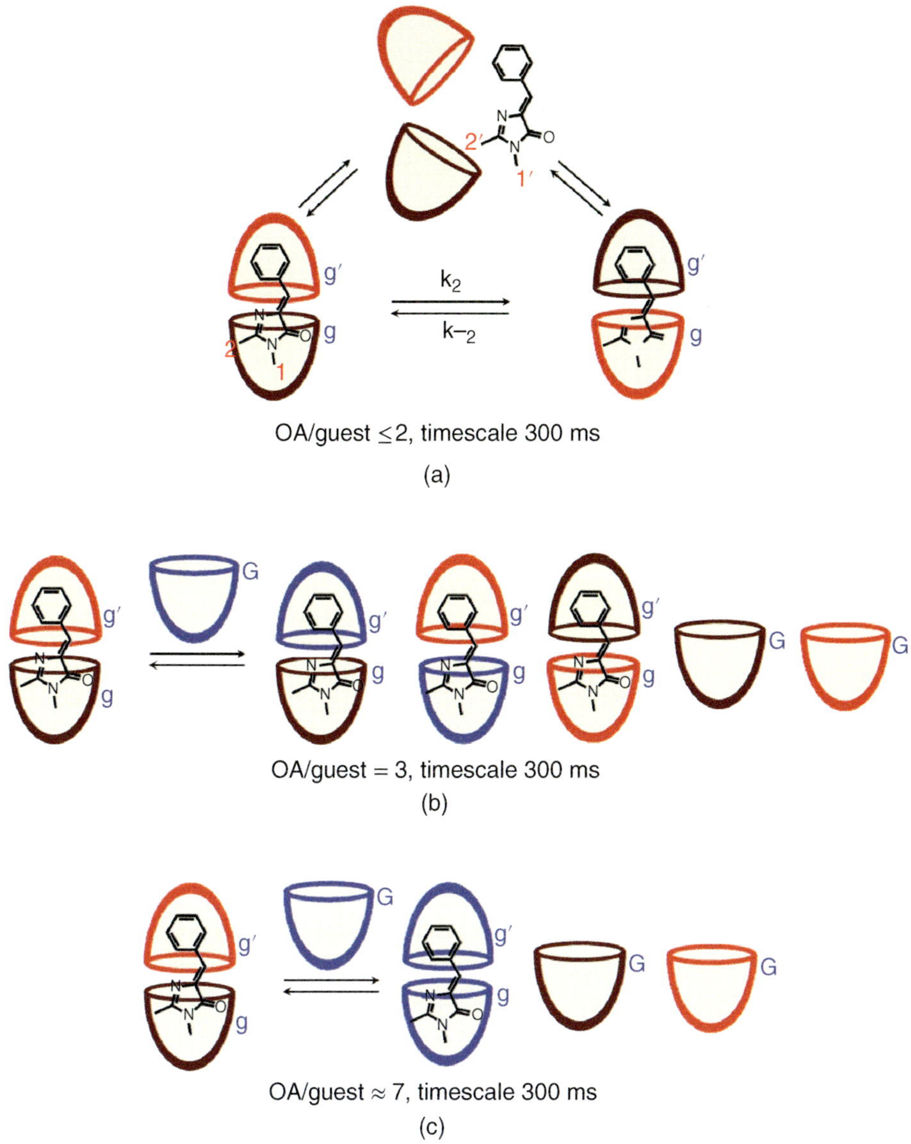

Figure 18.21 Cartoon representation of (a) exchange between two types of OA of 1:2 complex with each other when OA/guest $\leq$ 1:2, (b) exchange among three types of OA i.e. two from 1:2 complex and one unbound OA present in the solution when OA/guest =3:1, (c) exchange between unbound OA and one of the bound OAs from 1:2 complex separately when OA/guest $\approx$7:1.

and free OA are shown for a more hydrophobic guest (methyl is replaced by propyl). Interestingly, comparison of Figures 18.22 and 18.23, reveals the difference in exchange between capsules containing the methyl and propyl guests. In the case of propyl guest there is no correlation of g with either g′ or G (excess OA), suggesting that the capsule is stable and does not disassemble in 300 ms. Apparently with more hydrophobic guest the capsule is stable and does not fall apart in the NMR time scale.

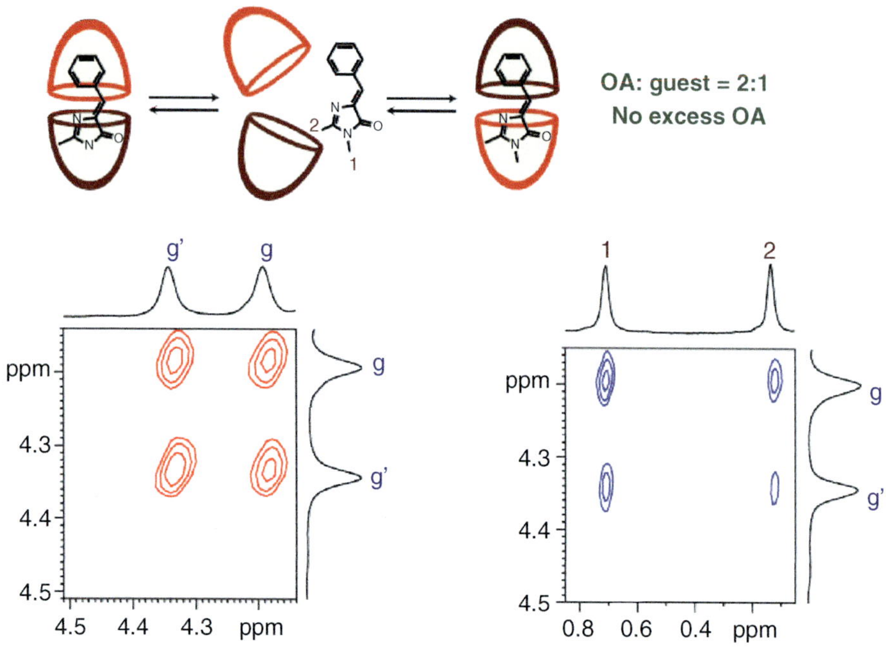

Figure 18.22 Top: A cartoon representation of intramolecular host exchange. Bottom left: Partial 2D ROESY (300 ms mixing time) spectra at guest:host ratio of 1:2. Correlation of the two cavitand protons are indicated. Note that g correlates with both g and g′. Bottom right: Correlation of the guest methyl groups with the host hydrogens. Note methyl groups correlate with both g and g′. The guest is substituted with two methyl groups.

Finally, an important point to note is that one could open and close the capsule by controlling the pH of the solution [70]. This is illustrated with coumarin-1 (7-diethylamino-4-methylcoumarin; same as C-152 where CF_3 is replaced by CH_3; Figure 18.19) as probe. Similar to other coumarins, C-1 exhibits medium polarity dependent emission maximum. In Figure 18.24a one could observe that upon addition of OA to a solution of C-1 in borate buffer the emission maximum shifts from 455 nm to 420 nm indicating C-1 is getting encapsulated within the hydrophobic OA capsule. However, upon increasing the pH from 8.9 to 4.9 by addition of HCl the emission maximum shifts to 455 nm, suggesting that the guest C-1 is in water (Figure 18.24b). This suggests that in the acidic medium the capsule is not stable and disassembles to release the probe. However, upon making the solution basic again (pH 4.9 to 8) the probe C-1 gets back into the capsule, confirming that the capsule is stable only in basic medium (Figure 18.24c). Thus by controlling the pH one can control the assembly and disassembly of the capsule.

Thus the photophysical, EPR, and NMR studies have revealed that the capsular stability depends on the size and hydrophobicity of guest molecules. With small hydrophobic guest molecules the capsule is stable and does not open and remains fully closed in 3 ns time scale, the excited singlet lifetime of most guest molecules. However, in > 10 μs time scale, the lifetime of excited triplet state, the capsule slightly opens and closes. In a longer time scale of 300 ms the capsule containing polar guests disassembles and reassembles. Dissociation of non-polar pyrene encapsulated OA capsule has been investigated by stopped flow technique and is reported to take place in much longer time scale of 2.7 s [99, 100]. The main message from the above studies is that the capsule is dynamic.

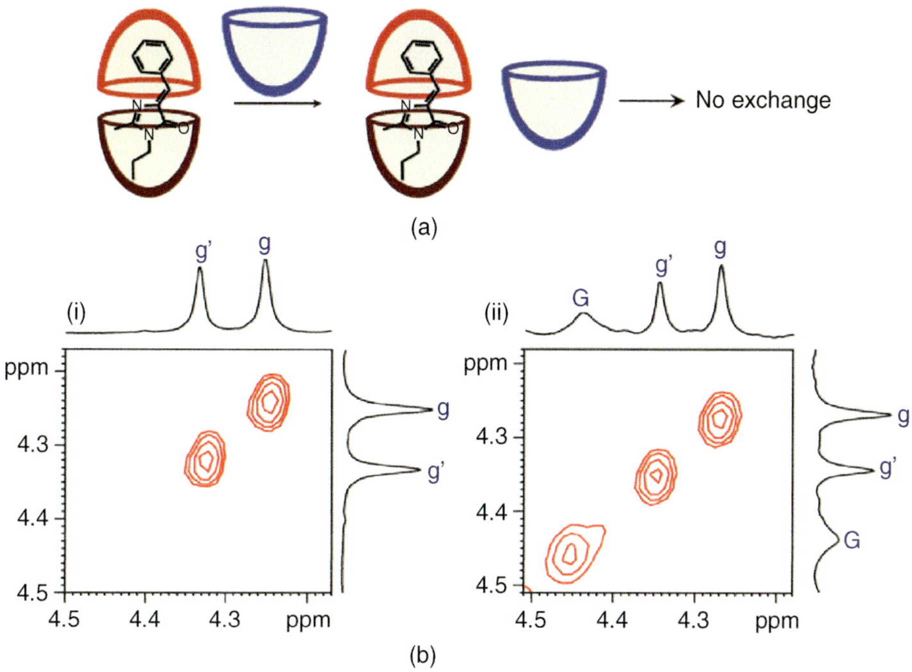

Figure 18.23 Top: Cartoon representation of non-exchanging host of complex of H/*N*-Pr with OA between two unsymmetrical host of 1:2 complex and with free host present in the solution. Bottom left: Partial 2D ROESY NMR spectra (300 ms mixing time) of H/*N*-Pr at (i) guest:host = 1:2, Note there is no correlation between g and g'. Bottom right: Correlation of the guest methyl groups with the host hydrogens. Note the absence of correlation between the cavitand hydrogens and free OA. The guest is substituted with a methyl and more hydrophobic propyl.

18.6 Room Temperature Phosphorescence of Encapsulated Organic Molecules

Organic molecules rarely phosphoresce at room temperature in solution [101]. In addition to low intersystem crossing (ISC) efficiency (S_1 to T_1) aromatic and olefinic molecules have low radiative rate constants for phosphorescence (k_P), making them susceptible for quenching by oxygen and other impurities (k_Q). As illustrated in Figure 18.25, several deficiencies need to be rectified to make such molecules phosphorescence at room temperature in solution: (a) increase ISC, (b) increase k_P, and reduce k_Q [101]. The capsular assembly that protects a guest molecule from external agents could eliminate the quenching by oxygen and impurities. Since OA capsule fully encloses a guest it should be much more effective than CD, CB and CA or any other similar three-dimensional hosts. Also, if one could co-encapsulate a heavy atom such as xenon, it should be possible to enhance the intersystem crossing from S_1 to T_1 as well as radiative rate constant from T_1 to S_0. Furthermore, since the guest molecules within the cage experience restricted motions, bimolecular collisions that quench the emissive T_1 could also be suppressed. These would favor the radiative rate constant from T_1 to S_0 to be competitive with channels that deactivate the T_1. Two examples shown below highlight the value of OA in enhancing room temperature phosphorescence from an aromatic molecule that has poor intersystem crossing rate constant and a thioketone that undergoes diffusion-limited self-quenching in solution.

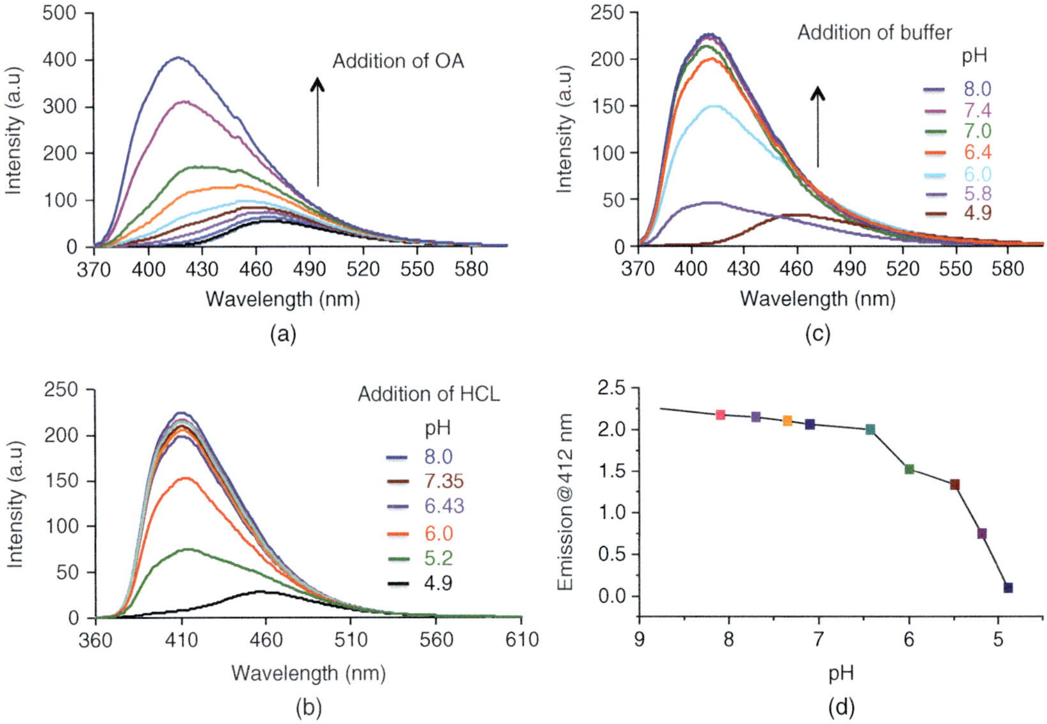

Figure 18.24 Fluorescence spectra of C-1 under various conditions. (a) Changes with addition of OA (b) Changes with addition of 0.025M HCl; solution made acidic. (c) Changes with addition of 10mM borate buffer; solution made basic. (d) Emission intensity of C-1at 412 nm versus pH.

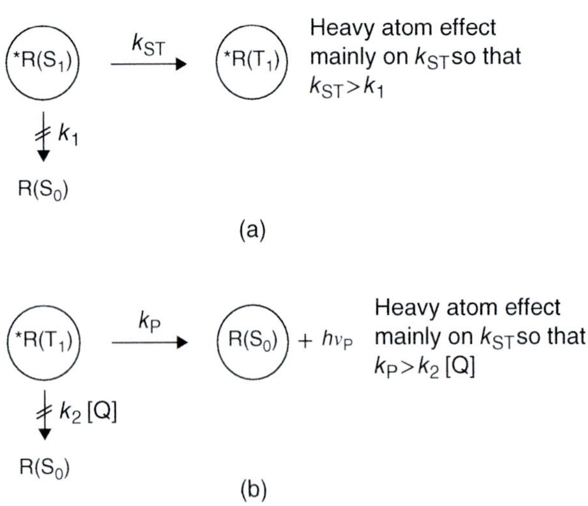

Figure 18.25 Strategy for the supramolecular control of room temperature phosphorescence.

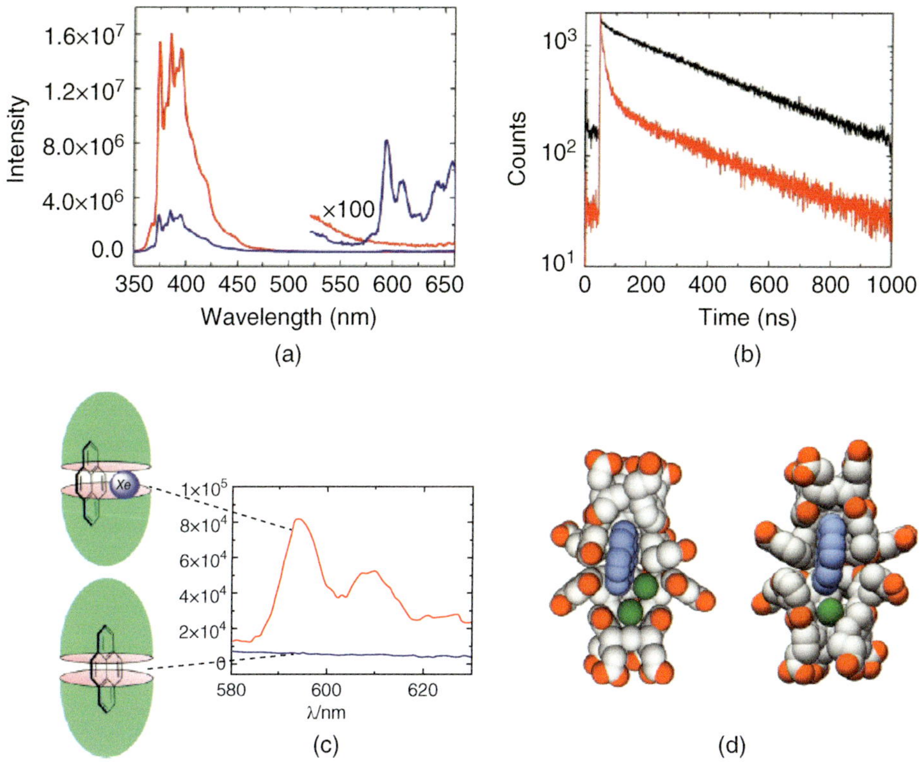

Figure 18.26 (a) Emission spectra of py @ OA2 complex (red) and py @ OA2 + Xe purged for 10 mins (blue) (λ exc = 337 nm); phosphorescence spectra around 580 to 650 nm of py @ OA2 observed (blue) after Xe purging was scaled up by 100 and shown in the figure (py = [5 μM], OA = [25 μM]). (b) Time-resolved fluorescence decay of py @ OA2 complex (black) and py @ OA2 + Xe purged for 10 mins (red). (c) Phosphorescence spectra of pyrene in presence and absence of Xe. (d) MD simulated possible structures of encapsulation of Xe with OA; the most representative structures of OA encapsulating pyrene + one Xe atom and two Xe atoms obtained from MD simulations.

In Figure 18.26 emission spectrum of pyrene included within OA capsule is shown. In solution and within OA there is no phosphorescence at room temperature (> 590 nm). However, when the capsule is co-included with Xe, phosphorescence could be recorded at room temperature [84, 102, 103]. While within OA phosphorescence from pyrene could be seen, it does not phosphoresce in Xe saturated isotropic solution. This example brings out the value of OA capsule in keeping the Xe and pyrene together. Without the presence of Xe adjacent to excited pyrene, no phosphorescence would be observed. Lack of change in fluorescene lifetime confirms that Xe enhances the emission through a static process and not via dynamic quenching. This example highlights the importance of both Xe and OA in enhancing the phosphorescence and also the need for both of them at the same time.

Thioketones are unique and undergo diffusion limited self-quenching [104]. Although they possess respectable rates of S_1 to T_1 intersystem crossing and radiative rate constants from T_1 to S_0, it is not easy to record phosphorescence at room temperature, even at 10^{-5} M, due to self-quenching [105, 106]. To record phosphorescence one should discourage the close encounter between two thioketone molecules. This could be carried out by enclosing a thioketone molecule within a supramolecular assembly (Figure 18.27). While this could be achieved with CD and CB, self-quenching is still likely if the C=S chromophore is projected outside. However, within

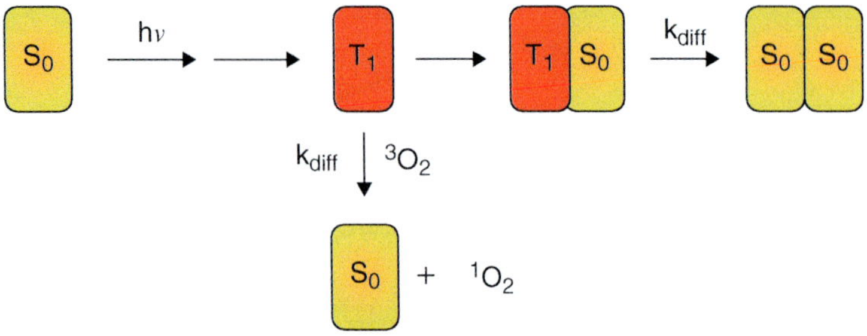

Figure 18.27 A cartoon representation of supramolecular strategy to arrest self-quenching and oxygen quenching by OA capsule. Inclusion within OA protects the guest thioketone.

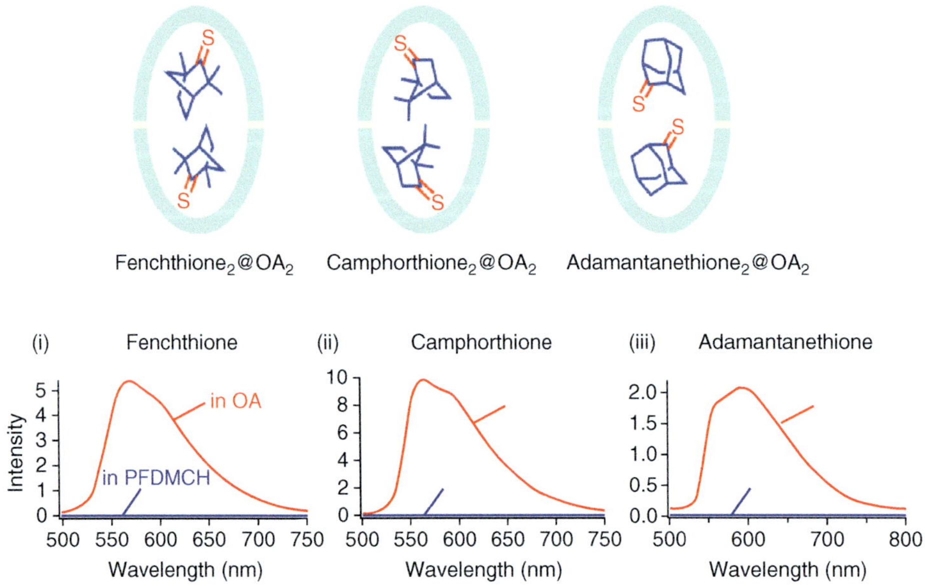

Figure 18.28 Top: A cartoon representation of the orientation of two thioketone molecules within the closed nano-container made up of OA cavitand, based on NOESY analysis. Bottom: Phosphorescence spectra of thioketones in perfluoro-1,3-dimethylcyclohexane (PFDMCH) (blue) and in 2:2 OA:thioketone capsular assemblies (red), [thioketone] = 0.1 mM, [OA] = 0.12 mM in 10 mM sodium tetraborate buffer (10 mM), $\lambda_{excitation}$ = 254 nm; under aerated conditions.

OA where two thioketone molecules' motion would be restricted self-quenching is not expected to occur. As illustrated in Figure 18.28, OA encapsulated thioketones emit phosphorescence at room temperature. It is important to note that the same molecules do not emit in a perfluorinated solvent at room temperature [52].

The two examples presented above bring out the value of an OA capsule in altering the excited state photophysics of enclosed guests.

18.7 Consequence of Confinement on the Photophysics of Anthracene

Studies within octa acid are providing excellent insight into what could be achieved within a nanocapsule and how it is different from cavitands such as CA, CB, and CD. In this section we

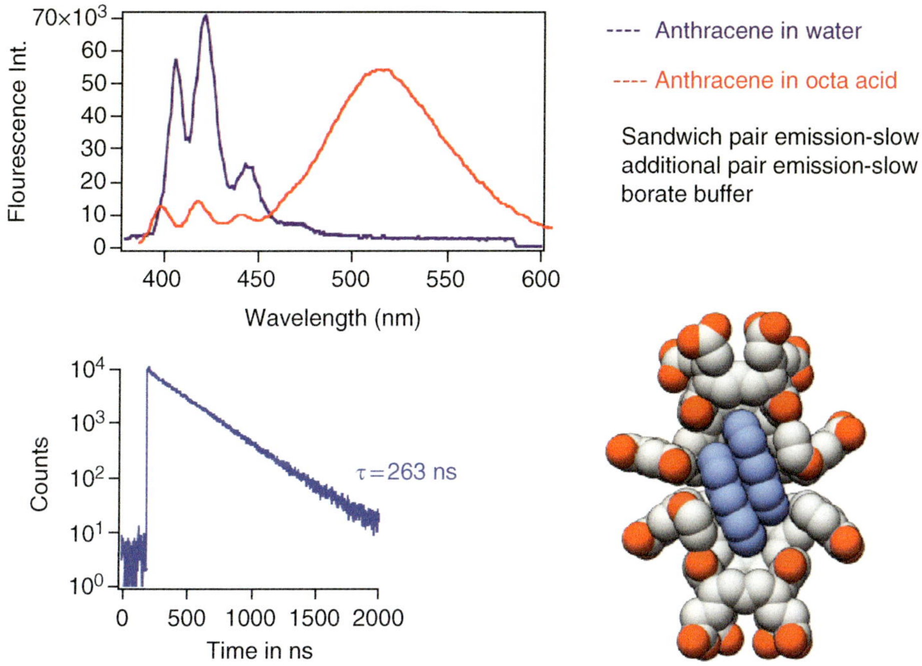

Figure 18.29 Top: Emission spectra of two anthracene molecules included within OA capsule. A control of anthracene in water is included. Bottom left: Excited state decay trace; lifetime 263 ns. Bottom right: MD simulated structure of 2:2 complex between anthracene and OA.

summarize our observations with anthracene, whose photochemistry is nearly a century old [107]. Anthracene is well known to dimerize with a limiting quantum yield of 1.0 in solution. On the other hand, although a large number of substituted anthracenes exhibit excimer emission, parent anthracene does not show excimer emission in solution. The excimer emission of anthracene can only be recorded by photofragmenting the synthetic dimer either in the crystalline state or in an organic glass at 77 K [108, 109].

Surprisingly, anthracene included within OA capsule do not dimerize but show excimer emission that has not been recorded earlier in solution. Two molecules of sparingly water-soluble anthracene could be included in OA capsule in a slipped geometry (Figure 18.29, bottom right). Irradiation of the 2:2 complex did not give either the expected dimer. The disappointment of not recovering dimer upon irradiation was replaced with excitement when the emission spectrum was recorded [54]. As shown in Figure 18.29, intense long-sought excimer emission was recorded. The long lifetime of this emission (τ = 263 ns) is consistent with the sandwich excimer (τ >200 ns) obtained by photofragmenting anthracene dimer in the crystalline state [109]. Apparently, dimerization is prevented by the confined space in which anthracene molecules reside in a slipped geometry. Such an arrangement does not permit bond formation between the two anthracene molecules across 9,10 positions. Furthermore, the confined space restrict the motion of anthracene molecules and does not allow them to realign to form bonds. Interestingly, closing the favored photochemical pathway opened the thus far unrealized photophysical pathway, namely excimer emission. This behavior is similar to that of a enzymes that favor a product by suppressing more favored pathways.

18.8 Selective Photo-Oxidation of Cycloalkenes

From the above examples it is clear that the OA capsule accommodates various molecules in defined geometries dictated by weak interactions between the walls of the host and the guest. Under these circumstances, a guest molecule with multiple reactive sites experiencing different extents of supramolecular steric hindrance is likely to react differently towards a reagent within the OA capsule from that in solution. Oxidation of olefins by one of the smallest oxidants, namely singlet oxygen, described below, exploits such a feature of the OA capsule.

Addition of singlet oxygen to olefin with several allylic hydrogens results in multiple allylic hydroperoxides via "ene" reaction [110]. The example we have considered, 1-alkyl cycloalkenes upon reaction with singlet oxygen in solution yield three allylic hydroperoxides in unequal amounts [111, 112]. In the example provided in Figure 18.30 there are three allylic hydrogens and ene reaction with singlet oxygen yields three different products. Because of the small size of singlet oxygen it is not easy to employ steric features to control the product distribution. However, remarkably within OA capsule the unexpected selectivity in product distribution is obtained with three 1-methylcycloalkenes (Figure 18.31). Analysis of the selectivity provides a clue to the importance of supramolecular steric features in controlling the product distribution.

Of the 20 olefins investigated with closely similar structures, product distributions in the case of only three olefins listed in Figure 18.30 were dramatically altered; the product that was minor in solution became major (> 90 % yield) within OA capsule (Figure 18.31) [43, 113]. First, the fact

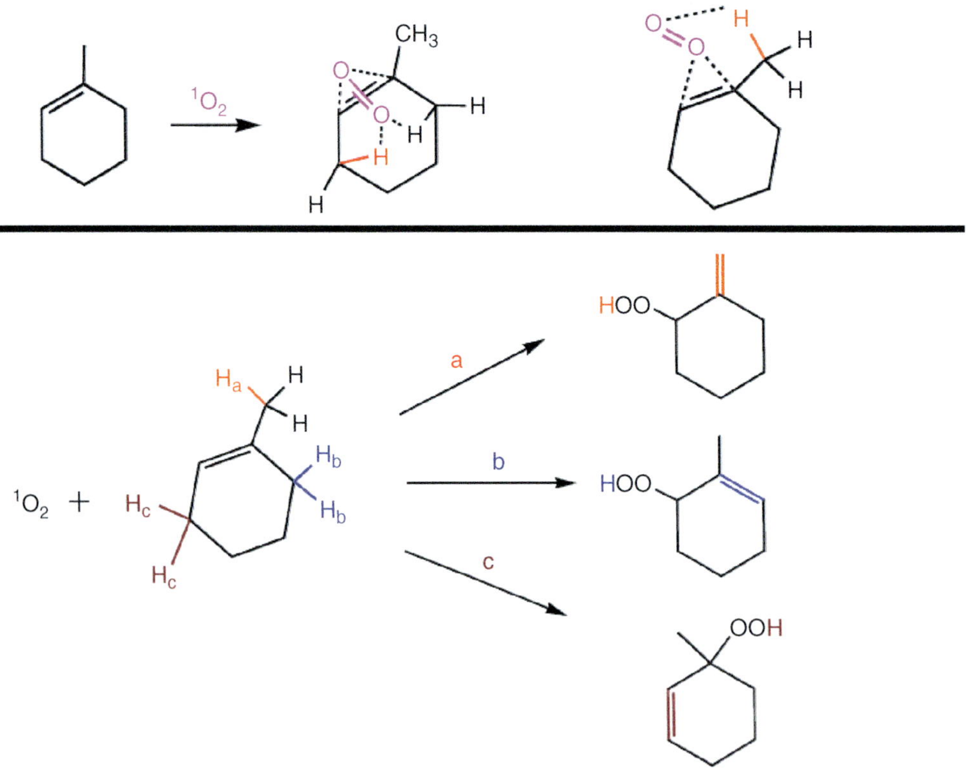

Figure 18.30 Possible products of ene reaction between singlet oxygen and methylcyclohexene. Addition from the two sides of the double bond shown.

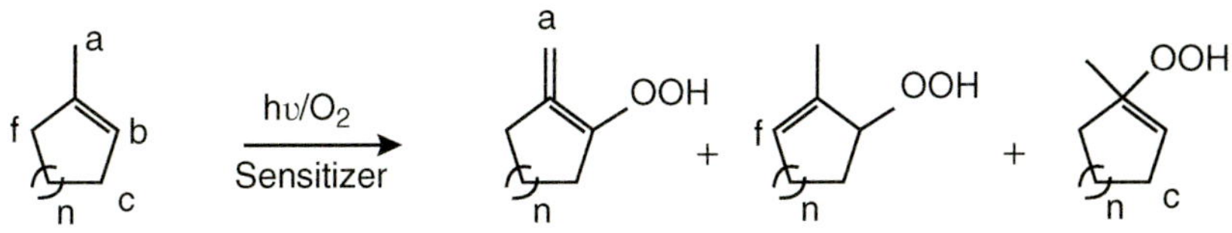

n=1	CH_3CN / Rose bengal	4%	43%	53%
	Octa acid / Rose bengal	-	5%	95%
n=2	CH_3CN / Rose bengal	44%	20%	36%
	Octa acid / Rose bengal	10%	-	90%
n=3	CH_3CN / Rose bengal	4%	48%	48%
	Octa acid / Rose bengal	4%	6%	90%

Figure 18.31 Product distribution upon addition of singlet oxygen to 1-methyl cycloalkenes in acetonitrile and within OA. Note the selectivity within OA. Sources: Natarajan, A., Kaanumalle, L. S., Jockusch, S., Gibb, C. L. D., Gibb, B. C., Turro, N. J., Ramamurthy, V. (2007) Controlling Photoreactions with Restricted Spaces and Weak Intermolecular Forces: Exquisite Selectivity during Oxidation of Olefins by Singlet Oxygen. J Am Chem Soc, 129, 4132-4133. ; Gupta, S., Ramamurthy, V. (2018) Characterization and Singlet Oxygen Oxidation of 1-Alkyl Cyclohexenes Encapsulated Within a Water-Soluble Organic Capsule. ChemPhotoChem, 2, 655–666.

that only 3 out of 20 olefins changed their behavior within OA capsule suggested that the capsule like enzyme is substrate-specific. Second, the obtained selectivity could be understood only on the basis of supramolecular structure of the host-guest complex and not on the molecular features of the guest alone. We consider one example, 1-methylcyclohexene, to highlight the point. ^{1}H NMR data suggested 1-methylcyclohexene$_2$@OA$_2$ complex to adopt the structure shown in Figure 18.32. The figure includes the $\Delta\delta$ values ($\Delta\delta$ is the chemical shift difference between within OA and in $CDCl_3$) which are a reflection of the location of the three allylic hydrogens within the capsule. From the $\Delta\delta$ values it is clear that of the three allylic hydrogens H_c (closer to the wider entrance of OA cavitand) would be most accessible and the H_a (methyl) (anchored at the tapered end of the capsule) would be least accessible to singlet oxygen that is expected to enter the capsule from the median. Thus while in solution singlet oxygen will not be able to distinguish between the three allylic hydrogens, within OA it would be able to attack only the H_c that is less supramolecularly hindered than the other two. As seen from the product distributions presented in Figure 18.31, this expectation is realized.

Additional interesting feature of this study deals with the process of activation of singlet oxygen by OA-encapsulated sensitizer. Having realized that the capsule opens and closes in microsecond time scale, we employed 4,4′dimethylbenzil (DMB) whose triplet lifetime is known to be long (~ 0.5 millisecond). We believed that this would allow oxygen to access the excited DMB. Thus OA-encapsulated DMB should be able to act as the triplet sensitizer and activate ground state oxygen. In fact, oxygen quenched the phosphorescence of DMB with a rate constant of $\sim 8 \times 10^7$ M^{-1} s^{-1}. The singlet oxygen that was generated within the OA capsule spent most of its lifetime in water as revealed by its photophysical properties. When D_2O was exchanged for H_2O, the lifetime increased from 5 μs to 41 μs, clearly indicating that singlet oxygen following generation within OA capsule exited to water and stayed there during its lifetime. We believe that the capsule opens-closes numerous times during the lifetime of encapsulated DMB triplet (596 μs) allowing oxygen to access the excited DMB. Interestingly, the energized oxygen with a lifetime of 41 μs in D_2O waits for the capsule with the olefin to open to enter and oxidize (Figure 18.33). This amazingly co-ordinated set

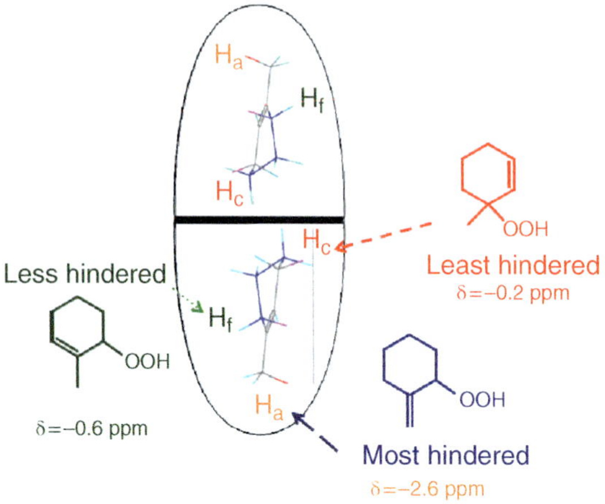

Figure 18.32 A cartoon representation of the orientation of 1-methylcyclohexene within OA. The orientation is arrived at based on NOESY correlations. The extent of change in chemical shift between $CDCl_3$ and within OA are indicated. The extent of shift is a reflection of how far the given hydrogen is from the median.

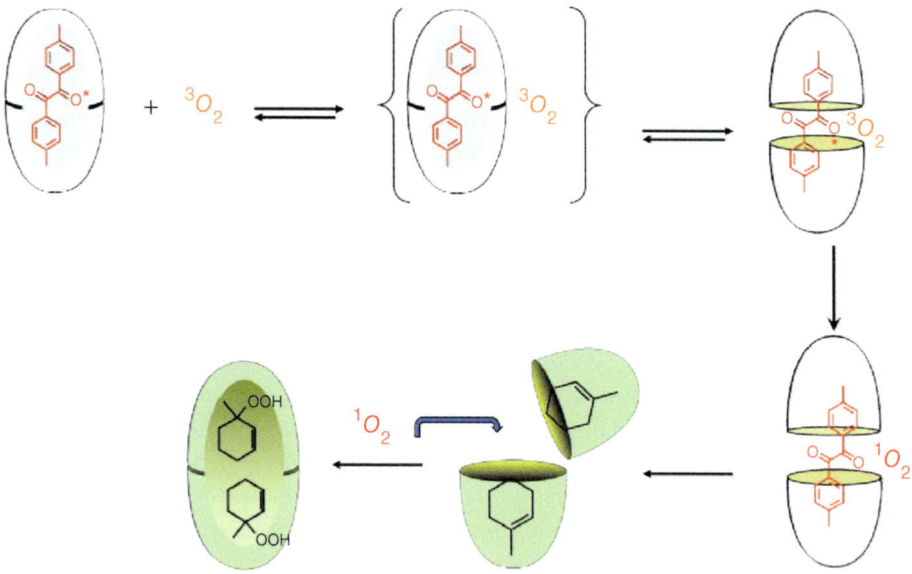

Figure 18.33 Schematic representation of energy transfer between encapsulated excited 4,4′-dimethylbenzil and oxygen and oxidation of encapsulated 1-methylcyclohexene by the generated singlet oxygen.

of cascading events leading to selective oxidation of olefins highlights the opportunities OA capsule offers to carry out selective bimolecular reactions in water. It is the "static" host-guest structure that provides selectivity in photoproducts while the "dynamics" of the capsule permit the oxygen to get activated in one capsule and oxidize the olefin in a different capsule.

18.9 Remote Activation of Encapsulated Guests: Electron Transfer Across an Organic Wall

One might wonder whether it would be possible to perform photoreactions that require activation from outside. For example, electron and energy transfer activated processes require a close

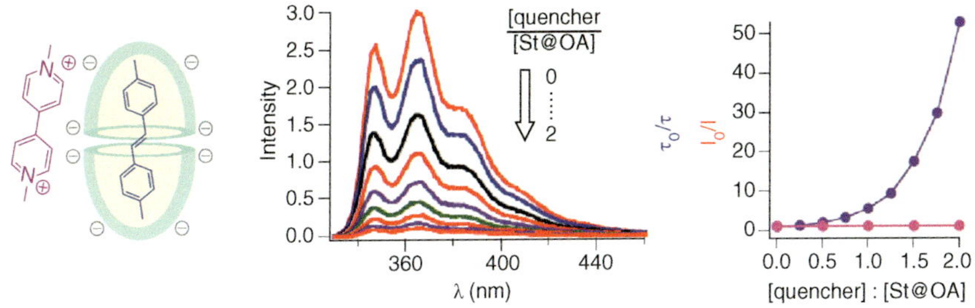

Figure 18.34 Top middle: Fluorescence spectra of 4,4′-dimetlstilbene@OA_2 at different amounts of MV^{2+}. Top left: Stern–Volmer plots of fluorescence quenching with MV^{2+} using steady-state fluorescence intensity and fluorescence lifetime. Top right: A cartoon representation of 4,4′-dimetlstilbene@OA_2 and Columbically attached MV^{2+}.

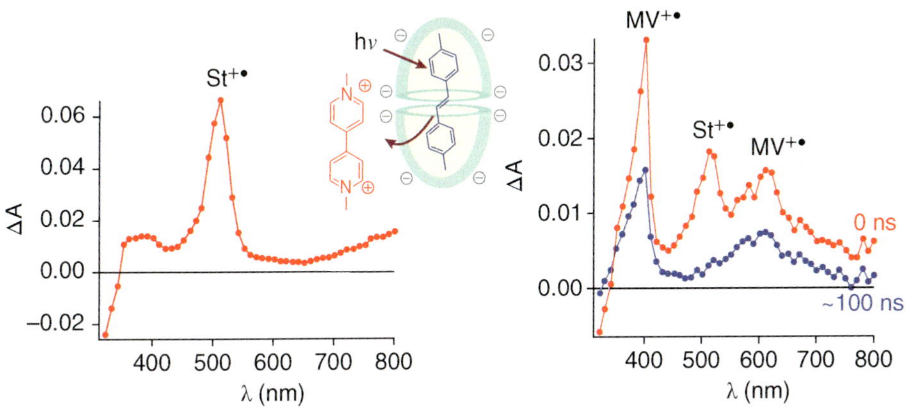

Figure 18.35 Transient absorption spectra after laser excitation of 4,4′-dimetlstilbene@OA_2 in the presence of MV^{2+}.

interaction between a donor and an acceptor. Once a molecule is encapsulated within an OA capsule the donor and acceptor would be separated by the capsular wall. The question now arises whether it would be possible to establish communication between the imprisoned photoactive active molecule and a free donor/acceptor molecule present in solution. We have established with *trans*-4,4′-dimethyl stilbene (DMS) as an example that encapsulated guest can be activated by an electron acceptor stationed outside the OA capsule [65]. In Figure 18.34 quenching of the emission of encapsulated DMS by methylviologen (MV^{2+}) is shown. Clearly increased addition of MV^{2+} results in decreased emission, suggesting that the two events are linked. The quencher MV^{2+} attached to the OA capsule through Coulombic interaction must be responsible for the quenching of DMS fluorescence. Consistent with this the Stern-Volmer plot shown in Figure 18.34 suggests that the quenching does not involve diffusion. The quenching is due to electron transfer from excited DMS to MV^{2+} is confirmed by recording the transient spectra of $DMS^{+\bullet}$ and $MV^{+\bullet}$ (Figure 18.35). While the rate constant for electron transfer in this pair has not been established, it is measured to be in the range of 0.4 to 4×10^9 s^{-1} for another donor-acceptor, namely coumarin–viologen pairs. Clearly electron transfer can occur through the walls of the OA capsule and activate the encapsulated guest molecule. For details readers are directed to published articles [58–62, 65].

Having established the feasibility of electronic communication between an encapsulated and a free molecule, we focused our attention on reactions of encapsulated radical ions generated by

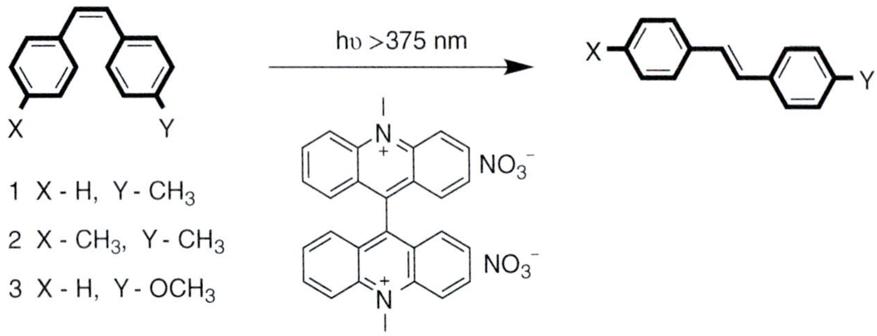

Figure 18.36 Photoisomerization of *cis*-stilbenes included within OA capsule sensitized by *bis*-N-methylacridinium nitrate. Reaction proceeds by electron transfer. In Table 18.1 the isomer ratio at the photostationary state is provided. Control means the irradiation is done without the sensitizer *bis*-N-methylacridinium nitrate.

Table 18.1

	Product distribution during electron transfer sensitization by BMAN[a, b]	
	Starting isomer: 100% *cis*	
Compound	*cis*	*trans*
1@$(OA)_2$	10	90
1@$(OA)_2$-control	90	10
2@$(OA)_2$	0	100
2@$(OA)_2$-control	85	15
3@$(OA)_2$	73	27
3@$(OA)_2$-control	100	0

electron transfer process. The example below deals with geometric isomerization of OA encapsulated *cis*-stilbenes triggered by an electron acceptor stationed outside the capsule through Coulombic interaction. In these experiments the excited *bis*-N-methylacridinium nitrate (BMAN) is used as the electron acceptor and encapsulated electron rich *cis*-stilbenes as donors. It is known that the *cis*-isomers of stilbenes are quantitatively converted to the *trans* isomers while the latter is stable under electron transfer conditions [114, 115]. As shown in Figure 18.36, the *cis* isomers of selected stilbenes are quantitatively converted to the *trans* isomer when BMAN was excited with light >375 nm. In the control experiments (without BMAN) there was very little isomerization. This unequivocally established that the radical cations of stilbenes can be generated by transferring an electron across the capsule wall. This opens up opportunities to manipulate reactions of guests not only through direct excitation but also via indirect excitation of a sensitizer.

18.10 Summary

Living systems are fastidious in terms of rate, and chemo-, regio-, and stereo-selectivities of the various reactions accomplished through transformations in an aqueous environment by confining molecules and restricting their mobility through weak interactions. Nature utilizes less reagents but more pre-organization and confinement to synthesize complex molecules with much less wastage. Opportunities synthetic confined spaces offer to modify and control the physical and chemical properties of molecules in ground and excited states are much less exploited. Much of the focus of organic chemists is to discover new reagents to accomplish complex transformations. The value of confined spaces in the context of making molecules is less recognized. This review has attempted to highlight the opportunities that exist for those who are interested in exploring the chemistry of molecules in synthetic confined spaces that has features similar to the space provided by proteins for molecules to reside and function. The principles delineated in this review are general and applicable to a wide variety of hosts that has become available in recent years [18] and hope the examples highlighted serve to elicit the interests of chemists to explore well-defined confined spaces as reaction cavities [116].

Acknowledgements

VR thanks the National Science Foundation (CHE-1807729) for continued financial support that helped to build the program. The author has benefited immensely from the experimental and intellectual contributions of co-workers (students and postdocs) and collaborators whose name appear in the cited references. Collaborators' willingness to share their expertise, knowledge, and time helped the author to explore new opportunities within confined spaces. VR is particularly thankful to N.J. Turro for introducing him to the exciting world of supramolecular photochemistry and encouraging him to pursue challenging problems.

References

1 Anslyn, E.V. and Dougherty, D.A. (2006). Modern Physical Organic Chemistry. Sacramento, CA: University Science Books.
2 Breslow, R. (ed.) (2005). Artificial Enzymes. Weinheim: Wiley-VCH.
3 Ramamurthy, V. (1986). Organic photochemistry in organized media. Tetrahedron 42: 5753–5839.
4 Ramamurthy, V. (ed.) (1991). Photochemistry in Organized & Constrained Media. New York: VCH.
5 Ramamurthy, V. and Inoue, Y. (eds.) (2011). Supramolecular Photochemistry. Hoboken: John Wiley.
6 Ramamurthy, V. and Venkatesan, K. (1987). Photochemical reactions of organic crystals. Chem Rev 87: 433–481.
7 Ramamurthy, V. and Gupta, S. (2015). Supramolecular photochemistry: from molecular crystals to water-soluble capsules. Chem Soc Rev 44: 119–135.
8 Ramamurthy, V. and Mondal, B. (2015). Supramolecular photochemistry concepts highlighted with select examples. J Photoch Photobio C 23: 68–102.
9 Ramamurthy, V., Eaton, D.F., and Caspar, J.V. (1992). Photochemical and photophysical studies of organic molecules included within zeolites. Acc Chem Res 25: 299–307.
10 Ramamurthy, V. and Eaton, D.F. (1988). Photochemistry and photophysics within cyclodextrin cavities. Acc Chem Res 21: 300–306.
11 Ramamurthy, V. (2015). Photochemistry within a water-soluble organic capsule. Acc Chem Res 48: 2904–2917.

12 Sivaguru, J., Natarajan, A., Kaanumalle, L.S. et al. (2003). Asymmetric photoreactions within zeolites: role of confinement and alkali metal ions. Acc Chem Res 36: 509–521.

13 Weiss, R.G., Ramamurthy, V., and Hammond, G.S. (1993). Photochemistry in organized and confining media: a model. Acc Chem Res 26: 530–536.

14 Ramamurthy, V. and Sivaguru, J. (2016). Supramolecular photochemistry as a synthetic tool: photocycloaddition. Chem Rev 116: 9914–9993.

15 Crini, G. (2014). Review: a history of cyclodextrins. Chem Rev 114: 10940–10975.

16 Cram, D.J. and Cram, J.M. (1997). Container Molecules and Their Guests. Cambridge: Royal Society of Chemistry.

17 Murray, J., Kim, K., Ogoshi, T. et al. (2017). The aqueous supramolecular chemistry of cucurbit[n]urils, pillar[n]arenes and deep cavity cavitands. Chem Soc Rev 46: 2479–2496.

18 Voloshin, Y., Belaya, I., and Kramer, R. (2016). The Encapsulation Phenomenon. Switzerland: Springer International Publishing.

19 Brinker, U.H. and Mieusset, J.-L. (eds.) (2010). Molecular Encapsulation. Chichester: John Wiley & Sons.

20 Ramamurthy, V., Jockusch, S., and Porel, M. (2015). Supramolecular photochemistry in solution and on surfaces: encapsulation and dynamics of guest molecules and communication between encapsulated and free molecules. Langmuir 31: 5554–5570.

21 Gibb, C.L.D. and Gibb, B.C. (2004). Well-defined, organic nanoenvironments in water: the hydrophobic effect drives capsular assembly. J Am Chem Soc 126: 11408–11409.

22 Jordan, J.H. and Gibb, B.C. (2015). Molecular containers assembled through the hydrophobic effect. Chem Soc Rev 44: 547–585.

23 Warmuth, R. (2010). Reactions inside carcerands. In: Molecular Encapsulation (eds. U.H. Brinker and J.-L. Mieusset), 227–268. Chichester: John Wiley & Sons.

24 Yoshizawa, M. and Catti, L. (2019). Bent anthracene dimers as versatile building blocks for supramolecular capsules. Acc Chem Res 52: 2392–2404.

25 Yoshizawa, M., Klosterman, J.K., and Fujita, M. (2009). Functional molecular flasks: new properties and reactions within discrete, self-assembled hosts. Angew Chem Int Ed 48: 3418–3438.

26 Yu, Y. and Rebek, J. (2018). Reactions of folded molecules in water. Acc Chem Res 51: 3031–3040.

27 Zhang, D., Ronson, T.K., and Nitschke, J.R. (2018). Functional capsules via subcomponent self-assembly. Acc Chem Res 51: 2423–2436.

28 Hong, C.M., Bergman, R.G., Raymond, K.N., and Toste, F.D. (2018). Self-assembled tetrahedral hosts as supramolecular catalysts. Acc Chem Res 51: 2447–2455.

29 Kobayashi, K. and Yamanaka, M. (2015). Self-assembled capsules based on tetrafunctionalized calix[4]resorcinarene cavitands. Chem Soc Rev 44: 449–466.

30 Liu, F., Helgeson, R.C., and Houk, K.N. (2014). Building on Cram's legacy: stimulated gating in hemicarcerands. Acc Chem Res 47: 2168–2176.

31 Sundaresan, A.K., Gibb, C.L.D., Gibb, B.C., and Ramamurthy, V. (2009). Chiral photochemistry in a confined space: torquoselective photoelectrocyclization of pyridones within an achiral hydrophobic capsule. Tetrahedron 65: 7277–7288.

32 Sundaresan, A.K., Kaanumalle, L.S., Gibb, C.L.D. et al. (2009). Chiral photochemistry within a confined space: diastereoselective photorearrangements of a tropolone and a cyclohexadienone included in a synthetic cavitand. Dalton Trans: 4003–4011.

33 Sundaresan, A.K. and Ramamurthy, V. (2007). Making a difference on excited-state chemistry by controlling free space within a nanocapsule: photochemistry of 1-(4-alkylphenyl)-3-phenylpropan-2-ones. Org Lett 9: 3575–3578.

34 Sundaresan, A.K. and Ramamurthy, V. (2008). Consequences of controlling free space within a reaction cavity with a remote alkyl group: photochemistry of para-alkyl dibenzyl ketones within an organic capsule in water. Photochem Photobiol Sci 7: 1555–1564.

35 Kulasekharan, R., Maddipatla, M.V.S.N., Parthasarathy, A., and Ramamurthy, V. (2013). Role of free space and conformational control on photoproduct selectivity of optically pure α-alkyldeoxybenzoins within a water-soluble organic capsule. J Org Chem 78: 942–949.

36 Kulasekharan, R., Choudhury, R., Prabhakar, R., and Ramamurthy, V. (2011). Restricted rotation due to the lack of free space within a capsule translates into product selectivity: photochemistry of cyclohexyl phenyl ketones within a water-soluble organic capsule. Chem Commun 47: 2841–2843.

37 Gibb, C.L.D., Sundaresan, A.K., Ramamurthy, V., and Gibb, B.C. (2008). Templation of the excited-state chemistry of α-(n-alkyl) dibenzyl ketones: how guest packing within a nanoscale supramolecular capsule influences photochemistry. J Am Chem Soc 130: 4069–4080.

38 Kaanumalle, L.S., Gibb, C.L.D., Gibb, B.C., and Ramamurthy, V. (2004). Controlling photochemistry with distinct hydrophobic nanoenvironments. J Am Chem Soc 126: 14366–14367.

39 Kaanumalle, L.S. and Ramamurthy, V. (2007). Photodimerization of acenaphthylene within a nanocapsule: excited state lifetime dependent dimer selectivity. Chem Commun: 1062–1064.

40 Parthasarathy, A. and Ramamurthy, V. (2016). Water-soluble octa acid capsule as a reaction container: Templated photodimerization of indene in water. J Photochem Photobiol 317: 132–139.

41 Parthasarathy, A., Samanta, S.R., and Ramamurthy, V. (2013). Photodimerization of hydrophobic guests within a water-soluble nanocapsule. Res Chem Intermed 39: 73–87.

42 Kaanumalle, L.S., Gibb, C.L.D., Gibb, B.C., and Ramamurthy, V. (2007). Photo-Fries reaction in water made selective with a capsule. Org Biomol Chem 5: 236–238.

43 Natarajan, A., Kaanumalle, L.S., Jockusch, S. et al. (2007). Controlling photoreactions with restricted spaces and weak intermolecular forces: exquisite selectivity during oxidation of olefins by singlet oxygen. J Am Chem Soc 129: 4132–4133.

44 Samanta, S.R., Choudhury, R., and Ramamurthy, V. (2014). Photoisomerization and photooxygenation of 1,4-diaryl-1,3-dienes in a confined space. J Phys Chem A 118: 10554–10562.

45 Parthasarathy, A. and Ramamurthy, V. (2011). Role of free space and weak interactions on geometric isomerization of stilbenes held in a molecular container. Photochem Photobiol Sci 10: 1455–1462.

46 Samanta, S.R., Parthasarathy, A., and Ramamurthy, V. (2012). Supramolecular control during triplet sensitized geometric isomerization of stilbenes encapsulated in a water soluble organic capsule. Photochem Photobiol Sci 11: 1652–1660.

47 Mohan Raj, A. and Ramamurthy, V. (2017). Volume conserving geometric isomerization of encapsulated azobenzenes in ground and excited states and as radical ion. Org Lett 19: 6116–6119.

48 Mohan Raj, A., Sharma, G., Prabhakar, R., and Ramamurthy, V. (2019). Space constrained stereoselective geometric isomerization of 1,2-diphenylcyclopropane and stilbenes in an aqueous medium. Org Lett 21: 5243–5247.

49 Baldridge, A., Samanta, S.R., Jayaraj, N. et al. (2010). Activation of fluorescent protein chromophores by encapsulation. J Am Chem Soc 132: 1498–1499.

50 Baldridge, A., Samanta, S.R., Jayaraj, N. et al. (2011). Steric and electronic effects in capsule-confined green fluorescent protein chromophores. J Am Chem Soc 133: 712–715.

51 Parthasarathy, A., Kaanumalle, L.S., and Ramamurthy, V. (2007). Controlling photochemical geometric isomerization of a stilbene and dimerization of a styrene using a confined reaction cavity in water. Org Lett 9: 5059–5062.

52 Jayaraj, N., Maddipatla, M.V.S.N., Prabhakar, R. et al. (2010). Closed nanocontainer enables thioketones to phosphoresce at room temperature in aqueous solution. J Phys Chem B 114: 14320–14328.

53 Mohan Raj, A., Sharma, G., Prabhakar, R., and Ramamurthy, V. (2019). Room-temperature phosphorescence from encapsulated pyrene induced by xenon. J Phys Chem A 123: 9123–9231.

54 Kaanumalle, L.S., Gibb, C.L.D., Gibb, B.C., and Ramamurthy, V. (2005). A hydrophobic nanocapsule controls the photophysics of aromatic molecules by suppressing their favored solution pathways. J Am Chem Soc 127: 3674–3675.

55 Gupta, S., Adhikari, A., Mandal, A.K. et al. (2011). Ultrafast singlet-singlet energy transfer between an acceptor electrostatically attached to the walls of an organic capsule and the enclosed donor. J Phys Chem C 115: 9593–9600.

56 Jockusch, S., Porel, M., Ramamurthy, V., and Turro, N.J. (2011). CIDEP from a polarized ketone triplet state incarcerated within a nanocapsule to a nitroxide in the bulk aqueous solution. J Phys Chem Lett 2: 2877–2880.

57 Jockusch, S., Zeika, O., Jayaraj, N. et al. (2010). Electron spin polarization transfer from a nitroxide incarcerated within a nanocapsule to a nitroxide in the bulk aqueous solution. J Phys Chem Lett 1: 2628–2632.

58 Porel, M., Chuang, C.-H., Burda, C., and Ramamurthy, V. (2012). Ultrafast photoinduced electron transfer between an incarcerated donor and a free acceptor in aqueous solution. J Am Chem Soc 134: 14718–14721.

59 Chuang, C.-H., Porel, M., Choudhury, R. et al. (2018). Ultrafast electron transfer across a nanocapsular wall: coumarins as donors, viologen as acceptor and octa acid as the mediator. J Phys Chem C 122: 328–337.

60 Mohan Raj, A., Porel, M., Mukherjee, P. et al. (2017). Ultrafast electron transfer from upper excited state of encapsulated azulenes to acceptors across an organic molecular wall. J Phys Chem C 121: 20205–20216.

61 Santos, F., Ramasamy, E., Ramamurthy, V., and Rodembusch, F.S. (2017). Photoinduced electron transfer across an organic molecular wall: octa acid encapsulated ESIPT dyes as electron donors. Photochem Photobiol Sci 16: 840–844.

62 Bhandari, S., Zheng, Z., Maiti, B. et al. (2017). What is the optoelectronic effect of the capsule on the guest molecule in aqueous host/guest complexes? A combined computational and spectroscopic perspective. J Phys Chem C 121: 15481–15488.

63 Chen, J.Y.C., Jayaraj, N., Jockusch, S. et al. (2008). An EPR and NMR study of supramolecular effects on paramagnetic interaction between a nitroxide incarcerated within a nanocapsule with a nitroxide in bulk aqueous media. J Am Chem Soc 130: 7206–7207.

64 Jagadesan, P., Mondal, B., Parthasarathy, A. et al. (2013). Photochemical reaction containers as energy and electron-transfer agents. Org Lett 15: 1326–1329.

65 Porel, M., Jockusch, S., Parthasarathy, A. et al. (2012). Photoinduced electron transfer between a donor and an acceptor separated by a capsular wall. Chem Commun 48: 2710–2712.

66 Varadharajan, R., Mohan Raj, A., and Ramamurthy, V. (2020). Remote electron and energy transfer sensitized photoisomerization of encapsulated stilbenes. ASAP: Photochem Photobiol Sci.

67 Jayaraj, N., Jockusch, S., Kaanumalle, L.S. et al. (2011). Dynamics of capsuleplex formed between octaacid and organic guest molecules – Photophysical techniques reveal the opening and closing of capsuleplex. Can J Chem 89: 203–213.

68 Das, A., Sharma, G., Kamatham, N. et al. (2019). Ultrafast solvation dynamics reveal that octa acid capsule's interior dryness depends on the guest. J Phys Chem A 123: 5928–5936.
69 Samanta, S.R., Baldridge, A., Tolbert, L.M., and Ramamurthy, V. (2020). Guest-host complexes of octa acid and amphiphilic benzylidene-3-methylimidazolidinones exchange hosts within the NMR time scale. ACS Omega 5: 8230–8241.
70 Mohan Raj, A., Talluri, S., Dubus, M. et al. (2018). Probing the pH dependent assembly-disassembly of water-soluble organic capsules with coumarins and anthracene. J Photochem Photobiol A: Chem 355: 398–407.
71 Porel, M., Jayaraj, N., Raghothama, S., and Ramamurthy, V. (2010). Chemistry in confined spaces: high-energy conformer of a piperidine derivative is favored within a water-soluble capsuleplex. Org Lett 12: 4544–4547.
72 Otolski, C.J., Mohan Raj, A., Ramamurthy, V., and Elles, C.G. (2019). Ultrafast dynamics of encapsulated molecules reveals new insight on the photoisomerization mechanism for azobenzenes. J Phys Chem Lett 10: 121–127.
73 Otolski, C.J., Mohan Raj, A., Sharma, G. et al. (2019). Ultrafast *trans* → *cis* photoisomerization dynamics of alkyl-substituted stilbenes in a supramolecular capsule. J Phys Chem A 123: 5061–5071.
74 Kulasekharan, R., Jayaraj, N., Porel, M. et al. (2010). Guest rotations within a capsuleplex probed by NMR and EPR techniques. Langmuir 26: 6943–6953.
75 Choudhury, R., Barman, A., Prabhakar, R., and Ramamurthy, V. (2013). Hydrocarbons depending on the chain length and head group adopt different conformations within a water-soluble nanocapsule: ^{1}H NMR and molecular dynamics studies. J Phys Chem 117: 398–407.
76 Jayaraj, N., Zhao, Y., Parthasarathy, A. et al. (2009). Nature of supramolecular complexes controlled by the structure of the guest molecules: formation of octa acid based capsuleplex and cavitandplex. Langmuir 25: 10575–10586.
77 Porel, M., Jayaraj, N., Kaanumalle, L.S. et al. (2009). Cavitand octa acid forms a nonpolar capsuleplex dependent on the molecular size and hydrophobicity of the guest. Langmuir 25: 3473–3481.
78 Choudhury, R. and Ramamurthy, V. (2018). Understanding the complexation of aliphatic and aromatic acids guests with octa acid. J Phys Org Chem 31: 1–9.
79 Jayaraj, N., Jagadesan, P., Samanta, S.R. et al. (2013). Release of guests from encapsulated masked hydrophobic precursors by a phototrigger. Org Lett 15: 4374–4377.
80 Jagadesan, P., Da Silva, J.P., Givens, R.S., and Ramamurthy, V. (2015). Photorelease of incarcerated guests in aqueous solution with phenacyl esters as the trigger. Org Lett 17: 1276–1279.
81 Kamatham, N., Mendes, D.C., Da Silva, J.P. et al. (2016). Photorelease of incarcerated caged acids from hydrophobic coumaryl esters into aqueous solution. Org Lett 18: 5480–5483.
82 Mohan Raj, A., Raymo, F.M., and Ramamurthy, V. (2016). Reversible disassembly–assembly of octa acid–guest capsule in water triggered by a photochromic process. Org Lett 18: 1566–1569.
83 Kamatham, N., Da Silva, J.P., Givens, R.S., and Ramamurthy, V. (2017). Melding caged compounds with supramolecular containers: photogeneration and miscreant behavior of the coumarylmethyl carbocation. Org Lett 19: 3588–3591.
84 Kamatham, N., Raj, A.M., Givens, R.S. et al. (2019). Supramolecular photochemistry of encapsulated caged ortho-nitrobenzyl triggers. Photochem Photobiol Sci 18: 2411–2420.
85 Ishida, Y., Kulasekharan, R., Shimada, T. et al. (2013). Efficient singlet–singlet energy transfer in a novel host–guest assembly composed of an organic cavitand, aromatic molecules, and a clay nanosheet. Langmuir 29: 1748–1753.

86 Ishida, Y., Kulasekharan, R., Shimada, T. et al. (2014). Supramolecular-surface photochemistry: supramolecular assembly organized on a clay surface facilitates energy transfer between an encapsulated donor and a free acceptor. Phys Chem C 118: 10198–10203.

87 Ishida, Y., Shimada, T., Ramasamy, E. et al. (2016). Room temperature phosphorescence from a guest molecule confined in the restrictive space of an organic–inorganic supramolecular assembly. Photochem Photobiol Sci 15: 959–963.

88 Tsukamoto, T., Ramasamy, E., Shimada, T. et al. (2016). Supramolecular surface photochemistry: cascade energy transfer between encapsulated dyes aligned on a clay nanosheet surface. Langmuir 32: 2920–2927.

89 Fujimura, T., Ramasamy, E., Ishida, Y. et al. (2016). Sequential energy and electron transfer in a three-component system aligned on a clay nanosheet. Phys Chem Phys 18: 5404–5411.

90 Ramasamy, E., Deshapriya, I.K., Kulasekharan, R. et al. (2014). Photophysical studies of an encapsulated neutral guest intercalated into the 2-dimensional space of α-Zr(IV) phosphate. Photochem Photobiol Sci 13: 301–309.

91 Ramasamy, E., Jayaraj, N., Porel, M., and Ramamurthy, V. (2012). Excited state chemistry of capsular assemblies in aqueous solution and on silica surfaces. Langmuir 28: 10–16.

92 Porel, M., Klimczak, A., Freitag, M. et al. (2012). Photoinduced electron transfer across a molecular wall: coumarin dyes as donors and methyl viologen and TiO_2 as acceptors. Langmuir 28: 3355–3359.

93 Samanta, S.R., Kulasekharan, R., Choudhury, R. et al. (2012). Gold nanoparticles functionalized with deep-cavity cavitands: synthesis, characterization, and photophysical studies. Langmuir 28: 11920–11928.

94 Mondal, B., Kamatham, N., Samanta, S.R. et al. (2013). Synthesis, characterization, guest inclusion, and photophysical studies of gold nanoparticles stabilized with carboxylic acid groups of organic cavitands. Langmuir 29: 12703–12709.

95 Gupta, S., Choudhury, R., Krois, D. et al. (2012). Cucurbituril adamantanediazirine complexes and consequential carbene chemistry. J Org Chem 77: 5155–5160.

96 Gupta, S., Choudhury, R., Krois, D. et al. (2011). Photochemical generation and reactivity of carbenes within an organic cavitand and capsule: photochemistry of adamantanediazirines. Org Lett 13: 6074–6077.

97 Choudhury, R., Gupta, S., Da Silva, J.P., and Ramamurthy, V. (2013). Deep-cavity cavitand octa acid as a hydrogen donor: photofunctionalization with nitrenes generated from azidoadamantanes. J Org Chem 78: 1824–1832.

98 Jones, G.I., Jackson, W.R., Choi, C.Y., and Bergmark, W.R. (1985). Solvent effects on emission yield and lifetimes for coumarin laser dyes. requirements for a rotatory decay mechanism. J Phys Chem 89: 294–300.

99 Tang, H., de Oliveira, C.S., Sonnatag, G. et al. (2012). Dynamics of a supramolecular capsule assembly with pyrene. J Am Chem Soc 134: 5544–5547.

100 Thomas, S.S., Tang, H., Gaudes, A. et al. (2017). Tuning the binding dynamics of a guest-octaacid capsule through non-covalent anchoring. J Phys Chem Lett 8: 2573–2578.

101 Turro, N.J., Ramamurthy, V., and Scaiano, J.C. (2010). Modern Molecular Photochemistry of Organic Molecules. Sausalito, CA: University Science Books.

102 Horrocks, A.R., Kearvell, A., Tickle, K., and Wilkinson (1966). Mechanism of fluorescence quenching in solution. Part 2 – Quenching by xenon and intersystem crossing efficiencies. Trans Faraday Soc 62: 3393–3399.

103 Ramamurthy, V. (1994). Organic guests within zeolites: xenon as a photophysical probe. J. Am. Chem. Soc. 116.

104 Ramamurthy, V. (1985). Thiocarbonyl photochemistry. In: A. Padwa, vol. 7 (ed. O. Photochemistry), 237–338. New York: Marcell Dekker.

105 Blackwell, D.S.L., Liao, C.C., Loutfy, R.O., de Mayo, P., Paszyc, S. (1972) On the luminescence of thiocarbonyl compounds. Mol Photochem, 4, 171–188.

106 Rajee, R. and Ramamurthy, V. (1979). Energy wastage in organic photochemistry: self-quenching in thiones. J Photochem 11: 135–138.

107 Bouas-Laurent, H., Castellan, A., Desvergne, J.-P., and Lapouyade, R. (2000). Photodimerization of anthracenes in fluid solution: structural aspects. Chem Soc Rev 29: 43–55.

108 Chandross, E.A., Ferguson, J., and McRae, E.G. (1966). Absorption and emission spectra of anthracene dimers. J Chem Phys 45: 3546–3553.

109 Ferguson, J. and Mau, W.-H. (1974). A spectroscopic study of the photodimerization of anthracene sandwich dimers in dianthracene. Mol Phys 27: 377–387.

110 Wasserman, H.H. and Murray, R.W. (1979). Singlet Oxygen. New York: Academic Press.

111 Foote, C.S. (1968). Photosensitized oxygenations and the role of singlet oxygen. Accounts of Chemical Research 1: 104–110.

112 Foote, C.S. (1971). Pure Appl Chem 27: 635–645.

113 Gupta, S. and Ramamurthy, V. (2018). Characterization and Singlet Oxygen Oxidation of 1-Alkyl Cyclohexenes Encapsulated Within a Water-Soluble Organic Capsule. ChemPhotoChem 2: 655–666.

114 Lewis, F.D., Petisce, J.R., Oxman, J.D., and Nepras, M.J. (1984). One-way photoisomerization of *cis*-stilbene via a cation radical chain mechanism. J Am Chem Soc 107: 203–207.

115 Lewis, F.D., Bedell, A.M., Dykstra, R.E. et al. (1990). Photochemical generation, isomerization, and oxygenation of stilbene cation radicals. J Am Chem Soc 112: 8055–8064.

116 Ramamurthy, V., Weiss, R.G., Hammond, G.S. (1993) A model for the influence of organized media on photochemical reactions. In: Advances in Photochemistry (eds D. Volman, G.S. Hammond, D.C. Neckers), vol. 18, John Wiley & Sons., pp. 67–234.

19

Effect of Confinement on the Physicochemical Properties of Chromophoric Dyes/Drugs with Cucurbit[*n*]uril: Prospective Applications

J. Mohanty[1,2,*], *N. Barooah*[1], *and A. C. Bhasikuttan*[1,2]

[1] *Radiation & Photochemistry Division, Bhabha Atomic Research Centre, Mumbai 400 085, India*
[2] *Homi Bhabha National Institute, Training School Complex, Anushaktinagar, Mumbai 400 094, India*

19.1 Introduction

Confinement of small guest molecules in cavitands or organized assemblies offers a convenient strategy to modulate the physicochemical properties of the guests and has proved very effective in influencing/controlling the course and outcome of the reaction path. The effect, which is mainly attributed to the net effect of various non-covalent interactions in the host-guest assembly, often respond specifically and external stimuli making them conveniently tunable. Such internment of a molecule achieved through supramolecular host-guest interaction has gained significant research interest due to their promising applications in sensor [1], drug delivery vehicles [2–4], novel catalysts for the construction of specific molecules [5, 6] and water-based dye laser [7–9]. In this regard, research literature has seen incredible contribution on the confinement effect on small guest/drug molecules through several macrocyclic receptors as they provide a unique hydrophobic cavity to rigidize and protect the included guest molecules from bulk media interaction.

Consequent to the light absorption, a chromophoric guest molecule gets excited to the higher energy states and this excited molecule is energetically unstable and gets deactivated by following different photophysical processes/pathways such as radiative processes (fluorescence and phosphorescence) and nonradiative processes (internal conversion, intersystem crossing, Forster resonance energy transfer, solvent-induced quenching or quenching by oxygen or additives, etc.) [10, 11]. This photophysical processes, especially the fluorescence behavior of the guest dye is greatly influenced/modulated by the macrocyclic encapsulation. In general, the effects of inclusion complexation are related to either the entrapment of the guest dye into the more hydrophobic environment provided by the host cavities [11] or the geometrical confinement of the fluorophore within the host, which restricts the rotational, vibrational and torsional motions of the guest thereby disfavoring undesirable nonradiative relaxation pathways [11, 12]. Furthermore, the fluorophore is mechanically protected from the external quenchers i.e. solvent and dissolved oxygen. As a result, the fluorescence quenching by oxygen or the proton or hydrogen transfer reactions caused by solvent on interaction with excited dye or exciplex formation gets prohibited. In some cases, due to encapsulation, aggregation of dye gets prevented, leading to significant changes in the photophysical properties of the dye solution [7, 13, 14]. In this book chapter, the modifications in the

*Corresponding author, Email: jyotim@barc.gov.in

Chemical Reactivity in Confined Systems: Theory, Modelling and Applications,
First Edition. Edited by Pratim Kumar Chattaraj and Debdutta Chakraborty.

photophysical properties of the guest dyes mainly observed with cucurbituril host molecules, especially, cucurbit[7]uril (CB7) and cucurbit[8]uril (CB8) have been discussed. CB8 macrocycle with large cavity size facilitates encapsulation of two homo- or hetero molecules which leads to the formation of emissive or nonemissive dimer/excimer or nonemissive exciplex [11, 15–18]. Single or multiphoton ionization processes, all of which can ultimately lead to a permanent degradation of the chromophore by irreversible chemical follow-up reactions, can be prevented by encapsulation [11, 14]. These modulations in the photophysical processes, especially, fluorescence behavior, photostability, aggregation behavior have been explored to design and develop fluorescence sensor, aqueous dye laser system, etc. [7, 14, 19]. Beside these photophysical processes, the inclusion complex formation also significantly alters the acid-base equilibrium. In the prototropic equilibrium, the protonated form of the guest dye gets stabilized upon complexation with the cucurbiturils. As a result, an upward supramolecular shift in the pK_a values is observed and the complexed dye becomes more basic in comparison to the free dye.

Catalysis occupies a pivotal role in synthetic chemistry, especially, in industrial chemistry. Catalytic performance can be described by three key parameters, activity, selectivity, and stability. In this context, the use of confined nanoreaction vessels to alter reactivity and control selectivity during chemical transformations represents a major paradigm shift from conventional methods [20, 21]. As a result, many fundamental challenges have emerged, in particular 1) incorporating reactive units/chromophores within confined cavities and 2) manipulating reaction conditions to induce product selectivity. Catalysts with preorganized structures and well-defined cavities, such as cyclodextrin, calixarenes and cucurbiturils can influence the catalytic performance. Some of the examples [5] that demonstrate the influence of confinement effects set by cucurbituril macrocyclic host that governs the activity and selectivity in catalysis through noncovalent host-guest interaction will be emphasized in this book chapter.

19.1.1 Confinement of Dyes/Drugs in Macrocyclic Hosts

Supramolecular noncovalent host-guest interactions are one of the methods to achieve confinement of the guest molecules as the macrocyclic hosts provide a unique hydrophobic and chemically inert cavity, besides the functional portals, to encapsulate the guest molecules. The prominent feature of macrocyclic hosts is the degree of rigidity of the ligand and the concomitant enforcement of a particular shape to the host-guest complex [22]. In this regard, various macrocyclic host molecules (both natural and synthetic) have been studied widely. The properties of the macrocycles, their characteristic binding behavior towards different guest molecules are discussed in this section.

19.1.1.1 Cyclodextrins

Cyclodextrins (CDs), a family of the classical cyclic oligosaccharides consist of (α-1,4)-linked α-D-glucopyranose units and contain a significantly hydrophobic central cavity and hydrophilic portals [23]. The central cavity of the truncated cone shaped cyclodextrins is lined by the skeletal carbons and ethereal oxygens of the glucose residues, which gives it a hydrophobic character and is instrumental for binding nonpolar alkyl and aryl residues. The polarity of the cavity has been estimated to be similar to that of ethanol solution [24]. The hydroxyl groups are orientated to the cone exterior with the primary hydroxyl groups of the sugar residues at the narrow edge of the cone and the secondary hydroxyl groups at the wider edge. This arrangement provides additional hydrogen bonding sites for the binding of organic guests, particularly anionic guest molecules. The natural α-, β-, and γ-cyclodextrins consist of six, seven, and eight glucopyranose units, respectively. CDs are the first receptor molecules whose binding properties towards organic guest molecules are in the range of 10 to 10^5 M^{-1} [25, 26] and require millimolar concentration to accomplish substantial complexation with them in aqueous medium, which have been widely investigated, yielding plentiful results on physical and chemical features of molecular complexation. Although the natural cyclodextrins and their complexes are hydrophilic, their aqueous solubility is rather moderate, especially that of β-cyclodextrin, which is about 16 mM. Practically, this restricts the large scale uses of cyclodextrin-based host-guest systems in various applications as quantitative complexation is often difficult to achieve with submillimolar concentration of guests [27]. However, it is worth mentioning here that the hydrophilic cyclodextrins are non-toxic at low to moderate oral dosages and they have several prospects in medicinal use, especially in drug formulation [23].

19.1.1.2 Calixarenes

Calix[*n*]arene (CX*n*) macrocycles, the homologues of cyclooligomeric phenols, have attracted much attention as basic molecular scaffolds for developing highly selective ionophores/ion sensors. CX*n* are synthesized from the base-catalyzed condensation of a *p*-substituted phenol with formaldehyde [28]. The name "calixarene" was devised by C. David Gutsche due to the resemblance of the bowl-shaped conformation of the smaller calixarenes to a Greek vase [29]. CX*n* are available in different cavity sizes of 4, 6, or 8 units of the monomeric unit (CX4, CX6, and CX8) [30]. The conformational variability of calixarenes is much larger than that of cyclodextrins [11]. Although several functionalized water-soluble calixarenes are known, the sulfonated analogues, introduced by Shinkai et al., are the most prominent water-soluble calixarenes having several potential applications [32]. The calixarenes such as CX6- and CX8-containing *p*-sulfonate groups (SCX6 and SCX8) have the same level of toxicity as glucose; however, *p*-sulfonato calix[4]arene (SCX4) shows slightly higher toxicity [31].

19.1.1.3 Cucurbiturils

Cucurbit[*n*]urils (CB*n*), a smart class of macrocyclic receptor molecules, consist of macrocyclic cavities made up of methylene-bridged glycoluril monomers. The macrocycle, initially exposed by Behrend in 1905 but fully characterized in 1981 [6, 11, 32–35], have a highly symmetrical hydrophobic cavity laced with two identical, highly polarizable negatively polarized carbonyl portals. Like cyclodextrins and calixarenes, different homologues of cucurbit[*n*]urils (commonly abbreviated as CB*n*; $n = 5$–14 represents the number of glycoluril units in the macrocycle; Figure 19.1) with varying cavity and portal dimensions are known [6, 32–35]. Of late, cucurbit[14]uril (remained in the twisted form and abbreviated as tQ [14]) has been synthesized which has two kinds of cavities (a central cavity and two side cavities) and adopts a folded, figure-of-eight confirmation [36]. Due to the portal interactions, the water solubility of the CB*n* is usually enhanced in the presence of salts/ionic guests or at higher H^+ concentration [32, 35]. Most of the cucurbituril homologues, except CB5, form stable inclusion complexes with guest molecules, especially, cationic guest molecules like diaminoalkanes, benzyl amines, adamantyl amine, methyl viologen cations, fluorescent dyes, surfactants, metal ions, metal nanoparticles, etc. through complete or partial encapsulation of the guest molecules [32–35, 37–41], whereas the electron-deficient carbon centers at the equatorial peripheral region of CB6, CB7, and CB8 interacts strongly with the redox-active polyoxovanadate, polyoxomolybdate, and polyoxotungstate anion clusters [42, 43]. On the contrary, the cavity of CB8 is large enough to accommodate more than one guest molecules to form multiple/higher order host-guest complexes [15, 18, 44]. They also exhibit low *in vitro* as well as *in vivo* toxicity, thereby facilitating applications in biology. Furthermore, the confinement provided by the hydrophobic cavity of cucurbit[*n*]uril to the guest molecules brings out significant

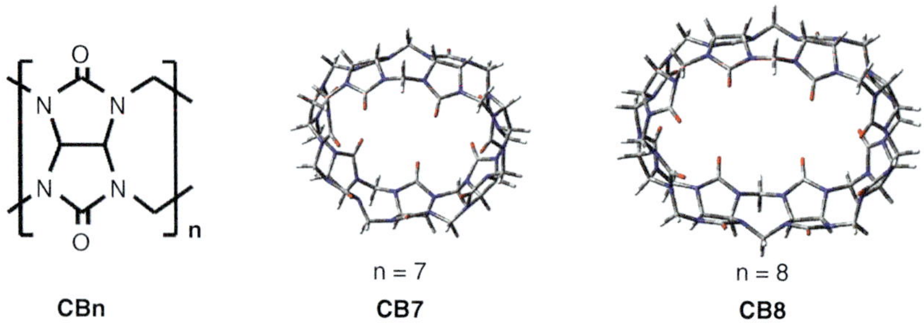

Figure 19.1 Structural formula of CB*n* (*n* = 5-10) and energy optimized structures of CB7 and CB8.

changes in the photophysical properties of the guest molecules and the catalytic performance in a chemical reaction and has activated a lot of research interest towards numerous applications. Aptly, their usages are attempted in fluorescence sensor, aqueous dye laser, drug delivery vehicles, anti-bacterial agents, catalysts, etc.

This book chapter will provide a brief account on the effect of confinement imposed by cucurbituril macrocyclic hosts on the physico-chemical properties of guest molecules such as fluorescence and photostability of the chromophoric dyes/drugs, supramolecular pK_a shift, solubility, as well as on its catalytic behavior and their potential technological and biological applications.

19.2 Confinement in Cucurbituril Hosts: Effects on the Physicochemical Properties of Guest Molecules – Advantages for Various Technological Applications

As discussed above, hitherto studies on the physicochemical properties due to the confinement exerted by cucurbituril have led to the demonstration of various applications such as in aqueous dye laser, molecular capsules as drug delivery vehicles, luminescent materials, sensors, antibacterial agents, etc. [7, 8, 18, 45–49]. In this section we describe the details of their fluorescence behavior, acid-base equilibrium and aggregation behavior of the molecular systems such as rhodamines, thioflavin T, thiazole orange, *p*-dimethylaminobenzonitrile, and fluoroquinolones in the presence of cucurbiturils.

Chromophoric dyes, especially those having technological applications, often face issues related to aggregation, stability, medium of usage, disposal etc. Increasing the photostability of fluorescent dyes either with additives or by structural modification is essential for their technological and biological applications [50]. In general, for most of the cationic dyes, photostabilization is deceptively low. In principle, this problem can be reduced by supramolecular encapsulation strategy that isolate the individual dye molecules and prevent self-aggregation or other destabilizing interactions with the surrounding environment due to confinement. However, in presence of CBs, a pronounced thermal stabilization due to deaggregation, desorption, and solubilization has been observed for xanthene dyes in aqueous solution through encapsulation/complexation which is supplemented by spectral shifts, enhanced fluorescence quantum yields, and lifetime characteristics. These additional features are attributed to the structural confinement of the dye in a less polar and low polarizability environment of CB7 cavity [13, 14, 51].

On the other hand, the solubility of the drug molecules can be tuned either through chemical modifications or through improved formulations. The encapsulation of drugs by macrocycles, provides an attractive method to increase their solubility by an appropriate formulation, i.e. by the formation of a host–guest complex. Among different host molecules, cyclodextrins have been extensively employed for this purpose and their advantages on drug solubility are well documented [52]. Using the similar method, calixarenes and cucurbiturils have also been reported to increase the solubility of several drug molecules [52].

19.2.1 Enhanced Photostability and Solubility of Rhodamine Dyes

Photostabilization of the rhodamine dyes are much required for their diverse applications in lasers and imaging techniques. The measurement of photobleaching quantum yields from the absorption data (Figure 19.2) reveals that CB7 increases the photostabilization of rhodamine 6G (Rh6G) by a factor of 1.5 at low irradiances (<1 kW/cm^2), but it becomes remarkably high (factor of 30) at high irradiances (ca. 10 MW/cm^2). Such irradiance situations are most pertinent for dye laser applications and also in confocal microscopy experiments [13].

The higher photostability of Rh6G is apparently caused by consolidation of several factors. The low polarizability in the host cavity should greatly disfavor unimolecular reactions like bond dissociations (for rhodamines, amino dealkylation is a common photodecomposition pathway) [51], which are known to be accelerated in more polarizable environments. Cucurbiturils also provide an extremely inert reaction medium; for example, it is very resistant towards oxidation and withstands even ozonolysis and imposes more rigidity to the dye structure and thereby, prevents the vibrational and rotational motion. Finally, the fluorescent dye is quite efficiently protected from water and oxygen, which may further reduce intermolecular follow-up (photo)chemistry, including photoionization from higher excited states relevant in the two-step two-photon decomposition [13, 14]. Alternatively, use of CB7 as an additive to Rh6G solutions improves the emission stability by greatly increasing the thermal stability of the dye solution, namely, by preventing unspecific adsorption

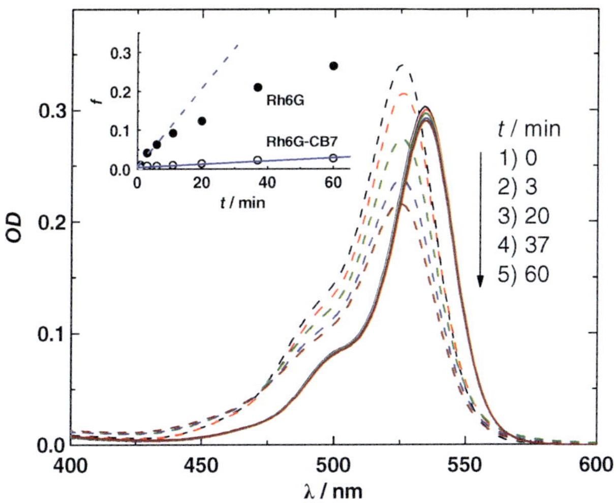

Figure 19.2 Photobleaching of Rh6G (ca. 4 μM) in aerated water in the absence (dotted lines) and presence (solid lines) of 1 mM CB7 followed through the decrease of the visible absorption with increasing time of pulsed 532-nm Nd-YAG laser irradiation. The inset shows the graph and correlation lines for the characteristic function f=log([10^A−1$_d$]/[10^A−1]) versus irradiation time for the determination of the relative quantum yield of photobleaching. For the unstabilized solution, the correlation line refers to the initial 10-min time period because of the formation of absorbing photoproducts at longer times. Reproduced with permission from [13], © 2005 John Wiley & Sons.

to glass and plastic surfaces at lower concentration and aggregation at very high concentration of the dye, a previously known problem for xanthene dyes [50]. The increased thermal as well as photochemical stability of CB7-containing Rh6G solution allows one to store aqueous Rh6G dye solutions, particularly in very dilute (nM) solutions, for an extended period of time. The fact that the enhanced photostabilization, prevention of aggregation and adsorption of the fluorescent dyes are achieved in an environmentally benign and biologically relevant solvent like water, the approach projects its candidature for several applications like in dye lasers, single molecule detection, dye-sensitized solar cells, optical storage devices, fluorescence-based assays, glow paints, etc. [13].

19.2.1.1 Water-Based Dye Laser

Aggregation/adsorption along with stability and solubility of organic dyes in aqueous solution are the major concerns for a dye laser system and is nicely demonstrated by the operation of a supramolecularly assisted aqueous dye laser system of rhodamine dyes with CB7 as the macrocyclic host additive [7, 8]. The effect of CB7 on the performance of aqueous kiton red S (KRS) dye solution has been investigated by both broad-band and narrow-band dye laser set-ups with respect to the practically relevant parameters like lasing efficiency, lasing stability and beam quality. A large increase in lasing efficiency is observed with addition of micromolar concentration of CB7 to the aqueous solution of kiton red dye (Figure 19.3), predominantly due to the deaggregating action of

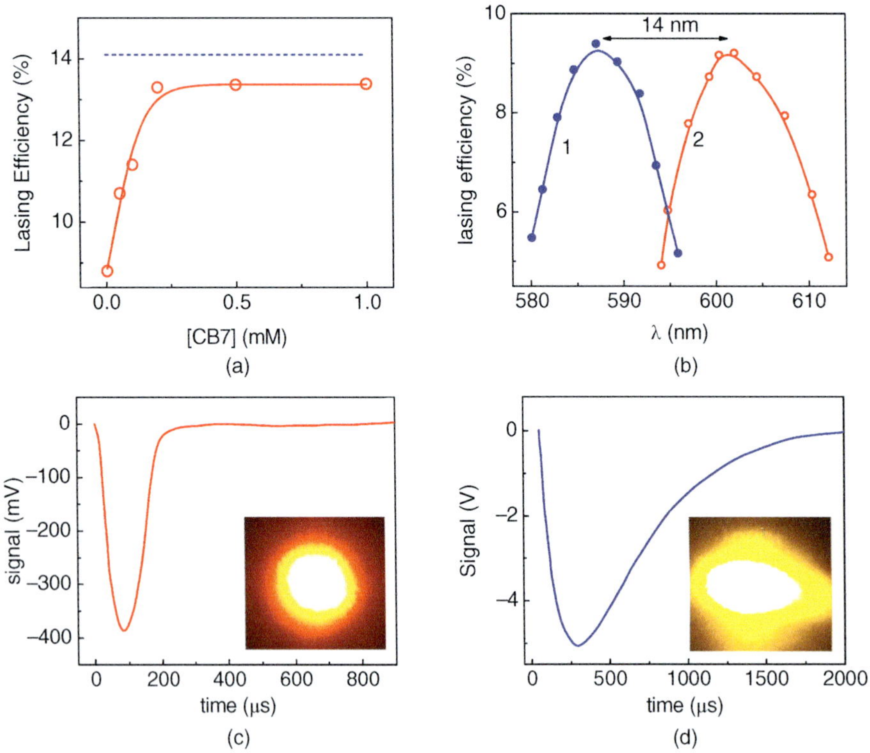

Figure 19.3 (a) Dependence of the lasing efficiency of KRS (200μM) on CB7 concentration. Shown for comparison (dashed line) are the lasing efficiencies of KRS in optically matched ethanol solutions. (b) Tuning curves of KRS in EtOH (1) and in the presence of 200μM CB7 (2) at a pump energy of 6.3 mJ. Thermal deflection signals of KRS in water-CB7 (c) and in EtOH (d) systems. Insets of C and D show the laser beam profiles for the respective medium. Reproduced with permission from [8], © 2010 Wiley-VCH Verlag GmbH & Co. KGaA, Weinheim.

CB7 on the dye [8, 19]. The resulting supramolecular dye lasers are environmentally more benign, more laboratory-safe, and less maintenance-sensitive than the presently employed dye laser systems based on organic solvents. The novel aqueous dye-CB7 systems are not only equally efficient with complementary tuning ranges, but also possess superior thermo-optic characteristics and unmatched beam profiles (Figure 19.3), which should jointly enable new and revived photonic applications based on such dye lasers [8, 19].

19.2.2 Enhanced Luminescence and Photostability of $CH_3NH_3PbBr_3$ Perovskite Nanoparticles

Recently, hybrid organic–inorganic lead halide perovskites are of great interest in photovoltaic devices and as luminescent materials for light-emitting devices. Their photoluminescence (PL) spectrum can be modified via controlled changes in their stoichiometry. There are many studies exploring the performance of these perovskites, in particular those with the $MAPbX_3$ (MA = CH_3NH_3, X = halide) stoichiometry, which address their optical gain, quantum yield, and their use as components of optical devices, among other features. However, there are concerns as to their stability and degradation, which affect their performance. In this contest, Gonzalez-Carrero et al. have demonstrated that the luminescence of $CH_3NH_3PbBr_3$ perovskite nanoparticles reaches its maximum by using CB7-adamantyl ammonium host-guest complexes as the capping ligand (Figure 19.4) on the nanocrystal surface [45]. These nanoparticles exhibit astonishing photostability and strong luminescence having a quantum yield of ~100% [45].

19.2.3 Enhanced Antibacterial Activity and Extended Shelf-Life of Fluoroquinolone Drugs with Cucurbit[7]uril

Chemical degradation of drug molecules is an apparent factor limiting their bioavailability, sustained medicinal activity, and shelf-life. In this regard, macrocyclic hosts can have beneficial effects on both their thermal and photochemical stability [52]. Amino groups of drug molecules

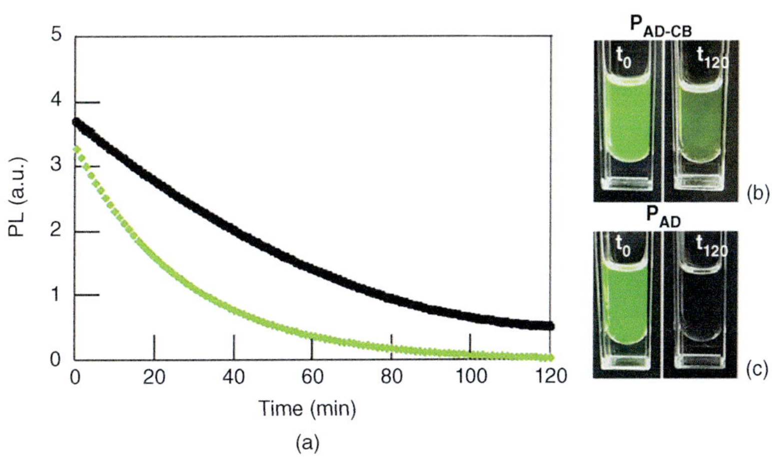

Figure 19.4 a) Photoluminescence of P_{AD-CB} (in black) and PAD (in green) dispersed in toluene and in contact with water as a function of the irradiation time; λ_{exc} = 350 nm, PL registered at 520 nm under air atmosphere; b, c) Images are of P_{AD-CB} and P_{AD} colloidal dispersions immediately after addition (left) of water and 120 min afterward (right). Reproduced with permission from [45], © 2016 John Wiley & Sons.

are prone to oxidation and nucleophilic reactions, and their conversion to ammonium groups provides a simple means to suppress both the reaction pathways. Saleh et al. have shown that the effectiveness and stability of the medicinally relevant benzimidazole-based molecules like lansoprazole and omeprazole increase by selectively encapsulating the benzimidazole moiety inside the cavity of CB7 [52, 53]. In another example, it has been shown that the stability of the histamine H_2-receptor antagonist ranitidine in the presence of CB7 increases significantly [54]. Decomposition of ranitidine leads to the formation of a dihydrothiazin-2-one oxime, and nucleophilic attack of water results in the (inactive) 5-N,N-dimethylaminomethyl-2-furylmethanol product. The increase in stabilization of ranitidine by CB7 is considered to be due to the prevention of the solvent nucleophilic attack and steric hindrance towards the formation of the cyclic intermediate [54]. In a recent study, Mohanty et al. have established the enhanced photostability and thermal stability of fluoroquinolone antibiotic drugs such as ofloxacin, norfloxacin, and danofloxacin [46] by confining the piperazine group/tertiary diazabicyclo[2.2.1]heptyl methylamino group in the CB7 cavity, as norfloxacin and ofloxacin undergo photochemical or thermal degradation of the piperazine side chain, either by degradation to an amino group or oxidation [55].

The "ideal" fluoroquinolone combines good clinical efficacy with low minimal inhibitory concentration (MIC) without any cytotoxicity. To achieve low MIC values, different generations of fluoroquinolones have been developed over the years as their substitution pattern modulates their antimicrobial activity [56]. Besides the covalent/synthetic modification of the basic chemical structure, noncovalently linked, externally controlled supramolecular host-guest interactions using macrocyclic hosts such as cyclodextrins, cucurbiturils, etc. [11, 19] are promising to improve and control the antibacterial activity of drugs. The significant enhancement in the antibacterial activity of second and third-generations fluoroquinolone drugs, namely danofloxacin (DOFL), norfloxacin (NRFL), and ofloxacin (OFL), in the presence of CB7 against four pathogenic bacteria at different pH conditions has been demonstrated [46]. In the case of all drugs, the minimum inhibitory concentration (MIC) was reduced by 3–5 fold in the presence of CB7 (Figure 19.5a). The extended shelf-life (antibacterial activity over time) of the fluoroquinone drugs in the presence of CB7 (Figure 19.5b), irrespective of four types of bacteria, can be attributed to the enhanced

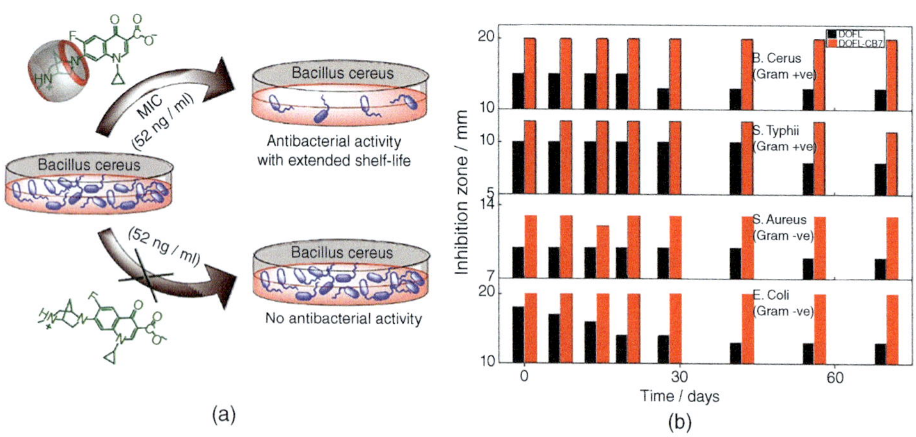

Figure 19.5 (a) Schematic representation of antibacterial activity of DOFL (52 ng/ml) in the absence and presence of CB7. (b) Bar chart representation of antibacterial activity (in terms of inhibition zone) of DOFL in the absence (black bar) and presence (red bar) of CB7 with time against four bacteria at pH 7.5. Adapted with permission from [46]. Licensed under CC BY 4.0.

thermal and photostability of their CB7 complexes [46]. All these studies are highly encouraging for the use of CB7 for the design and development of new long-acting antibiotic formulations.

19.2.4 Effect of Confinement on the Prototropic Equilibrium

The prototropic equilibrium of dyes can be significantly altered after confining the guest dyes in the cavity of macrocyclic hosts due to the differential binding affinity of the host molecules towards the different prototropic forms of the guests. This property is useful in medicinal and biological applications to achieve controlled binding and release of the active form of the drugs. Upward pK_a shifts are frequently observed for weakly basic molecules in the presence of cation-receptor or hydrogen bond-acceptor hosts, while downward pK_a shifts are expected for anion-receptor or hydrogen bond-donor host molecules, or those offering a nonpolar cavity [52].

Cucurbiturils, which act as cation receptors, preferentially stabilize the protonated form of guest molecules and therefore increase the pK_a values of the conjugate acids of amines. The effect of cucurbituril on the guest molecule is through affecting the protonation without the necessity of adjusting the pH. The binding constants of the guest molecules in their different protonation states with host mainly control/affect their complexation-induced pK_a shifts [12, 33].

19.2.4.1 Salt-Induced pK_a Tuning and Guest Relocation

Employing the competitive binding strategy, the pK_a value of the encapsulated guest shifted gradually towards that of the uncomplexed guest upon increasing the concentration of the salt as competitive binder, confirming a tunability essentially over the entire range of pK_a of the free dye to that of host:dye complex. Use of supramolecular pK_a shift towards drug delivery applications has been demonstrated in biological environment where the unprotonated dye would tend to be noncovalently bound to a protein, whereas the protonated dye prefers the CB7 host. In the presence of inorganic cations, the guest molecules (the dye, neutral red (NR) is used as a model for drug) could be actively released from the CB7 host cavity and relocated into the hydrophobic cavity of the protein, namely bovine serum albumin (BSA) (Figure 19.6) [2]. In another study, Mohanty et al. have also established the transfer of acridine dye from CB7 cavity to live cell lines by adding adamantly amine as the competitive binder [4]. Therefore, the presence of competitive binder provides a simple stimulus to release the drug from the macrocyclic cavity for biological/medicinal activity.

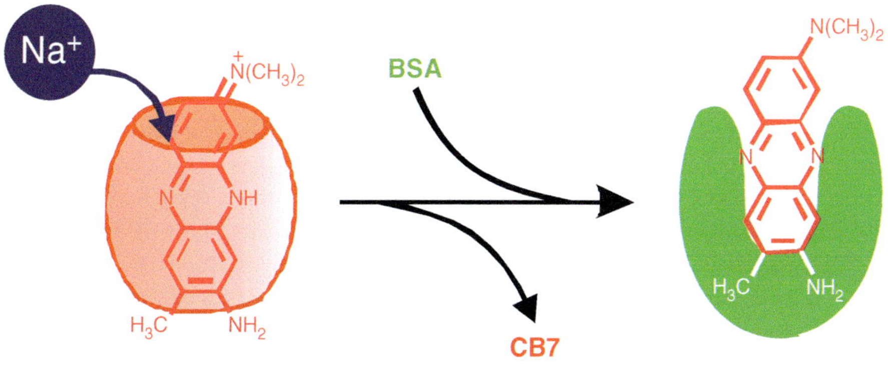

Figure 19.6 This scheme shows the salt-induced transfer of a dye from CB7 to BSA near pH 7.5. Reproduced from [2], © 2008, Royal Society of Chemistry.

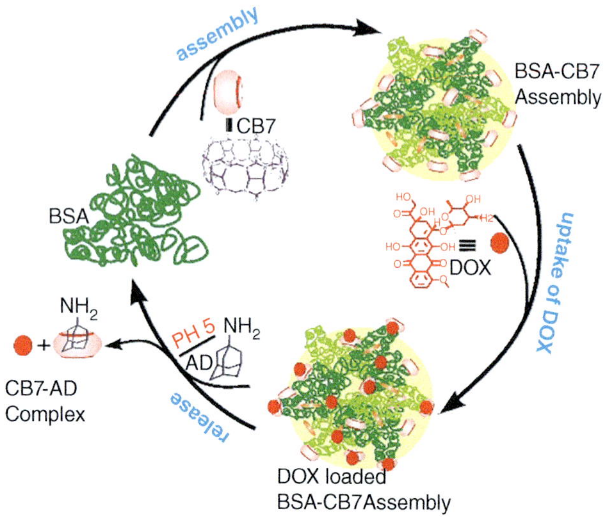

Figure 19.7 Schematic representation of the formation of CB7-BSA supra-biomolecular assembly, stimuli-responsive uptake and release of DOX. Reproduced with permission from [3], copyright © 2017 John Wiley & Sons.

19.2.5 Confinement in Cucurbit[7]uril-Mediated BSA: Stimuli-Responsive Uptake and Release of Doxorubicin

In a recent report, Mohanty et al. have established the construction of a non-toxic nanoassembly of bovine serum albumin (BSA) protein and the cucurbit[7]uril macrocycle as well as its stimuli-responsive breakage with adamantylamine or change in pH, which restores the protein structure and recognition properties [3]. The assembly showed efficient loading and controlled release of a standard anticancer drug, doxorubicin (DOX), and the same was validated in live cells (Figure 19.7) [3]. The cell viability studies documented that the DOX loaded assembly masks the cytotoxicity of DOX and the toxicity can be revived at the target on demand, triggering its therapeutic activation, and is more effective in the cancer cells [3]. In addition, they are highly promising for stabilizing/protecting the native protein structure, a viable approach to prevent/inhibit protein misfolding and aggregation, responsible for several neurodegenerative diseases.

19.2.6 Effect of Confinement on the Fluorescence Behavior of Chromophoric Dyes with Cucurbiturils

As revealed above, the guest molecules encapsulated in the cavity of CBs experience low polarity [13], an extremely low polarizability (close to the gas phase) [51], and more rigid environment, and consequently the complexed guest molecules show fundamentally different photophysical and chemical behaviors. In addition, CBs neither act as quencher nor display any absorption or fluorescence in the visible region, hence it is convenient to study the photophysical processes especially, fluorescence behavior of the guest molecules using CBs as the host. Moreover, the cavity of cucurbiturils also provides an inert reaction medium and protects the guests from the interaction with the surrounding water molecules [19]. The macrocyclic host-guest interactions impart structural rigidity, bringing out restrictions on the nonradiative processes such as vibrational and torsional motion and photo-isomerization, etc. in the dyes, thus modulating the fluorescence behavior.

19.2.6.1 Fluorescence Behavior of Chromophoric Dyes with Cucurbit[7]uril

Fluorescence Behavior of Thioflavin T with Cucurbit[7]uril: stimulus-responsive fluorescent molecular capsule and its breakage Thioflavin T (ThT) is a benzthiazolium dye which has been extensively applied in the early detection of amyloid fibril formation in tissues to diagnose neurodegenerative diseases such as Alzheimer's and Parkinson's diseases [57, 58]. Detection is mainly based on the characteristic fluorescence of ThT ($\Phi_f = 0.0003$) at 490 nm which increases intensely (~1000 fold) upon binding to the cavities of the amyloid fibrils [57, 58]. Though the application of ThT for the analysis of different aggregating systems is rapidly rising, the exact mechanism for binding of ThT into the cavities of the amyloid fibrils causing dramatic changes in its fluorescence properties is still largely elusive. Apart from the restriction towards the torsional motion between the benzthiazole and dimethylaminobenzene groups due to the confinement, excimer formation within the binding site is also proposed to cause further enhancement in the fluorescence for ThT on binding with amyloid fibrils [59]. In this regard, the cationic receptor behavior of cucurbit[*n*]uril (CB) hosts to introduce strong multiple binding and confinement to the cationic ThT dye has been examined. CBs can engulf ThT from either ends and can provide a large stabilization of the inclusion complex via ion-dipole interaction between the carbonyl portals of CBs and the cationic charge on the dye along with the hydrophobic interaction and confinement exerted by the host cavity [60]. Significant enhancements/modifications in the fluorescence yield (40-fold), lifetime and spectral features of ThT dye with CB7 have been observed due to the formation of 1:1 and 2:1 host–guest complexes between ThT and CB7, leading to the specific structural confinement [60].

Bhasikuttan et al. have further demonstrated the formation of a novel stimulus-responsive and highly *fluorescent supramolecular nanocapsule* through the cooperative metal ion binding to the stoichiometrically selected cucurbit[7]uril-thioflavin T complex (Figure 19.8a) [47]. The first

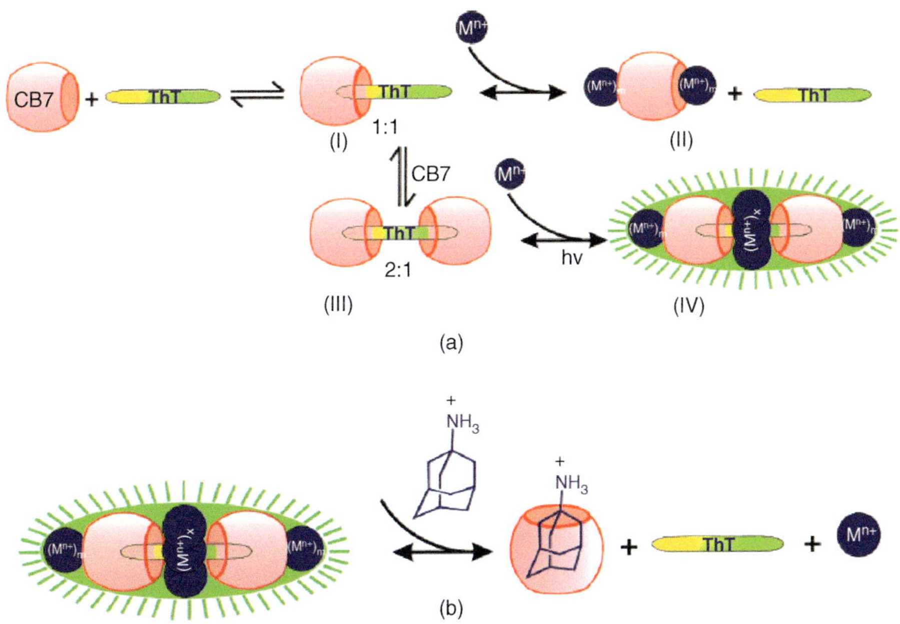

Figure 19.8 (a) Proposed binding interactions in the ThT, CB7 and metal ion system leading to the highly fluorescent supramolecular capsule and (b) release mechanism of ThT from the metal ion bound supramolecular capsule by amantadine hydrochloride as the stimulant. Adapted with permission from [47], © 2015, American Chemical Society.

example of such an unusual assembly became feasible due to the stoichiometry and the structural arrangement of the host-guest complex with two CB7 portals providing strong negative charge density for the metal ions to group and seal the complex, as a result, the dye experiences more confinement and protecting the incorporated dye, thus leads to the fluorescence intensity being enhanced dramatically [47]. The rupture of the capsular complex was demonstrated with a strong competitive guest, 1-amantadine hydrochloride, which helped in disrupting the capsule to release the dye (Figure 19.8b) [47]. By proper design criteria of the guest chromophores, the methodology can be explored for the binding and release of drug molecules and for application in on-off systems and will have immense potential as building blocks for molecular architectures displaying unique properties.

Fluorescence Behavior of Thiazole Orange with Cucurbit[7]uril: metal-ion mediated assemblies Thiazole orange (TO), is widely used as a fluorogenic probe for the detection of nucleic acids [61, 62]. The fluorescence intensity of TO in aqueous solutions is very weak (Φ_f =0.0002) [16, 49] which is attributed to the rapid nonradiative deactivation by photoisomerization as well as the torsional motion between the benzothiazole and quinoline groups around the monomethine bond. The absorption and fluorescence changes of TO in the presence of CB7 clearly indicates the formation of multiple complexes. At low CB7 concentrations, 1:4 CB7:TO complexes, consisting of TO dimers accumulated at each of the two CB7 portals are formed. With increasing CB7 concentrations, 2:1 host-guest complexes are formed where both the benzothiazole and the quinoline moieties of TO are encapsulated by 2 CB7 molecules (Figure 19.9) [49]. In the presence of Na^+ or Ca^{2+}, this structure is again disassembled to yield 1:1 CB7:TO complexes. The subsequent self-assembly of these 1:1 complexes at higher host concentrations is found to be sensitive to the nature of the metal ions. It is proposed that short dimeric structures composed of two 1:1 CB7:TO complexes are generated in the presence of Ca^{2+} ions, leading to fluorescence enhancement, whereas extended superstructures are formed by the assembly of many such 1:1 CB7:TO complexes in the presence of Na^+ ions, leading to the appearance of turbidity in the solution (Figure 19.9) [49]. This remarkable distinction between the supramolecular structures by changing the nature of the metal ions can have potential analytical applications.

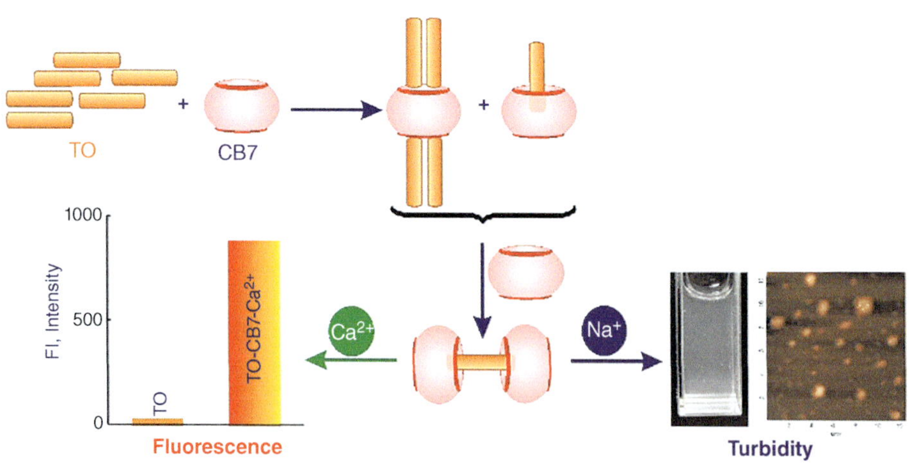

Figure 19.9 Schematic representation of various CB7:TO Structures that are proposed to be formed under different chemical environments. Adapted with permission from [49], © 2010, American Chemical Society.

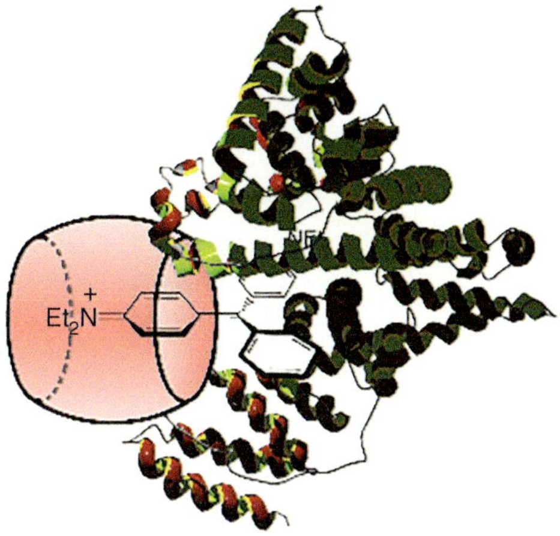

Figure 19.10 Sketch of the fluorescent ternary complex CB7•BG•BSA. Reproduced with permission from [64], © 2007, John Wiley & Sons.

Confinement Through Supra-Biomolecular Interaction: Fluorescence Behavior of Triphenyl Methane Dyes The effects of confinement in the radiative properties of fluorescent dyes by formation of supramolecular host-guest assemblies or association with biomacromolecules like proteins and oligonucleotides enables sensor applications and improves the understanding of the factors governing molecular recognition. Fluorescent dyes serve as probes for biomolecular systems and in selected cases, such as the triphenylmethane (TPM) dyes, as promising drugs for photodynamic therapy and as antimicrobial agents [63]. In this study, Bhasikuttan et al. demonstrated the dramatic enhancement (~300-fold) in the fluorescence properties of a representative TPM dyes, i.e. Brilliant Green (BG) with BSA by the addition of the macrocyclic host molecule CB7 (Figure 19.10) [64]. The resulting cooperative binding of a dye and potential drug with biological target molecules opens up a new approach to improve medicinal activity or the sensitivity of fluorescent sensor applications by a supramolecular enhancer strategy.

19.2.6.2 Fluorescence Behavior of Chromophoric Dyes with Cucurbit[8]uril

The confinement effect mainly depends upon the cavity size of the macrocyclic hosts as well as the dimension of the guest molecules. Cucurbit[8]uril (CB8), the higher homologue of CB7, is credited with forming higher stoichiometric complexes by co-localizing π-stacked or charge-transfer pairs in the cavity or 1:1 complex when the size of the guest molecule is large [19]. Depending upon the stoichiometry of the complexes, the confinement effect on the guest molecules exerted by the host molecule is different. As a result, the effect of confinement in the radiative properties of the guest molecules upon complexation varies.

Excimer Emission from Fluorogenic Dyes with Cucurbit[8]uril Mohanty et al. have demonstrated the host-guest assembly formation between thioflavin T and CB8 with 1:2 (CB8-ThT) and 2:2 stoichiometric ratios due to the strong ion–dipole interactions provided by the carbonyl portals of the CB8, which adequately support the stabilization of two π-stacked ThTs [15]. The noncovalently stabilized 2:2 (CB8-ThT) assembly brings out an intense novel excimer emission "turn on" at 570 nm and become, tunable and "turn off" in the presence of Ca^{2+} (Figure 19.11) [15]. Such tunable host–guest molecular assemblies find immense implications in activating nondiffusional- and stoichiometry-controlled photoinduced processes and excimer-based fluorescence on/off switches, and provide design criteria for tailor made molecular assemblies with CB8. Similarly, the CB8 recognized TO in its 2:2 stoichiometry exhibited spectacular fluorescence enhancement of the order of

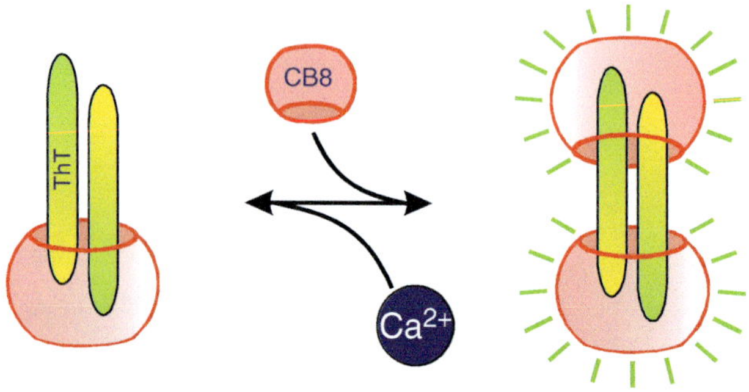

Figure 19.11 Schematic representation of a 2:2 CB8:ThT assembly formation and its dissociation in the presence of Ca^{2+}, demonstrating the on/off mechanism. Reproduced with permission from [15], © 2009, John Wiley & Sons.

1700 fold [16], which is the largest directly determined value so far reported for a dye in an organic macrocyclic system. This prospective 2CB8:2TO assembly responded to selected chemical stimuli such as metal ions, adamantylamine, and tryptophan, providing different dissociation mechanisms and demonstrating a controlled exchange and release action. This methodology can evolve into a general approach to deliver and operate intracellularly functional molecular components under chemical/thermal/optical trigger control, especially for therapeutic applications.

Excimer Emission from Coumarin Dyes with Cucurbit[8]uril Barooah et al. have established the dimer emission from CB8-templated bichromophoric coumarin fluorophores which exhibit ~150 fold increase in the excited-state lifetime. It has been reported, for the first time, the intrinsic mechanism of the dimer emission as due to the slipped spatial arrangement among the dimers within the CB8 cavity, as in the cases of H and J-type aggregates (Figure 19.12) [17]. The slip angles, θ, were found to correlate with the emissive behavior expected from the H and J-type configurations [17]. Besides their intriguing photophysical perspectives, the contrasting and strong modulations observed in the optical response of these dyes, having very subtle change in the guest chemical structure, are promising towards the design of spatially restricted, CB8-based optically-responsive, novel self-assembled systems.

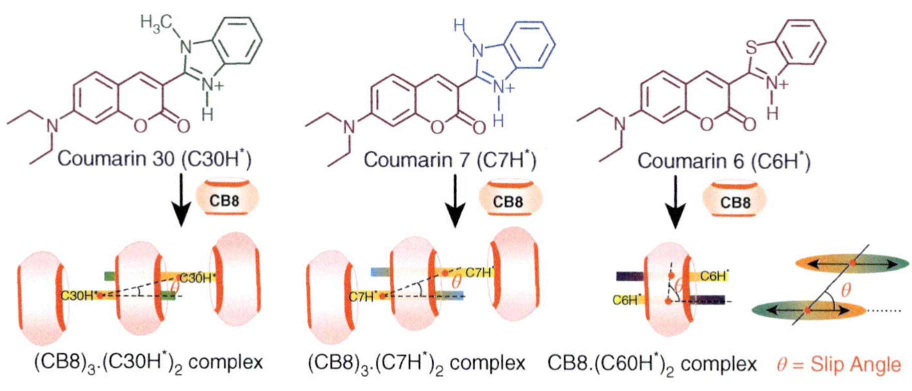

Figure 19.12 Structures of the protonated forms of C30, C7, and C6guests, and the proposed geometries of their host-guest complexes. Reproduced from [17], © 2015, Royal Society of Chemistry.

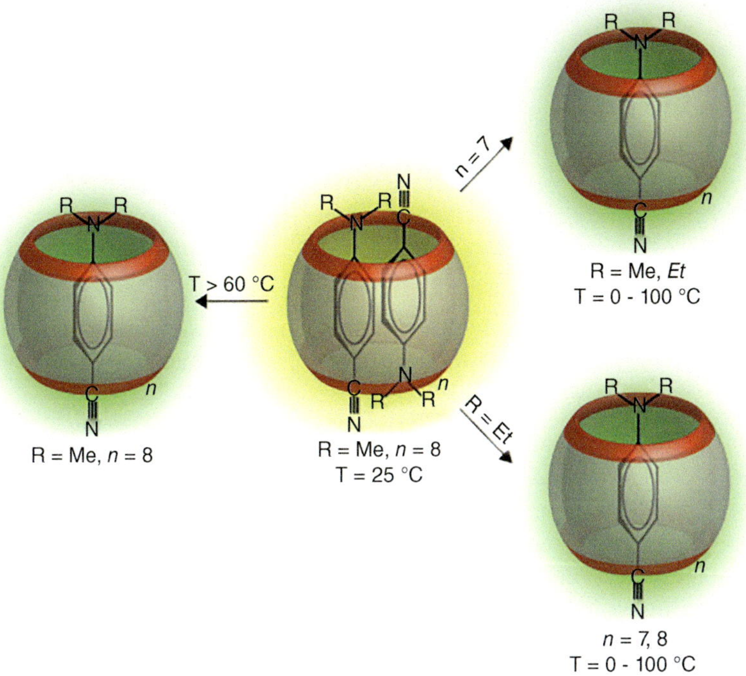

Figure 19.13 Schematic representation of the conversion of 1:2 CB8-DMABN complex into 1:1 complex at different stimuli-responsive conditions. Reproduced with permission from ref [18], © 2015, John Wiley & Sons.

Elusive Excimer Emission from* p*-Dimethylaminobenzonitrile Dyes with Cucurbit[8]uril The dual emission of *p*-dimethylaminobenzonitrile (DMABN) has fascinated photochemists for decades, but it is through the addition of cucurbit[8]uril (CB8) in water that its photophysics becomes truly multiemissive, because the host templates the dimer to allow the observation of the long sought for excimer emission [18]. The observed excimer emission band (third emission band) from *p*-dimethylaminobenzonitrile encapsulated inside the CB8 cavity as head-to tail dimer template, constitutes a completely new photophysical aspect for this famous chromophore. The properties of the excimer emission are intriguing, because it disappears in favor of monomer emission when a smaller host (CB7) is employed, when a slightly sterically modified chromophore is used (*p*-diethylaminobenzonitrile:DEABN), and when the temperature is increased slightly (to 60°C) (Figure 19.13) [18]. This tunable multiemissive behavior makes DMABN complexes with CB7 and CB8 for applications in differential and ratiometric sensing which includes its use as a sensitive optical supramolecular thermometer in the ambient temperature range, the construction of logic gates, etc.

Perylenediimide Derivative with Cucurbit[8]uril: Emission from Deaggregated Dye Perylenebis(diimide) (PDI) dyes, which are excellent fluorophores; have been used as pigments and colorants for decades on account of their excellent photophysical and thermal stability. Unfortunately, they tend to form non-fluorescent π-π stacks in water [65], which greatly limits biological applications based on confocal microscopic or single-molecule spectroscopic techniques. Scherman et al. have demonstrated a facile supramolecular approach for the use of PDI dyes as fluorophores in aqueous media through the elimination of self-aggregation using cucurbit[8]uril by encapsulating PDI inside its cavity

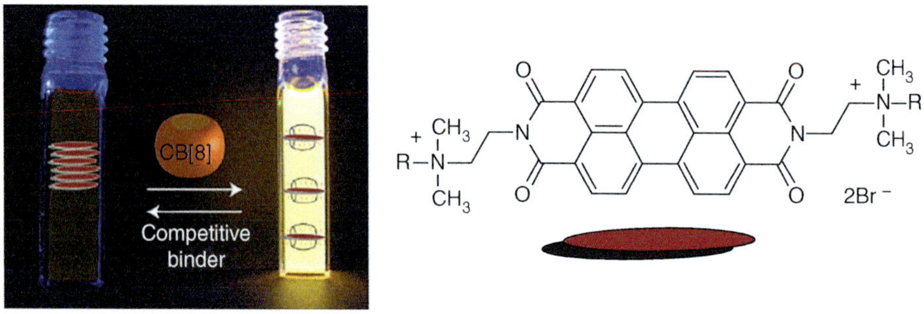

Figure 19.14 Schematic representation of the formation of 1:1 CB8-PDI complex and its dissociation in the presence of competitive binder. Reproduced with permission from [48], © 2012 Wiley-VCH Verlag GmbH & Co. KGaA, Weinheim.

(Figure 19.14), leading to a dramatic increase in the fluorescence quantum yield of the dye while ensuring high photochemical and chemical stability of the PDI dye [48]. Furthermore, these complexes were shown to be stimuli-responsive to competitive binders and can be (electro)chemically switched between the non-fluorescent reduced states and (re)oxidized fluorescent states. Moreover, addition of suitable dicationic or electron-rich second guests leads to the formation of 1:1:1 ternary complexes [48], which can be utilized for the reversible formation of higher-order supramolecular architectures in water.

19.2.7 Effect of Confinement on the Catalytic Performance within Cucurbiturils

The encapsulation of a catalyst in a molecular container with a well-defined confined space can influence the activity and selectivity of the catalytic reaction. In this section, some of the examples that illustrate the influence of confinement effects imposed by cucurbiturils in the catalytic performance. In a very early study, Mock and coworkers reported for the first time, the cycloaddition of alkynes and azides inside cucurbit[6]uril (CB6) cavity [66, 67]. The two substrates have been directly encapsulated inside the host cavity due to the presence of ammonium groups. The click reaction i.e. 1,3-dipolar cycloaddition is accelerated by a factor of 5.5×10^4 under the catalytic influence of cucurbituril, leading to the triazole product. This reaction was extensively used by Tuncel and co-workers for the synthesis of rotaxanes, pseudorotaxanes, and polyrotaxanes as well as for accessing new molecular switches, machines, and nanovalves [68].

So far, examples of CBs containing transition metal catalysts are limited. In this field, Nau et al. have demonstrated a chemoselective transformation of included guests promoted by transition-metal ions coordinated to the cucurbituril rim [69]. They have considered a two-phase system (Figure 19.15), in which the host CB7 acts as an inverse phase-transfer catalyst to bind a photoreactive substrate i.e. bicyclic azoalkanes 2,3-diazabicyclo[2.2.2]oct-2-ene (DBO), and 2,3-diazabicyclo[2.2.1]hept-2-ene (DBH) by hydrophobic interactions in the aqueous phase and allows the subsequent docking of transition metal ions to the carbonyl rim through ion–dipole interactions. The resulting ternary self-assembly is synergistically reinforced by weak metal–ligand bonding interactions, which affect the chemoselectivity in the spatially resolved laser photolysis of the aqueous phase. A new photoproduct i.e. cyclopentene (41%) was obtained from the DBH·CB7·Ag^+ ternary complex on photoirradiation [69]. The presence of the macrocycle is, indeed, essential to produce cyclopentene, since even the photolysis of DBH in the presence of 1M $AgNO_3$ did not lead to this unexpected cyclopentene (CP) product. It is evident that Ag^+ ions promote the photochemical formation of CP inside CB7. This reveals the potential for metal

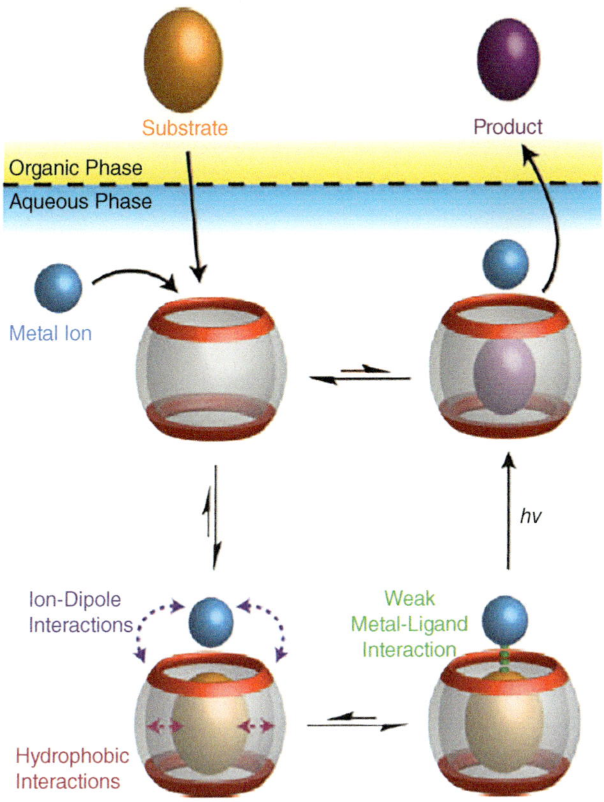

Figure 19.15 Dynamic self-assembly of a ternary guest/host/metal-ion complex and the transition-metal-promoted photoreaction. Reproduced with permission from [69], © 2011, John Wiley & Sons.

catalysis at the cucurbituril rim and the concomitant exploitation of the same macrocycles as inverse phase-transfer catalysts [69].

In a recent study, Zhang et al. have reported the use of cucurbit[7]uril in the biphasic oxidation of alcohols by using 2,2,6,6- tetramethylpiperidin-1-oxyl (TEMPO) as the catalyst and NaClO as the oxidant [70]. The proposed first step is the oxidation of TEMPO and the formation of key cationic intermediate ($TEMPO^+$), which can undergo unwanted side reactions in water. The use of a supramolecular TEMPO/CB7 complex resulted in higher conversion to the corresponding aldehyde as a consequence of the electrostatic stabilization of $TEMPO^+$ in its cavity [70].

In another study, Herrmann et al. have reported a novel design of a three-component supramolecular CB*n*-based system (amino acids, Cu^{2+} ions, and CB8) that creates a chiral nanoreactor [71]. The combination of amino acids as a source of chirality, Cu^{2+} as the catalytically active site and CB8 as the molecular container resulted in an asymmetric Lewis-acid catalyst for the Diels-Alder reaction of azachalcone with cyclopentadiene [71]. After optimization of the internal volume, this catalytic system is able to catalyze the Diels-Alder reaction with high enantio-selectivities (up to 92% ee), whereas the reactions in absence of cavity yields the racemic product. Apart from the high selectivities, improved activities were also obtained.

In a photochemical reaction, Sivaguru et al. have reported the use of CB8 for controlling the photoreactivity of coumarin derivatives such as 6-methylcoumarin [5, 72]. The photodimerization

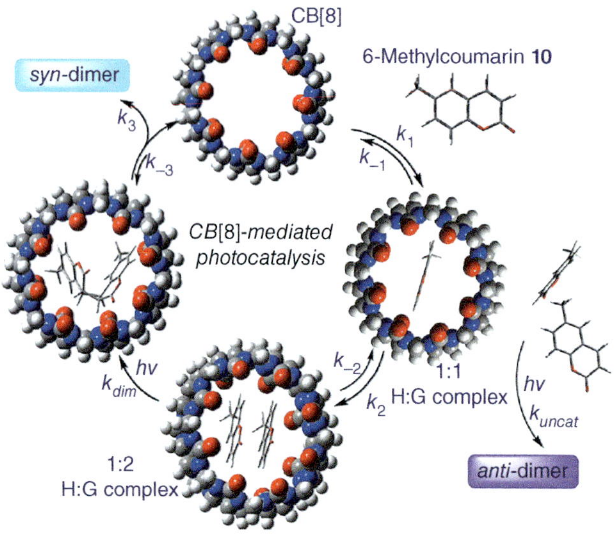

Figure 19.16 Catalytic cycle for photodimerization of 6-methylcoumarin mediated by CB8. Reproduced with permission from [5], © 2012, John Wiley & Sons.

was found to be inefficient (<10% conversion, 60 min irradiation) in water in the absence of CB8 with both syn- and anti-adducts as photoproducts (Figure 19.16). On the contrary, in the presence of CB8 the reaction (60 min irradiation) was found to be clean and efficient with syn adducts as the exclusive photoproduct with a syn-HH (head-to-head)/syn-HT (head-to-tail) ratio of about 70:30 (Figure 19.16) [5, 72]. In another photochemical reaction, Ramamurthy and co-workers have shown the controlled photodimerization of trans-1,2-bis(n-pyridyl)-ethylene E (n = 2, 3, or 4) and trans-stilbazole [73]. In the presence of 50 mol% of CB8, UV irradiation of both trans-1,2-bis(n-pyridyl)ethylenes (E) and stilbazoles (E) gave the corresponding syn-4 photodimer exclusively. On the other hand, UV irradiation of trans-1,2-bis(n-pyridyl)-ethylene (E) in solution leads to the formation of photoisomerized (E) isomer along with cyclized product and the hydration product.

19.3 Conclusion

Studies on cucurbituril-encapsulated molecular assemblies of small organic guest molecules has gained much attention because of the significant and stimuli-responsive modulation in their physico-chemical properties, which have tremendous prospects in the development of custom-designed materials for constructing sensors, on-off switches, drug delivery, therapeutics, catalysis, and other technological applications. In line with this, this chapter describes the effect of confinement on the physicochemical properties of certain chromophoric dyes/drugs within cucurbit[*n*]uril macrocycles and their prospective applications. Mechanistically, the macrocyclic confinement of the guest moiety and the stoichiometric compositional variations feasible within the host-guest assembly, brings out remarkable modulation in the molecular characteristics such as, excited state properties, fluorescence, aggregation behavior, solubility, stability, prototropic equilibrium, catalytic performance etc. Enhanced luminescence and photostability of $CH_3NH_3PbBr_3$ perovskite nanoparticles and enhanced antibacterial activity and extended shelf-life of fluoroquinolone drugs with cucurbit[7]uril have been achieved following convenient experimental procedures. Moreover, the subtle differences in the noncovalent interactions in the confined environment of the assembly have also been exploited to demonstrate various technological and biological applications such as aqueous dye laser, fluorescence-based sensor, molecular capsule, drug delivery vehicles, nanoreactors, catalysts etc. The variation of cavity dimensions available in the cucurbituril homologues and scope to introduce functionalized CB*n* or custom designed chromophoric guests offer new opportunities for controlled/targeted applications, particularly for cellular activation or therapeutics, modulate electronic/optical properties, molecular recognition and sensing, and other challenges in the field of supramolecular chemistry.

Acknowledgement

We acknowledge the co-authors of our published papers cited in the references. We thank Dr. A.K. Tyagi, Associate Director, Chemistry Group, BARC, India, for his constant encouragement and support.

References

1. Florea, M. and Nau, W.M. (2011). Strong binding of hydrocarbons to cucurbituril probed by fluorescent dye displacement: A supramolecular gas-sensing ensemble. Angewandte Chemie International Edition 50: 9338–9342.
2. Shaikh, M., Mohanty, J., Bhasikuttan, A.C. et al. (2008) Salt-induced guest relocation from a macrocyclic cavity into a biomolecular pocket: interplay between cucurbit[7]uril and albumin. Chemical Communications, 3681–3683.
3. Barooah, N., Kunwar, A., Khurana, R. et al. (2017). Stimuli-responsive cucurbit[7]uril-mediated BSA nanoassembly for uptake and release of doxorubicin. Chemistry–An Asian Journal 12: 122–129.
4. Khurana, R., Barooah, N., Bhasikuttan, A.C., and Mohanty, J. (2017). Modulation in the acidity constant of acridine dye with cucurbiturils: stimuli-responsive pK_a tuning and dye relocation into live cells. Organic & Biomolecular Chemistry 15: 8448–8457.
5. Pemberton, B.C., Raghunathan, R., Volla, S., and Sivaguru, J. (2012). From containers to catalysts: Supramolecular catalysis within cucurbiturils. Chemistry-A European Journal 18: 12178–12190.
6. Assaf, K.I. and Nau, W.M. (2015). Cucurbiturils: from synthesis to high-affinity binding and catalysis. Chemical Society Reviews 44: 394–418.
7. Mohanty, J., Pal, H., Ray, A.K. et al. (2007). Supramolecular dye laser with cucurbit[7]uril in water. ChemPhysChem 8: 54–56.
8. Mohanty, J., Jagtap, K., Ray, A.K. et al. (2010). Molecular encapsulation of fluorescent dyes affords efficient narrow-band dye laser operation in water. ChemPhysChem 11: 3333–3338.
9. Khurana, R., Agarwalla, S., Sridhar, G. et al. (2018). Ultra-bright rhodamines with sulfobutylether-β-cyclodextrin: A viable supramolecular dye laser in aqueous medium. ChemPhysChem 19: 2349–2356.
10. Lakowicz, J.R. (2006). Principles of Fluorescence Spectroscopy. New York: Springer.
11. Dsouza, R.N., Pischel, U., and Nau, W.M. (2011). Fluorescent dyes and their supramolecular host/guest complexes with macrocycles in aqueous solution. Chemical Reviews 111: 7941–7980.

12 Mohanty, J., Bhasikuttan, A.C., Nau, W.M., and Pal, H. (2006). Host-guest complexation of neutral red with macrocyclic host molecules: Contrasting pK_a shifts and binding affinities for cucurbit[7]uril and b-cyclodextrin. The Journal of Physical Chemistry B 110: 5132–5138.

13 Mohanty, J. and Nau, W.M. (2005). Ultrastable rhodamine with cucurbituril. Angewandte Chemie International Edition 44: 3750–3754.

14 Nau, W.M. and Mohanty, J. (2005). Taming fluorescent dyes with cucurbituril. International Journal of Photoenergy 7: 133–141.

15 Mohanty, J., Dutta Choudhury, S., Upadhyaya, H.P. et al. (2009). Control of the supramolecular excimer formation of thioflavin T within a cucurbit[8]uril host: A fluorescence on/off mechanism. Chemistry-A European Journal 15: 5215–5219.

16 Mohanty, J., Thakur, N., Dutta Choudhury, S. et al. (2012). Recognition-mediated light-up of thiazole orange with cucurbit[8]uril: Exchange and release by chemical stimuli. The Journal of Physical Chemistry B 116: 130–135.

17 Barooah, N., Mohanty, J., and Bhasikuttan, A.C. (2015). Cucurbit[8]uril-templated H and J dimers of bichromophoric coumarin dyes: origin of contrasting emission. Chemical Communications 51: 13225–13228.

18 Sayed, M., Biedermann, F., Uzunova, V.D. et al. (2015). Triple emission from *p*-dimethylaminobenzonitrile-cucurbit[8]uril triggers the elusive excimer emission. Chemistry-A European Journal 21: 691–696.

19 Bhasikuttan, A.C., Pal, H., and Mohanty, J. (2011). Cucurbit[*n*]uril based supramolecular assemblies: Tunable physico-chemical properties and their prospects. Chemical Communications 47: 9959–9971.

20 Vriezema, D.M., Comellas, A.M., Elemans, J.A.A.W. et al. (2005). Self-assembled nanoreactors. Chemical Reviews 105: 1445–1490.

21 Turro, N.J. (2005) Molecular structure as a blueprint for supramolecular structure chemistry in confined spaces. Proceedings of the National Academy of Sciences of the United States of America, 102, 10766–10770.

22 Comba, P. and Martin, B. (2005). Macrocyclic Chemistry: Current Trends and Future Perspectives, 303–325. Netherlands: Springer.

23 Loftsson, T., Jarho, P., Másson, M., and Järvinen, T. (2005). Cyclodextrins in drug delivery. Expert Opinion on Drug Delivery 2: 335–351.

24 Fromming, K.H. and Szejtli, J. (1994). Cyclodextrins in Pharmacy, Kluwer Academic. Dordrecht.

25 Rekharsky, M.V. and Inoue, Y. (1998). Complexation thermodynamics of cyclodextrins. Chemical Reviews 98: 1875–1918.

26 Houk, K.N., Leach, A.G., Kim, S.P., and Zhang, X. (2003). Binding affinities of host-guest, protein-ligand, and protein-transition-state complexes. Angewandte Chemie International Edition 42: 4872–4897.

27 Chatjigakis, A.K., Donze, C., Coleman, A.W., and Cardot, P. (1992). Solubility behavior of beta-cyclodextrin in water/cosolvent mixtures. Analytical Chemistry 64: 1632–1634.

28 Gutsche, C.D. (2008). Calixarenes: An Introduction. Cambridge: Royal Society of Chemistry.

29 Steed, J.W. and Atwood, J.L. (eds.) (2009). Supramolecular Chemistry. Chichester, U.K.: John Wiley & Sons Ltd.

30 Shinkai, S., Mori, S., Koreishi, H. et al. (1986). Hexasulfonated calix[6]arene derivatives: a new class of catalysts, surfactants, and host molecules. Journal of the American Chemical Society 108: 2409–2416.

31 Rodik, R.V., Boyko, V.I., and Kalchenko, V.I. (2009). Calixarenes in bio-medical researches. Current Medicinal Chemistry 16: 1630–1655.

32 Lee, J.W., Samal, S., Selvapalam, N. et al. (2003). Cucurbituril homologues and derivatives: new opportunities in supramolecular chemistry. Accounts of Chemical Research 36: 621–630.

33 Bhasikuttan, A.C., Pal, H., and Mohanty, J. (2011). Cucurbit[*n*]uril based supramolecular assemblies: tunable physico-chemical properties and their prospectus. Chemical Communications 47: 9959–9971.

34 Masson, E., Ling, X., Joseph, R. et al. (2012). Cucurbituril chemistry: a tale of supramolecular success. RSC Advances 2: 1213–1247.

35 Lagona, J., Mukhopadhyay, P., Chakrabarti, S., and Isaacs, L. (2005). The cucurbit[*n*]uril family. Angewandte Chemie International Edition 44: 4844–4870.

36 Cheng, X.-J., Liang, L.-L., Chen, K. et al. (2013). Twisted cucurbit[14]uril. Angewandte Chemie International Edition 52: 7252–7255.

37 Liu, S., Ruspic, C., Mukhopadhyay, P. et al. (2005). The cucurbit[*n*]uril family: prime components for self-sorting systems. Journal of the American Chemical Society 127: 15959–15967.

38 Bhasikuttan, A.C., Dutta Choudhury, S., Pal, H., and Mohanty, J. (2011). Supramolecular assemblies of thioflavin T with cucurbiturils: Prospects of cooperative and competitive metal ion binding. Israel Journal of Chemistry 51: 634–645.

39 Barooah, N., Mohanty, J., Pal, H., and Bhasikuttan, A.C. (2011). Supramolecular assembly of hoechst-33258 with cucurbit[7]uril macrocycle. Physical Chemistry Chemical Physics 13: 13117–13126.

40 Barooah, N., Bhasikuttan, A.C., Sudarsan, V. et al. (2011). Surface functionalized silver nanoparticle conjugates: demonstration of uptake and release of a phototherapeutic porphyrin dye. Chemical Communications 47: 9182–9184.

41 Dutta Choudhury, S., Barooah, N., Aswal, V.K. et al. (2014). Stimuli-responsive supramolecular micellar assemblies of cetylpyridinium chloride with cucurbit[5/7]urils. Soft Matter 10: 3485–3493.

42 Fang, X., Kogerler, P., Isaacs, L. et al. (2009). Cucurbit[*n*]uril#polyoxoanion hybrids. Journal of the American Chemical Society 131: 432–433.

43 Goel, T., Barooah, N., Mallia, M.B. et al. (2016). Recognition-mediated cucurbit[7]uril-heptamolybdate hybrid material: A facile supramolecular strategy for ^{99m}Tc separation. Chemical Communications 52: 7306–7309.

44 Shaikh, M., Dutta Choudhury, S., Mohanty, J. et al. (2010). Contrasting guest binding interaction of cucurbit[7–8]urils with neutral red dye: controlled exchange of multiple guests. Physical Chemistry Chemical Physics 12: 7050–7055.

45 Gonzalez-Carrero, S., Francés-Soriano, L., González-Béjar, M. et al. (2016). The luminescence of $CH_3NH_3PbBr_3$ perovskite nanoparticles crests the summit and their photostability under wet conditions is enhanced. Small 12: 5245–5250.

46 El-Sheshtawy, H.S., Chatterjee, S., Assaf, K.I. et al. (2018). A supramolecular approach for enhanced antibacterial activity and extended shelf-life of fluoroquinolone drugs with cucurbit[7]uril. Scientific Reports 8: 13925. (1–10).

47 Dutta Choudhury, S., Mohanty, J., Pal, H., and Bhasikuttan, A.C. (2010). Cooperative metal ion binding to a cucurbit[7]uril-thioflavin T complex: Demonstration of a stimulus-responsive fluorescent supramolecular capsule. Journal of the American Chemical Society 132: 1395–1401.

48 Biedermann, F., Elmalem, E., Ghosh, I. et al. (2012). Strongly fluorescent, switchable perylene bis(diimide) host-guest complexes with cucurbit[8]uril in water. Angewandte Chemie International Edition 51: 7739–7743.

49 Shinde, M.N., Dutta Choudhury, S., Barooah, N. et al. (2015). Metal-ion-mediated assemblies of thiazole orange with cucurbit[7]uril: A photophysical study. The Journal of Physical Chemistry B 119: 3815–3823.

50 Eggeling, C., Widengren, J., Rigler, R., and Seidel, C.A.M. (1998). Photobleaching of fluorescent dyes under conditions used for single-molecule detection: Evidence of two-step photolysis. Analytical Chemistry 70: 2651–2659.

51 Mohanty, J. and Nau, W.M. (2004). Refractive index effects on the oscillator strength and radiative decay rate of 2,3-diazabicyclo[2.2.2]oct-2-ene. Photochemical & Photobiological Sciences 3: 1026–1031.

52 Ghosh, I. and Nau, W.M. (2012). The strategic use of supramolecular pK_a shifts to enhance the bioavailability of drugs. Advanced Drug Delivery Reviews 64: 764–783.

53 Saleh, N., Koner, A.L., and Nau, W.M. (2008). Activation and stabilization of drugs by supramolecular pK_a shifts: Drug-delivery applications tailored for cucurbiturils. Angewandte Chemie International Edition 47: 5398–5401.

54 Wang, R. and Macartney, D.H. (2008). Cucurbit[7]uril host–guest complexes of the histamine H_2-receptor antagonist ranitidine. Organic & Biomolecular Chemistry 6: 1955–1960.

55 Albini, A. and Monti, S. (2003). Photophysics and photochemistry of fluoroquinolones. Chemical Society Reviews 32: 238–250.

56 Emmerson, A.M. and Jones, A.M. (2003). The quinolones: decades of development and use. Journal of Antimicrobial Chemotherapy 51: 13–20.

57 Naiki, H., Higuchi, K., Hosokawa, M., and Takeda, T. (1989). Fluorometric determination of amyloid fibrils in vitro using the fluorescent dye Thioflavin T. Analytical Biochemistry 177: 244–249.

58 Mathis, C.A., Bacskai, B.J., Kajdasz, S.T. et al. (2002). A lipophilic thioflavin-T derivative for positron emission tomography (PET) imaging of amyloid in brain. Bioorganic & Medicinal Chemistry Letters 12: 295–298.

59 Groenning, M., Olsen, L., Weert, M. et al. (2007). Study on the binding of Thioflavin T to beta-sheet-rich and non-beta-sheet cavities. Journal of Structural Biology 158: 358–369.

60 Dutta Choudhury, S., Mohanty, J., Upadhyaya, H.P. et al. (2009). Photophysical studies on the noncovalent interaction of thioflavin T with cucurbit[*n*]uril macrocycles. The Journal of Physical Chemistry B 113: 1891–1898.

61 Lubitz, I., Zikich, D., and Kotlyar, A. (2010). Specific high-affinity binding of thiazole orange to triplex and G-quadruplex DNA. Biochemistry 49: 3567–3574.

62 Nygren, J., Svanvik, N., and Kubista, M. (1998). The interactions between the fluorescent dye thiazole orange and DNA. Biopolymers 46: 39–51.

63 Kandela, I.K., Bartlett, J.A., and Indig, G.L. (2002). Effect of molecular structure on the selective phototoxicity of triarylmethane dyes towards tumor cells. Photochemical & Photobiological Sciences: 309–314.

64 Bhasikuttan, A.C., Mohanty, J., Nau, W.M., and Pal, H. (2007). Efficient fluorescence enhancement and cooperative binding of an organic dye in a supra-biomolecular host–protein assembly. Angewandte Chemie International Edition 46: 4120–4122.

65 Gorl, D., Zhang, X., and Wurthner, F. (2012). Molecular assemblies of perylene bisimide dyes in water. Angewandte Chemie International Edition 51: 6328–6348.

66 Mock, W.L., Irra, T.A., Wepsiec, J.P., and Adhya, M. (1989). Catalysis by cucurbituril. The significance of bound-substrate destabilization for induced triazole formation. The Journal of Organic Chemistry 54: 5302–5308.

67 Freeman, W.A., Mock, W.L., and Shih, N.-Y. (1981). Cucurbituril. Journal of the American Chemical Society 103: 7367–7368.

68 Tuncel, D. and Steinke, J.H.G. (2002). The synthesis of [2], [3] and [4]rotaxanes and semirotaxanes. Chemical Communications: 496–497.

69 Koner, A.L., Marquez, C., Dickman, M.H., and Nau, W.M. (2011). Transition-metal-promoted chemoselective photoreactions at the cucurbituril rim. Angewandte Chemie International Edition 50: 545–548.

70 Jiao, Y., Tang, B., Zhang, Y. et al. (2018). Highly efficient supramolecular catalysis by endowing the reaction intermediate with adaptive reactivity. Angewandte Chemie International Edition 57: 6077–6081.

71 Zheng, L., Sonzini, S., Ambarwati, M. et al. (2015). Turning cucurbit[8]uril into a supramolecular nanoreactor for asymmetric catalysis. Angewandte Chemie International Edition 54: 13007–13011.

72 Barooah, N., Pemberton, B.C., and Sivaguru, J. (2008). Manipulating photochemical reactivity of coumarins within cucurbituril nanocavities. Organic Letters 10: 3339–3342.

73 Maddipatla, M.V.S.N., Kaanumalle, L.S., Natarajan, A. et al. (2007). Preorientation of olefins toward a single photodimer: Cucurbituril-mediated photodimerization of protonated azastilbenes in water. Langmuir 23: 7545–7554.

20

Box-Shaped Hosts: Evaluation of the Interaction Nature and Host Characteristics of ExBox Derivatives in Host-Guest Complexes from Computational Methods

Giovanni F. Caramori[1,] and Alvaro Muñoz-Castro[2,*]*

[1] *Departamento de Química, Universidade Federal de Santa Catarina, Campus Universitário Trindade, CP 476, Florianópolis, SC, Brazil*

[2] *Laboratorio de Química Inorgánica y Materiales Moleculares, Universidad Autonoma de Chile, Llano Subercaceaux 2801, San Miguel, Santiago, Chile*

20.1 Introduction

The ability to incorporate guest species in appropriate molecular host offers a unique opportunity to explore and learn about confinement effects provided by sterically and electronically restrained environments [1]. It has been shown that the inner binding region of efficient hosts molecules triggers the formation of unexpected species [2, 3], enables molecular recognition capabilities [4–7], and modifies guest reactivity [8–10], among others [11, 12].

Cyclophane structures are prototypical examples of well-defined host species [13–16] providing efficient environments for guest inclusion, allowing the rational design of complex supramolecular structures such as catenanes, rotaxanes, and other inclusion complexes based on the stabilizing contribution of non-covalent interactions [17, 18], enabling the proposal of different molecular scale interconnection machines [13, 19]. In particular, the formation of two-dimensional molecular cages involving two viologen units connected by two *p*-xylylene linkers provides the interaction between π-electron-poor host and π-electron-rich guests in different ratios owing to the cavity size characteristics [20–24]. The so-called "blue box" (cyclobis(paraquat-*p*-phenylene) or $CBPQT^{4+}$, Ex^0Box^{4+}) is a classic building block, showing a tetracationic electron-deficient cavity developed by Stoddart and collaborators [25, 26]. The Ex^nBox^{4+} host series (n=0,1,2,3) [23] with *n* *p*-phenylene bridge between pyridinium rings (Scheme 20.1) provides different host sizes suitable for the encapsulation of a range of π-electron-rich guests from small to larger species owing to a ~7 Å width and a variable length from ~11 to ~24 Å [20, 21, 23].

Ex^0Box^{4+} is able to incorporate tetrathiafulvalene (TTF) units, phenyl-rings, and chains [27–33], and derivatives, proving a valuable template for redox-controlled modulation of intermolecular interactions enabling selectively from π-electron-rich to π-electron-poor guests. For $ExBox^{4+}$, extending the box size in one *p*-phenylene bridge increases the cavity size, making it possible to incorporate larger polycyclic aromatic hydrocarbons (PAHs) [20, 34] such as corannulene ($C_{20}H_{10}$) and decreasing the barrier for its bowl-to-bowl inversion [35] and perylene, among others, with enhanced affinity as the number of π-electron increases [20]. $ExBox^{4+}$ shows a more favorable binding affinity for planar species owing to the destabilization given by the introduction of strain in the molecular structure, as given by the five-membered ring on corannulene. The

* Corresponding Author

Chemical Reactivity in Confined Systems: Theory, Modelling and Applications,
First Edition. Edited by Pratim Kumar Chattaraj and Debdutta Chakraborty.

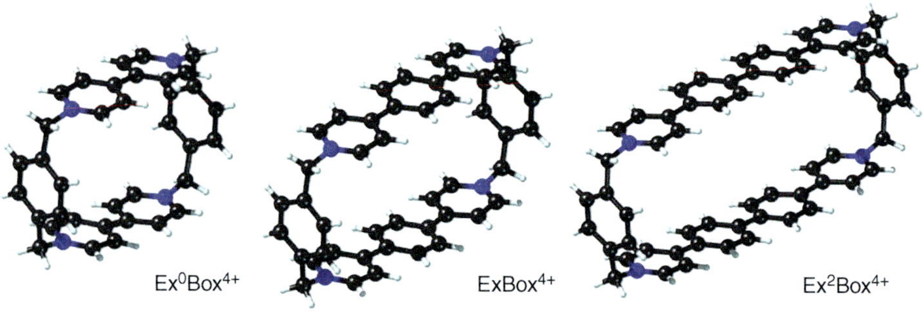

Scheme 20.1 Structures for isolated Ex^nBox^{4+} hosts (n=0, 1, 2).

addition of an extra *p*-phenylene bridge between pyridinium rings enables the incorporation of two guests simultaneously, besides the complete inclusion of more extended PAHs such as tetracene, tetraphene, and chrysene, among others [21]. This precludes the formation of larger ExBox derivatives, incorporating even C_{60} and C_{70} fullerenes as guests [36].

In this chapter, we will discuss the theoretical background of the binding nature involving noncovalent interactions between Ex^nBox^{4+} derivatives and different guest species, via density functional calculations, which are latter compared to other common host species.

20.2 Noncovalent Interactions Through Energy Decomposition Analysis

The role of noncovalent interactions in different chemical and biochemical processes is unquestionable [37, 38]. They appear in all kinds of matter (solid, liquid, and gas), controlling processes as recognition, transport, and catalysis. The understanding and predictions about how noncovalent interactions affect the behavior of chemical systems request to fathom their physical nature or the driving force behind them. In this sense, the use of energy decomposition analysis (EDA) schemes becomes crucial, since there is no quantum mechanical operator able to provide interaction energy values directly or still to decompose them into physical meaningful terms. It has been shown that EDA constitutes a very useful tool, able to provide computational insights about the physical nature of noncovalent interactions [39–42], since it decomposes intermolecular interaction energy values into terms such as electrostatic, polarization, repulsion, charge transfer, exchange, and correlation.

A plethora of energy decomposition schemes has been proposed [43, 44]. The most popular are variational approaches [45–47], derived from Morokuma's work [48] and the perturbation ones, based on SAPT (Symmetry Adapted Perturbation Theory) by Jeziorski and collaborators [49], which could be extended to large systems by the combination with DFT (SAPT-DFT) [50, 51]. Indeed, other alternative methods have been proposed as the Block-Localized Wave function (BLW) [52] and the Generalized Product Function (GPF) [53], both based on Valence Bond (VB) theory. It is important to note that the variational EDA schemes are not restricted to decompose DFT or HF interaction energies, but also MP2 [54, 55] or CCSD(T) [56–58] energies. All these different approaches stem from the efforts to overcome limitations of the original propositions and to improve both the physical meaning and chemical implications of the obtained terms, since they are not defined in a unique way.

A variational decomposition analysis, based on DFT, which has proved to be very useful to investigate interactions and bonding situations in large systems, containing more than 100 atoms, is the EDA-NOCV [59], which combines the traditional EDA [60] with natural orbitals for chemical valence (NOCV) [61]. It has become a workhorse for describing noncovalent interactions in diverse chemical systems [62–69]. For that reason, its theoretical basis is described below.

In the EDA-NOCV, the interaction energy between two fragments A and B of an interacting system AB is decomposed in a Ziegler-Rauk-like scheme [45]. The formation of the system AB with the corresponding wave function Ψ_{AB} and energy E_{AB}can be rationalized in different stages. Initially, the isolated fragments, in their geometric (A^0,B^0) and electronic (Ψ_A^0,Ψ_B^0) ground-states with energies (E_A^0,E_B^0), are placed close together (in the geometry of the interacting systems), it causes geometrical distortions ($A^0 \rightarrow A$, $B^0 \rightarrow B$) that lead to changes in wave functions ($\Psi_A^0 \rightarrow \Psi_A$,$\Psi_B^0 \rightarrow \Psi_B$) and energies ($E_A^0 \rightarrow E_A$,$E_B^0 \rightarrow E_B$), respectively. The amount of energy necessary to distort the isolated fragments to the final (interacting) stage is called preparation energy (Eq. 20.1).

$$\Delta E^{prep} = E_A - E_A^0 + E_B - E_B^0 \tag{20.1}$$

The EDA is focused on the analysis of instantaneous interaction ΔE^{int}(Eq. 20.2), which when added to the preparation energy leads to the dissociation energy (Eq. 20.3)

$$\Delta E^{int} = E_{AB} - E_A - E_B \tag{20.2}$$

$$\Delta E^{int} + \Delta E^{prep} = -D_e \tag{20.3}$$

In the first step of EDA, the distorted fragments are considered with frozen charge densities, and the electronic state is simply the product $\Psi_A\Psi_B$with energy E_{AB}^0. The interaction between prepared fragments with frozen densities leads to the quasiclassical Coulomb interaction (Eq. 20.4), which in general is attractive, but in some particular cases it can also be repulsive [70].

$$\Delta E^{elstat} = \sum_{\alpha \in A}\sum_{\beta \in B} \frac{Z_\alpha Z_\beta}{R_{\alpha\beta}} + \int dr V_B(r)\rho_A(r) + \int dr V_A(r)\rho_B(r) + \iint dr_1 dr_2 \frac{\rho_A(r_1)\rho_B(r_2)}{r_{12}} \tag{20.4}$$

In a second step, the product $\Psi_A\Psi_B$is renormalized and antisymmetrized to give an intermediate state Ψ^0 with energy E^0 that obeys the Pauli principle (Eq. 20.5). The difference ($E_{AB}^0 - E^0$) is called Pauli repulsion ΔE^{Pauli} (Eq. 20.6).

$$\Psi^0 = N\hat{A}\{\Psi_A\Psi_B\} \tag{20.5}$$

$$\Delta E^{Pauli} = E_{AB}^0 - E^0 \tag{20.6}$$

In a third step, the renormalized and antisymmetrized wave function is relaxed, leading to the molecular final state $\Psi_{AB}$of the interacting system AB with energy E_{AB}. This relaxation causes energy lowering that encompasses charge transfer (interactions between occupied molecular orbitals of one fragment with the unoccupied orbitals of the other) and polarization (empty and occupied orbital mixing on the same fragment) processes. It is called orbital interaction (Eq. 20.7). These contributions are summed up in ΔE^{int} (Eq. 20.8). When atom-pairwise empirical dispersion corrections are used, the difference in dispersion correction between the interacting system AB and its isolated fragments A and B can be obtained and included into ΔE^{int}, leading to a dispersion-corrected interaction energy ΔE_{dc}^{int} (Eq. 20.9).

$$\Delta E^{orb} = E_{AB} - E_{AB}^0 \tag{20.7}$$

$$\Delta E^{int} = \Delta E^{elstat} + \Delta E^{Pauli} + \Delta E^{orb} \tag{20.8}$$

$$\Delta E_{dc}^{int} = \Delta E^{elstat} + \Delta E^{Pauli} + \Delta E^{orb} + \Delta E^{disp} \tag{20.9}$$

The EDA-NOCV scheme not only provides an energy decomposition analysis, but also a charge decomposition with the differential density based on natural orbitals for chemical valence (NOCVs) [71]. The differential or deformation density $\Delta\rho(r)$ is the difference between the densities of the fragments before and after the interaction. The total charge deformation$\Delta\rho_k(r)$ can be expressed in terms of pairwise contributions $\Delta\rho_k(r)$ coming from complementary pairs of interacting NOCVs. The pictures of individual deformation densities, $\Delta\rho_k(r)$, is very useful since they show clearly the charge reorganization during the interaction. For instance, density channels can be graphically plotted showing inflow and outflow of electronic density during the formation of the interaction. The EDA-NOCV scheme also provides the energy stabilization associated with each pairwise orbital interactions ΔE_k^{orb} involved with the individual deformation densities, $\Delta\rho_k(r)$, which when added together recover the total orbital term (Eq. 20.10) Further information and a more detailed description of EDA-NOCV is found in the original work of Mitoraj [59].

$$\Delta E^{orb} = \sum \Delta E_k^{orb} \tag{20.10}$$

20.3 Ex^0Box^{4+} ($CBPQT^{4+}$)

The earlier formation of Ex^0Box^{4+} aggregates incorporating a tetrathiofulvalene (TTF) unit [32, 72, 73], and derivatives, [74] exemplifies the formation of π-electron-poor and π-electron-rich interaction between such host and guest species, respectively [20–24] (Figure 20.1). Olea-Ulloa [75] and coworkers, explore the nature of the host-guest interaction for $TTF \subset Ex^0Box^{4+}$ species via DFT calculations at the Triple-ξ Slater basis set plus two polarization functions (TZ2P) within generalized gradient approximation (GGA) according to the Perdew–Burke–Ernzerhof (PBE) functional [76] level (TZ2P/PBE) at different scenarios accounting for gas-phase model, given by: (i) $TTF \subset Ex^0Box^{4+}$ pair; for the explicit inclusion of counterions, (ii) $[TTF \subset Ex^0Box][(PF_6)_4]$, and, inclusion of counterions and solvation, via (iii) $[TTF \subset Ex^0Box][(PF_6)_4]$ embedded in a continuum approach for solvation with acetonitrile as solvent (COSMO) [77].

The formation of $TTF \subset Ex^0Box^{4+}$ shows small structural variations in comparison to the isolated host and guest parents, resulting in slight preparation energy (ΔE^{prep}) of about ~3 kcal/mol, in both gas-phase model (i), i.e. avoiding both PF_6^- counterions and solvation effects; and in the $[TTF \subset Ex^0Box][(PF_6)_4]$ model (ii) excluding just the solvation effects solely, which is given mainly by a planarization of the paraquat moieties. This ΔE^{prep} is given mostly from the slight planarization of viologen moieties. The inclusion of solvent effects increases to 13.79 kcal/mol the ΔE^{prep}, owing to the variation of the molecular cavity into the COSMO scheme.

The dispersion corrected interaction energy (ΔE_{dc}^{int}) obtained for the simplest model (i) amounts to −36.91 kcal/mol, which decreases when the counterions are taken into account (ii) to

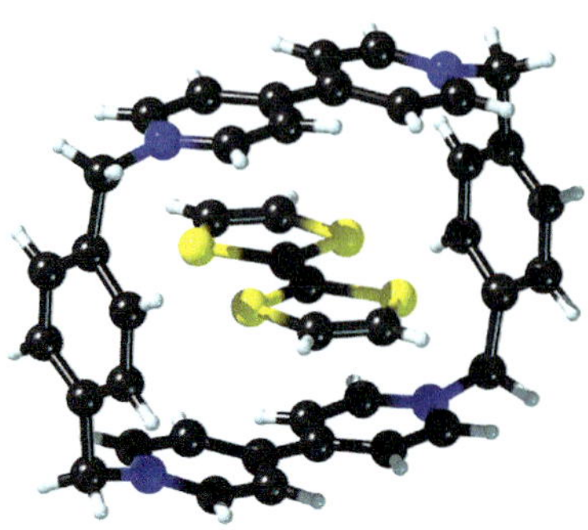

Figure 20.1 Structures for $TTF \subset Ex^0Box^{4+}$.

−24.97 kcal/mol, and to −22.77 kcal/mol when solvation effects are additionally involved, given mostly by the variation of the ΔE^{orb} term from charge transfer and polarization effects.

The total π-donor/π-acceptor interaction energy obtained as $\Delta E^{tot} = \Delta E^{prep} + \Delta E_{dc}{}^{int}$, vary along models (i), (ii) and (iii), according to −31.69, −22.00, and −8.98 kcal/mol, respectively. The ΔE^{disp} term is the most relevant term accounting from 39.5% in (i) to 50.1% in (iii) from the overall stabilizing energy (ΔE^{elstat}, ΔE^{orb}, and ΔE^{disp}). For comparison, the experimental formation energy determined by the UV–Vis dilution method indicates a complexation amounting to about −5 kcal/mol in acetonitrile [73], which is in the range of the calculated for model (iii). This suggests that in the evaluation of charged species the binding strength for this host-guest species, both solvent and counterions should be taken into account, and different models are able to account for different stabilizing contributions. The variation of such stabilizing terms along with different charge states for TTF⊂Ex^0Box, and related derivatives, have not been studied in detail to date, which is an interesting issue for further studies.

Jeppensen and coworkers, in a combined experimental and theoretical research effort, investigated the role of glycol chain functionalization for monopyrrolo-tetrathiafulvalene (MPTTF) species in the binding interaction with Ex^0Box^{4+} cage, leading to [2]pseudorotaxanes [78, 79]. Different lengths of glycol chains were involved as sidearms of the MPTTF-derivative backbone, providing from one to four oxygen atoms as external groups able to interact with the external region of the host (Figure 20.2). The experimental binding constants (K_a) for such [2]pseudorotaxanes were reported on the basis of UV–vis–NIR with structural characterization via ^{1}H NMR spectroscopy, denoting an increase along the series with its maximum when two oxygen atoms are included as sidearms [78, 79], which does not increase for cases with three and four glycol units. Aided by theoretical calculations at the meta-GGA 6-31G**/M06-L level embedded in an implicit solvent model [80], the authors explored the origin of this truncated trend, showcasing the formation of both C–H···π], and [C–H···O] interactions in the overall host-guest pair owing to TTF-moiety and glycol chain noncovalent interactions to the Ex^0Box^{4+} structure. For the shorter length chains, one and two oxygen atoms are allowed to interact efficiently towards one specific exo-hydrogen from the viologen section, namely the acidic Hα-protons of Ex^0Box^{4+}, whereas larger glycol chains are not flexible enough to rotate avoiding further [C–H···O] interactions for larger members.

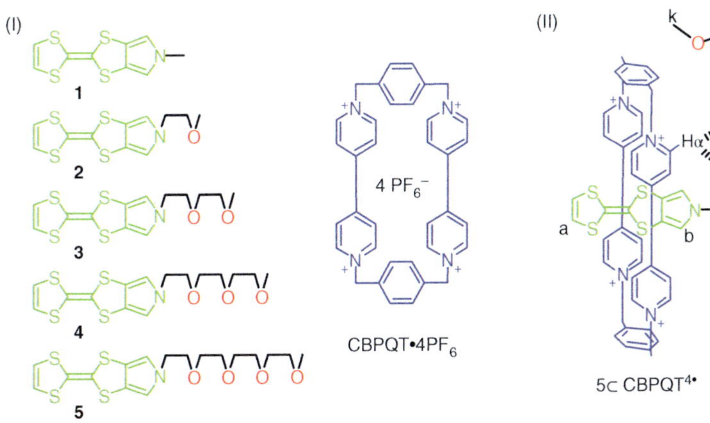

Figure 20.2 I) Different MPTTF derivatives (1-5) incorporated in the Ex^0Box^{4+} ($CBPQT^{4+}$) cavity, and 5⊂Ex^0Box^{4+} (II), denoting the [C–Hα···O] interaction. Reprinted (adapted) with permission from (Jeppesen et. al, J. Org. Chem. 2017, 82, 3, 1371-1379), © 2017, American Chemical Society.

As a result, the enhanced experimental binding constants observed along the series is attributed to the complementary stabilized given by the combination of charge transfer, [C–H···π] from the TTF-derivative, and extra [C–H···O] interactions involving glycol chains up to two oxygen atoms in such sidearm functionalization.

20.4 ExBox^{4+} and Ex^{2}Box^{4+}

The incorporation of PAHs guest species in ExBox^{4+} and Ex^{2}Box^{4+} given by the larger cavity size obtained by the incorporation of *p*-phenylene ring between bipyridinium moieties (Figure 20.3) shows the rational modification of cavity characteristics in the ExBox series, leading to the incorporation of guests with increased sizes and ratios [20–24].

The capabilities of ExBox^{4+} to incorporate several PAHs have been demonstrated by Stoddart and coworkers in 2013 [20, 35], by the introducing naphthalene, azulene, anthracene, phenanthrene, pyrene, tetracene, tetraphene, chrysene, helicene, triphenylene, perylene, coronene, and corannulenne, among others. For the corannulene case, its fit inside ExBox^{4+} induce a decrease in its bowl-depth from 0.87, in its isolated form, to 0.84 Å in the corannulene⊂ExBox^{4+}, leading to a decrease in the bowl-to-bowl inversion barrier from 11.5 to 7.9 kcal/mol, owing to the

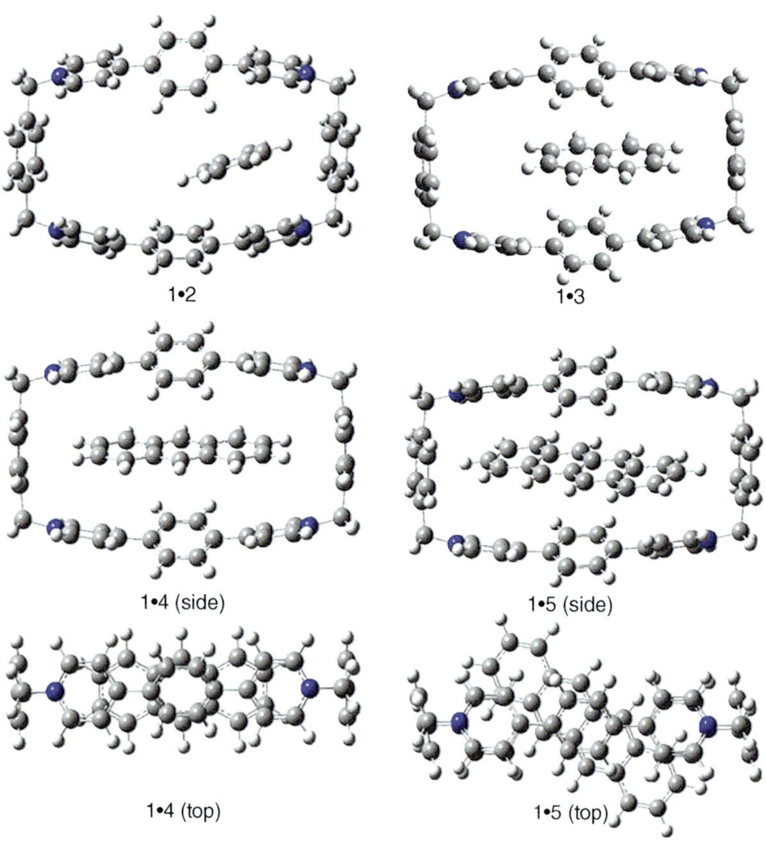

Figure 20.3 Most favorable optimized isomers for ExBox^{4+} (1) with benzene (2), naphthalene (3), anthracene (4), and, tetracene (5). Reprinted with permission from (Bachrach, J. Phys. Chem. A 2013, 117, 35, 8484-8491), © 2013, American Chemical Society.

destabilization of the relaxed structure and stabilization of the transition state [35] accounting for the modification of its behavior when is confined in the ExBox^{4+} cage.

Bachrach explored computationally the features involved in the ExBox^{4+} cage (**1**) and its related host capabilities towards linear acenes given by, benzene (**2**), naphthalene (**3**), anthracene (**4**), and, tetracene (**5**), at the ωB97X-D/6-311(d,p) level of theory [81] (Figure 20.3). It was found that the *o*-*o*'-CH interactions in the viologen unit give rise a ring strain energy of 20.2 kcal/mol, accounting for the non-planar structural arrangement of such groups in the ExBox^{4+} host. Interestingly, by removal of these *o*-*o*'-CH interactions in N-substituted viologen counterparts, the dihedral angle in such group is effectively reduced from ~38° in ExBox^{4+}, to ~10° and < 1° in the two proposed species, resulting in coplanar host with improved host-guest capabilities with more efficient π–π interactions [82].

The computational survey of host-guest binding energy of the ExBox^{4+} cage and benzene, reveals several conformations owing to the loose fit of C_6H_6 inside the cavity [81]. The different conformers lead to a wide range of binding free energies from two repulsive structures to stabilizing situations, from +6.53 to −7.41 kcal/mol, in gas-phase. When the solvation effects are incorporated via continuum solvation models (Conductor-like polarizable continuum model, CPCM), the binding energy decreases slightly to −4.88 kcal/mol for the most favored isomer with similar structure to that found for gas-phase calculations for the lower energy isomers (Figure 20.1).

In line with the increase of fused-ring number, the binding energy increases from −4.9 for benzene (**1•2**), to −13.7 naphthalene (**1•3**), −18.8 for anthracene (**1•4**), and −20.1 kcal/mol for tetracene (**1•5**) complex. This finding reveals that as the number of interacting π-electrons, the stabilizing energy is greater, and thus, the encapsulation processes is more exergonic. Hence, as the interacting π-surface increases, larger is the stabilization of the host-guest pair. The different orientations for such species reveal their preference to maximize the π–π surface interactions between host and guest [81].

For the naphthalene (**1•3**) and anthracene (**1•4**) cases, the acenes are located parallel to the viologen units denoting a better, showing an increase of about −5 kcal/mol in the stabilization of **1•3** in comparison to the benzene case. Moreover, **1•4** is stabilized by about −5 kcal/mol in comparison to the naphthalene case, owing to the larger extent of the π-surface. However, for tetracene, this stabilization increases by about of −2 kcal/mol in comparison to anthracene, due to that despite of the larger number of π-electrons, i.e. larger π-surface, the ExBox^{4+} cavity is not large enough to incorporate the four-fused rings in a completely parallel conformation, resulting in a decrease of the interacting π-surface given by the guest.

Das and Chattaraj via ωB97x-D/6-311G(d,p) calculations extend the insights into the host-guest capabilities of ExBox^{4+} towards hydrocarbons and azine rings [83]. The region compromising π–π interactions is showed by using noncovalent interaction index (NCI) [84], which is a useful approach to reveal the location of these interactions supporting the stabilization of the host-guest complex. The solvent effects were also incorporated via continuum solvation models. The authors observe a gradually decrease in the binding energy, as the number of nitrogen atoms is placed instead of C-H group in the ring, reaching in some cases, to a repulsive non-favorable situation, as the case for 1,2,4-triazine and tetrazine (Figure 20.4). The authors confirmed that electron-rich guests interact preferentially with pyridinium moieties, where benzene and pyridine are located in parallel displaced conformations, while guests like pyrimidine prefer a T-shape manner, discussing that the stability and aromaticity are important factors that may influence the extension of the interaction [83].

The obtained features are in line with the results provided by Bachrach [81], suggesting the decrease in the calculated binding energy when solvation effects are taken into account. The decrease in the favorable guest encapsulation by ExBox^{4+} as the N-substituted benzenes involves

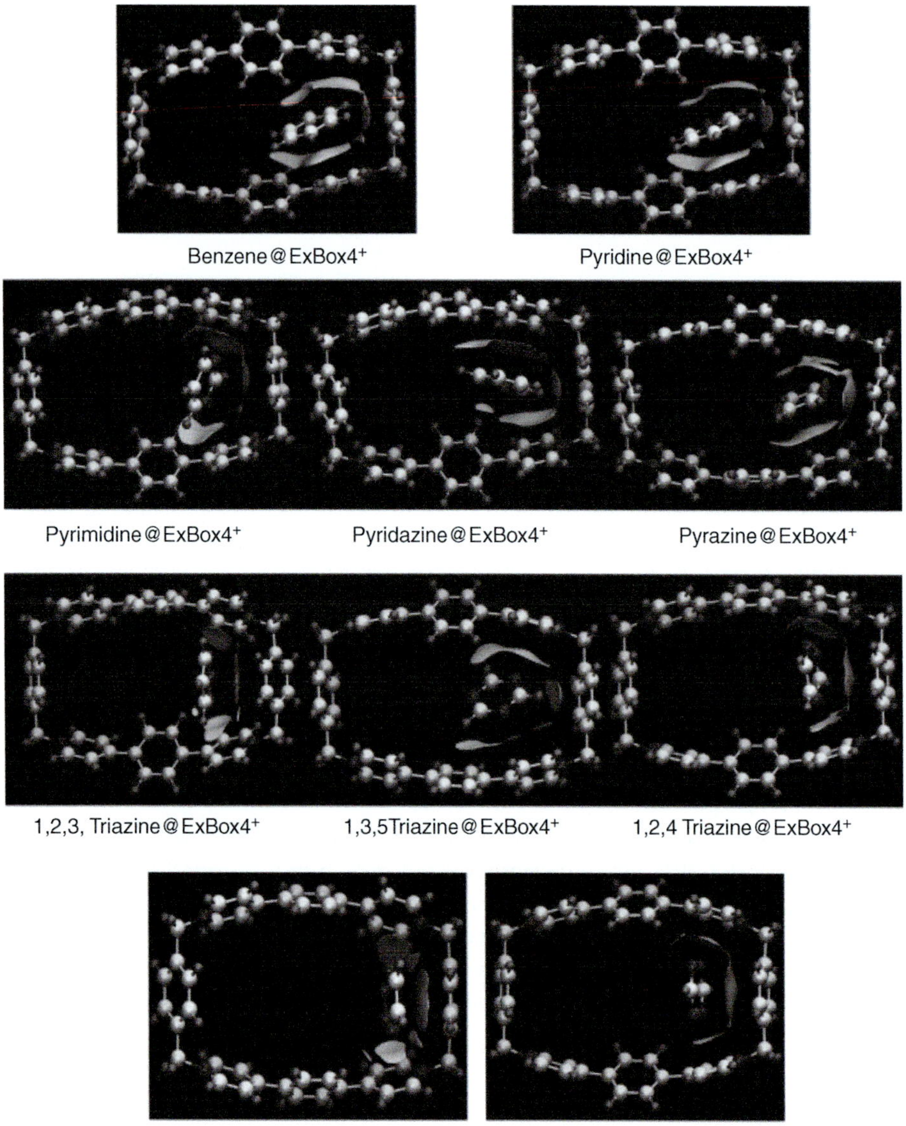

Figure 20.4 3D NCI plots for the guests and the ExBox^{4+} complex. Surfaces at gradient cutoff s=0.4 au. Source: Das, R., and Chattaraj, P.K. (2014) Host-Guest Interactions inExBox^{4+}. ChemPhysChem, 15 (18), 4108–4116. © 2016, John Wiley & Sons.

more nitrogen atoms from pyridine, pyrimidine, pyridazine, pyrazine, trizine, and tetrazines, is obtained in the calculations including solvation effects. For models incorporating counterions (Cl^-), the resulting binding energy given by the reaction free energy (ΔG) is more stabilized in comparison to solution-phase calculations (acetonitrile and water), ranging from −18.2 to −5.7 kcal/mol for the former, in contrast to the range between −4.9 to +1.9 kcal/mol, for the later. This suggest, that in solid-state, ExBox^{4+} derivative are prone to incorporate N-substituted benzenes, which can involve different antibiotic and antitumor agent, encouraging their use in medicinal chemistry. The use of the NCI analysis provides an intuitive and useful representation of the π–π interactions acting in the formation of the host-guest species, which have been employed in the understanding of different host-guest complexes [85].

Moreover, the authors explored the capabilities of ExBox^{4+} to incorporate more than one six-membered ring via a [2guest⊂ExBox]·4Cl model. Noteworthy, the two-dimensional cage offers room enough to retain a second guest member, which is calculated to be thermodynamically favorable, with a small decrease in their stabilizing interaction energy [81].

Furthermore, the potential to incorporate technological relevant small gas molecules, such as H_2 and CO, expanding the potential application of ExBox family, and their derivatives [86]. The evaluation involves a combined molecular dynamics simulations (MD) and quantum chemical calculations (ωB97x-D/6-311G(d,p)), showing that H_2 is prone to be located inside the ExBox^{4+} cavity, pointing towards the phenyl ring, where an overall of three H_2 molecules can be incorporated at such cavity showing a non-covalent dispersion character of the interaction. The exo-interaction is also denoted, which can bear up to 12 H_2 molecules, which are located above the eight six-membered rings from viologen and *p*-phenyl sections, and four nearby the N-CH_2 corners, showing an orbital character of the interaction owing to charge transfer behavior for exohedral H_2. As the number of H_2 molecules increases, the absorption energy per H_2 unit decreases from −2.4 to −1.7 kcal/mol for an overall total of 2 to 15 H_2 adsorbed units, resulting in a revealed hydrogen storage capacity of ~4.3 wt%, close to available efficient materials. The authors additionally describe the potential of ExBox^{4+} to encapsulate three CO molecules [86], with an exergonic nature of the processes, based on the electrostatic nature of the interaction. The presence of counterions in [nH_2⊂ExBox]·4Cl slightly improves the adsorption energy as well as reaction enthalpy. However, in the [nCO⊂ExBox]·4Cl case, the adsorption energy is decreased to a small extent.

Exploration of the lithium-decorated ExBox^{4+} suggests that Li_8ExBox^{4+} is a thermodynamically stable compound [86] which is able to support 24H_2 molecules, resulting in an increase of the hydrogen storage capacity to 6.23 wt%, suggesting a promising approach to adsorbate H_2 gas. Li_8ExBox^{4+} also extends the adsorption capabilities towards CO molecules, holding 14 units at 298 K, and up to 16 at a lower temperature (77 K). Thus, accounting for host improvement of ExBox^{4+} when decorated with lithium atoms.

Nagurniak and coworkers explored the situation with non-linear acenes, and larger PAHs with ExBox^{4+} at the BP86-D3/TZ2P level of theory [39], extending the observation that with an increase of the interacting π-surface comes a larger stabilization of the resulting host-guest complex. The studied PAHs were given by azulene, anthracene, phenanthrene, pyrene, tetracene, tetraphene, chrysene, benzo[*c*] phenanthrene, triphenylene, perylene, and coronene (Figure 20.5). It is found that the incorporation of larger PAHs increases the length between *p*-phenylene rings, with a subsequent decrease in the viologen-viologen unit distance, accounting for a certain degree of flexibility provided by ExBox^{4+} owing to the sp^3-CH_2 corners connecting *p*-phenylene rings and viologen fragments. This incorporation involves destabilization of ExBox^{4+} cage of about 2.2 to 4.5 kcal/mol along with the series, owing to the slight structural rearrangements of the cage. The stabilizing character of the interaction energy (ΔE^{int}) which ranges from −32.8 to −62.8 kcal/mol along the series, is mainly contributed by the dispersion term (58-63%) followed by the electrostatic character (ΔE^{elstat}, 25-27%), and lastly, the orbital contribution (ΔE^{orb}, 11-15%), where the dispersion term increases as the number of π-electrons in the guest PAH increases. The NCI analysis for Figure 20.6 reveals the greater extension of the van der Waals interaction regions, accounting for the π–π interaction in these host-guest complexes, which are larger than the found in benzene⊂ExBox^{4+} (Figure 20.4)

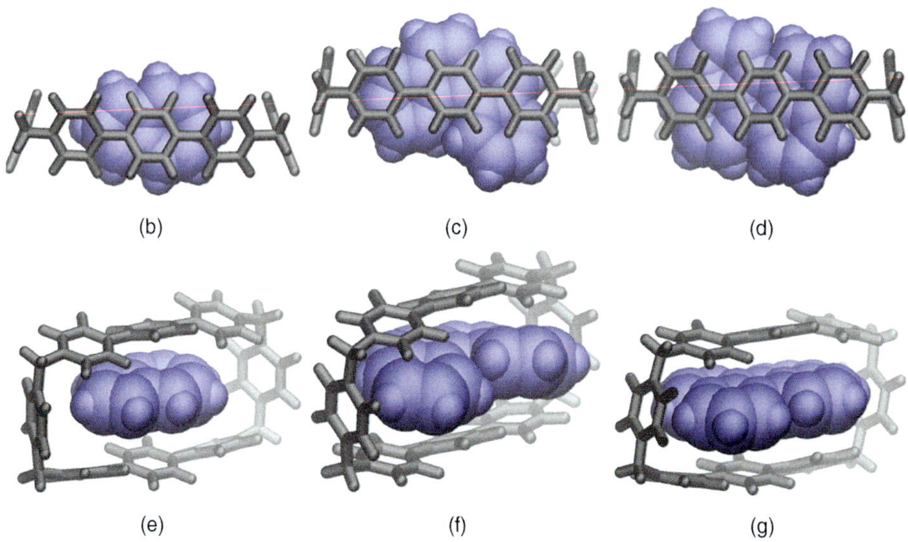

Figure 20.5 Two views from the relaxed structures for azulene⊂ExBox^{4+} (b and e), benzo[*c*]phenanthrene⊂ExBox^{4+} (c and f), and perylene⊂ExBox^{4+} (d and g). Source: Nagurniak, G.R., Caramori, G.F., Parreira, R.L.T., Bergamo, P.A.S., Frenking, G., and Muñoz-Castro, A. (2016) Shedding Light on the Nature of Host-Guest Interactions in PAHs-ExBox^{4+} Complexes. J. Phys. Chem. C, 120 (28), 15480–15487. © 2016, American Chemical Society.

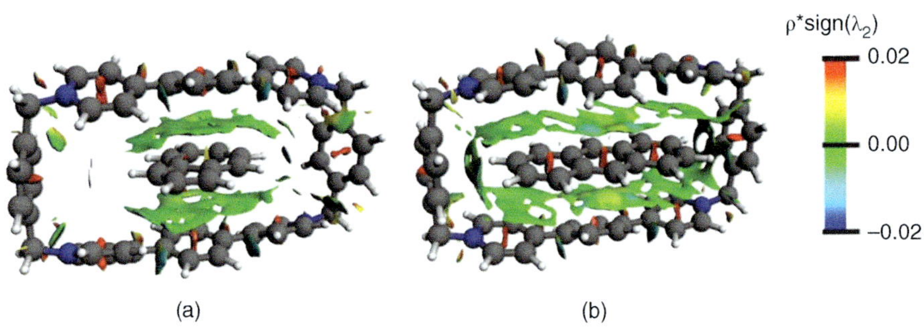

Figure 20.6 NCI analysis for azulene⊂ExBox^{4+} (a) and anthracene⊂ExBox^{4+} (b). Source: Nagurniak, G.R., Caramori, G.F., Parreira, R.L.T., Bergamo, P.A.S., Frenking, G., and Muñoz-Castro, A. (2016) Shedding Light on the Nature of Host-Guest Interactions in PAHs-ExBox^{4+} Complexes. J. Phys. Chem. C, 120 (28), 15480–15487. © 2016, American Chemical Society.

The next larger ExBox counterpart given by Ex^{2}Box^{4+} enables the incorporation of larger guests, where its interaction with π-electron-poor, π-electron-rich, and mixed-species was evaluated theoretically [40]. The resulting host-guest complexes between Ex^{2}Box^{4+} and π-electron-poor results in very small structural modifications of the parent host, similar to π-electron-poor and mixed cases. The possible incorporation of two guests is also estimated via the π-electron-poor 1,2,4-trichlorobenezene dimer, which also exhibits Cl···H-C interactions towards the *p*-phenyl fragments, resulting in a larger distortion of the host cage. The authors also discussed the inclusion of two benzene rings which is favorable, with the two guest being located towards the *p*-phenyl fragments, resulting in a larger distortion in comparison to the 1,2,4-trichlorobenezene dimer case, owing to the lack of Cl···H-C between the two guests in the benzene dimer case.

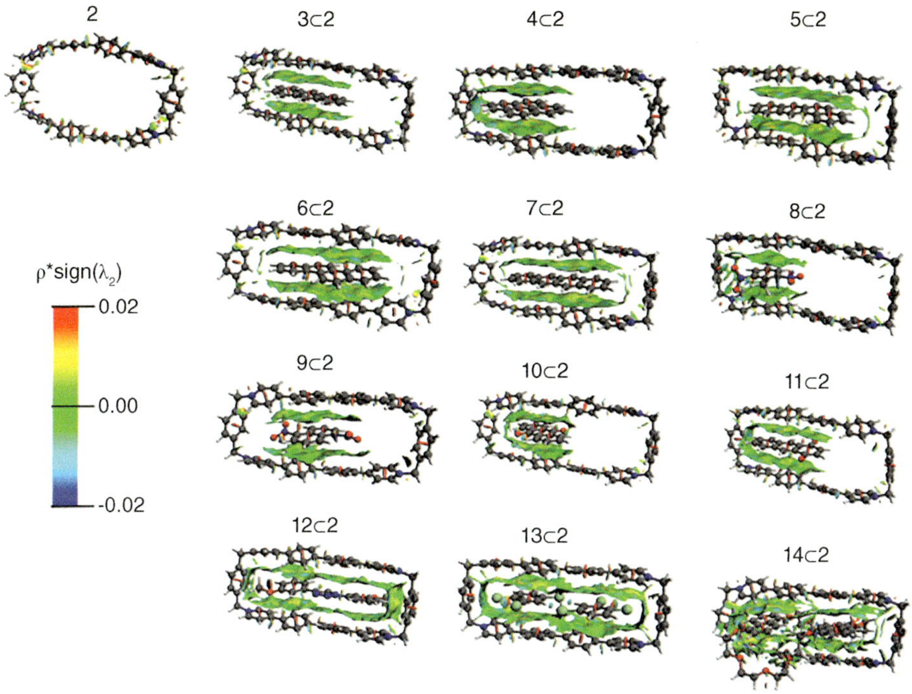

Figure 20.7 NCI analysis for NCI analysis for host-guest species between Ex^2Box^{4+} (2), and anthracene (3), pyrene (4), tetraphene (5), crysene (6), tetracene (7), 2,6-dinitrotoluene (8), 1,5-dinitronaphthalene (9), 9,10-Anthraquinone (10), 1,4-anthraquinone (11), 1,4-bis(3-methoxyphenyl)-1H-1,2,3-triazole (12), 1,2,4-trichlorobenzene (dimer) (13) and 1,5-Dinaphtho38crown-10 (14) (label employed in the original article). Source: Nagurniak, G.R., Caramori, G.F., Muñoz-Castro, A., Parreira, R.L.T., and da Silva, É.H. (2018) The ability of $Ex2Box^{4+}$ to interact with guests containing π-electron-rich and π-electron-poor moieties. Int. J. Quantum Chem., 118 (15), e25607. © 2018, John Wiley & Sons.

The interaction energy, for the studied species, shows an enhanced interaction between Ex^2Box^{4+} towards π-electron-rich guest in the range of −39.3 to −53.5 kcal/mol, with a lesser strength for the π-electron-poor case in the range between −29.8 to −39.0 kcal/mol. This interaction energy increases for the dimer guest cases with a value of −63.0 kcal/mol for the 1,2,4-trichlorobenezene dimer, and involving two bridged aromatic naphthalene rings, rising to −109.6 kcal/mol. For mixed cases, the interaction energy is located between the obtained for π-electron-poor and π-electron-rich guest. It is noteworthy that, for all the series, the stabilizing character of the interaction energy shows a contribution of ~25% from the electrostatic nature of the interaction (ΔE^{elstat}), followed by a ~14% from the orbital term (ΔE^{orb}), accounting for polarization and charge transfer effects, with larger contribution from the dispersion term (ΔE^{disp}), accounting for π–π interactions denoted by the NCI analysis on Figure 20.7 [40]. This is given by the guest species are planar PAHs based molecules, which requires efficient π–π interactions.

Different isomers were explored for 9,10-Anthraquinone⊂Ex^2Box^{4+}, which exhibits a mixed π-electron-poor and π-electron-rich character, resulting in a variation of the interaction energies of +12.3 kcal/mol, when the guest moves from one side of the host in its relaxed structure (point a, Figure 20.8), towards its center (point k, Figure 20.8) becoming gradually less stable. This is attributed to the electrostatic repulsion between the oxygen atoms of 9,10-Anthraquinone and the less electronegative region of the host at point k from Figure 20.8. Hence, for guests with mixed character, the location within the cavity requires to find a suitable match to ensure a stabilized conformation.

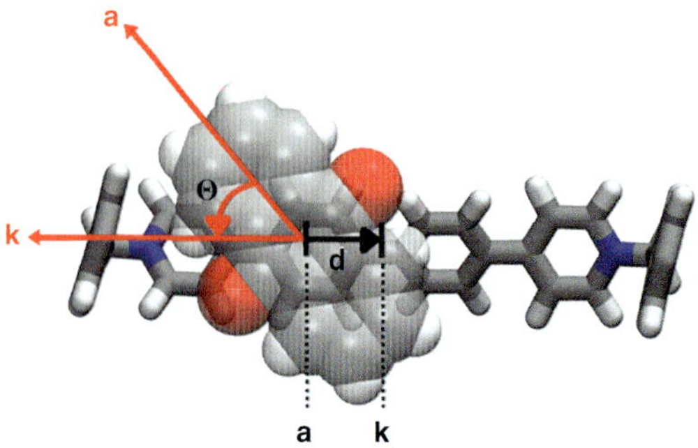

(a)

10⊂2	ΔE_{rel} (kcal·mol^{-1})	Θ (°)	d (Å)
a	0.00	50.7	2.77
b	0.98	45.0	2.40
c	2.59	39.5	2.10
d	4.30	34.0	1.80
e	5.40	28.5	1.50
f	6.17	23.0	1.20
g	7.65	17.5	0.90
h	9.51	12.0	0.60
i	11.21	6.50	0.30
j	12.32	1.50	0.00
k	12.33	0.00	0.00

(b)

Figure 20.8 Different isomers for 9,10-Anthraquinone⊂Ex^2Box^{4+}, accounting for the displacement of the guest in relation to Ex^2Box^{4+}, varying the angle Θ and the position d (black arrow) (a). (b) Relative change in energy from the variation of Θ and position d. Source: Nagurniak, G.R., Caramori, G.F., Muñoz-Castro, A., Parreira, R.L.T., and da Silva, É.H. (2018) The ability of $Ex2Box^{4+}$ to interact with guests containing π-electron-rich and π-electron-poor moieties. Int. J. Quantum Chem., 118 (15), e25607. © 2018, John Wiley & Sons.

20.5 Larger Boxes

Different derivatives of viologen units can be employed to further develop larger host and more confined cages towards functional materials [87], which are able to incorporate molecules such as C_{60} and C_{70} fullerenes [36]. Recently, ExBox molecular cages have been extended from two-dimensional cavities, standing on two viologen units, to three-dimensional hosts where porphyrin units serve as a standing group to held four viologens as sidearms [36]. The use of two tetraphenylporphyrin derivatives results in an efficient four-fold stand enabling the incorporation of four viologen leading to a rhomboid prismatic cage highlighting a large inner cavity taking advantage of the efficient supramolecular interaction between fullerene and porphyrins [88] and metalloporphyrins [89], resulting in an excellent fullerene receptor, named $TPPCage^{8+}$ by the authors.

This novel three-dimensional ExBox counterpart is able to incorporate spherical-like guests, namely, C_{60} and C_{70} fullerenes in a 1:1 ratio. The rationally designed cage shows particular selectivity towards C_{70} as observed by their association constants measured in dimtehylformamide with almost two-order of magnitude larger for C_{70} of $(2.4\pm0.2)x10^5$ M^{-1}, versus $(4.5\pm0.4)x10^3$ M^{-1} for C_{60} [36]. This extends the well-explored capabilities of ExBox derivatives to encapsulate polycyclic aromatic hydrocarbons (PAH) in a two-dimensional cavity.

The calculated models at the TZ2P/BP86-D with continuum solvation effects employing tetrahydrofuran as solvent [90], (Figure 20.9) shows similar structural features to the reported x-ray structures, where distances between side-arms viologen to fullerene amount to 3.261 and 3.139 Å, for C_{60} and C_{70} host-guest species respectively, which are close to the determined experimentally (3.30 and 2.94 Å) [36]. The calculated binding energy shows a favorable interaction between C_{60} and the three-dimensional cage amounting to −135.7 kcal/mol (ΔE_{int}) denoting that the cavity provides a proper fit [36]. The use of the packing coefficient (PC) introduced by Rebek and Mecozzi [91–93], provides an intuitive measurement of the ratio between host and guest volume, where an appropriate PC value range between 55-68%. The resulting packing coefficient measured for C_{60} and L

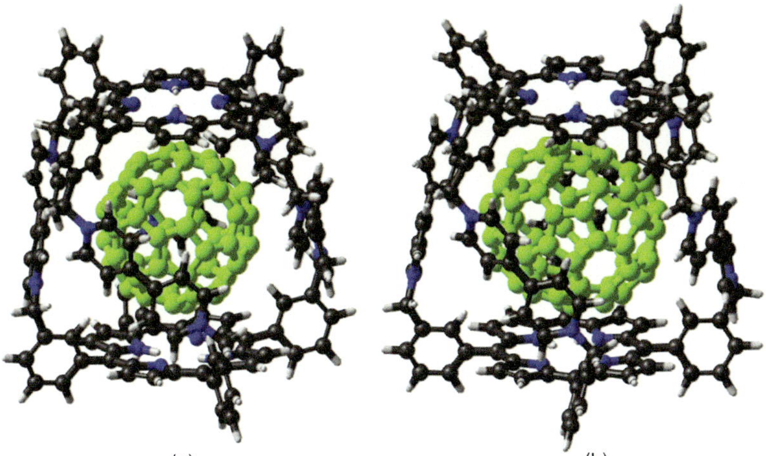

Figure 20.9 Optimized structures for $C_{60}\subset TPPCage^{8+}$ (a), and $C_{70}\subset TPPCage^{8+}$ (b). Source: Bachrach, S.M., and Andrews, A.E. (2014) All-Carbon, Neutral Analogue of $ExBox^{4+}$: A DFT Study of Polycyclic Aromatic Hydrocarbon Binding. J. Phys. Chem. A, 118 (31), 6104–6111. © 2014, American Chemical Society.

(calculated from the distances between C1-phenyl) amounts to 57%, denoting a favorable spatial fit accounting for the favorable ΔE_{int} estimated computationally.

The interaction involves contribution from electrostatic and orbital terms accounting for 21.5% and 28.1%, respectively of the stabilizing interaction, denoting the role of both electrostatic and the orbital polarization effects in the formation of $C_{60}\subset TPPCage^{8+}$. Moreover, the dispersion energy term (ΔE^{disp}) appears as the most relevant contribution (50.4%) denoting the non-covalent character as a result of strong $\pi\cdots\pi$ interactions which are remarkably shorter, about of 2.64 Å, in contrast to the typical $\pi\cdots\pi$ contacts found at ~3.40 Å. In comparison to the calculated benzene$\subset ExBox^{4+}$ structure by Das and Chattaraj [83], denoting a larger contribution from the ΔE^{elstat} (42.6%), followed by ΔE^{disp} (36.6%) and ΔE^{orb} (20.8%), the $C_{60}\subset TPPCage^{8+}$ formation stands more on noncovalent interactions owing to the larger π-surface/π-surface interaction.

For the $C_{70}\subset TPPCage^{8+}$ case, the different isomers indicate are given mostly by the dispersion energy term, suggesting that the preferred isomer is ruled by the formation of larger $\pi\cdots\pi$ interactions between host and guest. The calculated interaction energy (ΔEint) amounts to −139.74 kcal/mol in line to the preference to retain C_{70} over C_{60}, of the $TPPCage^{8+}$ [36]. Similar to $C_{60}\subset TPPCage^{8+}$, in the C_{70} case the energetic terms are in related contribution, with differences ascribed to the dispersion energy term (ΔE^{disp}). This result suggests that the available larger contact of π-surfaces may rule the favorable preference for C_{70}, also denoted by the more efficient packing given by the larger PC ratio of ~70%.

Moreover, the encapsulating capabilities of a hypothetical metalloporphyrin derivative of $TPPCage^{8+}$ are explored in literature, involving the coordination of Zn(II) in the N4-macrocycle ring, resulting in the related $Zn_2TPPCage^{8+}$, owing to the great co-crystallizing ability of Zn(II)porphyrin (ZnP) and several fullerenes [94, 95]. As result, the binding energy retaining the $Zn_2TPPCage^{8+}$ aggregate is reduced in approximately in 30 kcal/mol owing to the decrease of the stabilizing contribution from the dispersion term, owing to the lesser π–π interaction area, as revealed via non-covalent index analysis (NCI) [96, 97].

Interestingly, the origin of the electrostatic and orbital polarization character of the stabilization is given by a charge reorganization and charge accumulation at the porphyrin macrocycles regions

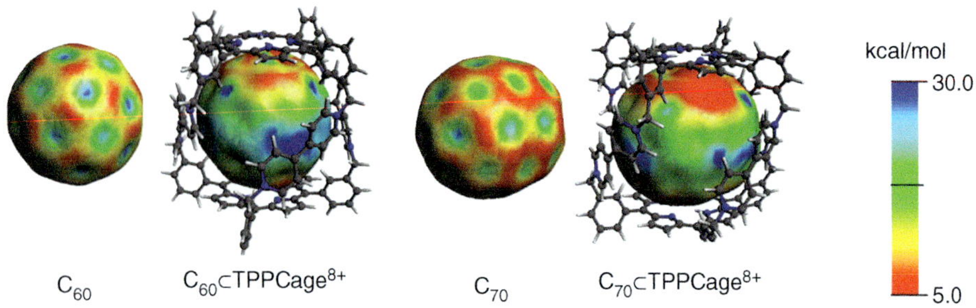

Figure 20.10 Electrostatic potential surface over fullerene electron density (isosurface at 0.001 a.u.) denoting the isolated and encapsulated species.

(Figure 20.10), with a charge depletion along the equator formed by the viologens side-arms. The π–π interactions are distributed in about ~80% at the porphyrin macrocycles regions and ~20% at the viologen section. Thus, the formation of stable and selective fullerene host involves the extension of the π···π and H···π stacking surface, and, charge reorganization over the fullerene surfaces, allowing a fine-tuning of the host characteristics by controlling the attached functional groups ranging from electron-deficient to electron-rich species in the host structure.

Noteworthy, the host-guest situation in C_{60}⊂$TPPCage^{8+}$ is calculated to be more favorable (-135.7 kcal/mol), than the recent host cage involving four $[M_3(3,5\text{-}(CF_3)_2Pz)_3]$ units, with M = Cu, Ag, Au, as depicted by Dias and coworkers [98]. The calculated interaction energy at a related level of theory ranges from −104.4 to −100.0 from Cu to Au species [99]. The C_{60}⊂$[M_3(3,5\text{-}(CF_3)_2Pz)_3]_4$ species, involves a stabilizing situation contributed by ~54% from the dispersion energy (ΔE^{disp}), ~29% from the electrostatic character and about of ~17% from the charge transfer term, being of more electrostatic character than C_{60}⊂$TPPCage^{8+}$. This denotes the different features of the appropriated host towards the effective encapsulation of the same guest species.

20.6 NMR Features

The formation of host pairs can be quantitatively followed by nuclear magnetic resonance experiments in both solution- and solid-state, owing to the high sensitivity of the technique, providing characteristic patterns (*fingerprint*) owing to the supramolecular aggregate formation [100–102]. Thus the chemical features of the host will modify the chemical environment of guests, and similarly, to the other way. This is related to the "anisotropy effect" concept in NMR spectroscopy textbooks, generally related to date to common organic functional groups [103–107], useful to explain the shielding or deshielding shift related to the aggregate formation. In this sense, the inherent shielding or deshielding regions within the ExBox series have been accounted, for Ex^0Box^{4+}, $ExBox^{4+}$, and Ex^2Box^{4+} (Figure 20.11) [39, 40, 75]. For such hosts, the n-pyridil ring contributes with a deshielding region at the inner section of the cavity which overcome the shielding contribution from the aromatic ring at the *p*-phenylene sides, whereas the central phenyl ring at the viologens units induces a shielding region both inner and external regions of the cavity. Thus guest atoms or groups will experience a shielding in the NMR spectra if the atoms are located towards the center of the structure, and more deshielding chemical shift, when they are located at one side of the host. This information is useful to determinate the dynamical processes involving the encapsulation and location of guest in the host-guest complex to be followed by regular NMR experiments.

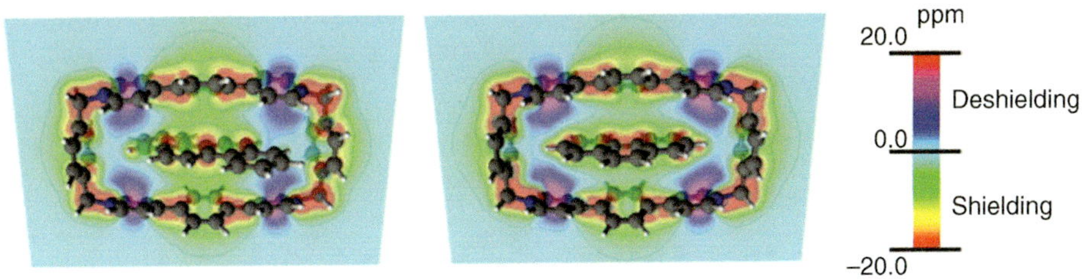

Figure 20.11 Magnetic response of the overall host-guest complex for chrysene⊂ExBox^{4+} and phenanthrene⊂ExBox^{4+} as examples, denoting the shielding contribution from the side *p*-phenyl groups, and the central phenyl ring in the viologen region. Moreover, the deshielding contribution from the pyridinium units. In addition, the shielding contribution from both PAHs is advised. Reprinted (adapted) with permission from (Nagurniak et. al, J. Phys. Chem. C, 2016, 120, 15480-15487). Copyright (2017) American Chemical Society.

20.7 All Carbon Counterpart

Bachrach proposed computationally, and interesting all-carbon backbone counterpart for ExBox^{4+} by replacing the nitrogen atoms by carbon atoms [108]. The resulting structure is a neutral derivative, showing a lesser *o*-*o*′-CH repulsive interactions decreasing the ring strain energy from +20.2 kcal/mol for ExBox^{4+} to +5.2 kcal/mol for the related to the all-carbon backbone species. This proposed counterpart retains the binding capabilities of ExBox^{4+}, denoting favorable incorporation of different PAHs from naphthalene, anthracene, and tetracene (Figure 20.12). However, for the proposed structure, a smaller binding energy is obtained which decrease in about 29% for naphthalene guest, 24% for the anthracene, 13% for tetracene, owing to the fact that a large *p*-surface from the guest enhance the interaction for the proposed host, decreasing the differences in the binding capabilities towards larger PAHs guest. In contrast, for when benzene is involved as a guest, the proposed host shows a more favorable binding energy (-8.2 kcal/mol), in contrast to the ExBox^{4+} case (-3.9 kcal/mol), owing to that a more efficient π–π and CH-π interaction are allowed in the former. Moreover, the structural distortions for the proposed counterpart are to a lesser extent in comparison to the ExBox^{4+} parent, when the host-guest complex is formed.

20.8 Conclusions

The use of versatile ExBox derivatives allows a rational design and fine-tuning of host species showcasing two- and three-dimensional cages of different sizes and characteristics. The modeling of such species is sensible to solvation effects owing to the large positive charges usually displayed by the ExBox hosts, which can be incorporated in order to gain information about the in-solution state behavior. The nature of the ExBox-guest interaction is of non-covalent dispersion interaction owing to π–π contacts for PAHs, which is dependent on the extension of the interactions well described by the non-covalent interaction index. From the analysis of H_2 and CO absorption capabilities, it is shown that the ExBox cage is able to sustain a stabilizing character based on electrostatic interactions for CO, and of charge-transfer interactions for H_2 molecules located outside the box. The use of the ExBox template is able to tune the nature and strength of the host-guest interaction allowing the adduct formation for different planar and curved π-surface guests, such as different PAHs and fullerene structures, which can depict a variable host-guest ratio. Thus, novel aggregates can be further explored by means of computational studies giving relevant information concerning the

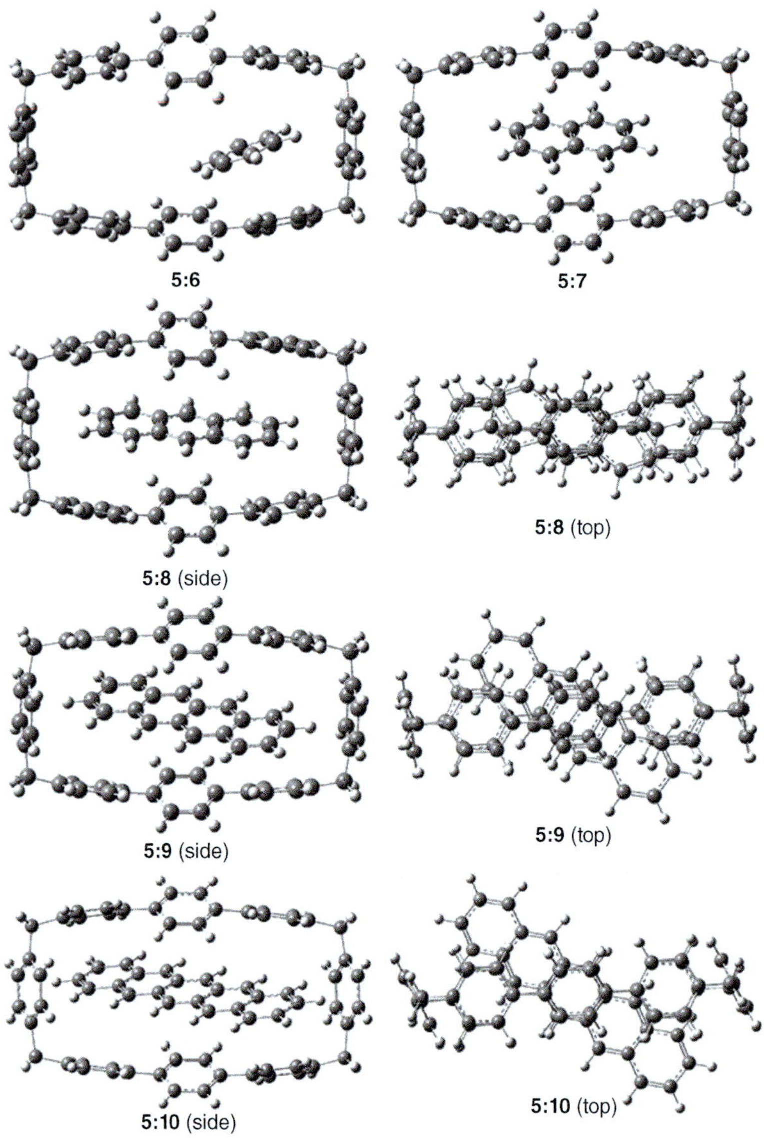

Figure 20.12 Optimized geometries for all-carbon backbone species related to $ExBox^{4+}$ and different guest. Reprinted with permission from (Bachrach, et.al, J. Phys. Chem. A 2014, 118, 6104–6111). Source: Nagurniak, G.R., Caramori, G.F., Parreira, R.L.T., Bergamo, P.A.S., Frenking, G., and Muñoz-Castro, A. (2016) Shedding Light on the Nature of Host-Guest Interactions in PAHs-$ExBox^{4+}$ Complexes. J. Phys. Chem. C, 120 (28), 15480–15487. © 2016, American Chemical Society.

impact of modifying the size and acceptor/donor capabilities of the cavity, towards the rational synthesis of efficient hosts, required for novel supramolecular arrays leading to extended networks based on pre-organized domains.

Acknowledgments

The authors thank the financial support from FONDECYT 1180683 and CNPq (grants: 311963/2017-0 and 311132/2020-0).

References

1 Bruns, C.J. and Stoddart, J.F. (2017). The Nature of the Mechanical Bond: From Molecules to Machines. Hoboken, New Jersey: John Wiley & Sons, Inc.

2 Cram, D.J., Tanner, M.E., and Thomas, R. (1991). The taming of cyclobutadiene. Angew. Chemie Int. Ed. English 30 (8): 1024–1027.

3 Mal, P., Breiner, B., Rissanen, K., and Nitschke, J.R. (2009). White phosphorus is air-stable within a self-assembled tetrahedral capsule. Science 324 (5935): 1697–1699.

4 Ariga, K. and Kunitake, T. The chemistry of molecular recognition – host molecules and guest molecules. In: Supramolecular Chemistry — Fundamentals and Applications, 7–44. Berlin/Heidelberg: Springer-Verlag.

5 Kollman, P.A. (1994). The nature of molecular recognition: examples from host/guest chemistry. In: Computational Approaches in Supramolecular Chemistry (ed. G. Wipff), 175–181. Dordrecht: Springer Netherlands.

6 Zhou, J., Yu, G., and Huang, F. (2017). Supramolecular chemotherapy based on host–guest molecular recognition: a novel strategy in the battle against cancer with a bright future. Chem. Soc. Rev. 46 (22): 7021–7053.

7 Yu, G., Jie, K., and Huang, F. (2015). Supramolecular amphiphiles based on host–guest molecular recognition motifs. Chem. Rev. 115 (15): 7240–7303.

8 Pluth, M.D., Bergman, R.G., and Raymond, K.N. (2009). Proton-mediated chemistry and catalysis in a self-assembled supramolecular host. Acc. Chem. Res. 42 (10): 1650–1659.

9 van Leeuwen, P.W.N.M. (2008). Supramolecular Catalysis. Verlag: Wiley-VCH.

10 Zhang, C., Wang, H., Zhong, J. et al. (2019). A mutually stabilized host-guest pair. Sci. Adv., 5 (11). In: eaax6707.

11 Diederich, F. and Dick, K. (1984). A new water-soluble macrocyclic host of the cyclophane type: host-guest complexation with aromatic guests in aqueous solution and acceleration of the transport of arenes through an aqueous phase. J. Am. Chem. Soc. 106 (26): 8024–8036.

12 Denis, P.A. (2019). On the estimation of the strength of supramolecular complexes of fullerenes. Int. J. Quantum Chem. 119 (2): e25670.

13 Rambo, B.M., Gong, H.-Y., Oh, M., and Sessler, J.L. (2012). The "Texas-sized" molecular box: a versatile building block for the construction of anion-directed mechanically interlocked structures. Acc. Chem. Res. 45 (8): 1390–1401.

14 Liu, Z., Nalluri, S.K.M., and Stoddart, J.F. (2017). Surveying macrocyclic chemistry: from flexible crown ethers to rigid cyclophanes. Chem. Soc. Rev. 46 (9): 2459–2478.

15 Chen, L., Lim, K.J.C., Babra, T.S. et al. (2018). A macrocyclic receptor containing two viologen species connected by conjugated terphenyl groups. Org. Biomol. Chem. 16 (27): 5006–5015.

16 Anelli, P.L., Ashton, P.R., Philp, D. et al. (1992). Molecular Meccano. 1. [2]Rotaxanes and a [2]catenane made to order. J. Am. Chem. Soc. 114 (1): 193–218.

17 Jean-Pierre, S. and Gaspard, P. (2010). From Non-Covalent Assemblies to Molecular Machines. KGaA, Weinheim, Germany: Wiley-VCH Verlag GmbH & Co.

18 Mena-Hernando, S. and Pérez, E.M. (2019). Mechanically interlocked materials. Rotaxanes and catenanes beyond the small molecule. Chem. Soc. Rev. 48 (19): 5016–5032.

19 Percec, V. (2013). Hierarchical Macromolecular Structures: 60 Years after the Staudinger Nobel Prize I. Cham: Springer International Publishing.

20 Barnes, J.C., Juríček, M., Strutt, N.L. et al. (2013). ExBox: A polycyclic aromatic hydrocarbon scavenger. J. Am. Chem. Soc. 135 (1): 183–192.

21 Juríček, M., Barnes, J.C., Dale, E.J. et al. (2013). Ex2Box: interdependent modes of binding in a two-nanometer-long synthetic receptor. J. Am. Chem. Soc. 135 (34): 12736–12746.

22 Dale, E.J., Ferris, D.P., Vermeulen, N.A. et al. (2016). Cooperative reactivity in an extended-viologen-based cyclophane. J. Am. Chem. Soc. 138 (11): 3667–3670.

23 Barnes, J.C., Juríček, M., Vermeulen, N.A. et al. (2013). Synthesis of Ex n Box cyclophanes. J. Org. Chem. 78 (23): 11962–11969.

24 Dyar, S.M., Barnes, J.C., Juricek, M. et al. (2014). Electron transfer and multi-electron accumulation in $ExBox^{4+}$. Angew. Chem., Int. Ed. 53 (21): 5371–5375.

25 Odell, B., Reddington, M.V., Slawin, A.M.Z. et al. (1988). Cyclobis(paraquat-*p*-phenylene). A tetracationic multipurpose receptor. Angew. Chemie Int. Ed. English 27 (11): 1547–1550.

26 Ashton, P.R., Odell, B., Reddington, M.V. et al. (1988). Isostructural, alternately-charged receptor stacks. the inclusion complexes of hydroquinone and catechol dimethyl ethers with cyclobis(paraquat-*p*-phenylene). Angew. Chemie Int. Ed. English 27 (11): 1550–1553.

27 Cai, K., Lipke, M.C., Liu, Z. et al. (2018). Molecular Russian dolls. Nat. Commun. 9 (1): 5275.

28 Fahrenbach, A.C., Sampath, S., Late, D.J. et al. (2012). A semiconducting organic radical cationic host–guest complex. ACS Nano 6 (11): 9964–9971.

29 Frasconi, M., Fernando, I.R., Wu, Y. et al. (2015). Redox control of the binding modes of an organic receptor. J. Am. Chem. Soc. 137 (34): 11057–11068.

30 Cao, J., Lu, H.-Y., Xiang, J.-F., and Chen, C.-F. (2010). Complexation between pentiptycene-based mono(crown ether)s and tetracationic cyclobis(paraquat-*p*-phenylene): who is the host or the guest? Chem. Commun. 46 (20): 3586.

31 Barnes, J.C., Frasconi, M., Young, R.M. et al. (2014). Solid-state characterization and photoinduced intramolecular electron transfer in a nanoconfined octacationic homo[2]catenane. J. Am. Chem. Soc. 136 (30): 10569–10572.

32 Philp, D., Slawin, A.M.Z., Spencer, N. et al. (1991). The complexation of tetrathiafulvalene by cyclobis(Paraquat-*p*-phenylene). J. Chem. Soc. Chem. Commun. (22): 1584.

33 Trabolsi, A., Khashab, N., Fahrenbach, A.C. et al. (2010). Radically enhanced molecular recognition. Nat. Chem. 2 (1): 42–49.

34 Dale, E.J., Vermeulen, N.A., Young, R.M. et al. (2016). Supramolecular explorations: exhibiting the extent of extended cationic cyclophanes. Acc Chem Res 49 (2): 262–273.

35 Juríček, M., Strutt, N.L., Barnes, J.C. et al. (2014). Induced-fit catalysis of corannulene bowl-to-bowl inversion. Nat. Chem. 6 (3): 222–228.

36 Shi, Y., Cai, K., Xiao, H. et al. (2018). Selective Extraction of C 70 by a Tetragonal Prismatic Porphyrin Cage. J. Am. Chem. Soc. 140 (42): 13835–13842.

37 Maharramov, A.M., Mahmudov, K.T., Kopylovich, M.N., and Pombeiro, A.J.L. (2016). Non-covalent Interactions in the Synthesis and Design of New Compounds. Chichester, West Sussex, UK: John Wiley & Sons, Inc.

38 Otero de la Roza, A. and DiLabio, G.A. (2017). Non-Covalent Interactions in Quantum Chemistry and Physics: Theory and Applications. Amsterdam: Elsevier.

39 Nagurniak, G.R., Caramori, G.F., Parreira, R.L.T. et al. (2016). Shedding light on the nature of host-guest interactions in PAHs-$ExBox^{4+}$ complexes. J. Phys. Chem. C 120 (28): 15480–15487.

40 Nagurniak, G.R., Caramori, G.F., Muñoz-Castro, A. et al. (2018). The ability of Ex 2 Box 4+ to interact with guests containing π-electron-rich and π-electron-poor moieties. Int. J. Quantum Chem. 118 (15): e25607.

41 Stasyuk, O.A., Sedlak, R., Guerra, C.F., and Hobza, P. (2018). Comparison of the DFT-SAPT and canonical EDA schemes for the energy decomposition of various types of noncovalent interactions. J. Chem. Theory Comput. 14 (7): 3440–3450.

42 Pecher, L. and Tonner, R. (2019). Deriving bonding concepts for molecules, surfaces, and solids with energy decomposition analysis for extended systems. Wiley Interdiscip. Rev. Comput. Mol. Sci. 9 (4): e1401.

43 Zhao, L., von Hopffgarten, M., Andrada, D.M., and Frenking, G. (2018). Energy decomposition analysis. WIREs Comput. Mol. Sci. 8 (3): e1345.

44 Phipps, M.J.S., Fox, T., Tautermann, C.S., and Skylaris, C.-K. (2015). Energy decomposition analysis approaches and their evaluation on prototypical protein–drug interaction patterns. Chem. Soc. Rev. 44 (10): 3177–3211.

45 Ziegler, T. and Rauk, A. (1977). On the calculation of bonding energies by the Hartree Fock Slater method. Theor. Chim. Acta 46 (1): 1–10.

46 Su, P., Jiang, Z., Chen, Z., and Wu, W. (2014). Energy decomposition scheme based on the generalized Kohn–Sham scheme. J. Phys. Chem. A 118 (13): 2531–2542.

47 Khaliullin, R.Z., Cobar, E.A., Lochan, R.C. et al. (2007). Unravelling the origin of intermolecular interactions using absolutely localized molecular orbitals. J. Phys. Chem. A 111 (36): 8753–8765.

48 Morokuma, K. (1971). Molecular orbital studies of hydrogen bonds. III. C=O···H–O Hydrogen Bond in H_2 CO···H_2 O and H_2 CO···$2H_2$ O. J. Chem. Phys. 55 (3): 1236–1244.

49 Jeziorski, B., Moszynski, R., and Szalewicz, K. (1994). Perturbation theory approach to intermolecular potential energy surfaces of van der Waals complexes. Chem. Rev. 94 (7): 1887–1930.

50 Podeszwa, R. (2010). Interactions of graphene sheets deduced from properties of polycyclic aromatic hydrocarbons. J. Chem. Phys. 132 (4): 044704.

51 Heßelmann, A. and Korona, T. (2014). Intermolecular symmetry-adapted perturbation theory study of large organic complexes. J. Chem. Phys. 141 (9): 094107.

52 Mo, Y., Gao, J., and Peyerimhoff, S.D. (2000). Energy decomposition analysis of intermolecular interactions using a block-localized wave function approach. J. Chem. Phys. 112 (13): 5530–5538.

53 Cardozo, T.M. and Nascimento, M.A.C. (2009). Energy partitioning for generalized product functions: The interference contribution to the energy of generalized valence bond and spin coupled wave functions. J. Chem. Phys. 130 (10): 104102.

54 Thirman, J. and Head-Gordon, M. (2015). An energy decomposition analysis for second-order Møller–Plesset perturbation theory based on absolutely localized molecular orbitals. J. Chem. Phys. 143 (8): 084124.

55 Thirman, J. and Head-Gordon, M. (2017). Efficient implementation of energy decomposition analysis for second-order Møller–Plesset perturbation theory and application to anion–π interactions. J. Phys. Chem. A 121 (3): 717–728.

56 Altun, A., Neese, F., and Bistoni, G. (2018). Local energy decomposition analysis of hydrogen-bonded dimers within a domain-based pair natural orbital coupled cluster study. Beilstein J. Org. Chem. 14: 919–929.

57 Schneider, W.B., Bistoni, G., Sparta, M. et al. (2016). Decomposition of intermolecular interaction energies within the local pair natural orbital coupled cluster framework. J. Chem. Theory Comput. 12 (10): 4778–4792.

58 Bistoni, G., Auer, A.A., and Neese, F. (2017). Understanding the role of dispersion in frustrated Lewis pairs and classical Lewis adducts: a domain-based local pair natural orbital coupled cluster study. Chem. – A Eur. J. 23 (4): 865–873.

59 Mitoraj, M.P., Michalak, A., and Ziegler, T. (2009). A combined charge and energy decomposition scheme for bond analysis. J. Chem. Theory Comput. 5 (4): 962–975.

60 Bickelhaupt, F.M. and Baerends, E.J. (2007). Kohn-Sham density functional theory: predicting and understanding chemistry. In: *Reviews in Computational Chemistry*, vol. 15, John Wiley & Sons Ltd, pp. In: 1–86.

61 Mitoraj, M. and Michalak, A. (2007). Natural orbitals for chemical valence as descriptors of chemical bonding in transition metal complexes. J. Mol. Model. 13 (2): 347–355.

62 Misturini, A., Ortolan, A.O., Caramori, G.F. et al. (2019). Tracking the absence of anion–π interactions in modified [2 3](1,3,5)cyclophanes: insights from computation. New J. Chem. 43 (33): 13271–13281.

63 Scorsin, L., Roehrs, J.A., Campedelli, R.R. et al. (2018). Cucurbituril-mediated catalytic hydrolysis: a kinetic and computational study with neutral and cationic dioxolanes in CB7. ACS Catal. 8 (12): 12067–12079.

64 Ortolan, A.O., Charistos, N.D., Guajardo-Maturana, R. et al. (2019). On the cation-π capabilities of small all sp 2 -carbon host structures. Evaluation of [6.8] 3 cyclacene from relativistic DFT calculations. Int. J. Quantum Chem. 119 (4): e25811.

65 Schneider, F.S.S., Segala, M., Caramori, G.F. et al. (2018). How do secondary phosphine oxides interact with silver nanoclusters? Insights from computation. J. Phys. Chem. C 122 (37): 21449–21461.

66 Ortolan, A.O., Caramori, G.F., Parreira, R.L.T., and Muñoz-Castro, A. (2018). Helicenes as molecular tweezers in the formation of cation–π complexes. bonding and circular dichroism properties from relativistic DFT calculations. ChemPhysChem 19 (18): 2321–2330.

67 Ortolan, A.O., Øestrøm, I., Caramori, G.F. et al. (2018). Anion recognition by organometallic calixarenes: analysis from relativistic DFT calculations. Organometallics 37 (13): 2167–2176.

68 Segala, M., Schneider, F.S.S., Caramori, G.F., and Parreira, R.L.T. (2016). Evaluation of electron donation as a mechanism for the stabilisation of chalcogenate-protected gold nanoclusters. ChemPhysChem 17 (19): 3102–3111.

69 Ortolan, A.O., Caramori, G.F., Matthias Bickelhaupt, F. et al. (2017). How the electron-deficient cavity of heterocalixarenes recognizes anions: insights from computation. Phys. Chem. Chem. Phys. 19 (36): 24696–24705.

70 Caramori, G.F. and Frenking, G. (2007). The nature of the Ru–NO bond in ruthenium tetraammine nitrosyl complexes. Organometallics 26 (24): 5815–5825.

71 Michalak, A., Mitoraj, M., and Ziegler, T. (2008). Bond orbitals from chemical valence theory. J. Phys. Chem. A 112 (9): 1933–1939.

72 Ashton, P.R., Balzani, V., Becher, J. et al. (1999). A three-pole supramolecular switch. J. Am. Chem. Soc. 121 (16): 3951–3957.

73 Witlicki, E.H., Hansen, S.W., Christensen, M. et al. (2009). Determination of binding strengths of a host–guest complex using resonance Raman scattering. J. Phys. Chem. A 113 (34): 9450–9457.

74 Deng, W.-Q., Flood, A.H., Stoddart, J.F., and Goddard, W.A. (2005). An electrochemical color-switchable RGB dye: tristable [2]catenane. J. Am. Chem. Soc. 127 (46): 15994–15995.

75 Olea Ulloa, C., Ponce Vargas, M., Guajardo Maturana, R., and Muñoz-Castro, A. (2013). Theoretical study of the binding strength and magnetical response properties involved in the formation of the $π$-donor/$π$-acceptor [TTF–CBPQT]4+ host–guest system. Polyhedron 54: 119–122.

76 Perdew, J.P., Burke, K., and Ernzerhof, M. (1996). Generalized gradient approximation made simple. Phys. Rev. Lett. 77 (18): 3865.

77 Klamt, A. and Jonas, V. (1996). Treatment of the outlying charge in continuum solvation models. J. Chem. Phys. 105 (22): 9972.

78 Kristensen, R., Andersen, S.S., Olsen, G., and Jeppesen, J.O. (2017). Probing the role of glycol chain lengths in π-donor–acceptor [2]pseudorotaxanes based on monopyrrolo-tetrathiafulvalene and cyclobis(paraquat-*p*-phenylene). J. Org. Chem. 82 (3): 1371–1379.

79 Nygaard, S., Hansen, C.N., and Jeppesen, J.O. (2007). Binding studies between triethylene glycol-substituted monopyrrolotetrathiafulvalene derivatives and cyclobis(paraquat-*p*-phenylene). J. Org. Chem. 72 (5): 1617–1626.

80 Wang, Y., Jin, X., Yu, H.S. et al. (2017). Revised M06-L functional for improved accuracy on chemical reaction barrier heights, noncovalent interactions, and solid-state physics. Proc. Natl. Acad. Sci. 114 (32): 8487–8492.

81 Bachrach, S.M. (2013). DFT Study of the ExBox·aromatic hydrocarbon host-guest complex. J. Phys. Chem. A 117 (35): 8484–8491.

82 Bachrach, S.M. and Nickle, Z.O.M. (2015). Designing a "flatter" $ExBox^{4+}$ analogue. J Phys Chem A 119 (42): 10613–10619.

83 Das, R. and Chattaraj, P.K. (2014). Host-guest interactions in ExBox 4+. ChemPhysChem 15 (18): 4108–4116.

84 Contreras-García, J., Johnson, E.R., Keinan, S. et al. (2011). NCIPLOT: A program for plotting noncovalent interaction regions. J. Chem. Theory Comput. 7 (3): 625–632.

85 González-Veloso, I., Rodríguez-Otero, J., and Cabaleiro-Lago, E.M. (2016). Carbon-nanorings ([10]CPP and [6]CPPA) as fullerene (C_{60} and C_{70}) receptors: a comprehensive dispersion-corrected DFT study. Phys. Chem. Chem. Phys. 18 (46): 31670–31679.

86 Das, R. and Chattaraj, P.K. (2014). Gas storage potential of ExBox 4+ and its Li-decorated derivative. Phys. Chem. Chem. Phys. 16 (40): 21964–21979.

87 Striepe, L. and Baumgartner, T. (2017). Viologens and their application as functional materials. Chem. – A Eur. J. 23 (67): 16924–16940.

88 Zhang, C., Wang, Q., Long, H., and Zhang, W. (2011). A highly C 70 selective shape-persistent rectangular prism constructed through one-step alkyne metathesis. J. Am. Chem. Soc. 133 (51): 20995–21001.

89 Tashiro, K., Aida, T., Zheng, J.-Y. et al. (1999). A cyclic dimer of metalloporphyrin forms a highly stable inclusion complex with C 60. J. Am. Chem. Soc. 121 (40): 9477–9478.

90 Camacho Gonzalez, J., Mondal, S., Ocayo, F. et al. (2020). Nature of C 60 and C 70 fullerene encapsulation in a porphyrin- and metalloporphyrin-based cage: Insights from dispersion-corrected density functional theory calculations. Int. J. Quantum Chem. 120 (3).

91 Mecozzi, S., Rebek, J., and Rebek, J. Jr., (1998). The 55% solution: a formula for molecular recognition in the liquid state. Chem. – A Eur. J. 4 (6): 1016–1022.

92 Assaf, K.I. and Nau, W.M. (2015). Cucurbiturils: from synthesis to high-affinity binding and catalysis. Chem. Soc. Rev. 44 (2): 394–418.

93 Nau, W.M., Florea, M., and Assaf, K.I. (2011). Deep inside cucurbiturils: physical properties and volumes of their inner cavity determine the hydrophobic driving force for host-guest complexation. Isr. J. Chem. 51 (5–6): 559–577.

94 Smalley, R.E. (1992). Self-assembly of the fullerenes. Acc. Chem. Res. 25 (3): 98–105.

95 Konarev, D.V., Khasanov, S.S., and Lyubovskaya, R.N. (2014). Fullerene complexes with coordination assemblies of metalloporphyrins and metal phthalocyanines. Coord. Chem. Rev. 262: 16–36.

96 Saleh, G., Gatti, C., and Presti, L.L. (2012). Non-covalent interaction via the reduced density gradient: Independent atom model vs experimental multipolar electron densities. Comput. Theor. Chem. 998: 148–163.

97 Saleh, G., Gatti, C., Lo Presti, L., and Contreras-García, J. (2012). Revealing non-covalent interactions in molecular crystals through their experimental electron densities. Chem. Eur. J. 18 (48): 15523–15536.

98 Jayaratna, N.B., Olmstead, M.M., Kharisov, B.I., and Dias, H.V.R. (2016). Coinage metal pyrazolates [(3,5-(CF 3) 2 Pz)M] 3 (M = Au, Ag, Cu) as Buckycatchers. Inorg. Chem. 55 (17): 8277–8280.

99 Ulloa, C.O.C.O., Ponce-Vargas, M., and Muñoz-Castro, A. (2018). Formation of coinage-metal···fullerene adducts. Evaluation of the interaction nature between triangular coinage metal complexes (M 3 = Cu, Ag, and Au) and C 60 through relativistic density functional theory calculations. J. Phys. Chem. C 122 (43): 25110–25117.

100 Pastor, A. and Martínez-Viviente, E. (2008). NMR spectroscopy in coordination supramolecular chemistry: A unique and powerful methodology. Coord. Chem. Rev. 252 (21–22): 2314–2345.

101 Cohen, Y. and Slovak, S. (2019). Diffusion NMR for the characterization, in solution, of supramolecular systems based on calixarenes, resorcinarenes, and other macrocyclic arenes. Org. Chem. Front. 6 (10): 1705–1718.

102 Weingarth, M. and Baldus, M. (2013). Solid-state NMR-based approaches for supramolecular structure elucidation. Acc. Chem. Res. 46 (9): 2037–2046.

103 Kleinpeter, E., Koch, A., and Seidl, P.R. (2008). Visualization and quantification of the anisotropic effect of C=C double bonds on 1H NMR spectra of highly congested hydrocarbons-indirect estimates of steric strain. J. Phys. Chem. A 112 (22): 4989–4995.

104 Field, L.D., Sternhell, S., and Kalman, J.R. (2013). Organic Structures from Spectra. Chichester, West Sussex, UK: John Wiley & Sons, Inc.

105 Klod, S., Koch, A., and Kleinpeter, E. (2002). Ab-initio quantum-mechanical GIAO calculation of the anisotropic effect of C–C and X–C single bonds – application to the 1 H NMR spectrum of cyclohexane. J. Chem. Soc. Perkin Trans. 2 (9): 1506–1509.

106 Klod, S. and Kleinpeter, E. (2001). Ab initio calculation of the anisotropy effect of multiple bonds and the ring current effect of arenes application in conformational and configurational analysis. J. Chem. Soc. Perkin Trans. 2 (10): 1893–1898.

107 Baranac-Stojanović, M. and Stojanović, M. (2013). Magnetic anisotropy of the CC single bond. Chem. – A Eur. J. 19 (13): 4249–4254.

108 Bachrach, S.M. and Andrews, A.E. (2014). All-carbon, neutral analogue of ExBox 4+ : A DFT study of polycyclic aromatic hydrocarbon binding. J. Phys. Chem. A 118 (31): 6104–6111.

Index

a
Adaptive Natural Density Partitioning (AdNDP) 242
antibacterial activity 377
Atom Centered Density Matrix Propagation (ADMP) 276
atoms in molecules 241

b
Baylis–Hillman reaction 301
Berkowitz-Parr relation 180
bichromophoric 384
bifunctional catalyst 310
bis-N-methylacridinium nitrate (BMAN) 362
Bishop-Kirtman perturbation theory 218
Block-Localized Wave function 396
blue box 395
bond critical point 242
bond paths 76
boron nitride nanotube 264
bovine serum albumin 380
breathing motion 346

c
C–H bond activation 298
calixarenes 373
carbon nanotubes 263
charge equilibriation methods 172
chemical potential 50
chemoselective 386
clathrate hydrates 274
click chemistry 274
coarse graining 173
compressibility 30
comproportionated 36
conceptual DFT 49
conductor-like polarizable continuum model 400
confined quantum gas 185
Coulomb's Law 113
cucurbit[n]uril 372
cycloalkenes 358
cyclodextrins 372
cylindrical 12

d
diamagnetic 36
disproportionated 36
doxorubicin 380
dual descriptor 52, 99
dual emission 385

e
Eigenmodes of reactivity 184
electric-dipole 215
electric-dipole coupling 3
electron affinity 57
Electron Transmission Spectroscopy 57
electronegativity 55
electrophilic 100
electrophilicity 52
electrophilicity index 87
enantiomeric excess 321
enantioselective catalysis 320
energy decomposition analyzes (EDA) 241
entanglement 4
entanglement induced selection rules 5
exchange-correlation functional 59

Chemical Reactivity in Confined Systems: Theory, Modelling and Applications,
First Edition. Edited by Pratim Kumar Chattaraj and Debdutta Chakraborty.

exciplex 372
excited states reactivity 107
external magnetic field 104

f
finite-field nuclear relaxation 218
first hyperpolarizability 215
Fisher information 231
fluoroquinolone 377
force field 174
4,4′dimethylbenzil 359
fragmentation models 94
Frontier Molecular Orbital theory 101
Fukui function 50
fullerene 1

g
Generalized Product Function (GPF) 396
Ghosh-Berkowitz-Parr, entropy 231
global descriptors 52
grand canonical ensemble 105, 179
group theory 12

h
hardness 51
heat capacity 199
Helmholtz potential 199
Herzberg-Teller Effect 5
Hirshfeld 105
Hohenberg-Kohn functional 51
homochiral 320
HSAB principle 52

i
impenetrable walls 69
inelastic neutron scattering 2
information gain 231
information-theoretic approach 231
insulators 37
interatomic charge transfer 169
interstellar medium 90
intrinsic reaction coordinate 81
invariant 5
inversion parity 10, 10
irreducible representations 3
isothermal calorimetry 339
isotropic pressure medium 26

j
J-couplings 232
Jahn-Teller effects 37

l
lattice model 170
Legendre-transformed resolutions 138
local descriptors 56
local electrophilicity 106
local hardness descriptors 106
local nucleophilicity 106
low-dosage-hydrate inhibitor 275
luminescence 377

m
magnetic field 137
magnetic moment 36
Marcus potential energy function 85
Markovnikov rule 109
Maxwell relation 194
metal nodes 301
metal-organic frameworks 293
metalloligands 306
metalloporphyrin 407
metastable 58
methane hydrates 278
methyl isobutyl ketone (MIBK) 319
Mg-CUK-1 75
micropolarity 347
minimal inhibitory concentration 378
MO degeneracy 102
Molecular Electrostatic Potential 50, 113
molecular orbital 240
momentum transfer vector 3
multipole decomposition 6

n
N-heterocyclic carbene 304
natural orbitals for chemical valence 242
negative pressure 26
neurodegenerative diseases 381
noble gas hydrates 279
NOESY 350
noncovalent interaction 229, 396
non-isotropic pressure 26
nonlinear chemical reactivity 179
nonradiative 371

nucleic acids 382
nucleophilic 100

o
octa acid 335
oligosaccharides 372
one pot three-component 314
open metal sites 293
OPLS-AA 338
orbital compression 214
orientational averaging 19

p
p-dimethylaminobenzonitrile 385
PAH 395
paramagnetism 36
perovskite 377
perylenebis(diimide) 385
phase transition 26
phosphorescence 353
photo-oxidation 358
π-Hole 118
PLATON 321
Poisson's equation 114
polarizability 52
post synthetic modification 304
prototropic equilibrium 379

q
quantization 273
Quasi degeneracy 102
quenching 346

r
radial distribution function 73
radiative 371
radical recombination 138
reaction electronic flux 82
reaction force analysis 82
real-space partitioning 105
recurrence relation 70
redox 36
regression relationship 120
remote activation 360
Rényi entropy 231
rhodamine dyes 375
ROESY 350
rotational level 1

s
Sanderson's equalization principle 89
scattering length operators 11
SCF 53
second-order transition moment 219
self-quenching 353
Shannon entropy 71, 231
σ-Hole 118
6-methylcoumarin 387
soft walls 72
softness 51
SP-DFT Fukui indices 141
spherical basis 6
spherical basis and eigenstates 7
spin-donicity 144
spin-electrophilicity 145
spin-philicity 144
spin polarization 103, 136
spin-polarized 171
spin potential equalization 144
spin transfer 137
stopped flow technique 352
Strecker's synthesis 90
stress tensor 26
superconductivity 39
Symmetry Adapted Perturbation Theory (SAPT) 396

t
TCSPC 348
Temperature-dependent response functions 195
2,2,6,6-tetramethylpiperidin-1-oxyl (TEMPO) 387
thiazole orange 382
thioflavin T 381
TPPCage^{8+} 406
transfer energy 20
translational level 1
transmetalation 299
two-photon absorption 219

v
valence bond 240
valence shell electron pair repulsion 35

volatile 297
volume collapse 32

W

water-based dye laser 376
water chains 265
Wigner 3-*j* symbol 8
Woodward Hoffmann rules 108

Z

zero-point vibrational averaging 218